Signal Transduction

Signal Transduction

Bastien D. Gomperts
Department of Physiology, University College London, UK

IJsbrand M. Kramer
Section of Molecular and Cellular Biology, European Institute of Chemistry and Biology, University of Bordeaux, France

Peter E.R. Tatham
Department of Physiology, University College London, UK

ELSEVIER
ACADEMIC
PRESS

AMSTERDAM BOSTON HEIDELBERG LONDON NEW YORK OXFORD
PARIS SAN DIEGO SAN FRANCISCO SINGAPORE SYDNEY TOKYO

Copyright © 2002, 2003, Elsevier (USA)
First published 2002
Second printing 2003
Third printing 2003
First published as a paperback edition 2003

Elsevier Academic Press
525 B Street, Suite 1900, San Diego, California 92101-4495, USA
http://www.elsevier.com

Elsevier Academic Press
84 Theobald's Road, London WC1X 8RR, UK
http://www.elsevier.com

Library of Congress Catalog Card Number: 2001096954

ISBN: 0-12-289631-9 hardback
ISBN: 0-12-289632-7 paperback

Typeset by J&L Composition, Filey, North Yorkshire, UK
Printed in Spain by Grafos SA Arte Sobre Papel, Barcelona

03 04 05 06 07 08 GF 9 8 7 6 5 4 3 2 1

Contents

Preface

Many people have asked for whom this book has been written and the most honest answer must be that it has been written for ourselves. It is our hope that it will also be both instructive and entertaining for students and professionals at many levels. We hope it avoids the worst excesses of the skimpily edited multi-author texts written by specialists, but necessarily this means that we have had to tread warily in areas in which none of us has first-hand experience, let alone expertise. Although we have touched the front edges of the subject, we have also endeavoured to provide an elementary basis with some historical background to all the topics covered. There has been no attempt to be comprehensive and we are aware that important topics that well qualify for inclusion in a book having the title *Signal Transduction* are conspicuous only by their absence. An obvious example must be the signalling processes that guide early embryonic development. Another is the field of signal transduction in plants. Throughout the period of writing, we have been the main beneficiaries as students of our own subject. As we learned more, we were encouraged to challenge, or at least to re-examine, some of the well-established dogmas. We have also learned to respect the wisdom of our forebears, whose freedom of thought and sometimes serendipitous discoveries in the 19th and early 20th centuries led to the creation of the modern sciences of physiology, pharmacology and cell biology, and related clinical fields, especially endocrinology and immunology.

The book conveniently divides in two parts. The first half (Chapters 1–9) provides the nuts and bolts of what might be termed 'classical' signal transduction. It concentrates mainly on hormones, their receptors, and the generation and actions of second messengers, particularly cyclic nucleotides and calcium. It was the advances in this area, particularly the discovery of the G-proteins, that originally gave rise to the expression 'signal transduction', although the term 'transduction' was stolen from elsewhere. In the second half of the book, introduced by Chapter 10, attention is concentrated on transduction processes set in action by growth factors and adhesion molecules, particularly the covalent modification of proteins (by phosphorylation) and of inositol-containing lipids and their roles in the initiation of intracellular signalling cascades. An important, though not exclusive, impetus to research in this area has been the quest to understand the cellular transformations underlying cancer, with the hope of devising effective therapeutic procedures. The gestation of the word 'cancer' has a long history, well revealed by its description in the dictionaries.

In preparing the book, we have had the benefit of advice and opinions given by friends and colleagues. These include Raj Patel (and his colleagues, particularly John Challiss, Jonathan Blank and David Critchley) at the Department of Biochemistry, University of Leicester) who, between them, read the complete script. Leon Lagnado (MRC Laboratory for Molecular Biology, Cambridge) read Chapters 1–6 and advised us particularly on the topic of visual transduction. Sussan Nourshargh (Imperial College School of Medicine at Hammersmith Hospital) advised on matters relating to cell adhesion and trafficking (Chapters 14 and 15). Greta Matthews read the entire script and gave much valuable advice. Alexander Levitzki (Department of Biochemistry, Hebrew University Jerusalem) encouraged us to disregard dogma when discussing the activation of G-proteins. We are also indebted to Peter ten Dijke (Netherlands Cancer Institute, Amsterdam) who read Chapters 10 and 16, Fergus McKenzie (Amersham Pharmacia, Cardiff) who read Chapters 11, 12 and 17, and the essay describing the cell cycle in Chapter 10, Lodewijk Dekker (Department of Medicine, University College London) who read part of Chapter 9, and Patricia Cohen (MRC Protein Phosphorylation Unit, University of Dundee) for her help and advice on the serine–threonine phosphatases. Elizabeth Genot (European Institute of Chemistry and Biology, Bordeaux) provided valuable advice on matters relating to the activation of T cells (Chapter 12). Steve Watson (Department of Pharmacology, University of Oxford) and Chris Richards (Department of Physiology, University College London) gave invaluable support and advice on matters relating to calcium. Geoffery Strachan advised on the translation of French and German texts into contemporary (19th century) English.

Many others, too numerous to name individually, have given us the benefit of their knowledge and understanding. In acknowledgement of their contribution we offer the following quotation from a paper dated 1878, entitled *On Pituri* by one of the pioneers ('A new second messenger is discovered', page 146) of signal transduction.

ON PITURI. By SYDNEY RINGER, M.D., and WILLIAM MURRELL, M.R.C.P., Lecturer on Practical Physiology at the Westminster School of Medicine, and Assistant Physician to the Royal Hospital for Diseases of the Chest

QUITE recently a student of University College, London, whose, name we have unfortunately forgotten, gave us a small packet containing a few twigs and broken leaves of the powerful and interesting drug, Pituri. These we placed in Mr Gerard's hands, and he kindly made first an extract from which he obtained a minute quantity of an alkaloid, and with this he made a solution containing one part of the alkaloid to twenty of water. Baron Mueller, from an examination of the leaves of pituri, is of opinion that it is derived from Duboisia Hopwoodii. Pituri is found growing in desert scrubs from the Darling River and Barcoote to West Australia. The natives, it is said to fortify themselves during their long foot marches, chew the leaves for the same purpose as Cocoa leaves are used in Bolivia. Dr G Bennett in the New South Wales Medical Gazette, May, 1873, says Pituri is a stimulating, narcotic and is used by the natives of New South Wales in like manner as the Betel of the East. It seems to be a substitute for tobacco. It is generally met with in the form of dry leaves, usually so pulverized that their character cannot be made out.

The use of pituri is confined to the men of a tribe called Mallutha. Before any serious undertaking, they chew these dried leaves, using about a tea-spoonful. A few twigs are burnt and the ashes mixed with the leaves. After a slight mastication the bolus is placed behind the ear (to increase it is supposed its strength), to be again chewed from time to time, the whole being at last swallowed. The native after this process is in a sufficiently courageous state either to transact business or to fight. When indulged in to excess, it is said to induce a condition of infuriation. In persons not accustomed to its use pituri causes severe headache.

Of course, the authors of this paper would themselves never have heard the expression 'signal transduction' and it would be a further 100 years before it made its appearance in the biological literature. The sensations brought about by pituri, an alkaloid that Ringer and Murrell described as sharing some of the pharmacological properties of atropine (courage, infuriation, frustration and headaches), are not dissimilar to those experienced by us in the writing of this book. However, we have never come to blows, never even felt the need to fight. Although it was drafted by three authors, the aim throughout has been to present a book written as if by one mind, one pair of hands. Of course, we all have our own particular fields of interest and (hopefully) expertise and we were individually responsible for the original drafting of particular chapters and sections. However, there is not a line of the book that has not been read, replaced, rewritten, expanded, cut or otherwise altered so that in the end, we hope that the text has a consistent style throughout.

Notes

For protein structural data we have made use of: the Protein Data Bank: Berman, H.M., Westbrook, J., Feng, Z. *et al*. The protein data bank. *Nucleic Acid Res.* 2000; 28: 235–42 (http://www.rcsb.org/pdb/).

For chemical structures we wish to acknowledge the use of the EPSRC's Chemical Database Service at Daresbury: Fletcher, D.A., McMeeking, R.F., Parkin, D. The United Kingdom Chemical Database Service. *J. Chem. Inf. Comput. Sci.* 1996; 36: 746–9.

Protein structures have been generated using Rasmol and CHIME: Sayle R., Milner-White, E.J. RasMol: Biomolecular graphics for all. *Trends Biochem. Sci.* 1995; 20: 374.

Martz, E. (University of Massachusetts, Amherst, MA, USA) and MDL Information Systems, Inc. (San Leandro, CA, USA).

■ References

We have tried to provide original text sources to nearly all the statements, experiments and discoveries discussed. The main reason for this is that we ourselves have necessarily had to extend the treatment of nearly all the topics presented far beyond the areas of our own experience or expertise. Thus, the comprehensive lists are there to provide us with some sort of reassurance that what we have written has not simply been conjured out of the air. Also, because we have made a particular feature of presenting original historical source material by quotation, which necessarily required referencing, it seemed logical also to include literature references to modern sources as well. Thus we hope that this book may serve as a valuable resource, in the manner of a basic literature review, for anyone wanting to explore further.

■ Abbreviations

The definitions of all the main abbreviations used in this book can be found in the index.

■ Genes and gene products

According to convention, acronyms printed in lower case indicate genes (e.g. ras), capitalized acronyms indicate their protein products (Ras). The genes of yeast are printed in upper case (RAS). The prefixes v- and c- indicate viral or cellular origin (v-Ras, c-Ras).

From the Shorter Oxford English Dictionary (3rd edition, 1944, with corrections 1977):

Transduction (trɒns₁dɐ·kʃən). *rare.* 1656. [ad. L. *tra(ns)ductionem*, *tra(ns)ducere*; see TRADUCE.] The action of leading or bringing across.

Traduce (trădiū·s), *v.* 1533. [ad. L. *traducere* to lead across, etc.; also, to lead along as a spectacle, to bring into disgrace; f. *trans* across + *ducere* to lead.] †1. *trans.* To convey from one place to another; to transport –1678. †b. To translate, render; to alter, modify, reduce –1850. †c. To transfer from one use, sense, ownership, or employment to another –1640. †2. To transmit, esp. by generation –1733. †b. *transf.* To propagate –1711. †c. To derive, deduce, obtain *from* a source –1709. 3. To speak evil of, esp. (now always) falsely or maliciously; to defame, malign, slander, calumniate, misrepresent 1586. †b. To expose (to contempt); to dishonour, disgrace (*rare*) –1661. †4. To falsify, misrepresent, pervert –1674.

1. b. Milton has been traduced into French and overturned into Dutch SOUTHEY. 2. Vertue is not traduced in propagation, nor learning bequeathed by our will, to our heires 1606. 3. The man that dares t., because he can With safety to himself, is not a man COWPER. b. By their own ignoble actions they t., that is, disgrace their ancestors 1661. 4. Who taking Texts .. traduced the Sense thereof 1648. Hence **Tradu·cement**, the, or an, action of traducing; defamation, calumny, slander. **Tradu·cingly** *adv.*

from The Oxford English Dictionary (2nd edition) On Compact Disc:

transduction (tra:ns'd□k∫ e n, træns-).
[ad. L. *transduction-em* (usually *traductionem*), n. of action f. *tra(ns)dfucere: see* TRADUCE.]

1. **The action of leading or bringing across.** rare.
    1656 BLOUNT *Glossogr., Transduction,* a leading over, a removing from one place to another.
    a1816 BENTHAM *Offic. Apt. Maximized, Introd. View* (1830) 19 In lieu of *adduction,* as the purpose requires, will be subjoined *abduction, transduction,...and* so forth.

2. **The action or process of transducing a signal.**
    1947 Jrnl Acoustical Soc. Amer. XIX. 307/1 It is rather interesting that the direct method of electronic transduction, instead of the indirect method of employing a conventional transducer and then amplifying the output with a vacuum tube, has not been developed.
    1970 J. EARL *Tuners & Amplifiers* iv. 87 Low impedance pickup cartridges using the moving-coil principle of transduction.
    1975 *Nature* 17 Apr. 625/1 The transduction of light energy into neural signals is mediated in all known visual systems by a common type of visual pigment.

3. *Microbiology.* **The transfer of genetic material from one cell to another by a virus or virus-like particle.**
    1952 ZINDER & LEDERBERG in *Jrnl. BacterioL* LXIV. 681 To help the further exposition of our experiments, we shall use the term transduction for genetically unilateral transfer in contrast to the union of equivalent elements in fertilization.
    1960 [see F Ill. 1 l].
    1971 *Nature* 18 June 46611 It has been suggested that transduction of genes by viruses was an important mechanism in evolution for spreading useful mutations between organisms not formally related.
    1977 *Lancet* 9 July 94/2 These were derived by selection of sensitive variants from gentamicin-resistant strains or by transduction of this resistance to sensitive strains.
    Hence
    trans'ductional a., of or pertaining to (genetic) transduction.
    1956 *Genetics* XLI. 845 (*heading*) Linear inheritance in transductional clones.
    1980 *Jrnl. Gen. Microbiol.* CXIX. 51 Transductional analysis revealed that one of the four mutations carried by strain T-693 was responsible for constitutive synthesis of both isoleucine and threonine biosynthetic enzymes.

Prologue: Signal transduction, origins and personalities

Transduction, the word and its meaning: one dictionary, different points of view

The expression *signal transduction* first made its mark in the biological literature around 1974,[1] and as a title word in 1979.[2–4] Physical scientists and electronic engineers had earlier used the term to describe the conversion of energy or information from one form into another. For example, a microphone transduces sound waves into electrical signals. Its widespread use in biospeak was triggered by an important review by Martin Rodbell, published in 1980 (Figure 1.1).[5] He was the first to draw attention to the role of

Alfred G Gilman and Martin Rodbell, awarded the Nobel Prize, 1994 "for their discovery of G-proteins and the role of these proteins in signal transduction in cells"

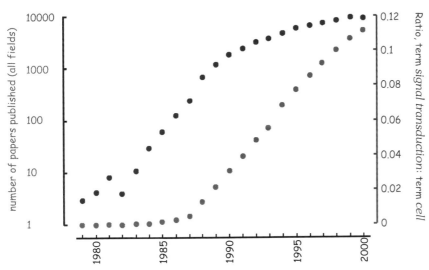

Figure 1.1 Occurrence of the term *signal transduction*. The left-hand axis records all papers using this term traced through the MedLine database. The right-hand axis records the proportion of papers using the term *cell* that also use the term *signal transduction*.

GTP and GTP-binding proteins in metabolic regulation and he deliberately borrowed the term to describe their role. By the year 2000, 12% of all papers using the term *cell* also employed the expression *signal transduction*.

■ Hormones, evolution and history

These chemical messengers . . . or 'hormones' (from ὁρμω, meaning I excite or I arouse), as we may call them, have to be carried from the organ where they are produced to the organ which they affect, by means of the bloodstream, and the continually recurring physiological needs of the organism must determine their repeated production and circulation throughout the body.[6]

■ The plasma membrane barrier

In the main, when we consider signal transduction we are concerned about how external influences, particularly the presence of specific hormones, can determine what happens on the inside of their target cells. There is a difficulty, since the hormones, being mostly hydrophilic (or lipophobic) substances are unable to pass through membranes, so that their influence must somehow be exerted from outside.

Steroids, prostaglandins, etc. are hydrophobic and do, of course, traverse cell membranes where they interact with intracellular receptors.

The membranes of cells, although very thin (3–6 nm) are effectively impermeable to ions and polar molecules. Although K^+ ions might achieve diffusional equilibrium over this distance in water in about 5 ms, they would take some 12 days (280 h) to equilibrate across a phospholipid bilayer (under similar conditions of temperature, etc.). Likewise, the permeability of membranes to polar molecules is low. Even to small molecules such as urea, the permeability of membranes is about 10^4 times lower than to water. So, for a hormone such as adrenaline the rate of permeation is too low to measure. The evolution of receptors has accompanied the development of mechanisms which permit external chemical signalling molecules, the first messengers, to direct the activities of cells in a variety of ways with high specificity and precise control in terms of extent and duration. With few exceptions (the steroid hormones, thyroid hormone), they do this without ever needing to penetrate their target cells.

■ Protohormones

The first messengers (which include the hormones), and their related intracellular (second) messengers, are of great antiquity on the biological timescale. It is interesting to consider which came first: the hormones or the receptors that they control? Substances recognized as having the actions of hormones in animals first made their appearance at early stages of evolution (Figure 1.2). Chemical structures closely related to thyroid hormones have been discovered in algae, sponges and many invertebrates. Steroids such as oestradiol are present in microorganisms and also in ferns and conifers. Catecholamines have been found in protozoa[7,8] and ephedrine, which is closely related, can be isolated from the stems and leaves of the Chinese herb Ma Huang (*Ephedra sinica*). It is still in use as an oral stimulant in the treatment of hypotension (low blood pressure) and in the relief of asthma. It is not only the low-molecular-weight hormones which are present in the cells of organisms which have no hormone/receptor responses; there are claims, based on immunological detection, for the presence of peptides related to insulin and the endorphins in protozoa, fungi, and even bacteria.[9] For these compounds no messenger-like function has been discerned in their species of origin and it is likely that the receptors that mediate their effects in animals evolved much later.

The a- and α-type mating factors of yeast which certainly act as messengers are very similar in structure to the gonadotrophin-releasing hormone (GnRH) that controls the release of gonadotrophins from the anterior pituitary in mammals.[11,12] Factors resembling the mammalian

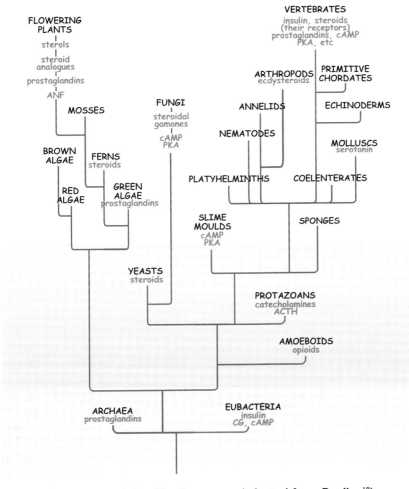

Figure 1.2 **Emergence of signalling hormones (adapted from Baulieu[10]).**

atrial natriuretic factors (ANF) are present in the cytosol of the single-cell eukaryote *Paramecium multimicronucleatum*[13] and in the leaves of many species of plants where they act as regulators of solvent and solute flow and of the rate of transpiration.[14] Adrenocorticotrophic hormone (ACTH) and β-endorphin are present in protozoa. These also contain high-molecular-weight precursors of these peptides reminiscent of the vertebrate pro-opiomelanocortin (POMC).[15,16] It is striking that pathways for the biosynthesis of these 'protohormones', often complex, were established early on, well before the evolution of membrane receptors.

Receptor-like proteins in non-animal cells have been much harder to identify. A recently described example is a protein expressed in the plant *Arabidopsis* that shares extensive sequence homology with the ionotropic glutamate receptor (iGluR) of mammalian brains.[17] A corollary arising from this is the possibility that the potent neurotoxins thought to be generated in defence against herbivores may have their origin as specific agonists, and were only later selected and adapted in some species as poisons. Molecules having a close relationship to the epidermal growth factor (EGF) and insulin receptors apparently evolved in sponges prior to the Cambrian Explosion, and it has been proposed that they may have contributed to the rapid appearance of the higher metazoan phyla.[18]

In general, however, it appears that many of the molecules that we regard as hormones arose long before the receptors that they control. An important consequence of this is that the responses to a given hormone can vary widely across different species and even within species. Numerous actions of prolactin have been identified.[19] It is the regulator of mammary growth and differentiation and of milk protein synthesis in mammals. In birds, it acts as a stimulus to crop milk production and in some species as a controlling factor for fat deposition and as a determinant of migratory behaviour.[20,21] It is a regulator of water balance in newts and salamanders (urodeles) and of salt adaptation and melanogenesis in fish.[19] Serotonin (5-hydroxy tryptamine), a neurotransmitter that controls mood in humans, is reported to stimulate spawning in molluscs, probably as a consequence of its conversion to melatonin (one wonders whether it affects their mood as well).

■ Protoendocrinologists

Despite excellent anatomical descriptions, almost nothing was known about the functions of the various organs which constitute the endocrine system (glands) until the last decade of the 19th century. Indeed, in the standard textbook of the period (*Foster's Textbook of Physiology*, 3 volumes and more than 1200 pages), consideration of the thyroid, the pituitary, the adrenals ('suprarenal bodies') and the thymus is confined to a brief chapter of less than 10 pages, having the title 'On some structures and processes of an obscure nature'.

The initial impetus prompting the systematic investigations which led to the discovery of the hormones can be ascribed to a series of papers that were much misunderstood.

The possibility of reversing the pain of old age by injecting the seminal fluid of dogs and guinea-pigs into human subjects might be considered the hopeless imaginings of a charlatan. However, here we are discussing the work of Charles Edouard Brown-Séquard, the successor of Claude Bernard at the Collège de France and also a member of leading scientific academies in England and the USA. He had held professorial appointments at both Harvard and Virginia; in London he was appointed physician at the National Hospital for the Paralysed and Epileptic (later the National Hospital for Nervous Diseases, now the National Hospital for Neurology and Neurosurgery). He was an associate of Charles Darwin and Thomas Huxley. He wrote over 500 papers relating to many diverse fields such as the physiology of the nervous system, the heart, blood, muscles and skin, the mechanism of vision, and much more. He was an outstanding experimentalist making fundamental contributions. Starting with his doctoral thesis, he described the course of motor and sensory fibres in the spinal cord, a field to which he returned many times. He was in constant demand as lecturer, teacher and physician on both sides of the Atlantic, crossing the ocean on more than 60 occasions. Of direct relevance to us must be his demonstration that the adrenal glands are essential to life.

In view of all this, it is curious that Brown-Séquard is now all but forgotten. On the rare occasions when he is recalled, it is generally in connection with a series of brief reports, published in 1889, in which he described the self-administration, by injection, of testicular extracts, which he considered had the effect of reinforcing his bodily functions.

Some brief quotations from his paper in the *Lancet* must suffice:[25]

In the margin:

In writing these paragraphs, we have relied heavily on the work of Victor Medvei,[22] Michael Aminoff[23] and Horace Davenport.[24]

I am seventy-two years old. My general strength, which has been considerable, has notably and gradually diminished during the last ten or twelve years. Before May 15th last, I was so weak that I was always compelled to sit down after about half an hour's work in the laboratory. Even when I remained seated all the time I used to come out of it quite exhausted after three or four hours of experimental labour . . .

The day after the first subcutaneous injection, and still more after the two succeeding ones, a radical change took place in me, and I had ample reason to say and to write that I had regained at least all the strength I possessed a good many years ago . . .

My limbs, tested with a dynamometer for a week before my trial and during the month following the first injection, showed a decided gain of strength . . .

I have measured comparatively, before and after the first injection, the jet of urine in similar circumstances – i.e., after a meal in which I had taken food and drink of the same kind and in similar quantity. The average length of the jet during the ten days that preceded the first injection was inferior by at least one quarter of what it came to be during the twenty following days. It is therefore quite evident that the power of the spinal cord over the bladder was considerably increased . . .

I will simply say that after the first ten days of my experiments I have had a greater improvement with regard to the expulsion of faecal matters than in any other function. In fact a radical change took place, and even on days of great constipation the power I long ago possessed had returned.

Finally, in a footnote:

It may be well to add that there are good reasons to think that subcutaneous injections of a fluid obtained by crushing ovaries just extracted from young or adult animals and mixed with a certain amount of water, would act on older women in a manner analogous to that of the solution extracted from the testicles injected into old men.

Possibly Brown-Séquard should be regarded as the father of HRT, but he was certainly not the inventor of organotherapy, the attempt to cure human disease by the introduction of glandular extracts from other animals.

Indeed, Gaius Plinius Secundus (known as Pliny the Elder, 23–79 CE) recorded his contempt of the Greeks who used human organs therapeutically although he did recommend the use of animal tissues, in particular that the testicles of animals should be eaten in order to cure impotence or to improve sexual function in men, and that the genitalia of a female hare should be eaten by women in order to achieve pregnancy.[24]

Ever mindful of the possibility of autosuggestion, Brown-Séquard pleaded that others should examine his claims and to consider them with objectivity. What a hope! But if any doubts remained, he went on record saying:[26]

Not only should one not be astonished that the introduction into the blood of principles deriving from the testicles of young animals be followed by an augmentation of vigour, but that this result should not have been foretold. Actually, everything demonstrates that the strength of the spinal cord and also, but to a lesser degree, that of the brain, has, in the male adult or the old, some variations linked to the activity of the testicles. To the facts which I have already mentioned in this regard in the meeting of 1 June, I believe one must add the following observations made a large number of times, over many years on two individuals aged 45 to 50 years old . . . On my advice, each time they had to undertake weighty activity, either physical or intellectual, they put themselves into a state of sexual excitement, but avoiding any ejaculation. The testicles thus acquired a strong temporary activity in function which was followed rapidly by the desired rise in the power of the nervous system. . . .

News travels fast and good news travels even faster. Within weeks this was the subject of an editorial paragraph in the *British Medical Journal* under the heading THE PENTACLE OF REJUVENESCENCE:[27]

On two occasions this month Dr. Brown-Séquard has made communication of a most extraordinary nature to the Societé de Biologie of Paris. The statements he made – which have unfortunately attracted a good deal of attention in the

---

SECONDE NOTE SUR LES EFFETS PRODUITS CHEZ L'HOMME PAR DES INJECTIONS SOUS-CUTANÉES D'UN LIQUIDE RETIRÉE DES TESTICULES FRAIS DE COBAYE ET DE CHIEN.
par M. BROWN-SEQUARD.
(*Communication faite le 15 juin*)

Non seulement il n'y a pas à s'étonner que l'introduction dans le sang de principes provenant de testicules de jeunes animaux soit suivie d'une augmentation de vigueur, mais encore on devait s'attendre à obtenir ce résultat. En effet, tout montre que la puissance de la moelle, épinière et aussi, mais à un moindre degré, celle du cerveau, a, chez l'homme adulte ou vieux, des fluctuations liées à l'activité fonctionnelle des testicules. Aux faits que j'ai mentionnés, à cette régard, dans la séance du 1re juin, je crois devoir ajouter que les particularités suivantes ont été observées un très grand nombre de fois, pendant plusieurs années, chez deux individus àgés de quarante-cinq à cinquante ans. Sur mon conseil, chaque fois qu'ils avaient à exécuter un grand travail physique ou intellectuel, ils se mettaient

public press – recall the wild imaginings of mediaeval philosophers in search of an elixir vitae. . . . MM Féré and Dumontpallier, in commenting on M. Brown-Séquard's statements, observed that they would require to be rigidly tested and fully confirmed by other self-experimenters before they were likely to meet with general acceptance, and in this opinion we fully concur.

This rather measured account by the editors of the BMJ contrasts with the reaction on the other side of the Atlantic, where it was taken up as:

THE NEW ELIXIR OF YOUTH. – Dr. Brown-Séquard's rejuvenating fluid is said to have been tried at Indianapolis on a decrepit old man with marvellous effect. Four hours after the injection the patient walked over a mile in twenty-five minutes. He declared that he felt more vigorous than he had done for twenty-five years. He read a newspaper without glasses, a thing he had not been able to do for thirty years. It should be added that no medical authority is given for this story.[28]

With all this happening, possibly he began to harbour some doubts, as he wrote:

In the United States especially, and often without knowing what I did or the most elementary rules regarding injections of animal materials, several physicians or rather the medicasters and charlatans have exploited the ardent desires of a great number of individuals and have made them run the greatest risks, if they have not done worse.

In England the reaction remained negative:

A MEMBER OF THE MEDICAL PROFESSION writes: You have not done me justice by your note on my circular relative to Dr. Brown-Séquard's experiments. Much as I disapprove of the animal torture in question, I should not have felt it my duty to print and post some 6,000 copies of my protest had that been all. I consider the idea of injecting the seminal fluid of dogs and rabbits into human beings a disgusting one, and when the treatment also involves the practice of masturbation, I think it is time for the medical profession in England to repudiate it. One may be a vivisector without also encouraging a loathsome vice, but as far as I could discover no word of protest was uttered by the leaders of the medical opinion, and I determined to take the thing into my own hands, with the effect, as I have good reason to believe, of making it very improbable that this method of pretended rejuvenescence will be introduced into England. I have a great number of encouraging letters from eminent Englishmen, who will take good care of this, at least. Vivisection may be an open question, but self abuse is not.[27]

Dr Edward Berdoe, who was also a member of the profession, left little doubt about his feelings on the matter:

the object of these abominable proceedings is to enable broken down old libertines to pursue with renewed vigour the excesses of their youth, to rekindle

dans un état de vive excitation sexuelle, en évitant cependant toute éjaculation spermatique. Les glandes testiculaires acquéraient alors temporairement une grande activité fonctionnelle, qui était bientôt suivie de l'augmentation désirée dans la puissance des centres nerveux.

Pentacle A five-pointed star, of supposed supernatural significance.

the dying embers of lust in the debilitated and aged, and to profane the bodies of men which are the temples of God, by an elixir drawn from the testicles of dogs and rabbits by a process involving the excruciating torture of innocent animals.... We may have also a new race of beings intermediate between man and the lower animals as a remoter consequence of the boon to humanity conferred by French physiology.[29]

Like it or not, many physicians, aware of Brown-Séquard's publications, were eager to test the possibilities of applying various organ extracts in their practice. Within a year or two, organotherapy was becoming respectable.[30,31]

In a letter to his colleague Jacques Arsene d'Arsonval (quoted by Borell[32]) Brown-Séquard said that '*the thing*' was '*in the air*'. What made it especially so was probably George Redmayne Murray's report[33] on the treatment of myxoedema. He described a patient who went on to survive with regular injections of sheep thyroid extract for a further 28 years.

Henry Dale's description of the discovery of adrenaline cannot be bettered:[34,35]

And respectable it remained, anyway in France, where in the 1920s Serge Voronoff was treating elderly gentlemen by implanting sections of chimpanzee or baboon testicles, the so-called 'monkey glands'.[30] R. V. Short[31] speculates that his patients may have been the first humans to become infected by HIV-1. Were it not for the fact that they remained impotent, the disease might have exploded on a world even less prepared than it was at the century's end. One might reasonably imagine that an AIDS epidemic originating at that time could have changed the course of European history.

Dr George Oliver, a physician of Harrogate, employed his winter leisure in experiments on his family, using apparatus of his own devising for clinical measurements. In one such experiment he was applying an instrument for measuring the thickness of the radial artery; and, having given his son, who deserves a special memorial, an injection of an extract of the suprarenal gland, prepared from material supplied by the local butcher, Oliver thought that he detected a contraction, or, according to some who have transmitted the story, an expansion of the radial artery. Which ever it was, he went up to London to tell Professor Schäfer what he thought he had observed, and found him engaged in an experiment in which the blood pressure of a dog was being recorded; found him, not unnaturally, incredulous about Oliver's story and very impatient at the interruption. But Oliver was in no hurry, and urged only that a dose of his suprarenal extract should be injected into a vein when Schäfer's own experiment was finished. And so, just to convince Oliver that it was all nonsense, Schäfer gave the injection, and then stood amazed to see the mercury mounting in the arterial manometer till the recording float was lifted almost out of the distal limb.

Thus the extremely active substance formed in one part of the suprarenal gland, and known as adrenaline, was discovered.

Schäfer's own account of his first meeting with George Oliver relates events that may have taken place on a another planet from Dales' version:[35]

'In the autumn of 1893 there called upon me in my laboratory in University College a gentleman who was personally unknown to me, but with whom I had a common bond of interest –

Within a few months, Oliver and Schäfer had demonstrated that the primary effect of the extract is a profound arteriolar constriction with a resulting increase in the peripheral resistance.[36,37] Their colleague Benjamin Moore reported that the activity could be transferred by dialysis through membranes of parchment paper, that it is insoluble in organic solvents but readily soluble in water, resistant to acids and to boiling, etc.[38] Curiously, nobody at that time seems to have remarked on its propensity to oxidation.

Hormones: a definition

Hormones are blood-borne 'first messengers', usually secreted by one organ (or group of cells) in response to an environmental demand to signal a specific response from another. The first such messenger to be endowed with the title of hormone was secretin, later shown to be a peptide released into the bloodstream from cells in the stomach lining, indicating the presence of food and alerting the pancreas. In the words of William Maddock Bayliss (co-discoverer with Ernest Henry Starling),

> There are a large number of substances, acting powerfully in minute amount, which are of great importance in physiological processes. One class of these consists of the hormones which are produced in a particular organ, pass into the blood current and produce effects in distant organs. They provide, therefore, for a chemical co-ordination of the activities of the organism, working side by side with that through the nervous system. The internal secretions, formed by ductless glands, as well as by other tissues, belong to the class of hormones.[39]

In fact, adrenaline, the signal for fright-fight-and-flight, is a much better candidate for the accolade of 'first hormone'.[40] Together with noradrenaline it is secreted into the bloodstream in consequence of emotional shock, physical exercise, cold, or when the blood sugar concentration falls below the point tolerated by nerve cells. Extracts having the activity of adrenaline, enhancing the force and volume of the heart output had been reported 10 years before the discovery of secretin, almost simultaneously, by George Oliver and Edward Schäfer in London,[36,37] and by Napoleon Cybulski and Szymonowicz in Krakow.[41]

What's in a name?

It has been customary in Europe to give the substance 4-[1-hydroxy-2-(methylamino)ethyl]-1,2-benzenediol, alternatively 3,4-dihydroxy-α-[(methylamino)methyl]benzyl alcohol, the name *adrenaline*. In the USA the same substance is called *epinephrine*. Why have the Europeans preferred the Latin root while the Americans go for the Greek? Behind this lie hints of scientific skulduggery.

John Jacob Abel, America's first professor of pharmacology, initially at the University of Michigan at Ann Arbor and then at the new Johns Hopkins Medical School in Baltimore, is credited with the isolation of the first hormone as a pure crystalline compound. As a part of his procedure, he treated the acidified and deproteinated extract of adrenal glands with alkaline benzoyl chloride and, after further steps including hydrolysis, he obtained a crystalline material which he reported as being very active. Chemical analysis yielded an elementary formula of $C_{17}H_{15}NO_4$. What is not so clear is whether the Japanese industrial chemist Jokichi Takamine, who had an arrangement with the Parke, Davis Company (Detroit, Michigan), gained some advantage from his visit to Abel's laboratory some time in 1899 or 1900. Certainly, Abel appears to have been quite candid with his visitor,

seeing that we had both been pupils of Sharpey, whose chair at that time I had the honour to occupy. I found that my visitor was Dr. George Oliver, already distinguished not only as a specialist in his particular branch of medical practice, but also for his clinical applications of physiological methods. Dr. Oliver was desirous of discussing with me the results which he had been obtaining from the exhibition by mouth of extracts of certain tissues, and the effects which these had in his hands produced upon the blood vessels of man, as investigated by two instruments which he had devised – one of them, the haemodynamometer, intended to read variations in blood pressure, and the other, the arteriometer, for measuring with exactness the lumen of the radial or any other superficial artery. Dr. Oliver had ascertained, or believed that he had ascertained, by the use of these instruments, that glycerin extracts of some organs produce diminution in calibre of the arteries, and increase of pulse tension, of others the reverse effect.'

The European Pharmacopoeia now also indicates the use of the term **epinephrine**. We justify our preference for adrenaline not only on its historical primacy but for reasons of logic and common usage. Who ever heard of epinephric receptors? Who ever used the expression 'that really gets my epinephrine up'?

whilst Takamine, used to the practices of commerce, never published reports of any sort until his preparation was well protected by patents and importantly in this case, a trademark with the name *Adrenalin* (no terminal '*e*'). What is clear is that Takamine used a simpler procedure for purifying the hormone, omitting the benzoylation step. T. B. Aldrich, also at the Parke, Davis Company but recently of the Department of Pharmacology at Johns Hopkins and therefore a late colleague of Abel's, determined the correct elementary formula of Takamine's preparation as $C_9H_{13}NO_3$.

The Parke, Davis Company lost little time in marketing the preparation, and continued to do so until 1975 when they replaced it with a synthetic product. It became evident that Abel's failure to provide the correct formula was due to incomplete hydrolysis of the benzoyl residues, and indeed, the compound to which he gave the name *epinephrin* (again, no terminal '*e*') was the monobenzoylated derivative, which nonetheless retained some biological activity. His revised formula appeared in the very first paper to be published in the first volume of the new *Journal of Biological Chemistry*. In a long and contentious account,[42] mainly devoted to explaining why he had got it right, and that the others had it wrong, he still managed to end up with a structure having 10 carbon atoms, $C_{10}H_{13}NO_3$. Of course, the material isolated from adrenal medulla is a mixture of adrenaline and noradrenaline and it is likely that the crystalline material was composed of both compounds, so that even an assignment of 9 carbon atoms would have been an overestimate. The correct structure and confirmation that adrenaline is a derivative of catechol was not long in coming. Independent reports from Dakin and from Stolz of complete chemical syntheses of the racemic mixture of the two optically active isomers came in 1904/5.[43,44]

The effects and the actions of adrenaline provide a pattern that has been and still remains useful in general discussions of hormone action.

■ Neurotransmitters

In comparison with the ready acceptance of the principle of blood-borne transmission of chemical signals between organs, the idea of chemical, as opposed to electrical, transmission of impulses between nerves and nerves, and between nerves and muscles had a long and fraught gestation. The phenomenon of vagal stimulation causing a slowing of the heart, had been described in 1845 by Weber (see reference[45]) and the possibility of chemical transmission of this signal was proposed as early as 1877 by Dubois-Reymond:

Of known natural processes that might pass on excitation, only two are, in my opinion, worth talking about: either there exists at the boundary of the contractile substance a stimulatory secretion in the form of a thin layer of ammonia, lactic acid, or some other powerful stimulatory substance; or the phenomenon is electrical in nature.

It took all of 60 years for the principle of chemical transmission to achieve acceptance as the general means of communication between nerves and muscles.[8]

Otto Loewi recorded[46] that he had conceived the idea of chemical transmission between nerves as early as 1903, but that at that time

> I did not see a way to prove the correctness of this hunch, and it entirely slipped my memory until it emerged again in 1920. . . . The night before Easter Sunday of 1920, I awoke, turned on the light, and jotted down a few notes on a tiny slip of paper. Then I fell asleep again. It occurred to me at six o'clock in the morning that during the night I had written down something most important, but I was unable to decipher the scrawl. The next night, at three o'clock, the idea returned. It was the design for an experiment to determine whether or not the hypothesis of a chemical transmission that I had uttered seventeen years ago was correct. I got up immediately, went to the laboratory, and performed a simple experiment on a frog heart according to the nocturnal design.

Loewi relates how, in his experiment, he induced contractions by electrical stimulation of the vagal nerve in an isolated heart. He then transferred some of the fluid from this heart into the ventricle of a second heart undergoing similar stimulation. The result was to slow it down and reduce the force of contraction. He gave the name '*vagusstoff*' to the inhibitory substance, later identified as acetylcholine.

One might have thought that this demonstration of chemical, not electrical communication between hearts might have settled the issue. However, it is doubtful whether the technique that he used at that time could have delivered the results that he described. There were problems of reproducibility which may relate both to the temperature of the laboratory and to seasonal variations in the response of the amphibian heart. In the winter months the inhibitory fibres predominate so that electrical stimulation suppresses the rate and force of contraction. In the summer the situation is reversed. Other problems arise from the transient nature of the pulse of neurotransmitter. Eventually, with the efforts of many others, these difficulties were overcome but, even so, in order to prove the role for acetylcholine in neurotransmission it remained necessary to demonstrate its presence in the relevant presynaptic nerve endings. Also, it was essential to establish that it is actually released when the nerve is stimulated electrically.

One of the main problems confronting all ideas concerning chemical transmission was the instability of the transmitter substance acetylcholine. This had been synthesized in the mid 19th century. René de M. Taveau, previously associated with J. J. Abel, showed that of 20 derivatives of choline, the acetyl ester is the most active in reducing heart rate and blood pressure, an effect opposed by atropine.[47] It was first isolated from natural sources in 1914 by Arthur Ewins, a member of the laboratory of Henry Dale as a component present in an extract of ergot. At the time of his appointment to a post at the Wellcome Physiological Laboratories in 1904, Dale remarks that his employer requested that

The fascinating history of the debates surrounding the issue of chemical versus electrical transmission at the synapse is excellently related by Horace Davenport.[24]

Henry Hallett Dale and **Otto Loewi** were awarded the Nobel prize in 1936 for their discoveries relating to chemical transmission of nerve impulses.

when I could find an opportunity for it without interfering with plans of my own, it would give him a special satisfaction if I would make an attempt to clear up the problem of ergot, the pharmacy, pharmacology and therapeutics of that drug being then in a state of obvious confusion. . . . I was, frankly, not at all attracted by the prospect of making my first excursion into (pharmacology) on the ergot morass.[48]

■ Ergot

Although ergot was recognized as the 'noxious pustule in the ear of grain' over 2500 years ago, written descriptions of ergot poisoning did not appear until the Middle Ages. It is a product of the fungal parasite *Claviceps purpurea* that affects grains, particularly rye. As described, the symptoms included burning pains in the limbs followed by a long and painful gangrene, the tissue becoming dry and being consumed by the Holy Fire, blackened like charcoal. Further complications included abortion and convulsions.

St Vitus' dance and the dance of death (*danse macabre*) were peculiar to the Middle Ages and it has been suggested[49] that the St Vitus dance may have been a manifestation of the seasonal starvation which occurred during the early summer months. This was the time of the 'hungry gap' when the grain stocks were at their lowest, awaiting replenishment with the August harvests. The peasants, forced to prepare meal from the most marginal sources inevitably consumed grains laced with ergot alkaloids and other mind-bending substances gathered from the fruit and flowers of wild plants. The St Vitus' dance became a public menace, and particularly in the Netherlands, Germany and Italy, during the 14th and 15th centuries, crazed mobs wandered from city to city. The scenes, and the psychedelic visions induced in the minds of the affected are well represented in the paintings of Pieter Bruegel (the Elder) (Figure 1.3).

St Vitus' dance, so called because sufferers sought to obtain relief at shrines dedicated to St Vitus, is also the name given to the unrelated condition later called **Sydenham's chorea**. This is mainly a disease of young people, generally associated with rheumatic fever.

It is in this social panorama, traversed by profound anxieties and fears, alienating frustrations, devouring and uncontrollable infirmities and dietary chaos that adulterated and stupefying grains contributed to delirious hypnotic states and crises, which could explode into episodes of collective possession or sudden furies of dancing. The forbidden zones, those most contaminated by the ambiguous, ambivalent magic of the sacred, seemed to emanate perverse influences and unleash dark energies. Psychological destitution, together with the torment of an ailing body, acted as detonator of the epileptic attacks and tumultuous and surging group fits, in which men were attracted and repelled by centres of sacrifice, like the altar.[49]

Relief, of course, was obtained at the shrine of St Anthony. Here, maybe fortuitously, the sufferers received a diet free of contaminated grain. Ergot was also known as a medicinal herb and was used by midwives to suppress postpartum bleeding though it did not find its way into regular clinical practice until the 19th century.

Figure 1.3 'One hundred and one Netherlandish proverbs', by Pieter Breugel the Elder (1559). It is high summer, though the hayfield seen at the top of the painting is not being harvested. Instead, it is being laid to waste by the pigs. The farmer and all the other villagers are beyond caring. In this hungry month of July, just before the staple crops come to fruition, sustenance is found in the 'bread of dreams', clearly laid out for display on the roof tiles. Containing a rich mixture of alkaloids derived from ergot-infested grains, this was the cause of communal madness and wild manifestations (St Vitus' dance) among the peasantry of the Middle Ages. Several hundred years later, Henry Dale and his colleagues isolated the neurotransmitter acetylcholine from that 'veritable cornucopia' that is ergot. Courtesy of the Gemäldegalerie Staatliche Museen zu Berlin Preußischer Kulturbesitz.

Today, the best-known products of ergot must be the hallucinogen lysergic acid diethylamide (LSD), and ergotamine which is used in the relief of the symptoms of migraine. However, for Henry Dale the extracts of ergot presented a veritable cornucopia of active substances, to which he returned repeatedly over several years. It was an impurity in a sample of ergot sent to him in 1913 for routine quality control, probably due to contamination by *Bacillus acetylcholine* (fresh ergot does not contain acetylcholine),[45] that led him back to the question of transmission at the contacts between nerves and cells. When injected into the vein of an anaesthetized cat, the extract caused profound inhibition of the heart beat and because it was obviously unsuitable for release as a drug, he obtained the whole batch for further investigation. The first thoughts were that the active constituent might be the stable compound muscarine, but on isolation it was found to be the profoundly labile acetylcholine.

At that time, and even later when it was identified as a chemical component in non-neural tissues, there was never a hint that acetylcholine might have physiological functions. This was not even suggested by the finding as late as 1930 that arterial injections of acetylcholine could induce contractures in denervated muscles. There were still a number of real problems to be overcome. Chief among these was the transient nature of the pulse of neurotransmitter. Also, it was not sufficient to show that acetylcholine applied from a pipette was capable of inducing a response. To prove its role in neurotransmission it still remained necessary to demonstrate its presence in the relevant presynaptic nerve endings and then to show that it is released upon electrical stimulation.

Feldberg describes his introduction of eserine and the use of an eserinized leech muscle preparation as a specific and sensitive device for measuring the acetylcholine present in the various effluents (blood, perfusate, etc.). This was the key that opened the way to the eventual conversion of that most obdurate sceptic, the electrophysiologist John Eccles.

Even so, without naming any names, Zenon Bacq[45] reports that even as late as 1950, certain eminent physiologists were still refusing to incorporate the theory of chemical transmission into their teaching.

Eserine or physostigmine, an alkaloid isolated from the Calabar bean, had previously shown by Loewi to be an acetylcholinesterase inhibitor.

John Eccles, awarded the Nobel Prize together with Andrew Huxley and Alan Hodgkin in 1963 for discoveries concerning the ionic mechanisms involved in excitation and inhibition in the peripheral and central portions of the nerve cell membrane.

Receptors and ligands

Among the numerous proteins inserted in the plasma membranes of cells are the receptors. These possess sites, accessible to the extracellular milieu, that bind with specificity soluble molecules, often referred to as ligands. The binding of a just a few ligand molecules may then bring about remarkable changes within the cell as it becomes 'activated' or 'triggered'. Although our knowledge of these interactions is quite extensive, it is not so very long ago that the very notion of a receptor was merely conceptual, indicating the propensity of a cell or tissue to respond in a defined manner to the presence of a hormone or other ligand. Nowadays, the receptors are familiar to us as products of the molecular biology revolution. They can be synthesized in the laboratory in milligram quantities as recombinant proteins; they can be modified by point mutations, deletions, insertions and by the formation of chimeric structures and they can be expressed in the membranes of cells from which they are normally absent. The concept of a specific binding site for a ligand certainly predates the discovery of the first hormones and can be ascribed to John Newport Langley.[50] On the basis of the mutual antagonism of the poisons atropine and pilocarpine (later found to be active at muscarinic cholinergic receptors), he proposed that these substances form 'compounds' in their target tissues to an extent based on the rules of mass action.

I think it is quite clear that if either atropin or pilocarpin is present in the blood of the submaxillary gland, then either pilocarpin or atropin respectively is able in sufficient quantity to produce the effects it produces when present alone in certain other quantity.

The greater the quantity of atropin the greater is this certain other quantity of pilocarpin; when a large quantity of pilocarpin overcomes the correspondingly large quantity of atropin it restores the effect of the secretory fibres less, and causes less secretion than when a smaller dose of pilocarpin overcomes a still paralysing but smaller dose of atropin.

Until some definite conclusion as to the point of action of the poisons is arrived at it is not worth while to theorise much on their mode of action; but we may, I think, without much rashness, assume that there is some substance or substances in the nerve endings or gland cells with which both atropin and pilocarpin are capable of forming compounds. On this assumption then the atropin and the pilocarpin compounds are formed according to some law of which their relative mass and chemical affinity for the substances are factors.

The structures of **atropine** and **pilocarpine** are shown in Figure 3.4, page 41.

Langley was careful to acknowledge the work of others, particularly Luchsinger who had already described the antagonistic actions of atropine and pilocarpine on the sweat glands of the foot of the cat in almost graphic terms:

there exists between pilocarpin and atropin a true mutual antagonism, their actions summing themselves algebraically like wave crests and hollows, like plus and minus. The final result depends simply and solely upon the relative number of molecules of the poisons present.

Twenty-eight years later, in his Croonian lecture of 1906, Langley (quoted in Davenport[24]) stated

Since neither curari nor nicotine, even in large doses, prevents direct stimulation of muscle from causing contraction, it is obvious that the muscle substance which combines with nicotine or curari is not identical with the substance which contracts.

Curari Old spelling of curare (*d*-tubocurarine), an arrow poison used by native South Americans. It is a competitive antagonist of acetylcholine at the neuromuscular junction.

It is convenient to have a term for the specially excitable constituent, and I have called it the receptive substance. It receives the stimulus and, by transmitting it, causes contraction. . . . The mutual antagonism of nicotine and curari on muscle can only satisfactorily be explained by supposing that both combine with the same radicle of the muscle, so that nicotine-muscle compounds and curari-muscle compounds are formed. Which compound is formed depends on the mass of each poison present and the relative chemical affinities for the muscle radicle.

References

1 Hildebrand, E. What does *Halobacterium* tell us about photoreception? *Biophys. Struct. Mech.* 1977; 3: 69–77.
2 Springer, M.S., Goy, M.F., Adler, J. Protein methylation in behavioural control mechanisms and in signal transduction. *Nature* 1979; 280: 279–84.
3 Koman, A., Harayama, S., Hazelbauer, G.L. Relation of chemotactic response to the amount of receptor: evidence for different efficiencies of signal transduction. *J. Bacteriol.* 1979; 138: 739–47.
4 Kenny, J.J., Martinez, M.O., Fehniger, T., Ashman, R.F. Lipid synthesis: an indicator of antigen-induced signal transduction in antigen-binding cells. *J. Immunol.* 1979; 122: 1278–84.
5 Rodbell, M. The role of hormone receptors and GTP-regulatory proteins in membrane transduction. *Nature* 1980; 284: 17–22.
6 Starling, E.H. On the chemical correlations of the functions of the body. *Lancet* 1905; 2: 339–41.
7 Janakidevi, K., Dewey, V.C., Kidder, G.W. Serotonin in protozoa. *Arch. Biochem. Biophys.* 1966; 113: 758–9.
8 Janakidevi, K., Dewey, V.C., Kidder, G.W. The biosynthesis of catecholamines in two genera of protozoa. *J. Biol. Chem.* 1966; 241: 2576–8.
9 Le Roith, D., Shiloach, J., Roth, J., Lesniak, M.A. Evolutionary origins of vertebrate hormones: substances similar to mammalian insulins are native to unicellular eukaryotes. *Proc. Natl. Acad. Sci. USA* 1980; 77: 6184–8.
10 Baulieu, E.-E. Hormones: A complex communication network. In: Baulieu, E.-E., Kelly P. A. (Ed) *Hormones: from molecules to disease.* Chapman & Hall, London, 1990; 3–169.
11 Le Roith, D., Shiloach, J., Berelowitz, M. *et al.* Are messenger molecules in microbes the ancestors of the vertebrate hormones and tissue factors? *Fed. Proc.* 1983; 42: 2602–7.
12 Hunt, L.T., Dayhoff, M.O. Structural and functional similarities among hormones and active peptides from distantly related eukaryotes. In: Gross, E., Meienhofer, J. (Eds) *Peptides: Structure and biological function.* Pierce Chemical Co, Rockford, IL, 1979; 757–60.
13 Vesely, D.L., Giordano, A.T. Atrial natriuretic factor-like peptide and its prohormone within single cell organisms. *Peptides* 1992; 13: 177–82.
14 Vesely, D.L., Gower, W.R. Jr., Giordano, A.T. Atrial natriuretic peptides are present throughout the plant kingdom and enhance solute flow in plants. *Am. J. Physiol.* 1993; 265: E465–77.
15 LeRoith, D., Shiloach, J., Roth, J. *et al.* Evolutionary origins of vertebrate hormones: material very similar to adrenocorticotropic hormone, β-endorphin, and dynorphin in protozoa. *Trans. Assoc. Am. Phys.* 1981; 94: 52–60.
16 Le Roith, D., Liotta, A.S., Roth, J. *et al.* ACTH and β-endorphin-like materials are native to unicelullar organisms. *Proc. Natl. Acad. Sci. USA* 1982; 79: 2086–90.
17 Lam, H.-M., Chiu, J., Hsieh, M.-H. *et al.* Glutamate-receptor genes in plants. *Nature* 1998; 396: 127–8.
18 Skorokhod, A., Gamulin, V., Gundacker, D., Kavsan, V., Muller, I.M. Origin of insulin receptor-like tyrosine kinases in marine sponges. *Biol. Bull.* 1999; 197: 198–206.
19 Nicoll, C.S. Prolactin and growth hormone: specialists on one hand and mutual mimics on the other. *Perspect. Biol. Med.* 1982; 25: 369–81.

20 Meier, A.H., Farner, D.S., King, J.R. A possible endocrine basis for migratory behaviour in the white-crowned sparrow, *Zonotrichia leucophrys gambelii*. *Anim. Behav.* 1965; 13: 453–65.

21 Meier, A.H. Daily rhythms of lipogenesis in fat and lean white-throated sparrows *Zonotrichia albicollis*. *Am. J. Physiol.* 1977; 232: E193–6.

22 Medvei, V.C. *The History of Clinical Endocrinology*. Parthenon Publishing Group, Carnforth, Lancs, 1993.

23 Aminoff, M.J. *Brown-Séquard: A visionary of science*. Raven Press, New York, 1993.

24 Davenport, H. Early history of the concept of chemical transmission of the nerve impulse. *Physiologist* 1991; 34: 129–39.

25 Brown-Séquard, C.E. The effects produced on man by subcutaneous injections of a liquid obtained from the testicles of animals. *Lancet* 1889; 2: 105–7.

26 Brown-Séquard, C.E. Seconde note sur les effets produits chez l'homme par les injection sous-cutanées d'un liquide retiré des testicules frais de cobaye et de chien. *C.R. Soc. Biol. Paris* 1889; 41: 420.

27 Editorial annotation. The new elixir of youth. *Br. Med. J.* 1889; 1: 1416.

28 Editorial annotation. The pentacle of rejuvenescence. *Br. Med. J.* 1889; 2: 446.

29 Berdoe, E. Circulated letter: A serious moral question. MS 980/67. Archives of the Royal College of Physicians, 1889.

30 Editorial annotation. Can old age be deferred? An interview with Dr Serge Voronoff, the famous authority on the possibilities of gland transplantation. *Sci. Am.* 1925; October: 226–7.

31 Short, R.V. Did Parisians catch HIV from 'monkey glands'? *Nature* 1999; 398: 657.

32 Borrell, M. Organotherapy, British physiology and the discovery of internal secretions. *J. Hist. Biol.* 1976; 9: 236–68.

33 Murray, G.R. Note on the treatment of myxodoema by hypodermic injections of an extract of the thyroid gland of a sheep. *Br. Med. J.* 1891; 2: 796–7.

34 Dale, H.H. Accident and opportunism in medical research. *Br. Med. J.* 1948; 2: 451–5.

35 Schäfer, E.A. Present condition of our knowledge regarding the functions of the suprarenal capsules. *Br. Med. J.* 1908; 1: 1277–81.

36 Oliver, G., Schäfer, E.A. On the physiological action of extract of the suprarenal capsules. *Proc. Physiol. Soc.* 1894; 9: i–iv.

37 Oliver, G., Schäfer, E.A. On the physiological action of extract of the suprarenal capsules. *Proc. Physiol. Soc.* 1895; 17: x–xiv.

38 Moore, B.M. On the chemical nature of a physiologically active substance occurring in the suprarenal gland. *Proc. Physiol. Soc.* 1895; 17: xiv–xvii.

39 Bayliss, W.M. *Principles of General Physiology*. Longmans, London 1924; 739.

40 Cannon, W.B. *Bodily Changes in Pain, Hunger, Fear and Rage. An account of researches into the function of emotional excitement* (reprinted 1963). Harper Torch Books, New York, 1929.

41 Bilski, R., Kaulbersz, J. Napoleon Cybulski, 1854–1919. *Acta Physiol. Pol.* 1987; 38: 74–90.

42 Abel, J.J., de Taveau, R. On the decomposition products of epinephrin hydrate. *J. Biol. Chem.* 1905; 1: 1–32.

43 Stolz, F. Über adrenalin und alkylaminoacetobrenzcatechin. *Ber. Dtsch. Chem. Ges.* 1904; 37: 4149–54.

44 Dakin, H.D. The synthesis of a substance allied to adrenalin. *Proc. Roy. Soc. Lond. B* 1905; 76: 491–7.

45 Bacq, Z.M. *Chemical Transmission of Nerve Impulses: An historical sketch*. Pergamon Press, Oxford, 1975.

46 Loewi, O. An autobiographical sketch. *Perspect. Biol. Med.* 1960; 4: 3–25.

47 Hunt, R., de Taveau, R. On the physiological action of certain cholin derivatives and new methods for detecting cholin. *Br. Med. J.* 1906; 2: 1788–91.

48 Dale, H.H. *Adventures in Physiology*. Pergamon Press, Oxford, 1953.

49 Camporesi, P. Bread of dreams. *Hist. Today* 1989; 39: 14–21.

50 Langley, J.N. On the physiology of the salivary secretion: Part II. On the mutual antagonism of atropin and pilocarpin, having especial reference to their relations in the sub-maxillary gland of the cat. *J. Physiol.* 1878; 1.

First messengers

The natural extracellular ligands that bind and activate receptors are best called first messengers, although they may be subdivided as hormones, neurotransmitters, cytokines, lymphokines, growth factors, chemoattractants, etc. Each of these terms attempts to define a class of agents that take effect in a particular setting. However, there can be much overlap, as none of them adheres to a strict definition. This is illustrated by Table 2.1, which provides examples of extracellular first messengers that function as hormones, growth factors and inflammatory mediators, and by Table 2.2 which lists a selection of substances that can act as neurotransmitters. Particular examples of multiple functions are:

- The coenzyme ATP and the cellular metabolite glutamate are neurotransmitters when they are secreted at synapses.

- The gut hormones gastrin, cholecystokinin and secretin are also present in the central nervous system (CNS), where they have diverse functions as neuromodulators (influencing the release of other neurotransmitters).

- Somatostatin, identified originally as a hypothalamic agent suppressing the secretion of growth hormone from the pituitary, also operates in the CNS as a neurotransmitter or neuromodulator. More than this, it is a paracrine agent in the pancreas and a hormone for the liver.

- The growth factor TGFβ also acts as a chemoattractant and as a growth inhibitor.

- Insulin acts not only to regulate glucose metabolism, but also as a growth factor.

- Noradrenaline (norepinephrine), depending on whether it is released at a synapse or from the adrenal medulla, can be considered as a neurotransmitter or as a hormone.

- Thrombin is a growth factor but is also involved in blood clotting as an activator of platelet function.

Hormones

Hormones are commonly released in small amounts at sites remote from the organs they target. On entering the circulation, they are diluted

Endocrine denotes the 'action at a distance' of hormones that may pervade the whole organism, searching out specific target tissues.
Paracrine denotes the action of an extracellular messenger that takes effect only locally. When a substance affects the same cell from which it has been released, the activity (perhaps part of a negative feedback pathway) is termed **autocrine**.

Table 2.1 A selection of first messengers found in the circulation

Class	Messenger	Origin	Target	Major effects
Messengers derived from amino acids	Adrenaline (epinephrine)	Adrenal medulla	Heart Smooth muscle Liver and muscle Adipose tissue	Increase in pulse rate and blood pressure Contraction or dilatation Glycogenolysis Lipolysis
	Noradrenaline (norepinephrine)	Adrenal medulla	Arteriolar smooth muscle	Vasoconstriction
	Serotonin (5-hydroxy tryptamine)	Platelets	Arterioles and venules	Vasodilatation and increased vascular permeability
	Histamine	Mast cells, basophils	Arterioles and venules	Vasodilatation and increased vascular permeability
Peptide hormones	Insulin	Pancreatic β-cells	Multiple tissues	Glucose uptake into cells Glycogenesis Protein synthesis Lipid synthesis (adipose tissue) NOTE: also a growth factor
	Glucagon	Pancreatic α-cells	Liver Adipose tissue	Glycogenolysis Lipolysis
	Gastrin	Intestine	Stomach	Secretion of HCl and pepsin
	Secretin	Small intestine	Pancreas	Secretion of digestive enzymes
	Cholecystokinin	Small intestine	Pancreas	Secretion of digestive enzymes
	Atrial natriuretic factor (ANF or ANP)	Heart	Kidney	Na^+ and water diuresis
	Adrenocorticotrophic hormone (ACTH)	Anterior pituitary	Adipose tissue Adrenal cortex	Lipolysis Production of cortisol and aldosterone
	Follicle stimulating hormone (FSH)	Anterior pituitary	Oocyte Ovarian follicles	Growth Oestrogen synthesis
	Luteinizing hormone (LH)	Anterior pituitary	Oocyte Ovarian follicles	Maturation Oestrogen and progesterone synthesis
	Thyroid-stimulating hormone (thryotropin or TSH)	Anterior pituitary	Thyroid	Generation and release of thyroid hormones
	Parathyroid hormone (PTH)	Parathyroid	Bone Kidney	Release of Ca^{2+} and phosphate Ca^{2+} reabsorption
	Vasopressin (antidiuretic hormone or ADH)	Posterior pituitary	Kidney Arteriolar smooth muscle	Water reabsorption from urine Vasoconstriction to increase blood pressure
	Oxytocin	Posterior pituitary	Uterine smooth muscle Milk ducts	Cervical dilation Milk ejection
Growth factors	Epidermal growth factor (EGF)	Multiple cell types	Epidermal and other cells	Growth
	Somatotropin (growth hormone or GH)	Anterior pituitary	Liver	Production of somatomedins
	Somatomedins 1 and 2 (insulin-like growth factors, IGF1 and 2)	Liver	Bone, muscle and other cells	Growth
	Transforming growth factor β (TGFβ)	Multiple locations	Multiple	Growth control
Eicosanoids	Prostaglandins PGA$_1$, PGA$_2$, PGE$_2$	Most body cells	Multiple	Inflammation, vasodilatation
Membrane permeant hormones	Thyroid hormone, T3 (tri-iodothyronine)	Thyroid	Multiple	Metabolic regulation
	Progesterone	Corpus luteum Placenta	Uterine endometrium	Preparation of endometrial layer Maintenance of pregnancy

Table 2.2 Neurotransmitters

Type	Transmitter	Structure
	Acetylcholine	
Tyrosine or tryptophan-derived	Adrenaline (epinephrine)	
	Noradrenaline (norepinephrine)	
	Dopamine	
	Serotonin	
Amino acids	Glutamate	
	Glycine	
	GABA	
Purine based	ATP	
Neuropeptides	Enkephalins	Tyr-Gly-Gly-Phe-Leu Tyr-Gly-Gly-Phe-Met
	Substance P	Arg-Pro-Lys-Pro-Gln-Gln-Phe-Phe-Gly-Leu-Met
	Angiotensin II	Asp-Arg-Val-Tyr-Ile-His-Pro-Phe
	Somatostatin	Ala-Gly-Cys-Lys-Asn-Phe-Phe-Trp-Lys-Thr-Phe-Thr-Ser-Cys

enormously and are subject to enzymes that break them down. Many of them circulate as complexes with specific binding-proteins, reducing their free concentration further. The result of all this is that their level in the vicinity of a target cell is likely to be extremely low and, accordingly, the cell receptors must possess high affinity. Another consequence is that although a target cell may react in milliseconds to hormone binding, overall response times are in the range of seconds to hours.

Examples of water-soluble hormones are included in the list of first messengers shown in Table 2.1. Adrenaline and noradrenaline (epinephrine and norepinephrine) are released from the adrenal gland and can act within seconds. The peptide hormones vary considerably in size, ranging from a few amino acids to full-scale proteins. For example, the hypothalamic factor that induces the release of thyroid stimulating hormone (TRH or TSH-releasing hormone) consists of only 3 amino acid residues, whereas FSH and TSH are heterodimeric proteins each of about 200 residues.

■ Growth factors

The first reports that fragments of biological tissue could be maintained in a living state *in vitro* appeared over 90 years ago,[1] but routine culturing of dispersed cells did not become established until the 1950s. The successful maintenance of dividing mammalian cells depends on the composition of the culture medium. This is traditionally a concoction of nutrients and vitamins in a buffered salt solution. A key ingredient is an animal serum, such as fetal bovine serum. Without such a supplement, most cultured cells will not replicate their DNA and therefore will not proliferate. An important early discovery was that the best mitogenic stimulus (i.e. inducing mitosis or cell division) is provided by natural clot serum and not plasma. A 30 kDa polypeptide released by platelets was later isolated and shown to have mitogenic properties. It was named platelet-derived growth factor, PDGF (see Chapter 10).

PDGF is a member of a class of messengers, the growth factors, comprising over 40 polypeptides, ranging in size from insulin (5.7 kDa) and EGF (6 kDa), to transferrin (78 kDa). As with the hormones, cells have high affinity, specific receptors for growth factors though binding may elicit multiple effects. Apart from stimulating (or inhibiting) growth, they may also initiate programmed cell death (apoptosis), differentiation and gene expression. The effects of growth factors, unlike those of hormones, may last for days. Also, unlike the hormones their effects are short range, generally influencing the growth and function of neighbouring cells.

■ Cytokines

In parallel with the discovery of growth factors, several extracellular signalling proteins were identified that interact with cells of the immune system. Because they activate or modulate the proliferative properties of this class of cells, they were initially termed 'immunocytokines'. As it became apparent that they also act on cells outside the immune or inflammatory systems, the name was shortened to 'cytokine'. The functions of growth factors and cytokines are so diverse that a clear distinction cannot be made. Thus cytokines include the growth factors already mentioned, along with such molecules as the interferons, tumour necrosis factor α (TNF α), numerous interleukins, granulocyte-monocyte colony stimulating factor (GM-CSF) and many others. Chemokines are cytokines that bring about local inflammation by recruiting inflammatory cells by chemotaxis and subsequently activating them (see Chapter 15).

■   Vasoactive agents

Physical damage to tissues or damage caused by infection generates an inflammatory response. This is a defence mechanism in which specialized cells (mostly leukocytes), recruited to the site, act in a concerted fashion to remove, sequester or dilute the cause of the injury. The process is complex, involving many interactions between cells and numerous extracellular messengers. Among these are cytokines that induce inflammation (pro-inflammatory mediators), or reduce it (anti-inflammatory mediators). To enable the recruitment of leukocytes and to retain a local accumulation of (diluting) fluid, there is vasodilatation and a regional increase in vascular permeability. Agents that bring this about include histamine (released by mast cells and basophils), serotonin (released by platelets) and pro-inflammatory mediators such as bradykinin.

   The eicosanoids are another important family of vasoactive compounds. They are derived from the polyunsaturated fatty acid, arachidonic acid (5,8,11,14-eicosatetraenoic acid) (Figure 2.1). Because arachidonic acid and many of its derivatives possess 20 carbon atoms, they are termed eicosanoids (from the Greek word for 20). They include the prostaglandins, thromboxanes and leukotrienes and they operate over short distances as potent paracrine or autocrine agents, controlling many physiological and pathological cell functions. In the context of inflammation, they cause plasma leakage, skin reddening and the sensation of pain. The drug aspirin possesses anti-inflammatory and pain-relieving properties because it inhibits a key enzyme in the pathway

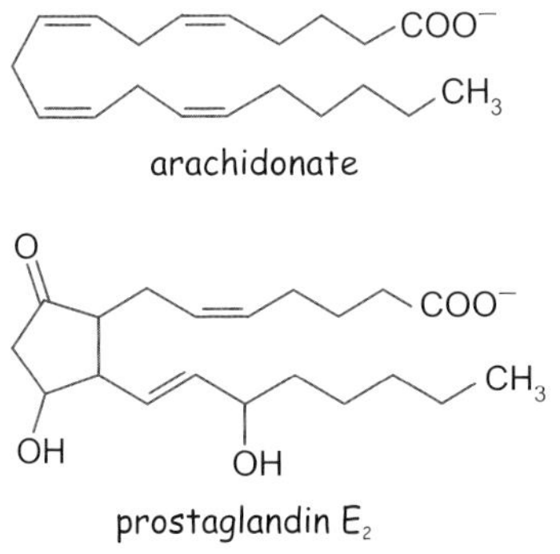

arachidonate

prostaglandin E$_2$

Figure 2.1 **Arachidonate and prostaglandin.**

generating prostaglandins, such as the vasoconstrictor prostaglandin E_2 (Figure 2.1).

■ Neurotransmitters and neuropeptides

Neurotransmitters are also first messengers, but their release and detection at chemical synapses contrasts strongly with that of endocrine signals. In the presynaptic cell, neurotransmitter-containing vesicles are organized to release their contents locally into the tiny volume defined by the synaptic cleft. Individual neurones contain only very small quantities of transmitter and this arrangement ensures that an effective concentration is quickly attained within the cleft following the secretion (by exocytosis) of the contents of relatively few vesicles. The released transmitter then diffuses across the cleft and binds to receptors on the post-synaptic cell. On the macroscopic scale, diffusion is a slow process, but across the short distance that separates the pre- and post-synaptic neurones (~0.1 μm or less), it is fast enough to allow rapid communication between nerves or between nerve and muscle at a neuromuscular junction. Table 2.2 shows the structures of some important neurotransmitters. In the central nervous system glutamate is the major excitatory transmitter, whereas GABA and glycine are inhibitory. Acetylcholine has its most prominent role at the neuromuscular junction where it is excitatory. Whether an excitatory or inhibitory effect is evoked depends on the nature of the ion channel that is regulated by the neurotransmitter receptor (see Chapter 3).

■ Lipophilic hormones

There is a class of hormones that differs in two major respects from the first messengers that we have examined so far. These are the steroid and the thyroid hormones (examples are shown in Figure 2.2). Their principle mode of action is to penetrate cells where they interact with the intracellular receptors which mediate their long-term effects.[2] These in turn penetrate the nucleus to provoke specific mRNA synthesis by binding to promoter elements on DNA. We shall give no further details of their function or mechanisms, other than to point out that progesterone, essential for establishing and then maintaining pregnancy in mammals, is an example of a steroid hormone having activities that emanate from two locations. Activation of intracellular receptors directs transcription and the synthesis of specific proteins. At the cell surface it interferes with the binding of the peptide hormone oxytocin to its receptor in rat uterine tissue.[3] It is possible that it is through inhibition of oxytocin binding that the progesterone steroids maintain uterine quiescence during pregnancy.

   Although we have described differences in the nature of selected first messengers, our aim in this book is to accentuate common aspects. This will become more apparent in later chapters, where it will be seen that similar intracellular mechanisms are used to promulgate an extracellular message. The ultimate effects of these first messengers are determined,

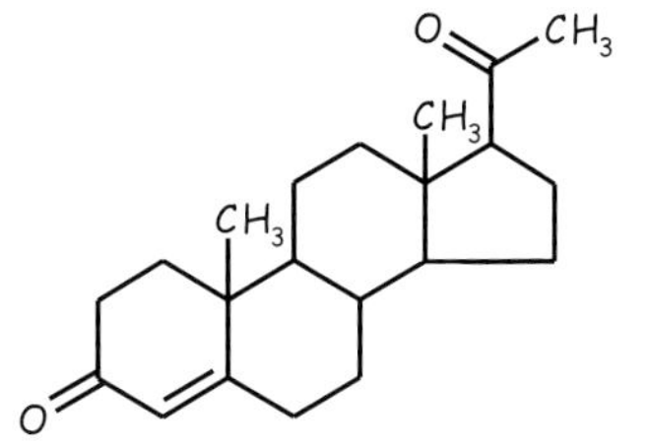

Figure 2.2 **Progesterone and triiodothyronine.**

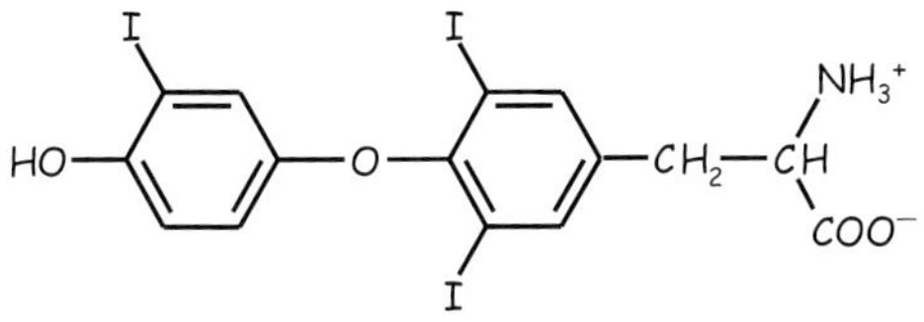

not by their chemical natures, but by the anatomical context within which they operate and by the characteristics of the receptor molecules that sense their presence in different classes of cells.

■ Intracellular messengers

Second messengers formed as a result of the activation of receptors are discussed in subsequent chapters. They include cAMP, Ca^{2+}, diacylglycerol and the 3-phosphorylated inositol lipids. We shall see that many, indeed most responses to stimulation of cell surface receptors are eventually mediated by protein phosphorylation and dephosphorylation. These are the major devices by which the binding of ligand is converted into a cellular consequence. In general, these second messengers interact with a specific set of enzymes (kinases and phosphatases) either directly or indirectly, to phosphorylate or dephosphorylate target proteins at specific sites. With a few special exceptions, phosphorylations due to these second messengers occur on serine and threonine residues.

■ Binding of ligands to receptors

A typical hormone may elicit physiological effects at concentrations as low as 10^{-10} mol/l ($10^{-10} \times 6 \times 10^{23} = 6 \times 10^{13}$ molecules per litre).

To get an idea of how many of these molecules an individual target cell might encounter, consider that a 12 μm diameter cell occupies a volume of 10^{-12} l. This volume of extracellular fluid will contain only 60 molecules of hormone. In spite of the necessary high affinity, it is important to

The value of 60 molecules may be compared with the number of ions of electrolyte present in the same volume – $\sim 1.8 \times 10^{11}$ (for isotonic saline). Furthermore, the same volume of intracellular water will contain about 10^{10} molecules of ATP (5–10 mmol/l) and after hormonal stimulation, a cell might contain as many as 10^6 molecules of the second messenger cAMP.

remember that the binding interaction is never covalent and is always reversible. Thus binding of hormone (H) to receptor (R) can be written:

$$H + R \rightleftharpoons HR$$

At equilibrium, the rates of the forward and backward reaction are equal (on-rate = off-rate) and by the law of mass action

$$k_{on}[H][R] = k_{off}[HR]$$

where k_{on} is the on-rate constant and k_{off} is the off-rate constant and the units of k_{on} and k_{off} are $mol^{-1} s^{-1}$ and s^{-1}, respectively. [H] and [R] represent the concentrations of free (unbound) hormone and free (unoccupied) receptors. A high affinity interaction means that the predominant species is HR. That is, the equilibrium is to the right or the equilibrium constant K is large, where

$$K = \frac{[HR]}{[H][R]} = \frac{k_{on}}{k_{off}}$$

and the rate constant k_{on} is much greater than k_{off}. The units of K are $l\ mol^{-1}$ (i.e. reciprocal concentration). Instead of an equilibrium constant for association, the expression above can be inverted:

$$K_D = \frac{[H][R]}{[HR]} = \frac{k_{off}}{k_{on}} \qquad\qquad \text{Equation 2.1}$$

K_D is the dissociation constant and has units of concentration (mol l^{-1}). It is also the concentration of hormone which, at equilibrium, causes 50% of the receptors to be occupied, i.e. [HR] = [R].

■ Binding heterogeneity

One might imagine that all receptors are equal in terms of the effects that they generate. This is far from the case. Binding site heterogeneity can be caused by the presence of receptor subtypes with different affinities or by receptors of a single class undergoing regulated conformational transitions between high and low affinity states. Since such a process would be driven by intracellular mechanisms, it could be tested for by inhibiting the cells metabolically (preventing ATP synthesis/depleting intracellular ATP). An alternative approach would be to compare the hormone binding affinities of intact cells with their isolated plasma membranes. An experiment of this kind should reveal the extent to which either the affinity or the number of exposed receptor sites are regulated by cellular metabolic status, which again might be modulated by events set in train by the initial hormone–receptor interaction itself.

■ Measurement of binding affinity

To measure the binding parameters, the cells, or their membranes, are exposed to hormone that is radioactively labelled. After equilibration the bound and free hormone pools are separated, generally by centrifugation or filtration. The supernate or filtrate contains the unbound material and the pellet or filtered material consists of cells or membranes bearing the bound hormone. These are briefly washed to remove residual and interstitial unbound hormone and then both samples are analysed for the presence of the label. The range of concentrations over which binding occurs and becomes saturated can be determined by varying the hormone concentration and plotting a graph of bound against free hormone concentration. For the very simplest systems, in which there is only one binding site per receptor and no interactions between receptors that might modify the affinity of binding at unoccupied sites, we can derive an equation for the binding curve, as follows.

If the total concentration of receptors in the preparation is R_{tot}, then

$$[R] = R_{tot} - [HR]$$

Substituting [R] in Equation 2.1 gives the single site equilibrium binding equation:

$$[HR] = R_{tot} \frac{[H]}{K_D + [H]}$$

Plotting [HR] against [H], we obtain a hyperbola (Figure 2.3). As with most representations of data ranging over several orders of magnitude, it is generally more useful to plot the concentrations on a logarithmic scale. By extrapolating the graph to an infinite concentration of hormone, we may then determine the total receptor concentration in the system $[R_{tot}]$ (Figure 2.3). In practice binding data are analysed by non-linear regression using computer software that takes account of multiple binding sites and other complications, yielding values of R_{tot} and K_D.

■ K_D and EC_{50}: receptor binding and functional consequences

Having derived R_{tot} by extrapolation or curve-fitting and knowing the number of cells, we can calculate the number of binding sites per cell. We can also ask how the estimated K_D relates to the likely hormone concentration *in vivo* and the biological responses that are evoked. More simply, is the magnitude of the tissue (or cell) response proportional to the fraction of occupied receptors? More often than not, there is no clear relationship.[4] In many cases, the maximal cellular response – whatever it is that the cell does as a consequence of stimulation – is achieved with only a tiny fraction of the receptors being occupied (Figure 2.4). We should therefore distinguish between K_D and the EC_{50}, the concentration of hormone at which the biological response or effect is half-maximal.

There are different ways of assessing a cellular response. An example is shown in Figure 2.4b in which an early response is estimated by measuring the generation of an intracellular second messenger (cAMP). Alternatively, more distal downstream events might be measured, such as secretion or contraction. Curiously, we find that the EC_{50} for each type of

(a)

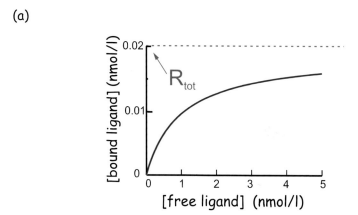

(b)

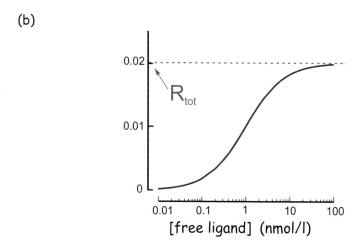

Figure 2.3 Saturation binding. Direct binding curves plotted using (a) linear and (b) semi-logarithmic scales.

response tends to be different. This is because K_D is a parameter describing ligand binding at equilibrium, whereas EC_{50} reflects consequent events that may have multiple kinetic components determined by the rates of enzyme catalysed reactions that vary independently of receptor occupation. The concentration of a hormone inducing a half-maximal response can be as much as two orders of magnitude (100-fold) smaller than the concentration required to saturate 50% of the receptors, as illustrated in Figure 2.4. Similarly, maximal stimulation of glucose oxidation by adipocytes is achieved by insulin when only 2–3% of its receptors are occupied.

(a)

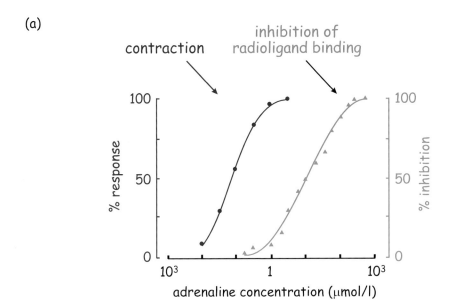

(b)

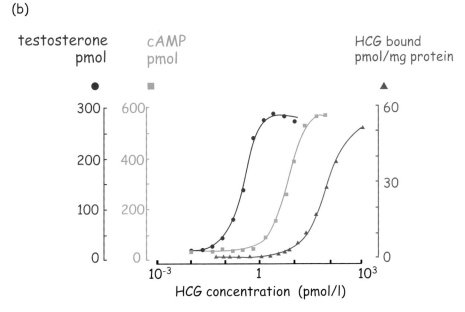

Figure 2.4 **Receptor occupancy and cell response.** (a) Correlation of adrenaline binding at α_1-adrenergic receptors and contraction of smooth muscle (rat vas deferens). In this experiment, receptor occupancy was determined by measuring the displacement of a specific α_1-antagonist, labelled with [125]I. Contraction occurred at ligand concentrations that were too low to have a measurable effect in the binding assay. Adapted from Minneman et al.[5] (b) Correlation of hormone binding (gonadotropin), cAMP generation and testosterone production from testicular Leydig cells. Note the 10^6-fold difference in affinity of the two hormone receptors. Adapted from Mendelson et al.[6]

■ Spare receptors

The receptors that remain unoccupied have unfortunately been termed 'spare receptors',[7] but this is not to suggest that their role is unimportant. This large apparent excess on most hormone-responsive cells helps to ensure that stimulation can occur at very low hormone concentrations. For example, imagine that an optimal cell response is only elicited when *all* of the available cell surface receptors are occupied. This condition could only be achieved at infinite hormone concentration, because the binding follows a saturation relationship as depicted in Figure 2.3. Conversely, if the cell response is optimal when only a small proportion of receptors are occupied, then a concentration of hormone much lower than the K_D will achieve this. The greater the receptor reserve, the lower the EC_{50} will be for a given level of occupancy. Figure 2.5 illustrates this.

The EC_{50} can be shifted by manipulations that alter the activity of the enzymes that take part in the stimulation pathway and that are well removed from events immediately set in train by the receptors. For instance, the drug theophylline is an inhibitor of cAMP phosphodiesterase, the enzyme that breaks down cAMP (Figure 2.6). Its effect is to augment the elevation of cAMP induced by receptor activation. Therefore the steady state concentration of cAMP due to stimulation of receptors can be induced by much lower concentrations of hormone.

■ Down-regulation of receptors

Cells with receptors that are subjected to regular or persistent activation, for example by drugs that are resistant to metabolic breakdown, may

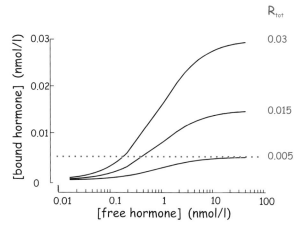

Figure 2.5 Spare receptors. Three saturation binding curves are shown for three different situations. In the upper curve the maximum extent of binding (R_{tot}) exceeds the level required for optimal stimulation (5 pmol/l, indicated by the dotted line) by 6-fold. Maximal activation is therefore achieved at the hormone concentration corresponding to the intersection of the line and the curve. If the receptor reserve is reduced (middle curve) so that R_{tot} is only three times greater than the required occupancy level, the intersection moves to the right and if the reserve approaches zero (R_{tot} = 5 pmol/l), the point of intersection moves towards infinity (lower curve).

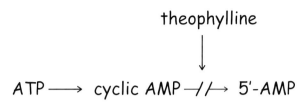

Figure 2.6 **Action of theophylline.**

become less amenable to stimulation. This down-regulation may be due to depletion of the number of exposed receptors, a reduction of their binding affinity or both. Either of these effects will increase the EC_{50} by reducing R_{tot} or increasing K_D. There is a familiar example. The instructions for use of the nasal decongestants based on xylometazoline give a warning not to continue usage beyond 7 days. After this time there is a tendency for the drug to cause 'rebound congestion' and further use then exacerbates instead of relieves the condition. Xylometazoline binds and activates α_2-adrenergic receptors that suppress production of cAMP. However, the specificity, though good, is not perfect and the drug has a low but nonetheless significant affinity for α_1-adrenergic receptors. The result is that when the preferred targets, the α_2-receptors, have eventually been desensitized and removed from the epithelial surface, action is transduced by the α_1-receptors which now predominate. The consequence is nasal congestion, most probably mediated, not by cAMP, but through the activation of phospholipase C, Ca^{2+} and protein kinase C.

An important mechanism of receptor down-regulation, to which we return in Chapter 4, is phosphorylation of the intracellular chains of receptor proteins by enzymes such as the protein kinases A and C and specific receptor kinases. This marks them as targets for removal from the cell surface by endocytosis.[8] It also allows the redirection of signals into different pathways.[9]

All in all, one may conclude that receptors are not static components of cells; they are in a dynamic state that is influenced by both exogenous and endogenous factors.

Discovery of the first second messenger, cAMP

Although experiments with radioactively labelled hormones and related reagents enabled the quantitative measurement of binding parameters of receptors, they told nothing of what receptors are, nor what they do. The critical advance was made in 1957 by Earl W. Sutherland. With his colleagues, he showed that the reactions stimulated by adrenaline or glucagon on liver which result in the activation of glycogen breakdown, occur in two distinct stages. The first of these results in the generation of a heat-stable and dialysable factor.[10] When applied together with ATP to '*dephospho liver phosphorylase*' (now called phosphorylase b), it promoted the phosphorylation reaction resulting in active phosphorylase. In a footnote they point out that

Earl W. Sutherland (1915–1974) awarded the Nobel Prize, 1971, "for his discoveries concerning the mechanisms of action of hormones"

The active factor recently has been purified to apparent homogeneity. From ultraviolet spectrum, the orcinol reaction, and total phosphate determination, the active factor appeared to contain adenine, ribose and phosphate in a ratio of 1:1:1. Neither inorganic phosphate formation nor diminution of activity resulted when the factor was incubated with various phosphatase preparations. . . . However, the activity of the factor was rapidly lost upon incubation with extracts from dog heart, liver and brain.

With the identification of the second messenger as cAMP in 1957, it became possible to link activation of specific classes of receptors with specific biochemical responses. Of course, elevation of cAMP is not the exclusive second messenger response to hormones. Nor is it the exclusive response to adrenaline and other closely related catecholamines. These bind, depending on the cell type or tissue, to a family of catecholamine or 'adrenergic' receptors, α_1, α_2, β_1, β_2, β_3, each of which has several distinct functions. These include the synthesis of cAMP (generally β), suppression of synthesis of cAMP (generally α_2) and elevation of Ca^{2+} (α_1).

References

1 Carrel, A. On the permanent life of tissues outside of the organism. *J. Exp. Med.* 1912; 516–28.
2 McEwan, B.S. Non-genomic and genomic effects of steroids on neural activity. *Trends Pharm. Sci.* 1991; 12: 141–7.
3 Grazzini, E., Guillon, G., Mouillac, B., Zingg, H.H. Inhibition of oxytocin receptor function by direct binding of progesterone. *Nature* 1998; 392: 509–12.
4 Stephenson, R.P. A modification of receptor theory. *Br. J. Pharm.* 1956; 11: 379–93.
5 Minneman, K.P., Fox, A.W., Abel, P.W. Occupancy of α1-adrenrenergic receptors and contraction of rat vas deferens. *Mol. Pharm.* 1983; 23: 359–68.
6 Mendelson, C., Dufau, M., Catt, K. Gonadotropin binding and stimulation of cyclic adenosine 3′:5′-monophosphate and testosterone production in isolated Leydig cells. *J. Biol. Chem.* 1975; 250: 8818–23.
7 Springer, M.S., Goy, M.F., Adler, J. Protein methylation in behavioural control mechanisms and in signal transduction. *Nature* 1979; 280: 279–84.
8 Goodman, O.B., Krupnick, J.G., Santini, F. *et al.* β-Arrestin acts as a clathrin adaptor in endocytosis of the β2-adrenergic receptor. *Nature* 1996; 383: 447–50.
9 Daaka, Y., Luttrell, L.M., Lefkowitz, R.J. Switching of the coupling of the β2-adrenergic receptor to different G proteins by protein kinase A. *Nature* 1999; 390: 88–91.
10 Rall, T.W., Sutherland, E.W., Berthet, J. The relationship of epinephrine and glucagon to liver phosphorylase: Effect of epinephrine and glucagon on the reactivation of phosphorylase in liver homogenates. *J. Biol. Chem.* 1957; 224: 463–75.

Receptors

Here we begin with a description of the receptors for the catecholamines (adrenaline, noradrenaline and dopamine) and for acetylcholine. These provide the basic paradigms by which the main classes of hormone, neurotransmitter and drug receptors have been defined.

Adrenaline (again)

Adrenaline (or epinephrine) is the hormone that is secreted in anticipation of danger, preparing the body for fight or flight.[1]

As Cannon has pointed out, this secretion (the internal secretion of the medulla of the suprarenal bodies) is poured into the blood during conditions of stress, anger, or fear, and acts as a potent reinforcement to the energies of the body. It increases the tone of the blood vessels as well as the power of the heart's contraction, while it mobilises the sugar bound up in the liver, so that the muscles may be supplied with the most readily available source of energy in the struggle to which these emotional states are the essential precursors or concomitants.[2]

The immediate actions are those of arousal: the sweat glands secrete and the body-hair stands on end; the pupils dilate to gather more light; the bronchi dilate to improve oxygen supply; the heart quickens and increases in its force of contraction. Then, through dilation of relevant vessels and constriction of others, the blood supply to the myocardium and selected skeletal muscles is increased, while the supply to the skin and visceral organs is reduced, postponing digestive and anabolic processes. Metabolic fuel stores are also mobilized (breakdown of liver and muscle glycogen and ultimately the catabolism of fat). Thus the secretion of adrenaline is familiar in the context of emergencies. However, it is important to remember that the adrenal gland continuously produces low levels of the hormone and that adrenaline is also released into the circulation during exercise.

Adrenaline belongs to the group of hormones and neurotransmitters classed as catecholamines. The other members are dopamine and noradrenaline. Their structures are shown in Table 2.2 (page 21). The precursor from which they are formed, by a sequence of steps, is the amino acid L-tyrosine. With a few exceptions, receptors are stereospecific and the catecholamines are active only as their L-enantiomers. Adrenaline differs from noradrenaline by the presence of a N-methyl group on the side chain. In humans, the chromaffin cells of the adrenal medulla store and secrete both adrenaline (~80%) and noradrenaline (~20%), and noradrenaline is also secreted by sympathetic neurones.

■ α- and β-adrenergic receptors

The receptors that respond to adrenaline or noradrenaline are called adrenergic receptors and they are present in a wide range of tissues. They may be divided into two major classes, α and β, although, as shown in Table 3.1, these are not defined in a precise way, either by their location or by the effects they elicit. Neither are they defined by their affinity for their natural first messengers: the differences in their sensitivity to either adrenaline or noradrenaline are not large.

Table 3.1 Properties that distinguish α- and β-adrenergic receptors

	α	β
Arteriolar smooth muscle	Vasoconstriction in viscera (other than liver)	Vasodilation in skeletal musculature and liver (β_2)
Bronchial smooth muscle	(not present)	Bronchodilation (β_2)
Uterine smooth muscle (myometrium)	Contraction (α_1)	Relaxation (β_2)
Heart		Inotropic effect (increased cardiac contractility) (β_1) Increased heart rate (β_1)
Eye	Pupillary dilation (α_1)	Lowering of intra-ocular pressure (β_2)
Skeletal muscle		Glycogenolysis (β_2)
Liver	Glycogenolysis (α_1)	Glycogenolysis (β_2)
Adipose tissue	Inhibition of lipolysis (α_2)	Lipolysis (β_1)
Pancreas	Inhibition of insulin secretion (α_2)	Insulin secretion (β_2)
Platelets	Inhibition of aggregation (α_2)	
Order of potency	adr ≥ nor ≫ iso*	iso > adr ≥ nor*

* Abbreviations: adr, adrenaline; nor, noradrenaline; iso, isoprenaline.

Adrenergic receptor agonists and antagonists

Agonists

What must characterize the different adrenergic receptors ultimately are the differences in their molecular structures. Knowledge in this area is still only partial, however, but we do have information from their interactions with synthetic ligands that are agonists or antagonists. In fact, they may be further subdivided on this basis, into four well-characterized subtypes α_1, α_2, β_1 and β_2 (Figure 3.1).

Although a particular subtype, by virtue of its dominance in a given tissue, may mediate a particular response, what really distinguishes the subtypes is their responsiveness to synthetic agonists such as isoprenaline

Molecular cloning has confirmed the existence of β_3 adrenergic receptors. Similar to β_1 and β_2 receptors, these are linked to adenylyl cyclase through the G-protein Gs, but differ in not being subject to phosphorylation by protein kinase A (PKA) or β-adrenergic receptor kinase (βARK). They possess their own characteristic profile of specific activators and inhibitors. They are expressed predominantly in adipose tissues where they play an important role in the regulation of lipid metabolism. For review, see Strosberg.[3]

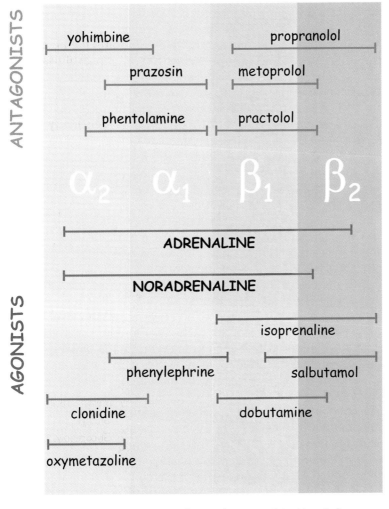

Figure 3.1 Adrenergic receptor agonists and antagonists. Lines indicate approximate specificities. Note that at high concentrations the selectivities tend to be reduced. Figure adapted from Hanoune.[4]

(isoproterenol) which causes blood vessels to constrict ($\beta > \alpha$), and phenylephrine ($\alpha > \beta$) which causes them to dilate. Adrenaline can do either, depending on the particular vessel. Clonidine (α_2) and salbutamol (β_2) are even more specific, since they can distinguish particular subtypes.[3]

Antagonists

As indicated in Figure 3.1, there are also synthetic antagonists. The structures of some selected examples are shown in Figure 3.2. Propranolol inhibits the action of β-receptors. This drug finds application in the treatment of certain types of hypertension because it inhibits the secretion of the peptide hormone renin from the juxtaglomerular apparatus of the kidney. Renin is required for the formation of angiotensin from angiotensinogen. There are other β-blockers, such as practolol and metoprolol, which at low concentrations are selective for β_1-receptors. Thus, in principle, it is possible to devise specific β_1 antagonists to suppress cAMP with which one could treat a heart patient without antagonizing lung β_2 receptors that would lead to bronchconstriction in patients at risk from asthma. Of course, the selectivity between receptor subtypes is rarely perfect, especially at higher dose levels. Also other factors such as the availability of the free form (e.g. not bound to plasma proteins), the tissue distribution, *in vivo* half-life and toxicity have to be taken into account before drugs can be used in clinical practice. Indeed, the first β-blockers failed to reach the clinic because they were found to induce tumours in mice.

How do receptors distinguish agonists from antagonists?

Why do some compounds, such as salbutamol, cause activation, whereas others, such as propranolol, are inhibitors of (β-)adrenergic receptor function? Certainly there are some systematic structural features that appear to determine their biological effects. In general, it is found that increasing the substitution on the amine nitrogen atom of catecholamines, increases the preference for β- over α-receptors. The β-blockers are all characterized by the absence of the catechol nucleus (aromatic ring with neighbouring hydroxyls) and by the presence of an $-OCH_2-$ group linking the side chain to the aromatic moiety.

An indication that the interaction of ligands with β-receptors differs for agonists and antagonists is given by the finding that reducing the temperature increases the binding affinity of agonists such as isoprenaline, but has little effect on the binding of antagonists such as practolol. A temperature-dependent increase in affinity indicates an equilibrium shift characteristic of an enthalpy-driven reaction, whereas an insensitivity to temperature changes suggests that a change in entropy is the main driving force. Indeed, estimates of the thermodynamic parameters reveal that the antagonists bind with an increase in entropy, which is to be expected when water molecules organized around a binding site are displaced. Surprisingly, the agonists mostly show a decrease in entropy on binding. This might mean that although the two classes of ligands may occupy the same

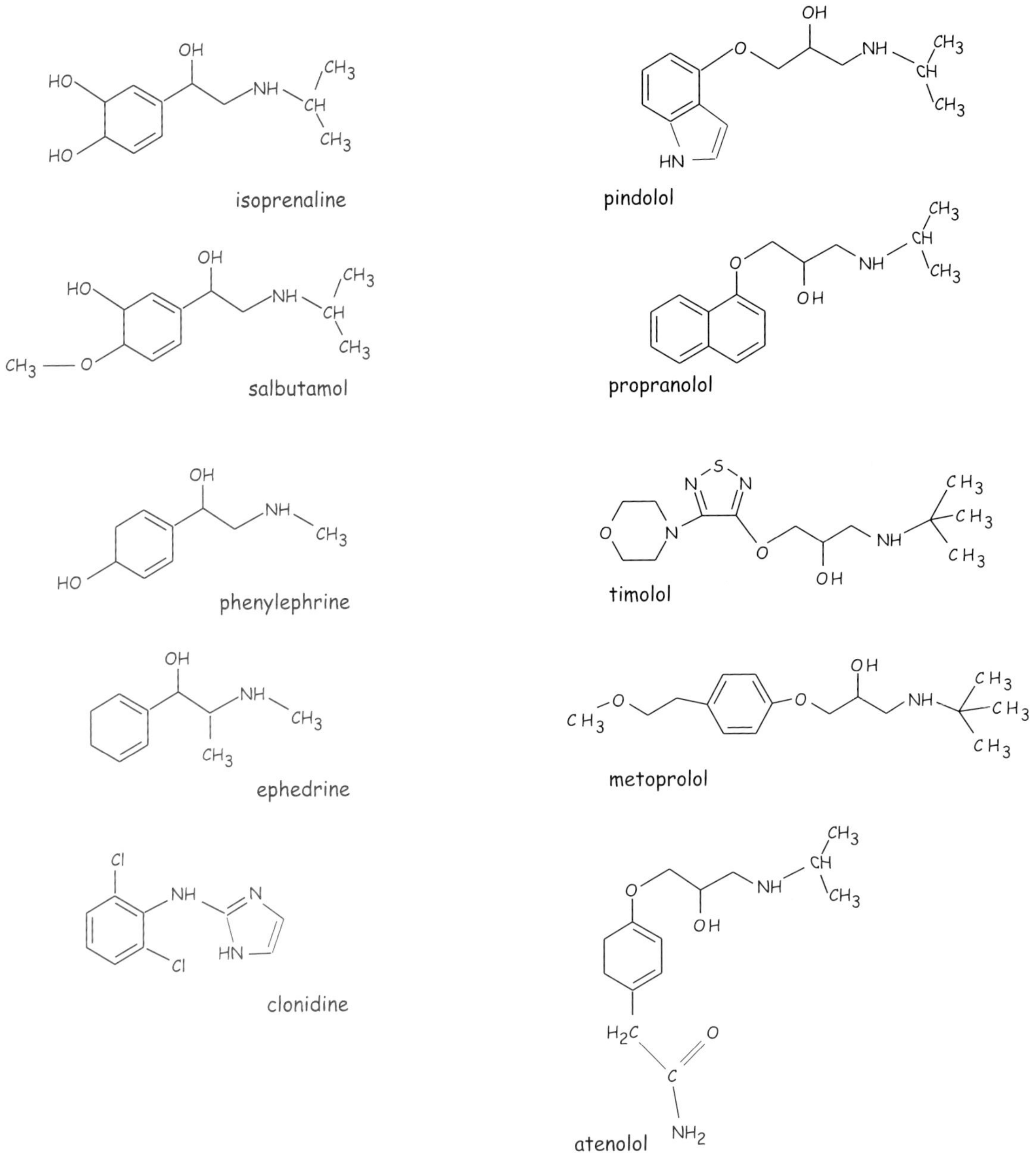

isoprenaline

pindolol

salbutamol

propranolol

phenylephrine

timolol

ephedrine

metoprolol

clonidine

atenolol

Figure 3.2 Structures of some drugs that interact with adrenergic receptors. A selection of the adrenergic drugs mentioned in this book. As the substitution of the amine nitrogen on the agonists (in red) is increased, their selectivity for β- over α-receptors increases. Some of the α-adrenergic agonists and all the β-blocking agents (in black) lack the catechol nucleus. The antagonists are also characterized by the insertion of an ether linkage (-O(CH$_2$)-) in the lateral chain. The structures of the physiological catecholamines adrenaline, noradrenaline and dopamine are shown in Table 2.2 (page 21).

binding pockets within the receptor molecule, which incidentally would explain the competitive nature of their binding, the actual points of attachment to the lining of the pocket (i.e. to particular amino acid residues) are different. The consequences of binding are alterations such as displacement of protons and water, breakage of hydrogen bonds, disturbance of van der Waals interactions and conformational changes in the receptor itself. In the case of agonists, this results in an overall increase in order (decreased entropy).[5] Importantly, only agonists induce the conformational alterations that enable the receptors to communicate with the ensuing components of the signal transduction apparatus.

■ Acetylcholine receptors

At the molecular level, all the effects of the catecholamines, including dopamine, are mediated through the single class of 7TM receptors (see below) and subsequently they all activate or inhibit an enzyme such as adenylyl cyclase or phospholipase C (see Chapter 5). This is always mediated through a GTP-binding protein (GTPase) (see Chapter 4). A large proportion (of the order of 60%) of the drugs used in clinical practice are directed at receptors of the 7TM class. In contrast, acetylcholine interacts with two very distinct classes of receptor that are quite unrelated to each other. These are the nicotinic receptors (ion channels) and muscarinic receptors (7TM receptors).

■ Acetylcholine

Although acetylcholine is a first messenger that interacts with receptors, it does not have the function of a hormone. It is confined to synapses between nerve endings and target cells. It is the primary transmitter at the neuromuscular junction (between motor nerve and muscle end plate). In the autonomic nervous system, it is also the transmitter at preganglionic nerve endings and in most parasympathetic postganglionic nerves. Parasympathetic stimulation through the vagus nerve causes the dilation of blood vessels, increases (fluid) secretions (as from the pancreas and the salivary glands) and slows the heart rate. At the neuromuscular junction, acetylcholine is released from the presynaptic terminal. It diffuses across the synaptic cleft and interacts with nicotinic receptors situated on the postsynaptic membrane. It is then removed from the synaptic cleft with 'flashlike suddenness', (in the words of Henry Hallett Dale). This occurs partly by diffusion, but mainly by the action of acetylcholine esterase, which converts it to choline and acetate within milliseconds. Since the affinity of nicotinic receptors for acetylcholine ($K_D \cong 10^{-7}$ mmol/l) is rather moderate (at least, in comparison with some other types of receptor for their respective ligands), the rate of dissociation is sufficiently fast to allow it to detach rapidly (recall that $K_D = k_{off}/k_{on}$), following the steep decline in the local concentration due to the activity of the esterase. Because of the extreme lability of acetylcholine, in experimental investigations it has been normal to work with stable, non-hydrolysable derivatives such as carbamylcholine (carbachol). As with the natural compound, carbachol is agonistic both at nicotinic and muscarinic receptors.

The compounds that inhibit the hydrolysis of acetylcholine are among the most toxic known (Figure 3.3). Thus, they may cause stimulation of cholinergic receptor sites throughout the CNS (central nervous system), depression at autonomic ganglia and paralysis of skeletal muscles. This is followed by secondary depression involving irreversible receptor desensitization, discussed below. In addition to these nicotinic functions, muscarinic responses to acetylcholine also tend to persist.[6,7]

Cholinergic receptor subtypes

Acetylcholine receptors fall into two classes, originally distinguished by their responses to the pharmacological agonists nicotine and muscarine. In this section we also give brief consideration to related receptors that respond to amino acid neurotransmitters.

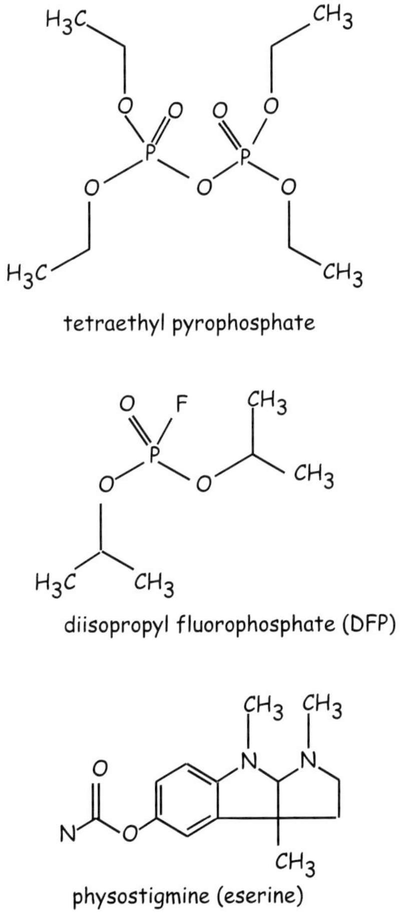

tetraethyl pyrophosphate

diisopropyl fluorophosphate (DFP)

physostigmine (eserine)

Figure 3.3 **Some inhibitors of serine esterases.**

The first synthesis of an irreversible inhibitor of cholinesterase was reported as early as 1854 by Clermont, predating the isolation of the alkaloid physostigmine (eserine) from Calabar beans by about 10 years (see Chapter 1). This was tetraethyl pyrophosphate. In his dedicated pursuit of science, Clermont even went so far as to comment on its taste and one may wonder that he ever survived to tell the tale. Further organophosphorus compounds were developed with the aim of preparing nasty surprises for flies (parathion, the most widely used insecticide of this class) and subsequently even nastier surprises for soldiers and, sadly, civilians too. These nerve gases include sarin (isopropyl-methylphosphonofluoridate), originally developed in Nazi Germany. Four years after the attack by Saddam Hussein's forces on the village of Birjinni on 25 August 1988, soil samples revealed traces of the sarin breakdown products. The LD$_{50}$ for humans is estimated (not tested!) as being of the order of 10 µg kg^{-1}. When sarin was released on the Tokyo metro system by zealots of the Aum cult, 12 people were killed and 5500 were injured.

All these compounds take their effects by alkylphosphorylation of active site serines, not only of acetylcholinesterase, but also of the serine esterases and serine proteinases (such as trypsin, thrombin, etc.). Clearly, the rapid removal of acetylcholine from the synapse is as essential for the process of neurotransmission as is its release. Interestingly however, some cholinesterase inhibitors of this class have found their way into clinical practice (e.g.

in the treatment of glaucoma).

With the threat of military and terrorist abuse of these noxious agents, attention has been turned to the matter of antidotes. Currently, atropine is administered to prevent over-occupation of the acetylcholine receptors. In addition, patients are treated with nucleophilic agents based on 2,3-butanedione monoxime with the aim of dephosphorylating and so reactivating the poisoned acetylcholine esterase.[6,7]

Tubocurarine is the main constituent of curare, a mixture of plant alkaloids obtained from *Chondrodendron tomentosum* or *Strychnos toxifera*, used as an arrow poison by some South American Indians. Claude Bernard (1856) showed that it causes paralysis by blocking neuromuscular transmission; Langley (1905) suggested that it acts by combining with the receptive substance at the motor endplate (see Chapter 1). Other related (synthetic) compounds such as gallamine differ mainly in their duration of action.

Nicotine Named after Jacques Nicot, the French ambassador to Lisbon who introduced the tobacco plant, *Nicotiana tabacum* into France in 1560.
Muscarine A product of the poisonous mushroom, *Amanita muscaria.*
Atropine From the Greek *a-tropos*, meaning no turning or inflexible, referring to the condition of the pupil when dilated by treatment with atropine. It is an alkaloid extracted from the berries of deadly nightshade, *Atropa belladonna*. The content of a single berry is sufficient to kill an adult human.

■ Nicotinic receptors

At the neuromuscular junction acetylcholine acts through nicotinic cholinergic receptors to stimulate the contraction of skeletal muscle. It also transmits signals in the ganglia of the autonomic nervous system. The receptors are rather non-specific cation channels, which conduct both Na^+ and K^+, the principal cations present in cells and extracellular fluid. Nicotinic receptors are antagonized by tubocurarine.

Neither nicotine nor tubocurarine has any significant effects on muscarinic receptors. By the same token, muscarine and atropine, compounds that activate and inhibit the muscarinic receptors are without significant effect on processes regulated by nicotinic receptors.

The nicotinic receptors are members of a superfamily of membrane proteins, that include the ionotropic receptors for serotonin (5-hydroxytryptamine, 5-HT), glycine and γ-aminobutyric acid (GABA). Note however that the glycine and GABA receptors are anion channels responding to ligands that cause electrical inhibition (see below). For all of these, the receptor and ion channel functions are intrinsic properties of the same protein.

■ Muscarinic receptors

Whereas acetylcholine acts through the nicotinic receptor to stimulate contraction of skeletal muscle, it decreases the rate and force of contraction of heart muscle through muscarinic receptors. These belong to the seven transmembrane-spanning (7TM) superfamily of G-protein linked receptors and they exist as five subtypes, M1, M2, M3, m4 and m5. All five subtypes are present in the CNS. M1 receptors are expressed in ganglia and a number of secretory glands. M2 receptors are present in the myocardium and in smooth muscle and M3 receptors are also present in smooth muscles and in secretory glands (Table 3.2).

Muscarine and related compounds have many actions, dependent on the tissues to which they are applied and the receptor subtypes that are expressed. The structures and the mechanisms of action of the receptors activated by muscarine are closely related to those activated by the catecholamines. Like the adrenergic receptors, the diversity of their functions first became apparent with the development of a specific antagonist. This was pirenzepine, which blocks M1 receptors and prevents gastric acid secretion, while being without effect on a number of other responses elicited by muscarinic receptors.

Although the muscarinic and nicotinic receptors for acetylcholine share a common physiological stimulus, and although the muscarinic receptors can regulate ion fluxes (for example in the heart), they are not in themselves ion channels. The difference goes much further than this. Not only are these receptors unrelated structurally and in evolutionary terms, but the conformation of the acetylcholine as it binds is likely to be different. Whereas nicotine and muscarine are very different from one another (Figure 3.4) and contain ring structures that lend rigidity, the acetylcholine molecule is more flexible and is able to adopt different

Table 3.2 Muscarinic receptor subtypes

By convention, upper-case letters (MI, etc.), are used to assign receptors that have been pharmacologically defined, whereas lower-case letters (m4, etc.) are used to assign those receptors that have been revealed only by molecular cloning.

Subtype	Antagonists	Tissue[a]	Transducer	Effector
MI	Atropine, pirenzipine	Autonomic ganglia	G_q	Phospholipase C (increased cytosol Ca^{2+})
M2	Atropine, AF-DX 384[b]	Myocardium Smooth muscle	G_i, G_o	Activation of K^+ channels, inhibition of adenylyl cyclase
M3	Atropine	Smooth muscle Secretory glands	G_q	Phospholipase C (increased cytosol Ca^{2+})
m4	Atropine, AF-DX 384[b]		G_i, G_o	Inhibition of adenylyl cyclase
m5	Atropine		G_q	Phospholipase C (increased cytosol Ca^{2+})

[a] Multiple subtypes occur in many tissues. All are present in the CNS. This column indicates tissues with high levels of expression.
[b] A benzodiazapine drug.

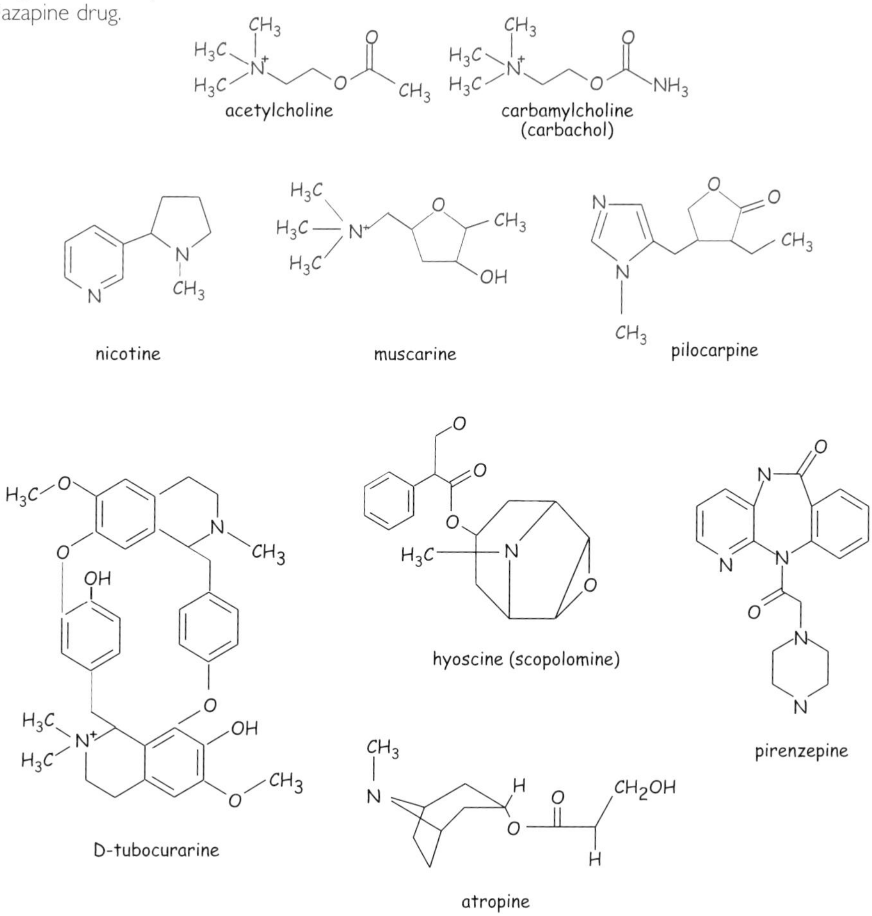

Figure 3.4 **Structures of nicotine, muscarine and some of their antagonists.**

conformations that may be stabilized as it slots into its binding sites in the two receptors.

As with the adrenergic receptors, the response pathways operated by the muscarinic receptors for acetylcholine are all mediated through GTP-binding proteins. In the case of M1, M3 and probably the m5 sub-type, the transduction is by G-proteins of the G_q family and this causes activation of phospholipase C (PLC), hydrolysis of phosphatidylinositol-4,5-bisphosphate (PIP_2) and release of inositol-1,4,5-trisphosphate (IP_3) which mobilizes intracellular Ca^{2+} stores (for detailed discussion, see Chapter 7). The M2 and m4 receptors address G-proteins of the G_i class with consequent inhibition of adenylyl cyclase and activation of K^+ channels (see Chapter 5).

■ Receptors related to the nicotinic receptor
   Amino acid receptors

In parallel with the receptors for acetylcholine, GABA, glutamate and serotonin also act at structurally and mechanistically distinct classes of receptor.

These are the metabotropic $GABA_B$ and glutamate (mGlu) receptors that communicate through G-proteins to activate PLC and mobilize intra-cellular Ca^{2+} on the one hand, and the ionotropic $GABA_A$, $GABA_C$ and glu-tamate (iGlu) receptors on the other, which are ligand-activated ion channels. Again, the ionotropic channel receptors and the metabotropic receptors for GABA and glutamate are antagonized by quite distinct classes of inhibitors.

> The term **metabotropic** is used to distinguish these G-protein linked receptors for glutamate, GABA, etc. from the **ionotropic** receptors that possess integral ion channels operated by the same ligands.

■ Ion channel-linked receptors

■ Nicotinic receptors are ion channels

Of all the receptors, it is probably true to say that the nicotinic cholinergic receptor is the most well-known and the best understood. There are sev-eral reasons for this:

*   Firstly, specialized tissues exist in which nicotinic receptors are pres-ent in huge quantities.

*   Secondly, there are toxins which bind specifically to the receptor with high affinity, enabling its isolation.

*   Finally, patch clamp electrophysiology has allowed the characteri-zation of the channel properties of individual nicotinic receptor molecules.

■ *Electric organs provide a source*

Tissues in which nicotinic receptors are abundant include the neuromus-cular junction, where there may be as many as 4×10^7 copies on a single motor endplate of skeletal muscle (Figure 3.5). When acetylcholine is secreted from the presynaptic membrane into the synaptic cleft, it binds

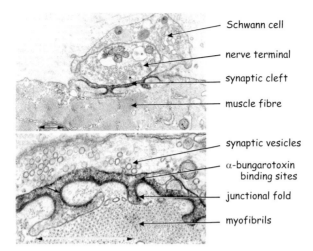

Schwann cell

nerve terminal

synaptic cleft

muscle fibre

synaptic vesicles

α-bungarotoxin
  binding sites

junctional fold

myofibrils

Figure 3.5 α-Bungarotoxin staining of nicotinic receptors at a neuromuscular junction. Thin section electron microscope photographs of a neuromuscular junction from frog cutaneous pectoris muscle. The nerve terminal containing some mitochondria and numerous synaptic vesicles is encased by a Schwann cell, and it abuts a muscle fibre. The preparation has been treated with α-bungarotoxin fused with horseradish peroxidase so that on treatment with 3,3′-diaminobenzidine, the acetylcholine receptors are revealed by the presence of the electron-dense reaction product. The label can be seen to be mainly confined at the extracellular surface of the post-synaptic membrane with some diffusion into the synaptic cleft. The scale bars are 0.5 μm (from Burden *et al.*[8]).

to receptors on the post-synaptic surface. This permits Na$^+$ (present in the extracellular space) to flow through the nicotinic channels, down its concentration and voltage gradient into the muscle cell, so causing a local depolarization that leads ultimately to muscle contraction. This abundance of receptors is vastly exceeded in the electric organs of the marine ray *Torpedo* and the fresh water electric eel *Electrophorus*. These electroplax organs are specialized developments of skeletal muscle and they express huge numbers of receptor molecules, about 2×10^{11} per endplate. As early as 1937, it was discovered that *Torpedo* electroplax could hydrolyse more than its own weight of acetylcholine in an hour. The first demonstration of the biosynthesis of acetylcholine by a cell-free electroplax preparation was reported in 1944.

α-Bungarotoxin blocks neuromuscular transmission

Because of their abundance in these specialized sources, it became possible to isolate, purify and reconstitute nicotinic receptors, albeit in small quantities before the advent of the cloning era. The isolation was achieved by affinity chromatography, using toxins that block neuromuscular transmission, in particular the peptide α-bungarotoxin isolated from snake venom. This peptide (8 kDa) binds specifically and with high affinity to nicotinic receptors ($K_D = 10^{-12}–10^{-9}$ mol/l), with consequent inactivation. In the laboratory, the toxin can be immobilized by attaching it covalently to a solid phase, such as Sephadex. By applying to this affinity column a solution of proteins released by detergent from muscle or electroplax, it is

α-Bungarotoxin binds at the agonist binding site of neuromuscular-type acetylcholine receptors. There are similar snake venom α-toxins that bind to the neuronal isoforms.

possible to isolate the receptor molecules.[9–11] Although X-ray crystallographic information is so far lacking, the quantity and organization of the protein has made structural determination by electron microscopy feasible (although at a lower level of resolution).

Patch clamp electrophysiology of single ion channels

With the introduction of the patch clamp technique by Neher and Sakmann,[12] it became possible to study individual ion channels *in situ* in a membrane.

Erwin Neher and **Bert Sakmann** shared the 1991 Nobel Prize for their discoveries concerning the function of single ion channels in cells.

It is hard to overestimate the contribution that this advance has made to our understanding of channel mechanisms. The idea itself is simple enough. A patch pipette, which is a glass microelectrode with a smooth (fire-polished) tip, 1–5 μm in diameter, is pressed gently against the outside of a cell. With gentle suction a very tight seal forms between the pipette tip and the plasma membrane. The seal resistance is typically more than a gigohm (10^9 Ω). Two alternative manoeuvres can then be used to remove a small area of membrane from the cell without breaking the 'gigaseal' (Figure 3.6). The end result is either an outside-out or an inside-out patch of membrane. These contain only a very few ion channels but, depending on the orientation, the ligand binding sites are either exposed to the exterior or occluded within the patch-pipette. The paucity of channels within the patch and the high-resistance seal then enable recordings of membrane current to be made, for example under voltage clamp conditions. These reveal individual channel openings and closures. Nicotinic receptors were among the first channels to be characterized in this way. Like most other ion channels they open and close abruptly in an all-or-none fashion and they do this spontaneously. The effect of acetylcholine is to increase the probability of a particular channel to be in its open state.

The nicotinic receptor of the neuromuscular junction is composed of five subunits of which two are identical, so that the complex comprises α_2, β, γ, δ (Figure 3.7). All the subunits are necessary for the functioning of the complex as a channel. The acetylcholine binds at pockets formed at the interfaces of the two α subunits with their δ and γ neighbours. Both binding sites must be occupied in order to activate the receptor. α-Bungarotoxin binds competitively, in the close vicinity of the acetylcholine binding sites. Chemical crosslinking experiments indicate that the subunits are organized around a central axis of fivefold symmetry, as shown in Figure 3.7. On the extracellular surface (synaptic space), the assemblage protrudes about 6 nm, in the form of a large funnel (see Figure 3.11). The central pore at the opening has a diameter of about 25 nm and this becomes narrower in the region where it traverses the membrane. On the obverse side of the membrane, the structure again protrudes about 2 nm into the cytoplasm.

The primary structure of the individual subunits indicates the presence of extensive stretches of sequence homology (35–40% identity) and in addition to this, there are many conservative substitutions. The four different peptides are understood to have evolved from a common ancestor

layout

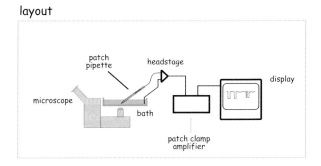

pipette/cell configurations

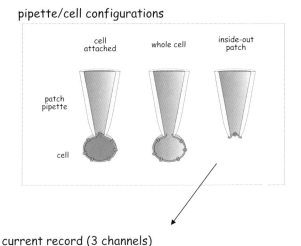

current record (3 channels)

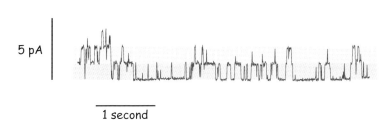

5 pA

1 second

Figure 3.6 Using membrane patches to record single channel opening and closure. Typically, cells are allowed to adhere to the bottom of a shallow dish. Individual cells are then attached to a patch pipette. A typical current record obtained from a patch of membrane detached from a cell is shown.

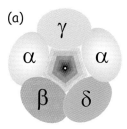

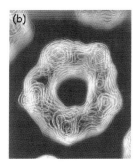

Figure 3.7 **The barrel structure of the nicotinic receptor:** (a) Organization of the five subunits that comprise the nicotinic receptor. According to Changeux, the subunits are shown in the order -αγαδβ-, but there is some uncertainty about this assignment: Unwin suggests αβαγδ. (b) Contour diagram illustrating the electron density of the nicotinic receptor. Successive sections at 0.2 nm intervals reveal the five subunits surrounding the central pathway which is about 2 nm wide at this level. The point of observation is from the outside of the membrane. The binding sites for acetylcholine are situated in clefts formed at the interfaces of the α and γ, and α and δ subunits. From Nigel Unwin, MRC Laboratory of Molecular Biology, Cambridge.

by a process of successive gene duplication and modification that has its origin more than 1.5×10^9 years ago. Inspection of four homologous sequences (M2) from each of the subunits α, β, α_2, β, γ, and δ (listed in Table 3.3), makes this evident. Within these short stretches of 26 amino acids, there are seven points of identity and then four more which represent conservative substitutions. Notice also that these four sequences each contain 15–17 strongly hydrophobic amino acids (leucine, valine, isoleucine, etc., having hydrophobicity assignments > 0: see Table 3.4).

Table 3.3 M2 sequences of the four subunits of the nicotinic receptor

	α7	β1	γ	δ	Identical (i) Conservative (c)
258	E	D	K	K	
	A	A	A	S	
	V	L	V	I	c
	L	L	L	L	i
254	L	L	F	L	
	M	L	L	L	
	F	F	F	F	i
251	V	V	V	V	i
	T	T	T	S	c
	L	L	Q	Q	
	S	T	A	A	
247	L	L	L	L	i
	L	L	L	L	i
	V	A	V	V	c
244	T	F	N	S	
	I	I	I	I	i
	G	S	A	A	
241	L	L	V	V	c
	S	G	T	S	
	I	M	C	T	
	K	K	K	K	i
237	E	E	Q	E	c
	-	-	G	G	
	G	G	G	S	
	G	A	A	D	
234	D	D	K	A	
	A	P	A	P	
	P	P	P	L	

M2 sequences of the four subunits of the nicotinic receptor. Numbers in the left-hand column refer to the residues in the α7-subunit that become labelled when exposed to water-soluble photoaffinity agents such as chlorprormazine (see Figure 3.10). Based on data for human α7, β1, γ and δ subunits published in the LGIC database[13] (see http://www.pasteur.fr/recherche/banques/LGIC/LGIC.html).

In order to understand how the receptor operates as an ion channel, we need to know how each individual component is organized as a transmembrane structure and then how they are arranged relative to each other.

by all five α-helical M2 segments together with contributions from M1. At the level of resolution provided by electron microscopy (9 Å, 0.9 nm), the entrance to the channel can be discerned (see Figures 3.11 and 3.12).

The narrowest section probably constitutes the selectivity filter. It is located at the border of the membrane with the cytoplasm and is composed solely of residues present on the M2 segments. At this point, the M2 probably forms an extended loop instead of the α-helical structure that lines the wider section of the channel. All aspects of ion selectivity are affected by mutations in this loop section (Figure 3.10). These include discrimination between cations and anions, the permeability to divalent cations and the relative permeability to different monovalent cations.[21–26]

■ Receptor desensitization

When a receptor has been exposed to acetylcholine for seconds or minutes (instead of milliseconds) it becomes unresponsive to further stimulation. This phenomenon, known as desensitization, is regulated by the β, γ and δ subunits. Although these subunits do not bind acetylcholine, it is apparent that they are not mere passengers riding on the back of the α-subunits. The extended intracellular loops linking M3 and M4 of each subunit contain sites for phosphorylation by protein kinases A and C (PKA and PKC). The longer acetylcholine is bound to the α-subunits, the more intense is the extent of phosphorylation. This explains why inhibitors of acetylcholine esterase, such as DFP (diisopropyl fluorophosphate) and sarin, are so toxic. By preventing the hydrolysis of acetylcholine, the stimulus is

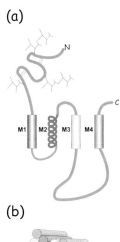

Figure 3.9 Peptide-chains and membrane organization of the human nicotinic receptor: (a) A single subunit showing the transmembrane segments. On the basis of their hydropathy profiles, the five subunits of the nicotinic acetylcholine receptor are each understood to traverse the membrane four times (M1–M4). As a consequence, both the N- and C-termini are exposed on the same (external) surface. For each subunit, the M2 segment is understood to contribute to the lining of the ion channel. (b) The transmembrane segments are packed so that the channel is lined by the M2 segments. The polar surfaces of the amphipathic M2 segments in each of the five subunits tend to orient towards each other, the non-polar surfaces seeking the non-polar environment offered by other subunits and the membrane bilayer. The ion channel is formed at the core of the structure.

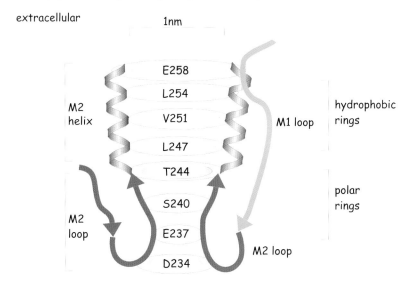

Figure 3.10 Positions of conserved side-chains that line the nicotinic channel. This simplified view of the pore just shows the M2 helices and loops of the two α-subunits. The amino acids that line the pore, contributed from these two segments, are indicated. Adapted from Corringer et al.[21]

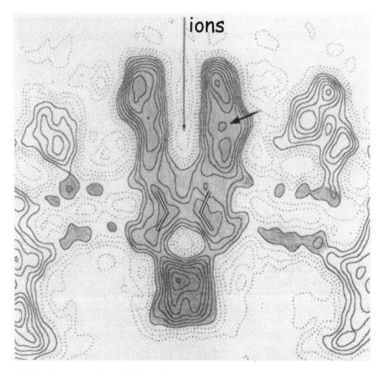

Figure 3.11 Cross-sectional view of acetylcholine receptors. The receptor is depicted in blue, the membrane phospholipid headgroups in yellow and attached cytoplasmic protein in pink. The receptor is about 12 nm long and it extends about 6.5 nm and 1.5 nm beyond the extracellular and intracellular surfaces respectively. The narrow pore (long arrow) is framed by two ~2 nm diameter entrances and shaped by a ring of five bent α-helical rods (indicated as bars). The gate of the channel is formed by amino acid side-chains projecting into the channel at the constriction point. The acetylcholine binding site appears to be a pocket (short arrow) in the α-subunit, located about 5 nm from the gate. From Nigel Unwin, MRC Laboratory of Molecular Biology, Cambridge.

initially prolonged, but when it does eventually diffuse away and the channels have closed, the extensive phosphorylation ensures that they cannot then be reopened. The receptors are said to be desensitized.

■ Other receptor-linked ion channels

Ion-channels regulated by serotonin (5-HT), GABA and glycine are clearly related to the nicotinic receptors. GABA and glycine regulate Cl⁻ channels present on the inhibitory neurones of the CNS. When these are opened there is a tendency for Cl⁻ ions to enter the cells causing membrane hyperpolarization and thus a decrease in the probability that a neurone attains its threshold for firing an action potential.[27]

The various channels regulated by glutamate, the major excitatory receptor of the CNS, comprise a separate family. Not only are their amino acid sequences unrelated, but the membrane topology is quite different. The subunits possess only three transmembrane segments, with the con-

Of course things are never as simple as one might wish. The direction of the ion flux must depend on the intracellular concentration of Cl⁻ but in the neurones of the hypothalamic suprachiasmic nucleus (SCN) this appears to vary diurnally, being higher during the daylight hours, probably due to the operation of a

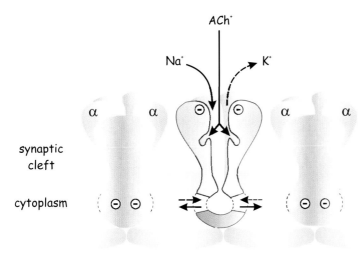

ACh⁺
Na⁺   K⁺

α   α   α   α

synaptic
cleft

cytoplasm

rapsyn

cyclically modulated ion pump.[27] The consequence of this is that, depending on the time of day, GABA acts either as an excitatory or as an inhibitory neurotransmitter. During the day, when the concentration of Cl^- in the SCN is in excess of about 15 mmol/l, GABA enhances the firing rate (8–10 Hz) and causes excitation. At night, opening of GABA$_A$ channels allows an influx of Cl^- which causes membrane hyperpolarization and a reduction in the firing rate (2–4 Hz).

Figure 3.12 The nicotinic ion channel: a schematic diagram of the nicotinic acetylcholine receptor. The narrowest part of the channel is occluded by amino acid side-chains contributed from the M2 segments of all five subunits. Two molecules of acetylcholine bind at sites on the two α-subunits. The receptor extends as a funnel some 6 nm into the synaptic cleft. On the cytoplasmic side, the acidic residues on surface of the receptor may determine the (cationic) selectivity of the channel. Rapsyn is a 43 kDa protein that anchors the channels to the cytoskeleton.

sequence that the N- and C-termini are expressed on either side of the membrane. In these receptors, the equivalent M2 segment presumed to line the channel is rudimentary, enters into the membrane and re-emerges on the same side (cytoplasm).[28]

■ The 7TM superfamily of G-protein linked receptors

The muscarinic receptors for acetylcholine, the GABA$_B$ receptors for γ-aminobutyrate and the mGlu receptors for glutamate, together with the adrenergic receptors, are all members of the same superfamily of receptor proteins. They are structurally related to the visual pigment rhodopsin and very distantly related to the bacteriorhodopsins which constitute ion pumps in the membranes of the extreme halophilic archaebacteria. The feature that relates all these structures is the topological organization of the single peptide chain, which in all cases is understood to traverse the membrane seven times (Figures 3.13 and 3.14). The seven membrane-spanning segments are linked by three exposed loops on either side of the membrane, with the N-terminus projected to the outside and the C-terminus in the cytosol. Only for the bacteriorhodopsins has the membrane topology been determined directly,[29] but although they share obvious features of structural organization, it is not certain whether the archaean 7TM proteins and the eukaryote receptors are derived from a common ancestor. The similarity between them could arise from conver-gent rather than divergent evolution. For the receptors and the visual pig-ment rhodopsin, the organization of the seven membrane-spanning

rhodopsin

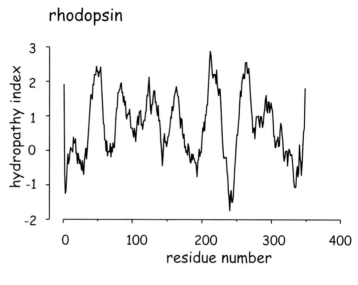

β₂-adrenergic receptor

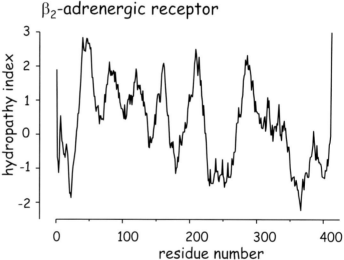

Figure 3.13 Hydropathy plots of rhodopsin and the β-adrenergic receptor. For the derivation of the hydrophobicity index, see legend to Figure 3.8. The sequence data were taken from the Swissprot database, accession numbers P08100 and P07550.

α-helices is deduced mainly by analysis of the hydropathy plots of the inferred amino acid sequences. The other essential feature that relates the 7TM receptors and the visual pigments is that they all interact, inside cells, with GTP-binding proteins. The bacteriorhodopsins are not linked to GTPases and they serve as ion pumps.

■   Categories of 7TM receptor

Of course, not only is the membrane organization of these proteins inferred, but so are the sequences (by analysis of the cDNAs). On many

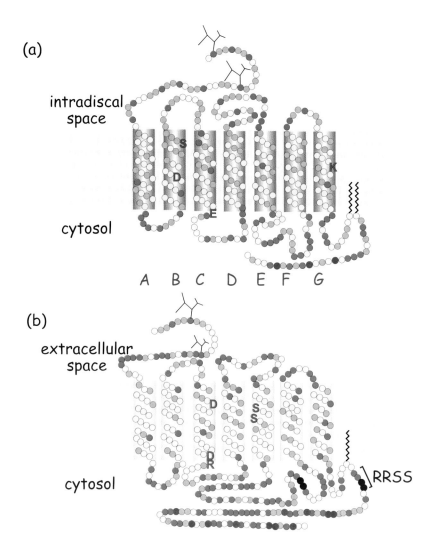

(a)

intradiscal space

cytosol

A B C D E F G

(b)

extracellular space

cytosol

RRSS

Figure 3.14 Membrane topology of rhodopsin and adrenergic receptors. The membrane organization of (a) rhodopsin and (b) the β_2-adrenergic receptor based on the hydropathy profiles of their amino acid sequences (see Figure 3.13). The circles indicate hydrophobic residues (yellow), non-polar residues (grey) and polar residues (blue). For both structures, the amino acids comprising the seven transmembrane spans (A–G) are predominantly hydrophobic, though non-polar and even charged residues are not entirely absent. The residues making contact with the ligands (11-*cis* retinal and adrenaline) are indicated (single-letter code). The incidence of polar and charged residues is much greater in the exposed loops and terminal domains. The residues shown as red circles are targets for phosphorylation by rhodopsin kinase or β_2-adrenergic receptor kinase (βARK), respectively[30,31]. The black circles are residues that are targets for phosphorylation by PKA (RRSS).

occasions cloning procedures have led to the identification of proteins, the products of individual genes and evidently members of this super-family, but which at the time had no known biological function. If we include the proteins expressed in olfactory epithelial cells and inferred to have the general topological organization of the 7TM receptors (probable receptors for odorant substances), then this family is truly enormous, probably approaching 2000 members, or ~5% of the mammalian genome.

Although the general structural plan of these proteins seems clear, there are of course important differences. If we simply consider the huge variety of activating ligands, encompassing ions (Ca^{2+}), small molecules and proteins, this must be self-evident (Table 3.5). One extreme example is that of rhodopsin which binds its ligand, 11-*cis* retinal, covalently. The advantage of this for the operation of the visual system is that the ligand is already in place, primed and ready to act as a trap for photons. It reacts by isomerization to its all-*trans* configuration within picoseconds and this perturbation is sensed by the rhodopsin causing it to interact with trans-ducin, a GTP-binding protein.[32, 33] This then activates a specific phospho-diesterase causing hydrolysis of intracellular cyclic GMP. The question of visual phototransduction is considered in further detail in Chapter 6.

Unlike rhodopsin, the receptors for hormones exist in both free and bound states. The rate at which they can be activated must be limited by the availability of the ligand, its rate of diffusion into the binding site and then its rate of attachment. The operation is necessarily slower. What happens next is now well recognized. All these receptors communicate with GTP-binding proteins and thence with intracellular effector enzymes to generate or mobilize second messengers (such as cAMP, Ca^{2+}, etc.).

Table 3.5 Categories of 7TM receptors

Receptor properties	Ligands
Ligand binds in the core region of the 7 transmembrane helices	11-*cis*-Retinal (in rhodopsin) Acetylcholine Catecholamines Biogenic amines (histamine, serotonin, etc.) Nucleosides and nucleotides Leukotrienes, prostaglandins, prostacyclins, thromboxanes
Short peptide ligands bind partially in the core region and to the external loops	Peptide hormones (ACTH, glucagon, growth hormone, parathyroid hormone, calcitonin)
Ligands make several contacts with the N-terminal segment and the external loops	Hypothalamic glycoprotein releasing factors (TRH, GnRH)
Induce an extensive reorganization of an extended N-terminal segment	Metabotropic receptors for neurotransmitters (such as GABA and glutamate) Ca^{2+}-sensing receptors, for example on parathyroid cells, thyroidal (calcitonin secreting) C-cells and kidney juxtaglomerular apparatus
Proteinase-activated receptors	Receptors for thrombin and trypsin

However, at the time when cAMP and its synthesizing enzyme adenylyl cyclase were discovered, understanding of the receptors themselves was purely conceptual. No receptor proteins had been isolated and there was a total lack of structural information. Indeed, even the understanding of membrane structure was far from being established. The possibility that receptors and the enzyme activities that they control might, or might not comprise a single entity was one that had not yet been contemplated.

■ Receptor diversity: variation and specialization

Among the 7TM receptors, there are only a few highly conserved residues, confined almost entirely to the hydrophobic membrane spanning regions. Although it is clear that these proteins are derived from a common ancestor, the divergence of their primary structures is enormous. More than this, if the divergence in the face of conserved molecular architecture is evident for the membrane spanning regions, then as we venture into the various binding domains for different classes of activating ligands, we find sequences that are appropriately specialized, bearing no discernible relationships with each other. The exposed segments vary in length from 7 up to 595 residues for the N-terminus, 12 to 359 residues for the C-terminus and 5 to 230 residues for the loops.[34]

Binding of low molecular mass ligands

The binding sites for most low molecular mass ligands such as acetylcholine, those derived from amino acids (catecholamines, histamine, etc.) and the eicosanoids (see Table 2.2, page 21) are located deep within the hydrophobic cores of their receptors (see Figure 3.16). Attachment of catecholamines is understood to involve hydrogen bonding of the two catechol OH groups to serine residues 204 and 207 separated by a single turn of the fifth membrane spanning (E) α-helix (Figure 3.15, also Figure 3.14) and by electrostatic interaction of the amine group to an aspartate residue (113) on the third membrane spanning (C) helix. The binding of acetylcholine similarly forms a cross-link between two membrane spanning helices of the muscarinic receptors. In this way, the bound hormone molecules are oriented in the plane parallel to the membrane. They form bridges between two transmembrane spans of their receptors, perturbing their orientation relative to each other. This is the origin of the signal that is conveyed to the cell. The β-blockers, such as propranolol, also bind at this location with high affinity and so impede the access of activating ligands. However, lacking the catechol group, they do not establish the critical link between membrane spanning segments. They fail to activate the receptor.

In contrast to the receptors for low molecular mass ligands, the binding sites for peptide hormones such as ACTH and glucagon are situated on the N-terminal segment and on the exposed loops linking the transmembrane helices (Figure 3.16b). The receptors for glycoprotein hormones have been adapted by elongated N-terminal chains that extend well out into the extracellular aqueous environment (Figure 3.16c).

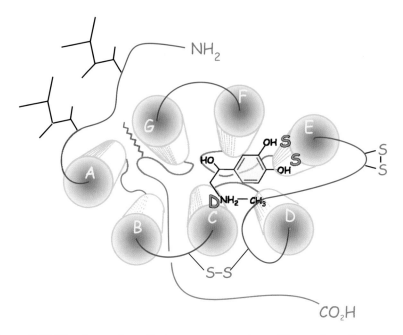

Figure 3.15 **Top-down view of a hormone receptor with an adrenaline molecule bound between the membrane-spanning segments.** The receptor is viewed from the extracellular surface. The N-terminal domain is glycosylated. The two catechol hydroxyls of adrenaline interact with the serine residues present in the membrane spanning α-helix E. The binding serines, S204 and S207, are separated by three residues (one turn) and so they are both projected towards the ligand, shifted by about 0.15 nm along the helix axis. The ligand amino group binds to an aspartate residue on the membrane spanning α-helix C. Adapted from Ostrowski et al.[35]

Calcium sensors and metabotropic receptors

The sites of attachment for the neurotransmitters (such as glutamate and γ-aminobutyrate) on metabotropic receptors, and for Ca^{2+} binding Ca^{2+}-sensors are situated externally, on specialized N-terminal extensions that can approach 600 residues in length[36] (Figure 3.16d). Binding of Ca^{2+} causes a pincer-like conformational change in the extended extracellular domain that exposes residues which then interact with the transmembrane core of the receptor. In this way, the extracellular domain of the receptor acts as its own auto-ligand. The Ca^{2+}-sensing receptor confronts a particular problem since it has to sense and then to respond to very small changes in Ca^{2+} concentration against a high basal concentration (± 0.025 mM in 2.5 mM). The cells of the parathyroid gland react by secretion of parathyroid hormone whenever the concentration of circulating Ca^{2+} dips below about 2.2 mM. Obviously, the affinity of the sensor has to be very low or it would be fully saturated at all times and under all conditions. On the other hand, the system must be sensitive to proportional changes in concentration that are minute compared with those sensed by conventional hormone receptors. These normally react to changes ranging over several orders of magnitude. The Ca^{2+} sensor is endowed with an

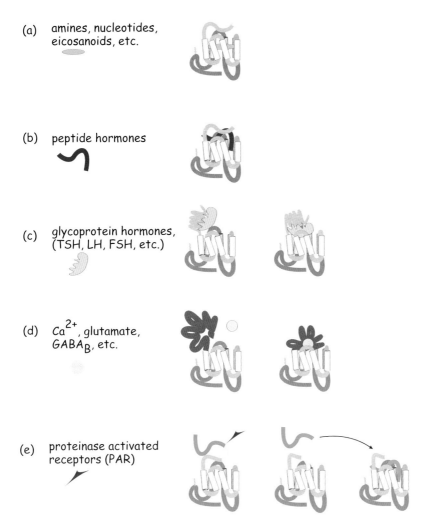

(a) amines, nucleotides, eicosanoids, etc.

(b) peptide hormones

(c) glycoprotein hormones, (TSH, LH, FSH, etc.)

(d) Ca^{2+}, glutamate, $GABA_B$, etc.

(e) proteinase activated receptors (PAR)

Figure 3.16 **Ligand binding sites**. The 7TM receptor is a jack of many trades, regulating a variety of different effectors and responding to ligands which come in many forms, having relative molecular masses in the range 32 (Ca^{2+}) to more than 10^5D. Most of the common low-mass hormones (such as adrenaline, acetylcholine) bind to sites within the hydrophobic core (a). Peptide and protein ligands are accommodated on the exterior face of the receptor (b, c). Although of low molecular mass, Ca^{2+} and the amino acids glutamate and GABA bind to extended N-terminal extensions, inducing new conformations which interact with the receptor (d). The proteinase-activated receptors are cleaved (e), the newly exposed N-terminus acting as an auto-ligand. The freed peptide may also interact separately with another receptor. Adapted from Ji et al.[34]

extracellular domain which acts as a low affinity chelator, binding or releasing Ca^{2+} as its concentration varies within the (extracellular) physiological range.[36] The Ca^{2+} sensor may operate as a dimer, the two components being joined by a disulphide bond between cysteine residues present in the extracellular domain.[37] A monomer–dimer equilibrium could underlie the special binding properties of this receptor.

Proteinase activated receptors (PARs)

Although blood platelets respond to many ligands (e.g. collagen exposed at sites of tissue damage, ADP released from damaged cells and, more importantly, secreted from activated platelets) the most potent activator is thrombin. Platelet-dependent arterial thrombosis triggers most heart attacks and strokes. Thrombin activates blood platelets causing them to aggregate within seconds but it also has numerous longer-term functions related to inflammation and tissue repair mechanisms requiring the stimulation of mitogenesis. These are mediated by a wide range of cell types. Thrombin is a serine proteinase enzyme related to trypsin and chymotrypsin and also to acetylcholine esterase. It has a unique specificity, cleaving peptide chains between arginine and serine, only as they are embedded in particular peptide sequences. As with the other serine esterases, the proteolytic activity of thrombin can be inhibited by the organophosphorus compounds mentioned earlier, and its action in stimulating blood platelets is also inhibited by compounds of this class. By cleaving the N-terminal exodomain of the thrombin receptor a new N-terminus is revealed which itself acts as a tethered ligand interacting with the exposed loops of the receptor (Figure 3.16e). A synthetic pentapeptide, equivalent to the five N-terminal amino acids revealed after the thrombin cleavage, also has agonistic activity for the thrombin receptor.[38] In addition to the tethered ligand, the cleaved 41-residue peptide acts as a strong agonist for platelets.[39]

The thrombin receptor is now recognized as a member of a larger family of protease activated receptors (PAR1–4), some of which are additionally activated by trypsin.[40–42] In the epithelia of the upper intestine the PAR2 receptors confer protection against self-digestion by proteolytic enzymes through the production of prostaglandins.[43] Similarly, in the bronchial airways, the presence of proteolytic enzymes released by inflammatory cells (mast cells) appears to be signalled by PAR receptors.[44] Activation results in the generation of prostaglandin E_2, causing bronchodilation. Although conventional in the sense that these PAR receptors are coupled to G-proteins, there is the particular problem that their activation by proteolytic cleavage is necessarily irreversible. Of course, for the functioning of platelets there is no problem since stimulation initiates a sequence of events which terminates in their demise. For the PAR receptors in other tissues, specialized mechanisms ensure their desensitization and removal.[45] In addition to the usual processes of receptor phosphorylation and endocytosis, these include cleavage of the tethered ligand in lysosomes. Because the receptors undergo cleavage both as a consequence of stimulation and again in the process of desensitization, resensitization of the system necessarily occurs by *de novo* protein synthesis.

Receptor–ligand interaction and receptor activation

A two-state equilibrium description of receptor activation

It is well recognized that ion channels are proteins that can exist in discrete states, typically 'open' or 'closed'. For ligand activated channels the open state probability is increased enormously when the ligand (e.g. acetylcholine) is bound.

A similar two-state equilibrium applies to the activation of the 7TM receptors. However, since Langley's first description of a 'receptive substance' towards the end of the 19th century,[46] it has been generally accepted that activation requires the actual binding of an agonist. Fundamental to this thinking is that the extent of the biological response is determined by the law of mass action linking the stimulus and the reactive tissue (i.e. the ligand and the empty binding site). The binding of a ligand is understood to induce a change in receptor conformation, such that only in this state does it communicate with its affiliated GTP-binding protein.

■ Receptors may be activated in the absence of ligand

The activity of adenylyl cyclase does not decline to zero in the absence of stimulating hormones. Of course, it is difficult to be quite sure that stimulating hormones are fully excluded even when a system has been extensively washed and then doped with inhibitors. Even so, there always remains a residual low level spontaneous (or constitutive) activity. Now we know that this is for real. However, even if activating ligands are not absolutely required, the presence of a receptor coupled to cyclase is obligatory. Quite simply, the unoccupied receptors themselves provide the necessary stimulus. The synthesis of cAMP in Sf9 insect cells, which are normally unresponsive to catecholamines, can be enhanced by transfecting them with β_2-adrenergic receptors. These cells now generate cAMP at a rate which is directly proportional to the level of receptor expression and they do this in the absence of any stimulating ligand[47] (see Figure 4.20, page 99) (though of course, the activity of the system is greatly increased if catecholamines are provided). We return to the matter of reconstituting the adrenergic receptor/cyclase system in Sf9 cells in Chapter 4.

This can be understood if the receptors exist in two conformational states, one of which can initiate downstream events (R*), the other cannot (r). The equilibrium between these two states exists regardless of the presence or absence of a stimulating ligand (Figure 3.17). Conventional agonists are those which bind to the active conformation (R*) and, as a result, increase the proportion of active receptors (R* + LIG.R*). Conversely, there are 'inverse' agonists (lig) that bind selectively to the receptor in its inactive conformation. These increase the proportion in the form (r + lig.r) in which they are unable to transmit signals. Thus the number of active receptors is reduced and the activity of the system is suppressed below its normal spontaneous or basal state. In between the extremes of full (conventional) and inverse agonists are those agonists that bind to receptors in both the active and inactive conformations. These are the partial agonists. Such compounds are unable to achieve maximal stimulation of effector enzymes (e.g. of adenylyl cyclase) even when all the receptor binding sites are occupied.

Sf9 insect cells and the baculovirus transfection system are used as an alternative to bacteria. Because they are eukaryotic cells and, moreover, animal cells, their newly synthesized proteins are post-translationally modified in the same way as in other multicellular organisms.

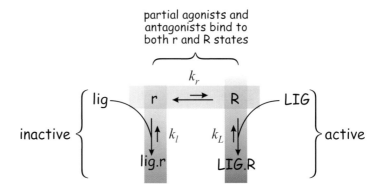

Figure 3.17 Receptor states and inverse agonists. This schematic diagram illustrates the equilibrium between the inactive (blue, r) and active (red, R) states of a receptor. Agonistic ligands (LIG) bind exclusively to the active form of the receptor and thus increase the proportion and the total amount (R + LIG.R) of the receptor in the form that it can transmit a signal. Conversely, inverse agonists (lig) bind exclusively to the inactive form of the receptor and thus increase the total amount of the receptor (r + lig.r) in the form that is incapable of transmitting signals.

The classification of drugs as agonists or antagonists, conventional, neutral or inverse, is a pharmacological minefield. In practice, many of the compounds in use in the clinic and in the laboratory, and depending on the circumstances, have activities that could enable them to be classified either as agonists or antagonists. A further complicating factor is that the spectrum of activities varies from tissue to tissue. For the right atrium of the (rat) heart, in which β_2-adrenergic receptors play a greater role than in other cardiac regions, it appears that almost all the β-blockers behave as inverse agonists.

Inverse agonism offers the potential of developing new drugs that attenuate the effects of mutant receptors that are constitutively activated. The neutral antagonists (β-blockers such as pindolol, etc.), bind to both the active and the inactive conformations and are better regarded as passive antagonists.

These impede the binding of both agonists and inverse agonists. Neutral antagonists prevent the stimulation of adenylyl cyclase by catecholamines, but they also oppose the inhibitory effect of inverse agonists. Examples of inverse agonists for β-adrenergic receptors are propranolol and timolol, originally classified as a β-blockers (see Figure 3.2) and used as such in the treatment of glaucoma and hypertension. However, unlike the β-blockade due to a neutral antagonist such as pindolol, which mainly affects the heart during exercise and stress (when sympathetic control is dominant), these compounds also suppress the resting heart rate.[48,49]

▪ Over-expression of receptors is sufficient

From this, one might imagine that if a receptor could be sufficiently over-expressed in a responsive tissue, the system would be fully activated even in the absence of a stimulating agonist. The equilibrium ratio r ↔ R* of the inactive and active species would be unaltered but the increased amount would ensure a sufficient quantity of the active form to induce activity. Indeed, 200-fold over-expression of the β_2-adrenergic receptor in the hearts of transgenic mice is sufficient to maximize cardiac function in the absence of any stimulus.[50] Under these conditions, no hormone is required to increase the level of active receptors to the point at which they can stimulate the tissue.

There are indications that receptor over-expression may play a role in the aetiology of some disease states and this offers the prospect that inverse agonists could provide more specific therapeutic approaches than the currently available neutral antagonists. Indeed, some forms of schizophrenia may be associated with an elevation of the D4 (dopamine) receptor in the frontal cortex.[51–53] Although the measured extent of the elevation is not large, about threefold compared with controls, it is possible that this could be a reflection of much greater changes in limited focal regions which could be of importance in determining the disease state. A number of compounds (raclopride, haloperidol) which find widespread application as antipsychotic drugs appear to possess inverse agonist activity at the dopamine receptors.[54–56] Similarly, the action of progesterone in reducing oxytocin-induced uterine excitability, essential for the maintenance of pregnancy (see Chapter 2), is also best understood as the action of an inverse agonist rather than as a conventional inhibitor.[57,58]

Of course, there is one receptor which is expressed at a level which is out of all proportion to all others. This is opsin, the major element of the visual pigment rhodopsin. It is legitimate to ask by what mechanism the photoreceptor system ensures that dark really means dark, that we see nothing when the lights are turned off? Even if the equilibrium would favour the inactive state to a degree much greater than any hormone receptor, the presence of such an enormous amount of rhodopsin must surely elicit some activating response, however minimal. Yet dark really does mean dark. This important question is discussed in connection with the mechanism of visual transduction in Chapter 6.

■ Receptor dimerization

Established assumptions are also being challenged by the realization that many receptors may function not as monomers, but as dimers or oligomeric clusters. As so often, there is little that is really new. The first hints concerning an oligomeric arrangment of β-adrenergic receptors emerged almost a quarter of a century ago.[59]

Heterodimers

A good example of receptor dimerization is the metabotropic GABA$_B$-R, because this is incapable of transmitting signals unless two subtypes of the receptor (splice variants, 1 and 2) are both present.[60–62] Both are normally present in the relevant neural tissues so there is no problem. However, when expressed in cells in which they are normally absent, successful reconstitution of the recombinant GABA$_B$-R with its associated K$^+$ channel and an appropriate GTP-binding protein requires expression of both receptor subtypes. The inference is that the GABA$_B$-R operates as an heterodimer or larger oligomer. Indeed, co-expression is not only a requirement for effective ligand binding and signal propagation but is also a prerequisite for maturation and transport of the receptor to the cell surface. The site of linkage between the partners has been proposed to reside in the C-terminal tail which is somewhat extended in comparison with other comparable receptors.

Other diseases that might be amenable to treatment with appropriate inverse agonists include inherited conditions such as retinitis pigmentosa, congenital night blindness (both due to mutations of rhodopsin), familial male precocious puberty and familial hyperthyroidism, among others. An inverse agonist active at the receptor for parathyroid hormone, recognized as being constitutively activated in patients suffering from Jansen-type metaphyseal chondrodysplasia, has recently been described.[58]

Homodimers

It is likely that the heterodimer $GABA_{B1}$–$GABA_{B2}$ exists as a stable entity and is unaffected by the state of receptor activation. Other receptors, such as those for glutamate (mGluR5[63]), vasopressin (V2[64]), adenosine, opiates (δ-opioid[65]) and even the well investigated $β_2$-adrenergic[64] and muscarinic receptors may be linked as homodimers. For some of these at least, the monomer-dimer equilibrium appears to reflect the state of receptor activation.

Stimulation of $β_2$-adrenergic receptors with catecholamines favours the formation of dimers[64] while, conversely, the application of an inverse agonist, timolol, favours monomer formation. When isolated, the dimers are stable even under the denaturing conditions applied for analysis by SDS gel electrophoresis. Of the seven transmembrane helices, the F segment may offer the site of intermolecular attachment determining dimer formation. A synthetic peptide based on the sequence of this segment, and which could therefore be expected to impede the access of the second protein molecule was found to reduce dimer formation.[64] It also inhibits the activation of adenylyl cyclase by isoprenaline. Does it follow that dimerization provides an alternative signal for activation of the $β_2$-adrenergic receptor? The evidence begins to support this idea, but a direct experiment is so far lacking.

Other receptors

For receptors of other types, represented generally by those with intrinsic tyrosine kinase activity (e.g. growth factor receptors such as those for EGF (Chapters 10 and 11)), those which are tightly linked to tyrosine kinases (cytokine receptors, Chapter 12) or those which interact with cytosolic tyrosine kinases (T-cell receptor, immunoglobulin receptors), dimerization is well established as the initiating step in the signal processes that they control.

Transmitting signals into cells

The receptor and the effector: one and the same, or are they separate entities?

When it was first realized that hormones induce enzymatic activity at the cell membrane, the receptors themselves were no more than concepts and ideas used to explain phenomena. It soon became apparent that a receptor and its catalytic function might not be two facets of a single entity. For example, adipocytes have receptors for many hormones of different classes. Adrenaline, ACTH, and glucagon all act to increase the concentration of intracellular cAMP. However, the maximal rate of cAMP formation due to a particular stimulus cannot be further enhanced by stimulation of a second receptor type.[66,67] Other agents such as prostaglandins (PGE) and clonidine (an $α_2$-adrenergic reagent), adenosine and insulin, acting through their own specific receptors, suppress the induced rate of cAMP synthesis. For these cells anyway, the inference is that all the receptors communicate with a common pool of adenylyl

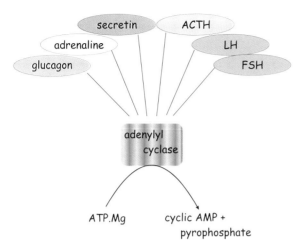

Figure 3.18 Multiple receptors coupled to a common pool of adenylyl cyclase.
There is a single pool of adenylyl cyclase which can be accessed by many disparate
hormone receptors. In general, when cyclase is maximally stimulated by one class of
receptors further activation by ligands binding at other receptors is not possible. (There
are, however, some exceptions to this in which maximal activation of cyclase requires two
classes of receptor to act together in synergy: see Chapter 5.) Adapted from Rodbell.[69]

cyclase (Figure 3.18). From this it follows that the receptors are likely to be
discrete entities, each capable of communicating independently with the
limited pool of enzyme.[68]

Support for this idea came from the finding that the rate of activation of
adenylyl cyclase in turkey red cell membranes by β-adrenergic receptors is
diffusion-controlled (see Figure 3.19). In membranes rendered less viscous
(or more fluid, by inclusion of an unsaturated fatty acid), the time taken to
establish the maximal rate of cAMP production can be substantially
reduced.[70] There are exceptions, however, since in the same membranes,
the time taken to establish the maximal rate following simulation through
the adenosine receptor is unaffected. The adenosine receptor (in these
cells anyway) appears to be permanently coupled to the effector enzyme.

To make things more complicated, however, there are situations in
which two hormones are required together to cause stimulation. For
example, in brain cortical slices the activation of adenylyl cyclase by
noradrenaline (α-adrenergic stimulation, blocked by phentolamine but
not by propranolol) requires the simultaneous application of adenosine
(blocked by theophylline).[72] Here, with full activation requiring the
involvement of two quite disparate receptors, one might want to infer that
these are integral with the catalytic unit itself (although there is a much
better explanation for this, which is discussed in Chapter 5: see page 113).

In the case that not all of the
cyclase is accessible to a
particular receptor, then
further increments of activity
can be obtained by
stimulating a second
receptor that has access to
the total cyclase pool. An
example of this is the partial
(70%) stimulation of turkey
red cell cyclase by adenosine.
This can then be augmented
by application of adrenaline.
There is no indication that
these receptors are
activating separate pools of
adenylyl cyclase.[68]

■ Mixing and matching receptors and effectors

A direct demonstration of the independent existence of receptors and
their downstream effector enzymes required a procedure by which they

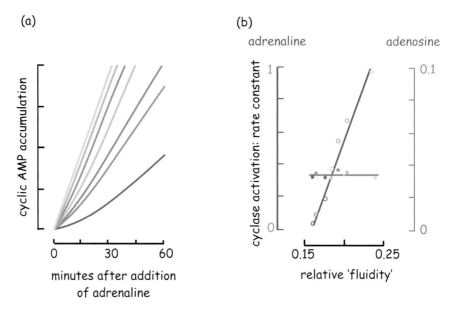

Figure 3.19 Diffusion control or permanent coupling in the activation of adenylyl cyclase by receptors? (a) Time-course of cAMP accumulation in response to adrenaline in turkey red cell membranes doped with increasing amounts of *cis*-vaccenic acid to enhance fluidity. As would be expected for a mechanism limited by the rate of diffusion, the time taken to establish the maximal activity of adenylyl cyclase is progressively curtailed as the membranes are rendered more fluid. An indication of membrane fluidity was obtained by measuring the polarization of fluorescence emitted by diphenylhexatriene, a probe molecule dissolved in the membranes. (b) The fluidity determines the rate constant for adenylate cyclase activation caused by adrenaline (open circles), but not by adenosine (filled circles). The inference is that the adenosine receptor is permanently coupled to the effector enzyme. Adapted from Hanski *et al.*[70] and Rimon *et al.*[71]

could be obtained separately and then reconstituted into a functioning unit. In pre-cloning days this presented a difficult problem, since the receptors and the catalytic units which they control always share a common membrane location. Rather than attempt to separate them physically, the solution was to inhibit the cyclase selectively, while leaving the receptors intact, and then to use a second cellular source of cyclase, bringing the two together in order to achieve activity. The alkylating agent *N*-ethylmaleimide was used to destroy the cyclase activity of turkey red blood cells while leaving their membrane receptors for adrenaline unimpaired. These cyclase-inactive cells were then fused to adrenal cortical cells as a source of the active enzyme (Figure 3.20). (Adrenal cortical cells lack catecholamine receptors but generate cAMP when treated with the peptide hormone ACTH.) Immediately after the fusion, cyclase could only be activated by ACTH, but over a period of a few hours, it once again became sensitive to adrenaline.[73] This delay is commensurate with the rate at which the surface proteins of the two component cells diffused laterally and merged with each other in the fused plasma membrane of the heterokaryon (the fused cell containing two nuclei). Since the cyclase to which adrenaline was normally coupled had been irreversibly

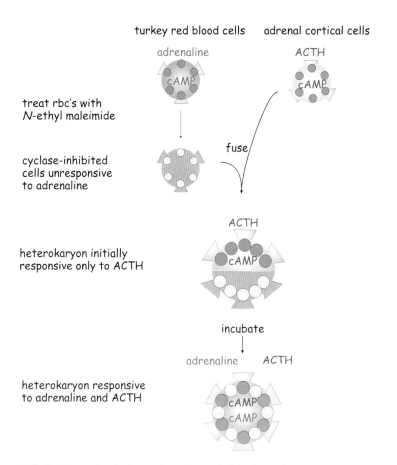

Figure 3.20 **Evidence for independent lateral diffusion of receptors and adenylyl cyclase.** By this experiment, illustrated schematically, it was shown that hormone receptors and their effector enzymes (adenylyl cyclase) are separate molecular entities capable of independent lateral diffusion in the plane of a cell membrane. For details see text.

inactivated before the fusion, this could only mean that the β-adrenergic receptor derived from the turkey red blood cells was now activating the adrenal cortical cell enzyme. It follows that the receptors and the catalytic units that they activate must be situated on different protein molecules. The experiment gave no indication that there is a third component, a GTP-binding protein, which intervenes between the two, communicating the message (Figure 4.1), from the activated receptors to the catalytic unit.

Intracellular 7TM receptor domains and signal transmission

The onward signals for all these receptors are carried forward by conformational perturbations conveyed to the loops and the C-terminal peptide exposed in the cytosol. Of these, the third intracellular loop (connecting the E and F chains), which is rather extended in most 7TM-receptor molecules, and also the C-terminal peptide are considered to be of particular importance. However, this definition of which residues and short stretches are involved gives no indication about how the signal is actually conveyed to the GTP-binding protein. Similarly, mutagenesis experiments have revealed amino acid residues and stretches on the GTP-binding proteins that are important in recognizing the activated receptors.

At this point, all reference to bacteriorhodopsin, the only 7TM molecule for which we have firm structural information, fails as a source of reference for comparison. The structural organization of all receptors and receptor-like molecules requires the two-phase environment provided by the membrane phospholipid bilayer stabilized by both structured and by bulk water phases on either side. So far, only very few intrinsic membrane proteins have provided crystals amenable for analysis by X-ray diffraction measurements. Anyway, even if we knew how the C-terminal peptide and the intracellular loops of bacteriorhodopsin are disposed, and how these alter with respect to each other when activated by light, it is hardly likely that this would contribute to our understanding of receptor mechanisms, since in comparison with membrane receptors these domains are very short. More importantly, bacteriorhodopsin does not transmit information through GTP-binding proteins.

◾ Adrenaline (yet again)

It was ever thus. The cab-choked street, the PR men clutching their ulcers, the jewellery displayed like medals on the chest of a Soviet general, the snoozing men from Wall Street, the Sardi's supper entrance. As always in any enterprise, Americans travel hopefully, fuelled by a thirst for adrenalin not experienced by most Europeans.

John Osborne, *Almost a Gentleman* (Faber & Faber, London, 1991)

◾ References

1 Cannon, W.B. *Bodily Changes in Pain, Hunger, Fear and Rage. An account of researches into the function of emotional excitement.* Harper, New York, 1929. Reprinted Torch Books, 1963.

2 Starling, E.H. *The Wisdom of the Body.* H. K. Lewis, London, 1923.

3 Strosberg, A.D. Structure and function of the β3-adrenergic receptor. *Annu. Rev. Pharm. Tox.* 2000; 37: 421–50.

4 Hanoune, J. The adrenal medulla. In: Baulieu, E.-E., Kelly P.A. (Ed) *Hormones, From Molecules to Disease.* Chapman & Hall, London, 1990; 309–33.

5 Weiland, G.A., Minneman, K.P., Molinoff, P.B. Fundamental difference between the molecular interactions of agonists and antagonists with the β-adrenergic receptor. *Nature* 1979; 281: 114–17.

6 Sellin, L.C., McArdle, J.J. Multiple effects of 2,3-butanedione monoxime. *Pharm. Toxicol.* 1994; 74: 305–13.

7 van Helden, H.P., Busker, R.W., Melchers, B.P., Bruijnzeel, P.L. Pharmacological effects of oximes: how relevant are they? *Arch. Toxicol.* 1996; 70: 779–86.

8 Burden, S.J., Sargent, P.B., McMahan, U.J. Acetylcholine receptors in regenerating muscle accumulate at original synaptic sites in the absence of the nerve. *J. Cell. Biol.* 1979; 82: 412–25.

9 Changeux, J.P., Kasai, M., Lee, C.Y. Use of a snake venom toxin to characterize the cholinergic receptor protein. *Proc. Natl. Acad. Sci. USA* 1970; 67: 1241–7.

10 Meunier, J.C., Sealock, R., Olsen, R., Changeux, J.P. Purification and properties of the cholinergic receptor protein from *Electrophorus electricus* electric tissue. *Eur. J. Biochem.* 1974; 45: 371–94.

11 Olsen, R.W., Meunier, J.C., Changeux, J.P. Progress in the purification of the cholinergic receptor protein from *Electrophorus electricus* by affinity chromatography. *FEBS Lett.* 1972; 28: 96–100.

12 Neher, E., Sakmann, B. Single-channel currents recorded from membrane of denervated frog muscle fibres. *Nature* 1976; 260: 799–802.

13 Le Novere, N., Changeux, J.P. The ligand-gated ion channel database. *Nucl. Acid Res.* 1999; 27: 340–2.

14 Kyte, J., Doolittle, R.F. A simple method for displaying the hydropathic character of a protein. *J. Mol. Biol.* 1982; 157: 105–32.

15 Segrest, J.P., Kahane, I., Jackson, R.L., Marchesi, V.T. Major glycoprotein of the human erythrocyte membrane: evidence for an amphipathic molecular structure. *Arch. Biochem. Biophys.* 1973; 155: 167–83.

16 Deisenhofer, J., Epp, O., Miki, K., Huber, R., Michel, H. Structure of the protein subunits in the photosynthetic reaction centre of *Rhodopseudomonas viridis* at 3Å resolution. *Nature* 1985; 318: 618–21.

17 Doyle, D.A., Cabral, J.M., Pfuetzner, R.A. *et al*. The structure of the potassium channel: Molecular basis of K conduction and selectivity. *Science* 1998; 280: 69–77.

18 Chang, G., Spencer, R.H., Lee, A.T., Barclay, M.T., Rees, D.C. Structure of the MscL homolog from *Mycobacterium tuberculosis*: A gated mechanosensitive ion channel. *Science* 1998; 282: 2220–6.

19 Hucho, F., Gorne Tschelnokow, U., Strecker, A. β-Structure in the membrane-spanning part of the nicotinic acetylcholine receptor (or how helical are transmembrane helices?). *Trends Biochem. Sci.* 1994; 19: 383–7.

20 Heidmann, T., Changeux, J.P. Time-resolved photolabeling by the noncompetitive blocker chlorpromazine of the acetylcholine receptor in its transiently open and closed ion channel conformations. *Proc. Natl. Acad. Sci. USA* 1984; 81: 1897–901.

21 Corringer, P.J., Le Novere, N., Changeux, J.P. Nicotinic receptors at the amino acid level. *Annu. Rev. Pharm. Toxicol.* 2000; 40: 431–58.

22 Utkin, Y.N., Tsetlin, V.I., Hucho, F. Structural organization of nicotinic acetylcholine receptors. *Membr. Cell. Biol.* 2000; 13: 143–64.

23 Bouzat, C., Barrantes, F.J. Assigning functions to residues in the acetylcholine receptor channel region. *Mol. Membr. Biol.* 1997; 14: 167–77.

24 Hucho, F., Tsetlin, V.I., Machold, J. The emerging three-dimensional structure of a receptor. The nicotinic acetylcholine receptor. *Eur. J. Biochem.* 1996; 239: 539–57.

25 Akabas, M.H., Kaufmann, C., Archdeacon, P., Karlin, A. Identification of acetylcholine receptor channel-lining residues in the entire M2 segment of the α subunit. *Neuron* 1994; 13: 919–27.

26 Leonard, R.J., Labarca, C.G., Charnet, P., Davidson, N., Lester, H.A. Evidence that the M2 membrane-spanning region lines the ion channel pore of the nicotinic receptor. *Science* 1998; 242: 1578–81.

27 Wagner, S., Castel, M., Gainer, H., Yarom, Y. GABA in the mammalian suprachiasmic nucleus and its role in diurnal rhythmicity. *Nature* 1997; 387: 598–603.

28 Bennett, J.A., Dingledine, R. Topology profile for a glutamate receptor: three transmembrane domains and a channel-lining reentrant membrane loop. *Neuron* 1995; 14: 373–84.

29 Henderson, R., Unwin, P.N. Three-dimensional model of purple membrane obtained by electron microscopy. *Nature* 1975; 257: 28–32.

30 Ohguro, H., Palczewski, K., Ericsson, L.H., Walsh K.A., Johnson, R.S. Sequential phosphorylation of rhodopsin at multiple sites. *Biochemistry* 1993; 32: 5718–24.

31 Fredericks, Z.L., Pitcher, J.A., Lefkowitz, R.J. Identification of the G-protein-coupled receptor kinase phosphorylation sites in the human β2-adrenergic receptor. *J. Biol. Chem.* 1996; 271: 13796–803.

32 Wheeler, G.L., Matuto, Y., Bitensky, M.W. Light-activated GTPase in vertebrate photoreceptors. *Nature* 1977; 269: 822–4.

33 Stryer, L. Cyclic GMP cascade of vision. *Annu. Rev. Neurosci.* 1986; 9: 87–119.

34 Ji, T.H., Grossmann, M., Ji, I. G protein-coupled receptors. I. Diversity of receptor-ligand interactions. *J. Biol. Chem.* 1998; 273: 17299–302.

35 Ostrowski, J., Kjelsberg, M.A., Caron, M.G., Lefkowitz, R.J. Mutagenesis of the β2-adrenergic receptor: How structure elucidates function. *Annu. Rev. Pharm. Toxicol.* 1992; 32: 167–83.

36 Brown, E.M., Gamba, G., Riccardi, D. *et al.* Cloning and characterization of an extracellular Ca^{2+}-sensing receptor from bovine parathyroid. *Nature* 1993; 366: 575–80.

37 Ward, D.T., Brown, E.M., Harris, H.W. Disulfide bonds in the extracellular calcium-polyvalent cation-sensing receptor correlate with dimer formation and its response to divalent cations in vitro. *J. Biol. Chem.* 1998; 273: 14476–83.

38 Scarborough, R.M., Naughton, M.A., Teng, W. *et al.* Tethered ligand agonist peptides. Structural requirements for thrombin receptor activation reveal mechanism of proteolytic unmasking of agonist function. *J. Biol. Chem.* 1994; 269.

39 Furman, M.I., Liu, L., Benoit, S.E. *et al.* The cleaved peptide of the thrombin receptor is a strong platelet agonist. *Proc. Natl. Acad. Sci. USA* 1998; 95: 3082–7.

40 Dery, O., Corvera, C.U., Steinhoff, M., Bunnett, N.W. Proteinase-activated receptors: novel mechanisms of signaling by serine proteases. *Am. J. Physiol.* 1998; *274C*: 1429–52.

41 Coughlin, S.R. Protease-activated receptors start a family. *Proc. Natl. Acad. Sci. USA* 1994; 91: 9200–2.

42 Kahn, M.L., Zheng, Y.W., Huang, W. *et al.* A dual thrombin receptor system for platelet activation. *Nature* 1998; 394: 690–4.

43 Hara, K., Yonezawa, K., Sakaue, H. *et al.* 1-Phosphatidylinositol 3-kinase activity is required for insulin-stimulated glucose transport but not for RAS activation in CHO cells. *Proc. Natl. Acad. Sci. USA* 1994; 91: 7415–19.

44 Cocks, T.M., Fong, B., Chow, J.M. *et al.* A protective role for protease-activated receptors in the airways. *Nature* 1999; 398: 156–60.

45 Trejo, J., Hammes, S.R., Coughlin, S.R. Termination of signaling by protease-activated receptor-1 is linked to lysosomal sorting. *Proc. Natl. Acad. Sci. USA* 1998; 95: 13698–702.

46 Langley, J.N. On the physiology of the salivary secretion: Part II. On the mutual antagonism of atropin and pilocarpin, having especial reference to thheir relations in the sub-maxillary gland of the cat. *J. Physiol.* 1878; 1: 339–69.

47 Chidiac, P., Hebert, T.E., Valiquette, M., Dennis, M., Bouvier, M. Inverse agonist activity of β-adrenergic antagonists. *Mol. Pharm.* 1994; 45: 490–9.

48 Varma, D.R., Shen, H., Deng, X.F. *et al.* Inverse agonist activities of β-adrenoceptor antagonists in rat myocardium. *Br. J. Pharm.* 1999; 127: 895–902.

49 Varma, D.R. Ligand-independent negative chronotropic responses of rat and mouse right atria to β-adrenoceptor antagonists. *Can. J. Physiol.* 1999; 77: 943–9.

50 Bond, R.A., Leff, P., Johnson, T.D. *et al.* Physiological effects of inverse agonists in transgenic mice with myocardial overexpression of the β-adrenoceptor. *Nature* 1995; 374: 272–6.

51 Seeman, P., Guan, H.C., Van-Tol, H. H. Dopamine D4 receptors elevated in schizophrenia. *Nature* 1993; 365: 441–5.

52 Marzella, P.L., Hill, C., Keks, N., Singh, B., Copolov, D. The binding of both [^{3}H]nemonapride and [^{3}H]raclopride is increased in schizophrenia. *Biol. Psychiatr.* 1997; 42: 648–54.

53 Stefanis, N.C., Bresnick, J.N., Kerwin, R.W., Schofield, W.N., McAllister, G. Elevation of D4 dopamine receptor mRNA in postmortem schizophrenic brain. *Brain Res. Mol. Brain Res.* 1998; 53: 112–19.

54 Griffon, N., Pilon, C., Sautel, F., Schwartz, J.C., Sokoloff, P. Antipsychotics with inverse agonist activity at the dopamine D3 receptor. *J. Neural Transm.* 1996; 103: 1163–75.

55 Malmberg, A., Mikaels, A., Mohell, N. Agonist and inverse agonist activity at the dopamine D3 receptor measured by guanosine 5′-γ-thio-triphosphate-^{35}S-binding. *J. Pharm. Exp. Ther.* 1998; 285: 119–26.

56 Hall, D.A., Strange, P.G. Evidence that antipsychotic drugs are inverse agonists at D2 dopamine receptors. *Br. J. Pharm.* 1997; 121: 731–6.

57 Grazzini, E., Guillon, G., Mouillac, B., Zingg, H.H. Inhibition of oxytocin receptor function by direct binding of progesterone. *Nature* 1998; 392: 509–12.

58 Gardella, T.J., Luck, M.D., Jensen, G.S. *et al.* Inverse agonism of amino-terminally truncated parathyroid hormone (PTH) and PTH-related peptide (PTHrP) analogs revealed with constitutively active mutant PTH/PTHrP receptors. *Endocrinology* 1996; 137: 3936–41.

59 Atlas, D., Levitzki, A. Tentative identification of β-adrenoceptor subunits. *Nature* 1978; 272: 370–1.

60 Jones, K.A., Borowsky, B., Tamm, J.A. *et al.* GABAB receptors function as a heteromeric assembly of the subunits GABABR1 and GABABR2. *Nature* 1998; 396: 670–4.

61 White, J.H., Wise, A., Main, M.J. *et al.* Heterodimerization is required for the formation of a functional GABAB receptor. *Nature* 1998; 396: 679–82.

62 Kaupmann, K., Malitchek, B., Schuler, V. *et al.* GABAB-receptor subtypes assemble into functional heteromeric complexes. *Nature* 1998; 396: 683–7.

63 Romano, C., Yang, W.L., O'Malley, K.L. Metabotropic glutamate receptor 5 is a disulfide-linked dimer. *J. Biol. Chem.* 1996; 271: 28612–16.

64 Hebert, T.E., Moffett, S., Morello, J.-P. *et al.* A peptide derived from a β2-adrenergic receptor transmembrane domain inhibits both receptor dimerization and activation. *J. Biol. Chem.* 1996; 271: 16384–92.

65 Cvejic, S., Devi, L.A. Dimerization of the δ opioid receptor: implication for a role in receptor internalization. *J. Biol. Chem.* 1997; 272: 26959–64.

66 Birnbaumer, L., Rodbell, M. Adenyl cyclase in fat cells. II. Hormone receptors. *J. Biol. Chem.* 1969; 244: 3477–82.

67 Bar, H.P., Hechter, O. Adenyl cyclase and hormone action. I. Effects of adrenocorticotropic hormone, glucagon, and epinephrine on the plasma membrane of rat fat cells. *Proc. Natl. Acad. Sci. USA* 1969; 63: 350–6.

68 Tolkovsky, A.M., Levitzki, A. Coupling of a single adenylate cyclase to two receptors: adenosine and catecholamine. *Biochemistry* 1978; 17: 3811–17.

69 Rodbell, M. The role of GTP-binding proteins in signal transduction: from the sublimely simple to the conceptually complex. *Curr. Top. Cell Regul.* 1992; 32: 1–47.

70 Hanski, E., Rimon, G., Levitzki, A. Adenylate cyclase activation by β-adrenergic receptors as a diffusion-controlled process. *Biochemistry* 1979; 18: 846–53.

71 Rimon, G., Hanski, E., Braun, S., Levitzki, A. Mode of coupling between hormone receptors and adenylate cyclase elucidated by modulation of membrane fluidity. *Nature* 2001; 276: 394–6.

72 Sattin, A., Rall, T.W., Zanella, J. Regulation of cyclic adenosine 3′,5′-monophosphate levels in guinea-pig cerebral cortex by interaction of α-adrenergic and adenosine receptor activity. *J. Pharm. Exp. Ther.* 1975; 192: 22–32.

73 Schulster, D., Orly, J., Seidel, G., Schramm, M. Intracellular cAMP production enhanced by a hormone receptor transferred from a different cell: β-adrenergic responses in cultured cells conferred by fusion with turkey erythrocytes. *J. Biol. Chem.* 1978; 253: 1201–6.

GTP-binding proteins and signal transduction

Nucleotides as metabolic regulators

In addition to providing the alphabet of the genetic code, nucleotides play many other roles. The pyrimidine bases act as identifiers for metabolites. For instance, CTP is a reactant in phospholipid biosynthesis in which the CMP moiety is transferred to phosphatidate in the formation of CDP-diglyceride (see Figure 8.7). Similarly, it reacts with choline phosphate in the formation of CDP-choline. Uridine nucleotides are involved in the assembly of polysaccharides:

$$\text{glucose-1-P} + \text{UTP} \xrightarrow{\text{UDP-glucose pyrophosphorylase}} \text{UDP-glucose} + \text{PP}_i$$

Purine nucleotides play their main regulatory roles in association with proteins, not metabolites. ATP acts as a link between metabolic processes and cellular activities. It is present in most cells at high concentration, between 5–10 mmol/l in the cytosol of muscle and nerve.

It is also subject to rapid turnover. The human body typically turns over 90% of its entire content of ATP within 90 seconds; thus, an amount equivalent to 75% of the body weight, about 40 kg, is turned over every day. In spite of this, the cellular concentration of ATP generally remains remarkably constant, even in muscle in the face of sustained dynamic work. In the extracellular environment ATP also acts as a neurotransmitter.

The possibility that GTP might act as a factor necessary for cellular processes first became apparent in connection with its requirement for gluconeogenesis (phosphoenolpyruvate carboxylase) and in the operation of the tricarboxylic acid cycle (succinyl CoA synthase). It is a necessary component, together with mRNA, activated amino acids (aminoacyl-tRNA) and ribosomes, in the initiation and elongation reactions of protein synthesis. In this sense, the initiation factors and elongation factors should be regarded as the original GTP-binding proteins.

Estimates of cytosol ATP concentration have been somewhat lower for other cell types (1–3 mmol/l). The determination of its concentration depends on a good estimate of cytosol volume.

ATP is not quite what it seems

In the early years following the discovery of cAMP, all that was required to detect the activity of adenylyl cyclase in cell membrane preparations was

(a)   (b)

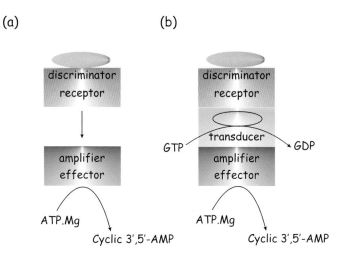

Figure 4.1 Introduction of the idea of a transducer in the relay of the receptor signal to the effector: (a) The early view. (b) A transducing GTP-binding protein relays the signal. Adapted from Rodbell.[5]

an activating hormone and the substrate ATP (as its Mg^{2+} salt) (Figure 4.1a). The membranes, containing an appropriate receptor and the catalytic unit, did the rest. At least, that was the general idea. The economical manufacture of ATP, in a high state of purity and in large quantities, has been a notable achievement of the commercial suppliers. It is now possible to purchase a gram of crystalline ATP, better than 99% pure, for less than the price of a pint of beer, but this was not always the case. Although commercially produced ATP was good enough as a substrate in the 1960s, sometime around 1970 it became erratic, occasionally registering zero activity. A similar impasse had been encountered previously in the investigation of fatty acid biosynthesis, and it had been found that an impurity had been lost as the quality of the ATP had been improved, and this was CTP. In the case of adenylyl cyclase experiments, it was the exclusion of contaminating traces of GTP that caused the loss of activity. Addition of GTP now allowed hormone-induced generation of cAMP to proceed.[1] Pertinently, it also had the effect of reducing the affinity of agonist (but not of antagonist) binding at receptors.[2–4] It was clear that GTP plays a central role in the cyclase reaction, not as a substrate but as a cofactor. The two established components, in Rodbell's terminology the discriminator (receptor) and the effector were now joined by a third, the transducer that binds GTP (Figure 4.1b).

■ GTP-binding proteins, G-proteins or GTPases

The current terminology of GTP-binding proteins embraces G-proteins and GTPases. All are capable of hydrolysing GTP, so, technically, all are GTPases. The term G-protein is generally reserved for the class of GTP-binding proteins that interact with 7TM receptors. All of these are composed of three subunits, α, β and γ, and so it is also common to refer to the class of heterotrimeric G-proteins. This distinguishes them from the other main class of GTP-binding proteins involved in signalling. These are monomeric and are related to the protein products of the ras proto-onco-genes. Collectively, the GTP-binding proteins are everywhere.[6] The basic cycle of GTP binding and GTP hydrolysis, switching them between active and inactive states is common to all of them and it is coupled to many diverse cellular functions.[7]

Historically, the G-proteins linked to the 7TM receptors were discovered before the Ras-related proteins and this is the order in which we consider them here.

G-proteins

Heterotrimeric structure

The structural organization of the components α, β and γ-subunits in a heterotrimeric G protein (G_i) is illustrated in Figure 4.2. The β-subunit, which has seven regions of β structure organized like the blades of a propeller, is tightly associated with the γ-chain and together they behave as a single entity, the $\beta\gamma$-subunit. The α-subunit contains the nucleotide binding site and it forms contacts with one face of the β-subunit. The whole assembly is anchored to the membrane by hydrophobic attachments, one at the N-terminal of the α-subunit and the other at the C-terminal of the

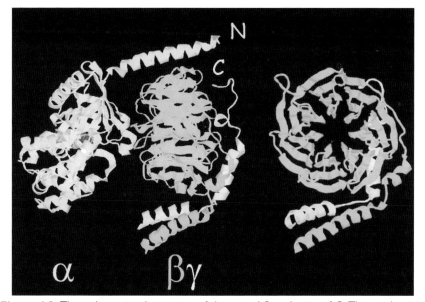

Figure 4.2 Three-dimensional structure of the α- and $\beta\gamma$-subunits of G_i. The α-subunit (left, cyan) has a molecule of GDP bound. The N-terminal helix is at top right. The $\beta\gamma$-subunits (β green, γ yellow) are in close apposition. The surface of the heterotrimeric structure that is in contact with the membrane is at the top of the figure. The hydrophobic attachments that are responsible are not shown. They involve the N-terminal of the α-subunit and the C-terminal of the γ-subunit. The separate $\beta\gamma$-subunit on the right has been rotated about a vertical axis to show the β propeller structure. (Data source: 1gp2.pdb[58]).

74

Here, we differ from most textbook accounts of G-protein activation. These state that the activated heterotrimer dissociates into two functional components, α-GTP and βγ. α-GTP then seeks out the effector enzyme and activates it.[8] However, with the notable exception of transducin (the G-protein that mediates visual transduction), the evidence for the physical separation of α-subunits from βγ-subunits is slight (though activated hetero-trimers are certainly more readily dissociated in test-tube experiments). A dissociation mechanism of G-protein activation may only be a hypothesis without a strong experimental basis but unfortunately it also represents the paradigm underlying the strategy of most experimental investigations in this area. It may be fortuitous that exogenous G-protein α-and βγ-subunits retain the activities of activated heterotrimeric G-proteins.[9]

γ-subunit. The structures of these lipid modifications are shown in Figure 4.9 and specific modifications are listed in Table 4.2).

The GTPase cycle: a monostable switch

The cycle of events regulated by all GTP-binding proteins starts and ends with GDP situated in the guanine nucleotide binding site of the α-sub-units (Figure 4.3). Throughout the cycle, the G-protein remains associated with the effector enzyme through its α- or βγ-subunits, but its association with the activated receptor is transient[8,11] ('collision coupling'[10]). Upon stimulation, and association of the receptor with the G-protein–effector complex, the GDP dissociates. It is replaced by GTP. The selectivity of binding and hence the progress of the cycle is determined by the 10-fold excess of GTP over GDP in cells (GTP is present at about 100 µmol/l). After activation, the contact between the G-protein and the agonist–receptor complex is weakened, so allowing the receptor to detach and seek out fur-ther inactive G-protein molecules. This provides an opportunity for signal amplification at this point.

The irreversibility of the cycle is determined by cleavage of the terminal phosphate of GTP (the GTPase reaction). With the restoration of GDP in the nucleotide binding site of the α-subunits, the G-protein, and with it the effector enzyme, return to their inactive forms. G-proteins are often referred to as molecular switches. Actually, this may not be the best description, because they do not necessarily switch off abruptly. A better analogy might be the light switch on the landlady's staircase, the kind

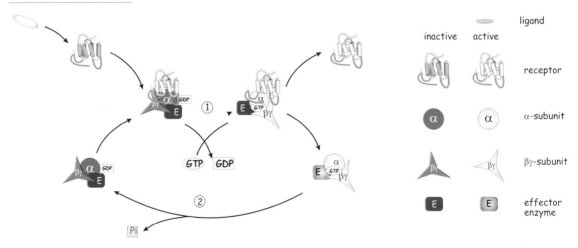

Figure 4.3 Activation and deactivation of G-proteins by guanine nucleotide exchange and GTP hydrolysis. Activation and deactivation of the G-protein α-subunit by (1) guanine nucleotide exchange and (2) GTP hydrolysis. The receptor binds ligand and is activated to become a catalyst, enhancing the rate of detachment of bound GDP from the α-subunit of the G-protein. The vacant site is rapidly occupied by GTP and this causes activation of the α- and βγ-subunits, enabling them in turn to activate specific effector enzymes. The interaction between receptor and G-protein is transient, allowing the receptor to catalyse guanine nucleotide exchange on a succession of G-protein molecules. The system returns to the resting state following hydrolysis of the bound GTP on each of these.

which, when pressed (GDP/GTP exchange) switches on the light for just the time it takes to make the dash up to the next landing (when GTP hydrolysis occurs). In electronic parlance this would be called a mono-stable switch. The activation is thus kinetically regulated, positively by the initial rate of GDP dissociation and then negatively by the rate of GTP hydrolysis. In this way, the state of activation can be approximated by the ratio of two rate constants, each of which is under independent control:[12]

active/inactive = on/off
$$= [G\alpha \cdot GTP]/[G\alpha \cdot GDP]$$
$$= k_{diss}/k_{cat} \qquad\qquad \text{Equation 4.1}$$

where k_{diss} is the rate constant for the dissociation of GDP and k_{cat} is the rate constant for the hydrolysis of GTP.

The GTPase cycle, first perceived in the processes of visual phototrans-duction[13] and β-adrenergic activation of adenylyl cyclase, is central to many important signal transduction mechanisms.[7] The GTP-bound form of the α-subunit is required to activate effector enzymes such as adenylyl cyclase and phospholipase C. The duration of this interaction lasts only as long as the GTP remains intact. As soon as it is hydrolysed to GDP communication ceases and the effectors revert to their inactive state.

■ Switching off activity: switching on GTPase

It was initially surprising to find that the rate of GTP hydrolysis catalysed by isolated G-proteins is far too slow to account for the transient nature of some of the known G-protein-mediated responses. The extreme example must be the visual transduction process which, of necessity, must be turned off rapidly. When the lights go off, we perceive darkness promptly. If transducin (G_t, the G-protein responsible for visual transduction) was able to persist in its active form ($\alpha_T \cdot GTP$), the illumination would appear to dim gradually as the hydrolysis of GTP proceeds. It is now clear that the rate of the GTPase reaction (k_{cat}) is stepped up very considerably as soon as activated α-subunits come into contact with effector enzymes. This is certainly the case for the GTP hydrolysis catalysed by α_T[14,15] (see Chapter 6). Likewise, the GTPase activity of the α-subunit of G_q is accelerated as a result of its association with its effector enzyme, phospholipase Cβ,[16,17] so ensuring that the α-subunit retains its activity just for as long as it takes to make productive contact. More generally, the GTPase activity of the main classes of G-proteins, (with the exception of the G_s subgroup) are subject to regulation through the family (at least 20 members) of RGS proteins (regulator of G-protein signalling). These interact with specific α-subunits to accelerate the rate of GTP hydrolysis and act as acute regulators of a wide variety of physiological processes.[18] The smaller RGS proteins (<220 residues) appear to have no particular homologies apart from their conserved catalytic (RGS) domain. Others, however, are much larger (370–1400 residues), and possess multiple structural domains (DH, PH, PTB, PDZ, etc: see Chapter 18) that might allow them to link up with other proteins involved in signalling. An example is the desensitization of a Ca^{2+}

The lower maximal activation (for example, of adenylyl cyclase) achieved by the so-called partial agonists can be ascribed to lower rates of GDP dissociation.[12] Unlike full agonists, partial agonists fail to distinguish perfectly between the active and the inactive conformations of the receptor, thus they have some of the character of inverse agonists (see page 59).

channel (N-type) which is mediated through GABA$_B$ receptors and G$_O$ and opposed by RGS12.[19] This provides yet another level of control over the processes regulated by G-proteins. Also, there is the possibility that the RGS proteins might themselves be subject to regulation through the level of their expression.

There are other proteins having similar activity that act to enhance the GTPase activity of the small monomeric GTP-binding proteins such as Ras. These are known as GTPase activating proteins or GAPs (see below).

Modulation of receptor affinity by G-proteins

One of the earliest indications of a role for GTP in the activation of receptor-mediated-processes was the lowering of the affinity for activating ligands by stable chemical analogues of GTP (GTPγS, GppNHp, etc.; Figure 4.4).[20] Because these analogues are not readily hydrolysed they ensure that the α-subunits remain persistently activated. This, coupled with the observation that application of agonistic (but not antagonistic) ligands accelerates the rate of GTP hydrolysis,[21] gave the clue that the communication between receptors and GTP-binding proteins is a two-way affair. The receptors speak to the G-proteins and the G-proteins speak to the receptors. Following activation, the contact between the G-protein and the agonist–receptor complex is weakened, so allowing the receptor to seek out another G-protein molecule. The situation for inverse agonists is the converse (Figure 4.5).[22–25] We referred earlier (Chapter 3) to the example of the oxytocin receptor, for which progesterone acts as an inverse agonist, binding to the receptor in its inactive conformation. Here GTPγS characteristically causes the affinity for the peptide hormone to decline while the affinity for steroid hormone increases.[26]

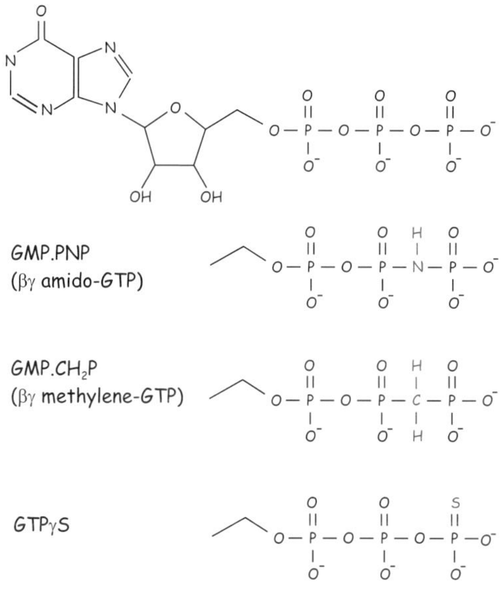

Figure 4.4 **Structures of stable analogues of GTP.**

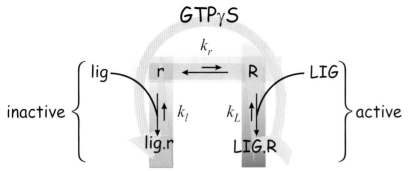

Figure 4.5 Opposing effects of guanine nucleotides on the binding of agonists and inverse agonists to receptors. The diagram illustrates the equilibrium between the inactive (cyan, r) and active (magenta, R) states of a receptor. Agonistic ligands (LIG) bind exclusively to the active form of the receptor and thus increase the proportion and the total amount (R + LIG.R) of the receptor in the form that can transmit a signal. Conversely, inverse agonists (lig) bind exclusively to the inactive form of the receptor and thus increase the total amount of the receptor (r + lig.r) in the form that is incapable of transmitting signals. The stable analogue, GTPγS, depresses the affinity of receptors for agonists while enhancing their affinity for inverse agonists. There is no effect on the affinity for the common competitive inhibitors. Compare this figure with Figure 3.17.

▪ α-Subunits

Historically, an experimental challenge has been to manipulate receptors, G-proteins and effector enzymes independently of each other. Because the receptors and the effectors are intrinsic membrane proteins, they cannot readily be added or withdrawn in an experimental system. Even though they can be mixed and matched in cell fusion experiments, as discussed in Chapter 3, this system does not allow the manipulation of the different subunits of the heterotrimeric G-proteins. Instead, it was the use of the lymphoma cell line S49 cyc⁻ that led to the identification of the α-subunits as the GTP-binding component relaying signals from receptors to adenylyl cyclase.[27]

In the wild-type S49 cells, elevation of cAMP has the effect of arresting the cell cycle in the G1 phase (for details, see Chapter 10) and then promoting cell death[28,29]. By growing the cells in the presence of isoprenaline to elevate cAMP and then rescuing the few survivors, and then repeating this procedure, a resistant line, cyc⁻, was obtained[30] (Figure 4.6).

Although endowed with β₂-adrenergic receptors and an adenylyl cyclase which could be activated by the terpenoid forskolin (a direct activator of adenylyl cyclase: see Chapter 5), these cells are unresponsive to agonists binding at the β₂-adrenergic receptor. The signal transduction pathway may be said to have become uncoupled. Something stands in the way (or is missing), so that the communication between the receptor and the effector enzyme is blocked.

Communication between the receptor and the cyclase can be established by provision of the GTP-binding component (α-subunit of the G-protein G_s) to the membranes of these cyc⁻ cells.[31–33] By failing to express the GTP binding α_s-subunit, the cyc⁻ cells avoid the possibility of lethal elevation of cAMP. This system has been widely used as a test-bed to study

Of course, **cyc⁻** is a misnomer, based on the initial thought that these cells might be devoid of adenylyl cyclase.

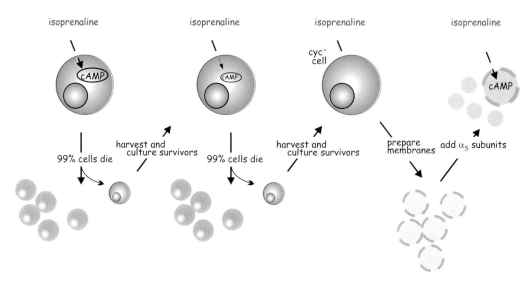

Figure 4.6 Reconstitution of hormone stimulated cyclase activity in S49 cyc⁻ lymphoma cells by addition of α_s-subunits. The figure illustrates the principle underlying the selection of a line of cells (S49 cyc⁻) that fail to generate cAMP in response to stimulation by ligands that bind to receptors linked to G_s. Activity can be restored to the isolated membranes by addition of G-protein α_s-subunits.

Table 4.1 Some functions of G-protein α-subunits

	Generation of second messengers	Other processes
α_s	Stimulation of adenylyl cyclase	
α_{olf}	Stimulation of adenylyl cyclase	
α_{i1}	Inhibition of adenylyl cyclase	
α_{i2}	Inhibition of adenylyl cyclase	Mitogenic signalling[34,35] Positive regulation of insulin action[36]
α_{i3}	Inhibition of adenylyl cyclase	
α_O	Inhibition of type I adenylyl cyclase[37]	Potentiation of EGF-R and Ras signals to MAP kinase[38] Sequestration of Rap1GAP[39], activation of STAT3[40], inhibition of cardiac L-type currents[41]
α_t	Stimulation of cyclic GMP phosphodiesterase	
α_z	K⁺ channel closure	
α_q	Stimulation of PLCβ[42,43] Stimulation of Btk[44]	Affects migratory responses of fibroblasts to thrombin Platelet function disrupted in $\alpha_q^{-/-}$ (knock-out) mice[45]
α_{12}	Stimulation of Btk[46], RasGAP (Gap1m)[46] Suppresses α_{13} on RhoGEF	
α_{13}	Stimulates activity of RhoGEF[47]	Regulation of developmental angiogenesis[48] Affects migratory responses of fibroblasts to thrombin[48]

the effects of different GTP-binding proteins, for example, proteins with selected point mutations or chimeric variants containing components of different G-proteins such as G_s and G_i (inhibitory). This opened the way to understanding the roles and interactions of G-proteins in the sequence of steps leading from receptor occupation to cyclase activation.

α-Subunits determine G-protein diversity

The diversity of the heterotrimeric G-proteins is principally a function of their α-subunits (Table 4.1). Unlike the receptors, of which there are probably thousands, there are only 16 genes specifying the α-subunits in animals (with alternative splice variants of G_s and G_o, these provide about 20 gene products in all) (Figure 4.7).

Most animal cells express 9 or 10 of the 16 possible α-subunit gene products. Some, such as the α_t, α_{olf} and α_{gust} genes are expressed only in single classes of sensory cells (photoreceptors, olfactory epithelial neurones and taste receptors). Others are found in cells that share a common embryonic origin. Thus, α_o-subunits are expressed in cells derived from

Possibly as much as 1% of the human genome codes for G-protein coupled receptors, the odorant receptors of the olfactory epithelium alone contributing several hundred.

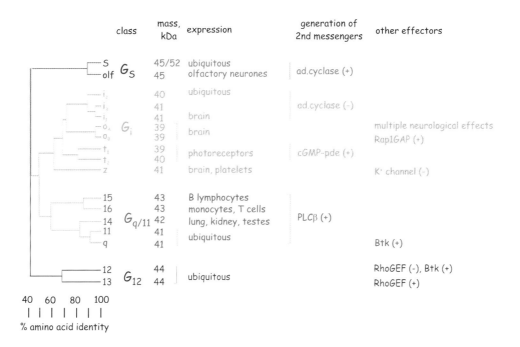

Figure 4.7 **Evolutionary relationship of G-protein α-subunits.** All proteins that bind GTP have related sequences in those segments that interact directly with the nucleotide (see Figure 4.12). The sequence identity of the α-subunits of heterotrimeric G-proteins is greater than 40%. In the figure, examples of downstream effectors that generate second messengers are given. Some other effector proteins that are discussed elsewhere in the book are also listed. Note that G_s is expressed in short and long forms as a result of alternative splicing. Btk, Bruton's tyrosine kinase,[44,46] is a non-receptor protein tyrosine kinase (see Chapter 12). It is likely that G_q and G_{12} also regulate other related protein tyrosine kinases. RhoGEF[47,49] see Chapter 14. Rap1GAP interaction with G_o[39] see Chapter 9. The dendrogram is adapted from Simon et al.[50]

the neural crest and in endocrine tissues such as the pancreatic β-cells and the pituitary gland. α_z is found primarily in neurones and α_{16} is expressed exclusively in cells of haematopoietic origin. Only α_s, α_{i2} and α_{11} are universally expressed in animal cells and most cells also express α_q and either α_{i1} or α_{i3}.

When the amino acid sequences of the various α-subunits are aligned, it can be seen that about 40% of the residues are invariant. Moreover, the proteins are strongly conserved in evolution. It is likely that this high degree of conservation has been driven by the need to maintain the very specific, yet multiple physiological functions of each of these protein gene products.

We can classify the family of α-subunits in four main classes, each of which contains a number of closely related isotypes (Figure 4.7). Thus, the α-subunits of mammalian α_s and α_{olf} differ only in 12% of their amino acids and α_{i1} and α_{i3} differ in only 6%.

Within these groups, the functions are, in the main, closely related. All the three α_i proteins inhibit adenylyl cyclase; all α-subunits of the $G_{q/11}$ family activate PLCβ. Other α-subunits function not in the regulation of second messengers, but to modulate the activity of proteins in other signal transduction pathways. The opposing effects of α_{12} and α_{13} on the activity of RhoGEF is an example of this[47,49] (see Chapter 14).

Also, α_q and α_{12} are able to activate Btk (Bruton's tyrosine kinase) and possibly other related non-receptor protein tyrosine kinases (see Chapter 12).

The downstream effectors of some of the α-subunits still remain very uncertain. In particular, the functions of G_o have so far evaded any sort of precise description. A problem here is that G_o appears to have multiple effects. One might have thought that ablation of the gene coding for such an abundant protein, as much as 2% of membrane protein in brain, would generate a drastically affected phenotype that would offer a ready diagnosis. Yet $\alpha_O^{-/-}$ (knockout) mice survive some weeks. They are hyperactive, tending to run in circles for hours on end, and display hypersensitivity in standard pain tests.[41] There appears to be some involvement of G_o in the regulation of voltage-sensitive (L-type) Ca^{2+} channels, but this is likely to be indirect. Among other functions of G_o, it regulates the activity of Rap1, a Ras-related protein[39] (see Chapter 9), it enhances signalling from the EGF receptor, potentiating the activity of ERK (extracellular signal regulated protein kinase)[38] and it activates signalling by STAT3[40] (see Chapter 11).

Sites on α-subunits that interact with the membrane and with other proteins

The C-terminal region of α-subunits dictates the specificity of interaction with receptors. It is the site of ADP-ribosylation by pertussis toxin (see Chapter 5). The N-terminal sequence is the site of interaction with $\beta\gamma$-subunits. When removed, interaction with $\beta\gamma$-subunits is prevented. The N-terminal glycine of the α-subunit of G_i and G_o is the site of attachment of myristic acid ($CH_3(CH_2)_{12}COOH$) on the α-subunits of G_o and G_i, a

The α_s subunits of rats and humans differ by only a single amino acid out of a total of 394; α_{i1} is identical in cows and humans and $\alpha_{i2}, \alpha_{i3}, \alpha_o$ and α_z retain more than 98% identity among all mammalian species.

RhoGEF A guanine nucleotide exchange catalyst for the Ras-related GTPase Rho.

$\mathbf{G_{olf}}$ regulates adenylyl cyclase in olfactory epithelial neurones.

modification that facilitates heterotrimer formation. Antibodies raised against synthetic peptides having the same sequences that are present in the C-terminal domain of α-subunits block the line of communication from receptors.[51,52] Point mutations in this region also determine an 'uncoupled' (*unc*) phenotype[53] in which downstream processes can be stimulated by stable analogues of GTP (see Figure 4.4). They also fail to respond to receptors.

It has been much harder to determine the sites at which α-subunits communicate with effectors. This is because the protein interfaces are not composed of just a few residues on a single stretch of the chain. Instead, they are distributed as clusters that are present on different loops, separated by long distances in the sequence. After an extensive analysis of chimeric proteins constructed from α_{i2} and α_s, three distinct regions were perceived to be necessary for the interaction with adenylyl cyclase.[54] More recently, detailed structural information based on X-ray crystallography of α_t and α_{i1} has revealed the structural changes that occur when GDP is exchanged for GTP.[55] Here, the lessons learned from crystallographic investigations of the small GTPases, such as Ras, provide a valuable key to the understanding of the mechanics of the more complex α-subunits. There may yet be more to learn as molecular models are obtained in the same way for complexes of Ras with the Ras-binding domains of their effector proteins (Raf-1, see Chapter 11) and modulators (RasGAP: see below).

Crystallographic studies of α_t and α_i indicate that these proteins are folded as two independent domains[56–58] (Figure 4.8). One of these (rd) has a structure closely related to Ras, and contains the guanine nucleotide binding site. The other is a compact helical domain (hd, a bundle of six helices) that is absent from Ras and other monomeric GTP-binding

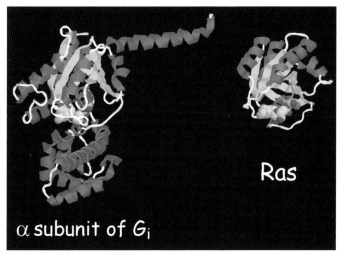

Figure 4.8 The α-subunits of G proteins and the monomeric GTPases exhibit structural similarities. α_i (left) has an rd domain (ras domain, upper half) that resembles the small GTPase Ras (right). The lower half of the α_i structure is the hd domain (helical domain). Each molecule has a bound GDP. (Mg^{2+} is only indicated in the RasGDP structure, green sphere.) Data source: α_i: 1gp2.pdb[58], RasGDP: 4q21.pdb.[125-128]

proteins. All the sites of interaction linking the α-subunits to other proteins had been mapped to the conserved Ras-like domain and the amino-terminal sequence. However, the sequences of the hd domains in the different α-subunits vary considerably and so it was reasonable to think that they might be linked to function. What this function might be, if any, was largely a matter for speculation. Among the ideas considered were that it might be involved in effector recognition, that it might modulate the affinity of GTP-binding or act as an intrinsic GTPase-activating protein (GAP: see below). Now it now appears that the hd domain of the transducin α-subunit (α_t) acts as an allosteric modulator, enhancing the efficiency of the interaction of the activated α_t with its effector, cyclic GMP phosphodiesterase[59] (see Chapter 6). The helical domains of the α-subunits of the other main G-protein classes may also be allosteric modulators of effector enzymes.

■ βγ-Subunits

β and γ subtypes

There are 5 β subtypes sharing 50–90% identity (with molecular masses of either 35 kD or 39 kD) and at least 12 subtypes of γ (5–10 kDa). The sequences of γ are more diverse. Not all of the potential βγ pairs exist in nature; for example, there is $\beta_1\gamma_2$, but not $\beta_2\gamma_1$. Regardless of this, in reconstitution experiments, the βγ-subunits seem to be interchangeable in their association with α-subunits, with one clear exception: the α-subunit of transducin, α_t, associates with $\beta_1\gamma_1$, uniquely expressed in photoreceptors. Mammalian γ-subunits are post-translationally modified at the C-terminus by the addition of the 20-carbon geranylgeranyl group (four isoprene units) or the 15-carbon farnesyl group (three isoprene units) (Table 4.2; Figure 4.9) . This hydrophobic adduct ensures that they are tethered to the membrane and with them, their associated β-subunits. Although mutated γ-subunits unable to undergo the prenylation reaction still associate with β-subunits, the resulting heterodimers are soluble and unable to interact with α-subunits such as α_s.

Table 4.2 Heterotrimeric G-protein post-translational lipid modifications

Subunit	Modification
α_q	Palmitoyl
α_s	Palmitoyl
α_i	Myristoyl
α_o	Myristoyl
α_t	Myristoyl
α_{olf}	Myristoyl
α_z	Myristoyl, palmitoyl
$\gamma_1, \gamma_8, \gamma_{11}$	Farnesyl
$\gamma_{2-7}, \gamma_{9-11}$	Geranylgeranyl

prenylation of GTPases

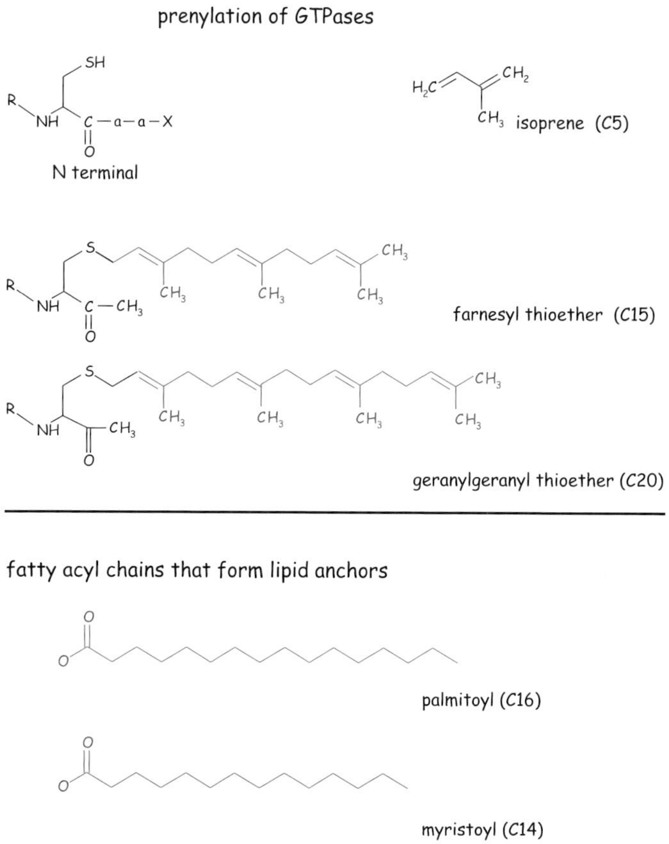

SH

R—NH—C—a—a—X
        ‖
        O
N terminal

H_2C=C—CH_2
      |
      CH_3 isoprene (C5)

farnesyl thioether (C15)

geranylgeranyl thioether (C20)

fatty acyl chains that form lipid anchors

palmitoyl (C16)

myristoyl (C14)

Figure 4.9 Membrane tethering of βγ-subunits through post-translational modification.

βγ-Subunits as signalling proteins

The first indication that βγ-subunits can transmit information independently of α-subunits and of second messengers (cAMP, Ca^{2+}, etc.) was in connection with the process by which acetylcholine reduces cardiac output through parasympathetic stimulation.[60] As related in Chapter 1, this very effect had provided the first firm evidence in favour of the idea of chemical transmission at synapses. Some 70 years later, it was found that this muscarinic response is mediated by a G-protein.[61] No soluble second messenger is involved and the G-protein, likely to be G_i or G_o (sensitive to pertussis toxin: see Chapter 5), is understood to interact directly with a cardiac K^+ channel. As with the early idea of chemical transmission, the proposal that βγ-subunits are capable of signal transduction was regarded as heretical and generated much fractious debate.[62] It then emerged that purified βγ-dimers induce the opening of K^+ channels when applied to the inside surface of isolated membrane patches.[60,63] Counter-arguments came from those who insisted that the phenomenon was more likely due to the presence of contaminating α-subunits or even to the detergents used to maintain the G-proteins in solution. However, similar experiments using purified α-subunits produced erratic results at best. The dust

only settled when it was shown that co-expression in *Xenopus* oocytes of βγ-subunits together with the muscarinic K$^+$ channel results in constitutive (persistent) activity. In contrast, application of α_i-subunits containing GTPγS or co-expression of α-subunits with the K$^+$ channel is without effect.[64] By the time the argument had been finally resolved (at least to the satisfaction of most people), it had also been shown that βγ-subunits are activators of phospholipase A$_2$[65] and some β-isoforms of phospholipase-C.[66]

It is now clear that at the functional level, G$_s$ and all the other heterotrimeric G-proteins behave as if they were a dimeric combination of an α-subunit and an inseparable βγ pair. βγ-Subunits have the following functions:

- They ensure the localization, effective coupling and deactivation of the α-subunits.
- They regulate the affinity of the receptors for their activating ligands.
- They reduce the tendency of GDP to dissociate from α-subunits (thus stabilizing the inactive state).
- They are required for certain α-subunits (G$_i$ and G$_o$) to undergo covalent modification by pertussis toxin (ADP-ribosylation; see Chapter 5).
- They interact directly with some downstream effector systems.
- They regulate receptor phosphorylation by specific receptor kinases.

Receptor phosphorylation, down-regulation and pathway switching

The βγ-subunits also serve to present negative feedback signals as demonstrated first through the activation of rhodopsin kinase[67] and later with the β-adrenergic receptor kinase (βARK).

> The class of enzymes, including βARK and rhodopsin kinase, that phosphorylate 7TM receptors are called **G-protein-coupled receptor kinases** (GRKs).

Receptors such as muscarinic and adrenergic receptors, phosphorylated at serines and threonines on the C-terminal domain (see Figure 3.14, page 53) become targets for the binding of arrestins and this prepares them for removal by endocytosis.[68,69] It is thought that this mechanism is of particular importance in locations, such as sympathetic synapses, where agonist concentrations have a tendency to soar (the transient concentration of catecholamines within a synaptic cleft can reach 10^{-4} mol/l). Moreover, while the phosphorylation prevents the normal line of communication to G-proteins (desensitization), the bound arrestin can also act as a docking site for the soluble protein tyrosine kinase Src. This initiates a new signalling pathway resulting in the activation of ERK (see Chapter 11).[70]

As the concentration of the stimulus is increased, two phosphorylating mechanisms, catalysed by protein kinase A (PKA, cAMP activated protein kinase, see Chapter 9), and by βARK are called into action. Since these two kinases have different substrate specificities (consensus sequence selectivities), reacting with different residues in the C-terminal domain of the receptor, the pattern of phosphorylation is different (see Figure 3.14, page 53). It depends on whether there has been mild (low stimulus concentration, inducing PKA activity) or strong activation (high stimulus concen-

Figure 4.10 Heterologous and homologous desensitization by phosphorylation of receptors:
(a) Heterologous desensitization. Under mild stimulating conditions, the generation of cAMP results in the phosphorylation of all proteins (including receptor molecules) having the PKA substrate consensus motif (-RRSS-). The phosphorylation of receptors occurs regardless of their state of occupation and so the consequence is a generalized down-regulation of all the receptors that regulate cAMP production. (b) Homologous desensitization. Under strong stimulation, the βγ-subunits associated with the receptor act as an anchor for the soluble receptor kinase (βARK in the case of $β_2$-adrenergic receptors) which phosphorylates only this receptor molecule.

tration, additionally activating βARK). The consequence of phosphorylation at the PKA-reactive sites provides a classic feedback mechanism, which acts to disrupt the line of communication with the effector, adenylyl cyclase, and so shuts down the production of cAMP. The phosphorylation due to PKA, which may be maximally activated when less than 10% of the receptors are occupied (see Chapter 3), will also act to down-regulate other receptors having appropriate phosphorylation sites. These also will then be prevented from activating cyclase (heterologous desensitization, Figure 4.10). Phosphorylation by βARK or other receptor specific kinases affects only those receptors that are in contact with βγ-subunits (homologous desensitization). In this way, we can see that the responsiveness of activated receptors can be suppressed either specifically, by individual receptor kinases (homologous desensitization: *same* receptor), and also more generally through the action of second messenger activated, broad spectrum kinases such as protein kinase A and protein kinase C (heterologous receptor desensitization).

It must be evident that the whole sequence of control, from the first interaction of a hormone with its receptor right through to the generation of second messengers, and then back again to the receptors, is tightly regulated at all stages. Throughout, it remains flexibly sensitive to the needs of the cell.

◼ Ras proteins

◼ Monomeric GTP-binding proteins discovered as oncogene products

The Ras proteins are often referred to as proto-oncogene products. This is because they were first discovered as the transforming products[71,72] of a group of related retroviruses, including the Harvey murine (H) virus[73] and Kirsten sarcoma (K) virus.[74] The transforming genes are fusions of the viral

gag gene (see page 262) and one of the Ras genes derived from the rats through which the virus had been passaged.

N-Ras was discovered as a transforming gene product having sequence homology to the other Ras proteins, present in a neuroblastoma cell line.[75,76] They are all single-chain polypeptides, 189 amino acids in length, bound to the plasma membranes of cells by lipidic post-translational attachments at their C-termini (Table 4.3). They all bind guanine nucleotides (GTP and GDP) and they are GTPases.[77,78] Evidence for a link to human tumours came with the finding that cultured fibroblasts transfected with DNA derived from a human tumour cell line contain a mutated form of Ras.[79]

In a high proportion of human tumours one of the three endogenous cellular forms of Ras is altered by somatic mutations that inhibit the rate of GTP hydrolysis.[80] This ensures that they are in a persistently activated state. On the other hand, non-oncogenic forms, c-Ras, are present in all cells. They are regulators of cell growth and differentiation (see Chapter 11).

■ Subfamilies of Ras

The sequences of the Ras proteins are closely related (see Figure 4.11).

The first 164 amino acids of human H-Ras and chicken Ras differ in only two positions, and the sequences of the first 80 amino acids of human N-Ras and *Drosophila* D-Ras are identical. These close similarities are supported by many conservative substitutions. As with the α-subunits, it is likely that this high degree of conservation has been driven by the need for these proteins to communicate with a large number of other components, including the activators, inhibitors and effectors (see Figure 4.15).

The Ras proteins are regarded as the archetypes of a large superfamily. All members share some sequence homology to Ras and then fall into distinct groups, called Ras, Rho, Rab, Ran and Arf. Within each subfamily, the homologies are rather strong. Beyond the more immediate subgroups of the Ras superfamily, all these proteins share some limited sequence homologies with short 'fingerprint' sequences present in the bacterial elongation factors (involved in protein synthesis) and also in the α-subunits of the heterotrimeric G-proteins (Table 4.4, Figure 4.12).

Gag (glycosylated antigen) is the gene encoding the internal capsid of the viral particle crk (C10 regulator of kinase).

The incidence of mutated Ras proteins varies among different types of tumour: 90% of human pancreatic adenocarcinomas and 50% of colon adenocarcinomas are associated with Ras mutations. They are rarely found in adenosarcoma of the breast.

Yeast RAS activates adenylyl cyclase.[81] The proteins RAS1 and RAS2, having 321 and 309 residues, are longer than their mammalian counterparts, though the first 80 residues still maintain a high (80%) degree of homology. Yeast cells lacking both RAS1 and RAS2 are non-viable, but viability can be restored if they are induced to express the homologous mammalian H-Ras.[82,83] (By convention, capital letters are used to indicate yeast genes.)

Table 4.3 Monomeric GTPase post-translational lipid modifications

GTPase	Target sequence	Modification
H-Ras	CVLS	Farnesyl
N-Ras	CVVM	Palmitoyl
K-Ras	CVIM	Farnesyl
Rho, Rac	CXXL	Geranylgeranyl
Rab	CC/CxC	Geranylgeranyl
ARF	N-terminal glycine	Myristoyl

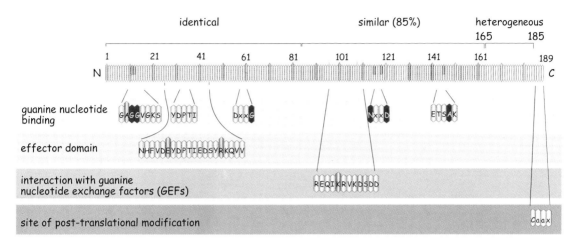

Figure 4.11 Main features of the Ras primary sequence. The mammalian Ras proteins (H, K and N) share very close identity for the first 164 amino acids (green indicates identity, mauve indicates a conservative substitution). With the exception of the cysteine residue (186), the C-terminal segment is highly divergent (yellow). The lower sections of the figure illustrate the details of sequence motifs involved in binding to the guanine nucleotide, the effector domain and the C-terminal-Caax box that forms the substrate for post-translational modification by isoprenylation. The residues marked in red are associated with oncogenic mutations.

Indeed, the presence of these short sequence motifs, appropriately distributed along the chain of a protein, can be taken as a fairly sure indication that it will be a GTPase. More than this, the presence of β-strands immediately adjacent to these highly conserved motifs is an invariant feature, whether they are Ras-related, elongation factors or α-subunits. Not too surprisingly, these conserved motifs constitute the sites of contact with guanine nucleotides.

■ Structure

A stereo image of the three-dimensional structure of Ras is shown in Figure 4.13. This indicates the elements of secondary structure and those connecting regions, G1–G5, that form the nucleotide binding in colours corresponding to Table 4.4 and Figure 4.12. These motifs are also present in the elongation factors and the α-subunits of heterotrimeric G-proteins. The residues 26–45 encompassing the G2 contact, comprise the effector region that communicates with downstream proteins. Within this region, residues 30–40 are conserved in all forms of Ras, from yeast to mammals, and when mutated, the products are generally inactive (as measured in cell transformation assays: see 'Cancer and transformation', page 246). Ras proteins mutated in the effector domain (switch region, see below) retain the ability to bind and catalyse the hydrolysis of GTP. Even when combined with a second (transforming) mutation that suppresses GTP hydrolysis and therefore prolongs the lifetime of the activated GTP bound state, the effector mutants remain biologically inactive.[84] The sequence of amino acids between codons 97 and 108 are responsible for interaction with guanine nucleotide exchange proteins (GEFs).[85]

Table 4.4 Conserved nucleotide binding motifs in H-Ras and bovine α_i. Binding contacts in red bold; non-conserved residues in lower case

Contact	Residues	Sequence	
G1	Ras *(10 - 17)*	GaggyGKS	Binds to α- and β-phosphates
	α_i *(40 - 47)*	GagesGKS	
G2	Ras *(32 - 36)*	YpdTi	Mg^{2+} coordination, effector loop
	α_i *(178 - 182)*	rvkTt	NB: The equivalent arginine (R201) in the G2 loop of α_s is the site of ADP-ribosyl attachment by cholera toxin
G3	Ras *(57 - 60)*	DtaG	Binds to Mg^{2+}, glycine binds to the γ-phosphate of GTP
	α_i *(200 - 203)*	DvgG	
G4	Ras *(116 - 119)*	NKcD	Confers specificity for guanine over other nucleotides
	α_i *(269 - 272)*	NKkD	
G5	Ras *(145 - 147)*	tsA	Buttresses guanine base recognition site
	α_i *(324 - 326)*	tcA	

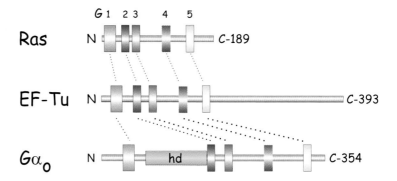

Figure 4.12 Conserved nucleotide binding motifs in Ras, G_i and EF-Tu. Three families of GTPases have generally divergent sequences but possess short stretches (G1–G5) that are similar to each other. When folded these segments are almost superimposable and comprise the guanine nucleotide binding pocket. The motifs G2 and G3 (in blue) lie within the switch regions 1 and 2 that undergo a conformational change when GDP is exchanged for GTP. The extended sequence marked hd on the α-subunit represents the helical domain which is absent from the other GTPases. Adapted from Bourne.[6]

Two regions of Ras, called the switch regions, change their conformation when the GDP is exchanged for GTP. These are indicated in Figure 4.14; here the nucleotide binding site is occupied by the non-hydrolysable GTP analogue GppNHp. Both switch-1 and switch-2 regions are implicated in the binding to effectors (such as the serine-threonine kinase Raf-1: see Chapter 11) and to the GTPase activating protein RasGAP. Switch-1 corresponds to the G2 loop that forms a part of the binding site for Mg^{2+}

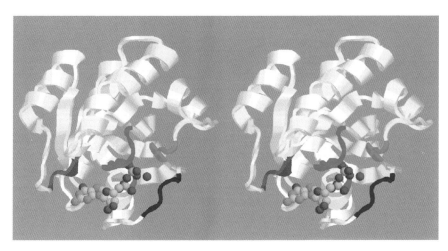

Figure 4.13 **Three-dimensional structure of RasGDP.** In this stereo image, the conserved motifs G1–G5 that make contact with the nucleotide are depicted in the following colours to match Table 4.4: G1 green, G2 blue, G3 cyan, G4 purple, G5 yellow. GDP is shown in CPK colours. The Mg^{2+} ion is coloured dark green. Instructions for viewing stereo images can be found on page 396. (Data source 4q21.pdb[125–128].)

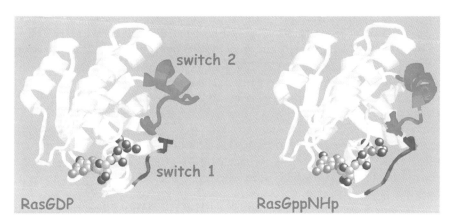

Figure 4.14 **Structural differences in the switch regions of GDP- and GTP-bound Ras.** Structures of the N-terminal residues of H-Ras complexed with GDP and with GppNHp (an analogue of GTP). The switch regions are indicated. The nucleotides are depicted as ball-and-stick structures. The green atom is Mg^{2+}. (Data sources 4q21.pdb[125–128] and 5p21.pdb[129] respectively.)

and the switch-2 region includes G3 which contacts the γ-phosphate of GTP, and the following helix (α_2-helix) (Figure 4.14 and Table 4.4). The conformational transitions in these two switch regions are coupled so that binding of GTP brings about an ordered helix $\rightarrow$ coil transition at the N-terminus (closest to the nucleotide) of switch-2. Subsequent changes to the $\alpha2$ helix effectively reorganize the components of the effector binding interface. All this is reversed when the bound GTP is hydrolysed to GDP.

■ Post-translational modifications

Beyond residue 165, the sequences become highly divergent. However, at the C-terminal extremities there is again homology in the form of -Caax (-Cysteine, aliphatic, aliphatic, anything: –CVLS in human H-Ras, –CVVM in N-Ras and –CIIM in K-Ras; see table 4.3). The –Caax motif is post-translationally modified by the addition of a farnesyl (C15, 3 isoprene units) extension, through stable thioether linkage to the cysteine. This is followed by carboxymethylation, cleavage of the final three amino acids and methylation of the newly exposed C-terminus (now cysteine) (Figure 4.9). A second lipid modification occurs by palmitoylation of another cysteine residue within the C-terminal hypervariable region. These modifications ensure strong attachment to the internal surface of the plasma membrane. When the post-translational prenylation reaction is prevented, either by mutation of the -Caax sequence or by farnesyl transferase inhibitors, normal Ras functions are suppressed.[86] The non-farnesylated Ras accumulates in the cytoplasm and as a result it scavenges the associated kinase Raf (see Chapter 11), preventing it from coming into contact with the membrane.[87]

■ GTPases everywhere!

We now know of more than 70 monomeric GTPases. With few exceptions they have been identified by screening cDNA libraries for Ras-like proteins and their functions have necessarily been assigned (and are still being assigned) subsequent to their discovery. This is an ongoing activity. The normal strategy is to express (or introduce) mutant forms, expected to possess altered functions which might be revealed as an altered cell phenotype. Thus, the sequence comprising the effector domain of Ras was established long before any effectors were identified. By comparison, the first discovered heterotrimeric G-proteins – G_s, G_i, G_o, transducin (G_t), etc. – were isolated by classical purification strategies (seeking proteins that bind and hydrolyse GTP and which interact with appropriate receptors and downstream effectors known to be affected by GTP).

To date, representative functions have been assigned for members of all the groups of Ras-related proteins in vertebrates and also in other eukaryotic organisms such as yeasts, flies, nematodes and slime moulds (Table 4.5). Some of the Ras-like proteins appear to have more than one function,

K-Ras-2B is different. It is not palmitoylated but instead there is a positive charge cluster which is thought to assist its association with negatively charged phospholipids on the membrane inner surface.

Farnesyl transferase inhibitors such as **simvastatin** or **lovastatin** are used to suppress cholesterol biosynthesis in patients having a history of stroke or heart disease. These drugs inhibit the conversion of 3-hydroxy-3-methylglutaryl CoA to mevalonate (catalysed by HMG CoA reductase). This is the key step in the pathway to the formation of isoprene units which then polymerize to form C_{10} geranyl, C_{15} farnsesyl, C_{20} geranylgeranyl and C_{30} squalene. The last of these is the precursor in the biosynthesis of cholesterol. Farnesyl transferase inhibitors are potential anticancer agents.

Table 4.5 Functions of Ras-related proteins

GTPase	Regulatory function
Ras	Cell proliferation and differentiation
Rho	Actin cytoskeleton modification Cell proliferation
Rab	Intracellular vesicle trafficking
Ran	Cell cycle transitions (S and M phases)
Arf	Activation of PLD Vesicle formation

possibly determined in some cases by the cells in which they are expressed or their subcellular localization. Doubtless, more will become evident.

Mutations of Ras promoting cancer

As with the heterotrimeric G-proteins, the proportion of Ras in the activated form is given by the ratio k_{diss}/k_{cat} (see Equation 4.1, page 75). Some of the oncogenic viral gene products (vRas) differ from the wild-type cellular forms by single point mutations which inhibit the GTPase activity (reducing k_{cat}), and this prolongs the active state. For this reason, a higher proportion of the activated GTP-bound form exists in vRas transformed cells. The most common mutations promoting the transformed phenotype occur at residues 12, 13 and 61 (see figure 4.11).

Other mutations promoting spontaneous activation of Ras are those which enhance the rate of dissociation (k_{diss}) of bound nucleotide (GDP). In the unmodified protein, this occurs at a very slow rate indeed (about 10^{-5} mol s^{-1} per mole of protein), but there are numerous mutations which increase this rate significantly. Many of these are at or in the close vicinity to the residues interacting with the nucleotide. In particular, substitution of Ala-146 increases the rate of guanine nucleotide exchange 1000-fold. As with the GTPase mutants, these mutations ensure that a higher proportion of the protein exists in the activated GTP-bound form. In consequence these too, are transforming mutations.

Functions of Ras

Although the identification of receptors and their catalytic activities generally preceded the discovery of the heterotrimeric G-proteins that link their operations, the proteins operating immediately upstream or downstream of Ras at first eluded definition. Even so, a place for Ras in the regulation of mitogenesis was not in doubt. Cellular expression of the transformed *ras* genes[88,89] or direct introduction of their products into single cells by microinjection[90,91] promotes cell detachment and loss of contact inhibition, characteristics of the transformed phenotype (see Box: Cancer and cell transformation, page 246). Conversely, microinjection of antibodies (to neutralize the native cellular Ras), or of a dominant inhibitory mutant form (S17N), prevents normal growth and cell division.[92–94]

S17N Ser replaces Asn at position 17.

However, mitogenesis is not the exclusive outcome of activating Ras. Other cell types undergo differentiation as the consequence of Ras activation. In PC12 cells the introduction of the Ras oncogene product provides a signal for neurite outgrowth.[95,96]

PC12 cells are an adrenal medullary cell line derived from a phaeochromocytoma.

How Ras controls all this has been far from clear. It is true to say that Ras lies at the centre of a network of interacting pathways, is activated and modulated directly and indirectly by several receptors and, in turn, has its influence on a large number of downstream processes (see Figure 4.15 and Chapter 11). Indeed, it is amazing that a protein as small as Ras can interact with so many other proteins. Among functions directly linked to

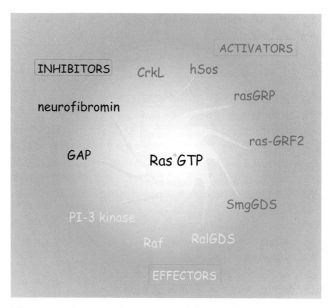

Figure 4.15 **Some of the many influences of, by and for Ras.**

Ras are the activation of the protein kinase Raf and the activation of phosphatidylinositol 3-kinase (PI 3-kinase: see Chapter 13). Raf is the first member of the sequence of phosphorylating enzymes (kinases) that leads to the activation of ERK and thence to the transcription of genes controlled through the serum response element, SRE (Chapter 11).

▪ Ras-GAPs

In cell-free preparations, the half-life of the GTP bound to Ras is between 1 and 5 hours. With a rate constant for hydrolysis of about 4.6×10^{-4} s^{-1}, wild-type Ras hardly functions as a GTPase. Once in this form, one might expect it to remain almost permanently activated. It would also be hard to understand why the differences in the rates of GTP hydrolysis by the wild-type and the oncogenic mutants should be so crucial. However, when the rate of GTP hydrolysis in cells is measured, things are very different.[97] Ras-GTP injected into *Xenopus* oocytes is converted to its inactive GDP form within 5 minutes. Cells contain a protein or proteins that accelerate the rate of GTP hydrolysis enormously. By contrast, GTP bound to GTPase-defective transforming mutants of Ras remains intact. These observations led to the discovery and isolation of the GTPase activating protein RasGAP (also known as p120^{GAP}) from *Xenopus* eggs and from mammalian tissue. The GTPase activating proteins (or GAPs) were the first proteins found to interact directly with Ras.

▪ Ras-GAP

This GAP protein interacts specifically with wild-type Ras, accelerating the GTPase activity by up to five orders of magnitude. When over-expressed,[98,99] or injected into fibroblasts at a concentration in excess of the normal level,[100] RasGAP inhibits the mitogenic action of growth factors (see Chapter 11). In this respect it acts, as one might expect, as a negative regulator of Ras. However, in T- and B-lymphocytes and in adipocytes a mitogenic signal is mediated by a decrease in the activity of p120[GAP].[101–103]

In addition to its (Ras-interacting) catalytic function, RasGAP contains a number of identifiable domains[104,105] (see Figure 4.16a). These include a PH domain,[106] a C2 domain,[107] two SH domains and one SH3 domain[108] (these are described in Chapter 18). The presence of such domains, a characteristic of many proteins involved in signal transduction, implies the potential for a multiplicity of interactions with other signalling proteins and other molecules.

Mechanism of GTPase activation

In trying to understand the mechanism of activation of the intrinsic GTPase activity of Ras by GAP proteins, two main ideas have been examined.[110] First, there is the possibility that the GAP acts simply by driving the Ras protein into a conformation active for GTP hydrolysis, without itself forming a part of the active site. In this case, the action of GAP on Ras would be catalytic and hence non-stoichiometric. Now, however, it is thought that the RasGAP interacts with the Ras stoichiometrically, contributing cationic residues (arginine) that stabilize the transition state of

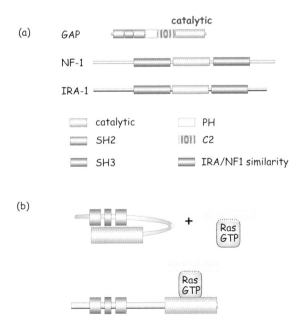

Figure 4.16 **The domain organization of RasGAP and related GTPase activating proteins.**

GAPs and effectors
Does RasGAP do more than regulate GTPase activity? Is it, like the effectors of heterotrimeric G-proteins, also an effector of downstream processes?[109] The fact that RasGAP interacts only with Ras.GTP, not with Ras.GDP, is certainly consistent with the idea of an effector function. Furthermore, the interaction is mediated through its contact with the effector domain of Ras (Figure 4.16b). Some (though not all) mutations in this domain prevent the interaction with GAP. Against this is the finding that RasGAP suppresses the transformation of fibroblasts induced by over-expression of normal wild-type Ras. Here, the GAP appears to play the role of a negative regulator, simply accelerating the rate of GTP hydrolysis.
An alternative possibility would be that Ras GAP might be the mediator of just a subset of Ras functions, the rest of which are linked to other effector proteins. For instance, it can stimulate transcription of the c-*fos* promoter[110] but here its catalytic domain plays no part since deletion mutants comprising only the SH2 and SH3 domains are equally effective.

However, Ras still appears to play a part since the induction of c-*fos* by the GAP deletion mutant is prevented by N17Ras, a dominant inhibitory mutant. It would appear that the GAP deletion mutant cooperates with another signal emanating from activated Ras. Also, as a result of complex formation with RhoGAP (specific for the Ras related GTPase Rho), RasGAP acts to inhibit the formation of stress fibres and focal adhesions, both of which are Rho-dependent processes[111] (see Chapter 14). The assignment of GAP functions will only be resolved when we can identify the relevant protein–protein interactions in normal cells.

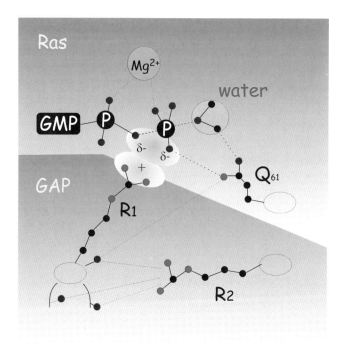

Figure 4.17 Organization of the active site of Ras and its interaction with RasGAP. The point of contact between the catalytic GTPase site of Ras and its GTPase activating protein (Ras GAP). A primary arginine finger extends across the cleft between the two proteins, and neutralizes the negative charge (pink) that develops in the transition state of the reaction. A second arginine residue (R2) stabilizes the primary arginine finger. The departing phosphate is depicted as a pentacoordinate intermediate. The Ras Q61 also contributes to the transition state. Adapted from Scheffzek *et al.*[110]

the reacting GTP (Figure 4.17). This allows access of Gln-61 (Q61) which directs an incoming water molecule for the hydrolysis of GTP. Mutations of this residue, frequently found in human tumours, prevent the enhancement of GTPase activity by GAP and lock the protein in its active conformation. A similar interaction of Rho with RhoGAP defines the catalytic centre of the Rho GTPase. For both Rho and Ras, the interaction with their respective GAPs is transient. They form active heterodimeric enzymes and then separate as the GTPases take up the conformations determined by the presence of GDP in the nucleotide binding pocket.

Other molecules having GAP activity are more likely to be pure negative regulators of Ras. One of these is neurofibromin, the product of the *NF1* gene, which is associated with neurofibromatosis type 1 (or von Recklinghausen's neurofibromatosis).

Neurofibromin contains a segment clearly related to the catalytic domain of RasGAP. In neurofibroma-derived cell lines deficient in this protein, 30–50% of the Ras is present in the GTP-bound form (as opposed to less than 10% in normal cells). Unlike RasGAP, the NF1 gene product possesses no identifiable domains or motifs (beyond its catalytic domain)

von Recklinghausen's neurofibromatosis is one of the most common of human hereditary disorders, having an estimated incidence of 1 in 3500

and shares no sequence similarity with any other mammalian signalling proteins (see Figure 4.16). These findings are all consistent with the idea that neurofibromin acts as a GTPase activator, pure and simple.

■ GEFs: guanine nucleotide exchange factors

The identification of proteins that catalyse GTP/GDP exchange on Ras (called guanine nucleotide exchange factors, or GEFs) was achieved by studying mutations in yeast and other simple eukaryotes, in which an exchange factor is deficient.

In yeast (*S. cerevisiae*), the RAS gene products regulate adenylyl cyclase,[111] mediating the response to starvation conditions and leading to spore formation. By mutating non-RAS genes that destroy the ability to activate cyclase, it is possible to identify other proteins that contribute to this signalling pathway. Some of these mutations can be overridden (bypassed) by the presence of constitutively activated RAS and are therefore likely to lie upstream of the GTPase. The CDC25 gene product is such a protein.[112] Deletion is lethal, but can be compensated by expression of constitutively activated RAS, strongly suggesting that the CDC25 gene product acts as an upstream activator, probably a RAS-GEF.[113] In addition, the effect of deletion can be overcome by expression of the C-terminal portion of CDC25, which is understood to house its catalytic activity. The catalytic domain of a human GEF (hSos-1: see Chapter 11) can restore cyclase activity in yeast lacking CDC25.[114]

We return to the question of activating guanine nucleotide exchange on Ras in Chapter 11.

Essay: Activation of G-proteins without subunit dissociation

It is a fact that the activation of G-proteins is generally discussed in terms of subunit dissociation. However, the evidence for the idea that G-protein heterotrimers actually separate into two independent active components, α and βγ, is not strong.

The conventional description (which can be found in most textbooks) is, at best, an over-simplification. The βγ-subunits, which certainly remain inseparable from each other under all physiological conditions, can be counted as a single entity. We need to ask whether α truly separates from βγ? Some arguments favouring dissociation may be summarized:

• Over-expression, or provision of purified α-subunits to isolated membrane preparations, causes activation of downstream effector enzymes such as adenylyl cyclase or phospholipase C.

• Provision or over-expression of βγ-subunits tends to oppose the activation due to α-subunits in some experimental systems such as platelet membranes.

• Provision of fluoride ions, or stable analogues of GTP (GTPγS, etc.), to G-proteins in detergent solution, and in the presence of high concentrations of Mg^{2+} ions, enhance their tendency to dissociate. These

individuals worldwide. Almost half of the patients have no previous family history of the disease, so with 1 in 10 000 individuals harbouring a new mutation, it follows that NF1, of all human genes, must be one of the most prone to mutation. It predisposes to benign and malignant tumour formation especially in cells derived from the neural crest.

Most work in this field has been confined to the investigation of the activation (and inhibition) of adenylyl cyclase, initially in turkey red blood cell membranes and later in other cell-derived and reconstituted systems. The turkey red cell membrane cyclase is particularly amenable since it is totally dependent on the presence of a stimulating hormone and is rather slow, allowing the collection of many data points. There have been no equivalent attempts to elucidate the mechanism of activation of phospholipase C.

manipulations are an essential component in the strategies applied in subunit purification.

- The α- and $\beta\gamma$-subunits of the retinal G-protein transducin certainly dissociate from each other and from the membrane when activated by rhodopsin.[115] However, about 8% remain attached to the membrane as the intact heterotrimer and it is possible that it is this fraction that activates the effector, cyclic GMP phosphodiesterase.

- X-ray crystallographic studies of GDP-bound, heterotrimeric G-proteins and of complexes of activated α-units with their effectors have shown (1) that the switch regions are obscured by the $\beta\gamma$-subunits in the inactive G-protein and (2) that contact with the effectors is made through the same regions.

Although it is clear that G-protein subunits can dissociate under conditions promoting activation and certainly do so when pressed hard enough (fluoride plus detergent plus Mg^{2+}), it is far from clear that this is what actually happens in cell membranes.

Against these arguments is the finding that the process of adenylyl cyclase activation is a first-order reaction, dependent only on the concentration of the activated receptor. From this it follows that the Gs is permanently coupled to adenylyl cyclase. This coupling can be maintained through a 3000-fold purification of adenylyl cyclase. Indeed, when carried out in the presence of an activating guanine nucleotide (GppNHp), a 1:1:1 stoichiometry of cyclase to α_s to β is retained.[116,117] There is no need to invoke the dissociation of the $\beta\gamma$-subunits from the complex in order for α_sGTP to cause activation even though this might deliver a stronger stimulus. The important possibility that arises is that the presence of both subunits may be required in order for the G-protein to fulfil its authentic physiological role. What would happen if the subunits are prevented from dissociating?

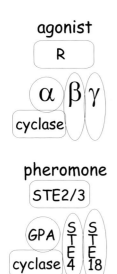

agonist

pheromone

Figure 4.18
Equivalence of pheromone and hormone receptors. In *S. cerevisiae*, STE2 and STE3 genes code for 7TM receptors specific for α- and **a**-mating factors. *GPA, STE4* and *STE18* encode proteins homologous to mammalian G-protein α-, β- and γ-subunits. Ste18 shares 35% homology with (retinal) γ_1.

 Pheromone-induced mating response in yeast

An excellent system in which to challenge the ideas concerning the dissociation of G-protein subunits and their relationship to receptors is provided by the mating response of yeast (*S. cerevisiae*). Here, in response to **a**- and α-type mating factors (peptides of 12 and 13 residues, similar to gonadotrophin releasing hormone [GnRH][118]) the α- and **a**-type haploid cells undergo cell cycle arrest at the late G_1 phase (see Chapter 10), attach to each other, fuse and eventually give rise to diploid $\alpha/$**a** cells (Figure 4.19). The secreted pheromones bind to the **a**/α-factor receptors, products of the genes *STE2* and *STE3*. The signals are transduced by the products of the genes *GPA, STE4* and *STE18* (Figure 4.19).

Deletion of GPA is lethal because the free $\beta\gamma$-subunits then activate a pathway leading to growth inhibition. However, mammalian α-subunits (of any class, even chimeric mammalian/yeast α-subunits) that can bind to the $\beta\gamma$-subunits are able to restore viability, while not restoring the signal transduction pathway leading to the generation of diploid cells.

(a)

(b)

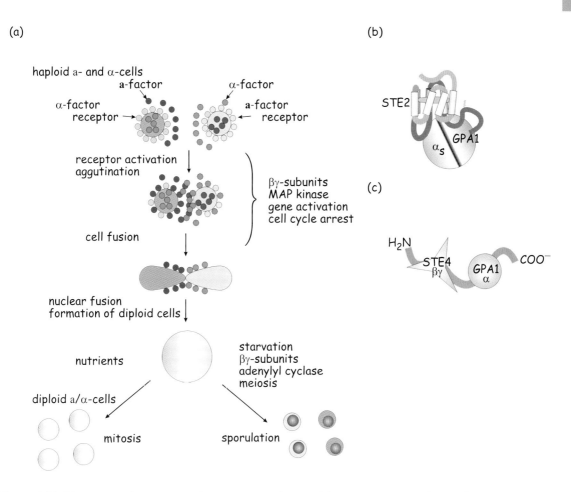

(c)

Figure 4.19 Pheromone-induced mating response of yeast: (a) Haploid yeast cells (**a** and α) generate **a** and α mating factors (pheromones) that bind to specific receptors on and α and a cells. This is the stimulus for a series of events that involves agglutination, cell cycle arrest, and then the mating reaction involving fusion and the generation of diploid cells. Given the presence of nutrients, the cells undergo multiple divisions, growing as diploids. In starvation they undergo meiosis, forming four haploid spores. Since α and **a** pheromone receptors are G-protein linked receptors coupled to adenylyl cyclase, the production of diploid cells can be used to read out the integrity of the signal transduction pathway. (b) Representation of the construct containing Ste2 (receptor) fused to a chimera of Gpa1 (yeast, N-terminal segment) and α_s (rat, C-terminal segment). This fused receptor–α_s construct is capable of efficient signal transduction. (c) Representation of the construct containing Ste4 (β-subunit) fused to Gpa1 (α-subunit). This fusion construct transmits signals as efficiently as the native G-protein subunits.

Indeed, the presence of all of the genes coding for G-protein subunits is required. By the use of mutants in which any of these genes are absent, and replacing these by fusion constructs, effectively chimeras composed of the N- and C-terminal sections of the absent proteins, it is possible to ask searching questions concerning the status and interactions of the various components of the signal transduction pathway.

Using such an approach, the mating response of cells made sterile by the presence of a null mutation (knock-out) in *STE4* (β-subunit) was restored by introduction of genes coding for chimeras composed of the

N-terminus of Ste4 coupled to the C-terminus of Gpa1.[9] These were fully functional in the transduction of pheromone signals, in promoting growth arrest and mating. Since the chimeric construct allows no possibility of subunit dissociation it is likely that the activated receptor acts to induce conformational changes in the heterotrimer, exposing hidden binding interfaces that allow communication with downstream effectors. The implication is that since the fusion construct is fully competent to convey the signals ascribed to both the α- and the βγ-subunits, then possibly the wild-type proteins which are demonstrably capable of dissociating may not actually do so. If they do, then it is likely that they remain in very close proximity to each other throughout the cycle. Indeed, it begins to appear that most of the molecules involved in signal transduction, receptors, G-proteins, effectors, are close neighbours at all times and under all conditions of activation.

■ Constructing the mammalian β-adrenergic transduction system in insect cells

Earlier we described how the use of S49 cyc⁻ lymphoma cells was instrumental in determining the role of the α_S-subunit of the signal transduction process. More versatile are the Sf9 insect cells.[119] With these cells and the baculovirus vector it is possible to manipulate the expression of all the components, mammalian receptors, α-subunits and βγ-subunits independently of each other with regard to identity, specified mutations and levels of expression. The native adenylyl cyclase can be used to read out the information transfer from the activated receptors.

When expressed in Sf9 cells, the affinity of the mammalian β-adrenergic receptor for its ligands is not only very low but also insensitive to the presence of guanine nucleotides (Figure 4.20a–c). The response of the native (insect) cyclase to the mammalian receptor is also low, and in general, the transduction of the signals from the mammalian receptor by the native insect G_s is inefficient. This system displays features of an 'uncoupled' (unc) phenotype.[120] All this changes when the Sf9 cells express mammalian α_S-subunits in addition to the mammalian receptor (Figure 4.20d). The affinity of ligand binding (isoprenaline) is now high but, importantly, it is also sensitive to the presence of guanine nucleotides, just as in mammalian cells. In this aspect of receptor function, it appears that the insect βγ-subunits are able to cooperate with the mammalian α-subunits. However, the GTPase activity remains insensitive to the presence of stimulating ligands. It requires the additional expression of mammalian βγ-subunits for the system to take on the full characteristics of the mammalian signalling pathway (Figure 4.20e). β_2-Adrenergic receptors expressed together with mammalian G_s heterotrimers allow high signal throughput (high V_{max}), high ligand binding affinity and now, ligand-enhanced GTPase activity. The conclusion from all this is that the α- and the βγ-subunits communicate with the receptor throughout the period of activation. They never really lose sight of each other, nor of the receptors.

That this might really be the case finds support in the frequent reports of the co-immunoprecipitation of receptors together with G-proteins and of

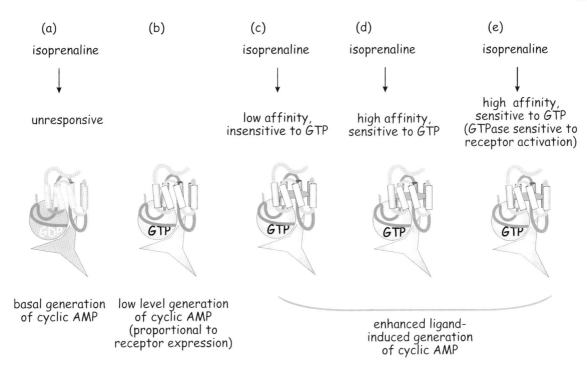

Figure 4.20 **Reconstitution of the β-adrenergic response in insect cells:** (a) Although they possess G-protein (grey) coupled to adenylyl cyclase, Sf9 (insect) cells lack β-adrenergic receptors and are therefore unresponsive to catecholamines. (b) When transfected with the human β-adrenergic receptor (pale green), Sf9 cells generate cAMP at a low rate that is proportional to the extent of receptor expression. No activating ligand is needed for this low-level activity. (c) The rate of cAMP generation is greatly enhanced when the transfected cells are stimulated with isoprenaline (yellow). However, in membrane preparations, the affinity of the receptor is not sensitive to the presence of GTP. The native G-protein does not transmit a feedback signal to the mammalian receptor. Nor is the GTPase activity of the insect α-subunit sensitive to the presence of the agonist. (d) By additionally transfecting the cells with mammalian α_s (blue), the affinity of the receptor becomes sensitive to the activation state of the G-protein. However, GTPase activity remains insensitive to the presence of the hormone. (e) To establish the complete pathway of forward and backward control, it is necessary to express both the mammalian α- and βγ-subunits (pink). The GTPase activity of the α-subunit is now sensitive to the presence of the activating hormone. For details of this experiment, see reference 120.

G-proteins with their effector enzymes. Similarly, the purification of 7TM receptors on affinity supports can result in co-purification of G-protein subunits, frequently of several different classes. As examples, the receptors for somatostatin and for opioids (δ-opioid) co-purify with α- and βγ-subunits to an extent that depends on the state of their activation.[121–123] This raises the possibility that receptors, rather than communicating in a linear fashion through one G-protein to one catalytic effector, may communicate, depending on their state of activation, with different G-proteins and hence with different effector enzymes (Figure 4.21).

For yeast cells devoid of the endogenous *STE2* (pheromone receptor) and *GPA1* (α-subunit) genes, transduction of the mating response can be achieved by a fusion protein generated from the N-terminus of Ste2 linked to a chimera composed of the N-terminus of Gpa1 and the C-terminus of rat α_s (see figure 4.19b).[124] The presence of the Gpa1/α_s

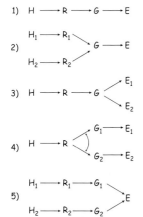

Figure 4.21 Pathways of information flow through receptors and G-proteins. The communication of signals from receptor to effector is not necessarily a simple linear sequence of steps (1). There are other more complex modes by which receptors, transducers and effectors are linked. Some of these may seem self-evident: different receptors accessing one class of G-proteins (2) and a single class of G-protein regulating more than one type of effector enzyme (3). The regulated switching of attention of adrenergic receptors between G_s and G_i (4) is considered in Chapters 9 and 11. The synergistic interaction of two receptors and two G-proteins in the activation of some isoforms of adenylyl cyclase (5) is discussed in Chapter 5.

construct (i.e. not linked to the receptor), although capable of restoring viability to haploid cells lacking Gpa1, fails to restore mating competence. This is due to its inability to recognize the receptor, but there is no problem when the two are fused together as Ste2-Gpa1/α_s. It follows that the C-terminus of the α-subunit operates mainly to bring the G-protein into the proximity of the receptor, allowing guanine nucleotide exchange and ensuring efficient coupling. Questions related to the communication of signals to downstream effectors are not raised in this experiment since, in yeast, this is a function of the $\beta\gamma$-subunits.

Contrary to the conventional descriptions of G-protein activation, couched in terms of fleeting interactions between receptors and dissociating subunits, the evidence now begins to point to the conclusion that receptors, G-proteins and their effectors, comprise loose operational ensembles. For sure, G-proteins can dissociate, and more readily so when activated. Whether they actually do this to any significant extent under physiological conditions remains uncertain.

References

1 Rodbell, M., Birnbaumer, L., Pohl, S.L., Krans, H.M. The glucagon-sensitive adenyl cyclase system in plasma membranes of rat liver: An obligatory role of guanylnucleotides in glucagon action. *J. Biol. Chem.* 1971; 246: 1877–82.

2 Rodbell, M., Krans, H.M., Pohl, S.L., Birnbaumer, L. The glucagon-sensitive adenyl cyclase system in plasma membranes of rat liver: Effects of guanylnucleotides on binding of ^{125}I-glucagon. *J. Biol. Chem.* 1971; 246: 1872–6.

3 Maguire, M.E., Arsdale, P.M. Van, Gilman, A.G. An agonist-specific effect of guanine nucleotides on binding to the β-adrenergic receptor. *Mol. Pharm.* 1976; 12: 335–9.

4 Lefkowitz, R.J., Mullikin, D., Caron, M.G. Regulation of β-adrenergic receptors by guanyl-5'-yl imidodiphosphate and other purine nucleotides. *J. Biol. Chem.* 1976; 251: 4686–92.

5 Rodbell, M. Nobel Lecture. Signal transduction: evolution of an idea. *Biosci. Rep.* 1995; 15: 117–33.

6 Bourne, H.R. GTPases everywhere! In: Dickey, B.F., Birnbaumer, L. (Eds) *GTPases in Biology*, Vol. 1. Springer-Verlag, Berlin, 1993; 3–15.

7 Bourne, H.R., Sanders, D.A., McCormick, F. The GTPase superfamily: a conserved switch for diverse cell functions. *Nature* 1990; 348: 125–32.

8 Gilman, A.G. G proteins and dual control of adenylate cyclase. *Cell* 1984; 36: 577–9.

9 Klein, S., Reuveni, H., Levitzki, A. Signal transduction by a nondissociable heterotrimeric yeast G protein. *Proc. Natl. Acad. Sci. USA* 2000; 97: 3219–33.

10 Arad, H., Rosenbusch, J.P., Levitzki, A. Stimulatory GTP regulatory unit Ns and the catalytic unit of adenylate cyclase are tightly associated: mechanistic consequences. *Proc. Natl. Acad. Sci. USA* 1984; 81: 6579–83.

11 Levitzki, A. From epinephrine to cAMP. *Science* 1988; 241: 800–6.

12 Arad, H., Levitzki, A. The mechanism of partial agonism in the beta-receptor dependent adenylate cyclase of turkey erythrocytes. *Mol. Pharm.* 1979; 16: 749–56.

13 Liebman, P.A., Parker, K.R., Dratz, E.A. The molecular mechanism of visual excitation and its relation to the structure and composition of the rod outer segment. *Annu. Rev. Physiol.* 1987; 49: 765–91.

14 Arshavsky, V.Y., Dumke, C.L., Zhu, Y. *et al.* Regulation of transducin GTPase activity in bovine rod outer segments. *J. Biol. Chem.* 1994; 269: 19882–7 .

15 He, W., Cowan, C.W., Wensel, T.G. RGS9, a GTPase accelerator for phototransduction. *Neuron* 1998; 20: 95–102.

16 Biddlecombe, G.H., Berstein, G., Ross, E.M. *et al.* Regulation of phospholipase C-β1 by Gq and m1 muscarinic receptor. *J. Biol. Chem.* 1996; 271: 7999–8007.

17 Chidiac, P., Ross, E.M. Phospholipase C-β1 directly accelerates GTP hydrolysis by Gαq and acceleration is inhibited by Gβγ subunits. *J. Biol. Chem.* 1999; 274: 19639–43.

18 Vries, L. de, Farquhar, M.G. RGS proteins: more than just GAPs for heterotrimeric G proteins. *Trends Cell. Biol.* 1999; 9: 138–44.

19 Schiff, M.L., Siderovski, D.P., Jordan, J.D. *et al.* Tyrosine-kinase-dependent recruitment of RGS12 to the N-type calcium channel. *Nature* 2000; 408: 723–7.

20 Rodbell, M., Krans, H.M., Pohl, S.L., Birnbaumer, L. The glucagon-sensitive adenyl cyclase system in plasma membranes of rat liver: Binding of glucagon: method of assay and specificity. *J. Biol. Chem.* 1971; 246: 1861–71.

21 Cassel, D., Selinger, Z. Catecholamine-stimulated GTPase activity in turkey erythrocyte membranes. *Biochim. Biophys. Acta* 1976; 452: 538–51.

22 Burgisser, E., Lean, A. De, Lefkowitz, R.J. Reciprocal modulation of agonist and antagonist binding to muscarinic cholinergic receptor by guanine nucleotide. *Proc. Natl. Acad. Sci. USA* 1982; 79: 1732–6.

23 Green, R.D. Reciprocal modulation of agonist and antagonist binding to inhibitory adenosine receptors by 5′-guanylylimidodiphosphate and monovalent cations. *J. Neurosci.* 1984; 4: 2472–6.

24 Westphal, R.S., Sanders Bush, E. Reciprocal binding properties of 5-hydroxytryptamine type 2C receptor agonists and inverse agonists. *Mol. Pharm.* 1994; 46: 937–42.

25 Sundaram, H., Newman Tancredi, A., Strange, P.G. Characterization of recombinant human serotonin 5HT1A receptors expressed in Chinese hamster ovary cells. [³H]Spiperone discriminates between the G-protein-coupled and -uncoupled forms. *Biochem. Pharm.* 1993; 45: 1003–9.

26 Grazzini, E., Guillon, G., Mouillac, B., Zingg, H.H. Inhibition of oxytocin receptor function by direct binding of progesterone. *Nature* 1998; 392: 509–12.

27 Haga, T., Ross, E.M., Anderson, H.J., Gilman, A.G. Adenylate cyclase permanently uncoupled from hormone receptors in a novel variant of S49 mouse lymphoma cells. *Proc. Natl. Acad. Sci. USA* 1977; 74: 2016–20.

28 Coffino, P., Bourne, H.R., Tomkins, G.M. Mechanism of lymphoma cell death induced by cAMP. *Am. J. Pathol.* 1975; 81: 199–204.

29 Bourne, H.R., Coffino, P., Tomkins, G.M. Somatic genetic analysis of cAMP action: characterization of unresponsive mutants. *J. Cell. Physiol.* 1975; 85: 611–20.

30 Bourne, H.R., Coffino, P., Tomkins, G.M. Selection of a variant lymphoma cell deficient in adenylate cyclase. *Science* 1975; 187: 750–2.

31 Musacchio, A., Cantley, L.C., Harrison, S.C. Crystal structure of the breakpoint cluster region-homology domain from phosphoinositide 3-kinase p85 α subunit. *Proc. Natl. Acad. Sci. USA* 1996; 93: 14373–8.

32 Ross, E.M., Howlett, A.C., Ferguson, K.M., Gilman, A.G. Reconstitution of hormone-sensitive adenylate cyclase activity with resolved components of the enzyme. *J. Biol. Chem.* 1978; 253: 6401–12.

33 Northup, J.K., Sternweis, P.C., Smigel, M.D. *et al.* Purification of the regulatory component of adenylate cyclase. *Proc. Natl. Acad. Sci. USA* 1980; 77: 6516–20.

34 Gupta, S.K., Gallego, C., Lowndes, J.M. *et al.* Analysis of the fibroblast transformation potential of GTPase-deficient gip2 oncogenes. *Mol. Cell. Biol.* 1992; 12: 190–7.

35 Lyons, J., Landis, C.A., Harsh, G. *et al.* Two G protein oncogenes in human endocrine tumors. *Science* 1990; 245: 655–9.

36 Malbon, C.M., Moxham, C.C. Insulin action impaired by deficiency of the G-protein subunit Giα2. *Nature* 1996; 379: 840–4.

37 Sunahara, R.K., Dessauer, C.W., Gilman, A.G. Complexity and diversity of mammalian adenylyl cyclases. *Annu. Rev. Pharm. Toxicol.* 1996; 36: 461–80.

38 Antonelli, V., Bernasconi, F., Wong, Y.H., Vallar, L. Activation of B-Raf and regulation of the mitogen-activated protein kinase pathway by the G(o)β chain. *Mol. Biol. Cell.* 2000; 11: 1129–42.

39 Jordan, J.D., Carey, K.D., Stork, P.J., Iyengar, R. Modulation of rap activity by direct interaction of Gα(o) with Rap1 GTPase-activating protein. *J. Biol. Chem.* 1999; 274: 21507–10.

40 Ram, P.T., Horvath, C.M., Iyengar, R. Stat3-mediated transformation of NIH-3T3 cells by the constitutively active Q205L Gαo protein. *Science* 2000; 287: 142–4.

41 Jiang, M., Gold, M.S., Boulay, G. *et al.* Multiple neurological abnormalities in mice deficient in the G protein Go. *Proc. Natl. Acad. Sci. USA* 1998; 95: 3269–74.
42 Singer, W.D., Brown, H.A., Sternweis, P.C. Regulation of eukaryotic phosphatidylinositol-specific phospholipase C and phospholipase D. *Annu. Rev. Biochem.* 1997; 66: 475–509.
43 Taylor, S.J., Chae, H.Z., Rhee, S.G., Exton, J.H. Activation of the β1 isozyme of phospholipase C by α subunits of the Gq class of G proteins. *Nature* 1991; 350: 516–18.
44 Bence, K., Ma, W., Kozasa, T., Huang, X.Y. Direct stimulation of Bruton's tyrosine kinase by G(q)-protein α-subunit. *Nature* 1997; 389: 296–9.
45 Offermanns, S., Toombs, C.F., Hu, Y.H., Simon, M.I. Defective platelet activation in Gα(q)-deficient mice. *Nature* 1997; 389: 183–6.
46 Jiang, Y., Ma, W., Wan, Y. *et al.* The G protein Gα12 stimulates Bruton's tyrosine kinase and a rasGAP through a conserved PH/BM domain. *Nature* 1998; 395: 808–13.
47 Hart, M.J., Jiang, X., Kozasa, T. *et al.* Direct stimulation of the guanine nucleotide exchange activity of p115 RhoGEF by Gα13. *Science* 1998; 280: 2112–14.
48 Offermanns, S., Mancino, V., Revel, J.P., Simon, M.I. Vascular system defects and impaired cell chemokinesis as a result of Gα13 deficiency. *Science* 1997; 275: 533–6.
49 Kozasa, T., Jiang, X., Hart, M.J. *et al.* p115 RhoGEF, a GTPase activating protein for Gα12 and Gα13. *Science* 1998; 280: 2111.
50 Simon, M.I., Strathmann, M.P., Gautam, N. Diversity of G proteins in signal transduction. *Science* 1991; 252: 802–8.
51 McKenzie, F.R., Kelly, E.C., Unson, C.G., Spiegel, A.M., Milligan, G. Antibodies which recognize the C-terminus of the inhibitory guanine-nucleotide-binding protein (Gi) demonstrate that opioid peptides and foetal-calf serum stimulate the high-affinity GTPase activity of two separate pertussis-toxin substrates. *Biochem. J.* 1988; 249: 653–9.
52 McKenzie, F.R. and Milligan, G. δ-opioid-receptor-mediated inhibition of adenylate cyclase is transduced specifically by the guanine-nucleotide-binding protein Gi2. *Biochem. J.* 1990; 267: 391–8.
53 Sullivan, K.A., Miller, R.T., Masters, S.B. *et al.* Identification of receptor contact site involved in receptor-G protein coupling. *Nature* 1987; 330: 758–60.
54 Berlot, C.H., Bourne, H.R. Identification of effector-activating residues of Gsα. *Cell* 1992; 68: 911–22.
55 Iiri, T., Farfel, Z., Bourne, H.R. G-protein diseases furnish a model for the turn-on switch. *Nature* 1998; 394: 35–8.
56 Lambright, D.G., Noel, J.P., Hamm, H.E., Sigler, P.B. Structural determinants for activation of the α-subunit of a heterotrimeric G protein. *Nature* 1994; 369: 621–8.
57 Coleman, D.E., Berghuis, A.M., Lee, E. *et al.* Structures of active conformations of Gi alpha 1 and the mechanism of GTP hydrolysis. *Science* 1994; 265: 1405–12.
58 Wall, M.A., Coleman, D.E., Lee, E. *et al.* The structure of the G protein heterotrimer Gi alpha 1 beta 1 gamma 2. *Cell* 1995; 83: 1047–58.
59 Liu, W., Northup, J.K. The helical domain of a G protein α subunit is a regulator of its effector. *Proc. Natl. Acad. Sci. USA* 1998; 95: 12878–83.
60 Logothetis, D.E., Kurachi, Y., Galper, J., Neer, E.J., Clapham, D.E. Purified subunits of GTP-binding proteins regulate muscarinic K$^+$ channel activity in heart. *Nature* 1987; 325: 321–6.
61 Pfaffinger, P.J., Martin, J.M., Hunter, D.D., Nathanson, N.M., Hille, B. GTP-binding proteins couple cardiac muscarinic receptors to a K channel. *Nature* 1985; 317: 536–8.
62 Wickman, K., Clapham, D.E. Ion-channel regulation by G-proteins. *Physiol. Rev.* 1995; 75: 865–85.
63 Neer, E.J., Clapham, D.E. Roles of G protein subunits in transmembrane signalling. *Nature* 1988; 333: 129–34.
64 Reuveny, E., Slesinger, P.A., Inglese, J. *et al.* Activation of the cloned muscarinic potassium channel by G protein βγ subunits. *Nature* 1994; 370: 143–6.
65 Jeselma, C.L., Axelrod, A. Stimulation of phospholipase A2 activity in bovine rod outer segments by the βγ subunits of transducin and its inhibition by the α-subunit. *J. Biol. Chem.* 1987; 84: 3623–7.
66 Camps, M., Hou, C., Sidiropoulos, D. *et al.* Stimulation of phospholipase C by guanine-nucleotide-binding protein βγ subunits. *Eur. J. Biochem.* 1992; 206: 821–31.

67 Wilden, U., Hall, S.W., Kuhn, H. Phosphodiesterase activation by photoexcited rhodopsin is quenched when rhodopsin is phosphorylated and binds the intrinsic 48-kDa protein of rod outer segments. *Proc. Natl. Acad. Sci. USA* 1986; 83: 1174–8.

68 Menard, L., Ferguson, S.S., Barak, L.S. *et al.* Members of the G protein-coupled receptor kinase family that phosphorylate the β2-adrenergic receptor facilitate sequestration. *Biochemistry* 1996; 35: 4155–60.

69 Goodman, O.B., Krupnick, J.G., Santini, F. *et al.* β-Arrestin acts as a clathrin adaptor in endocytosis of the β2-adrenergic receptor. *Nature* 1996; 383: 447–50.

70 Daaka, Y., Luttrell, L.M., Lefkowitz, R.J. Switching of the coupling of the β2-adrenergic receptor to different G proteins by protein kinase A. *Nature* 1999; 390: 88–91.

71 Scher, C.D., Scolnick, E.M., Siegler, R. Induction of erythroid leukaemia by Harvey and Kirsten sarcoma viruses. *Nature* 1975; 256: 225–6.

72 Zheng, B., Vries, L. de, Farquhar, M.G. Divergence of RGS proteins: Evidence for the existence of six mammalian RGS subfamilies. *Trends Biochem. Sci.* 2001; 24: 411–14.

73 Harvey, J.J. An unidentified virus which causes the rapid production of tumours in mice. *Nature* 1964; 204: 1104–5.

74 Kirsten, W.H., Carter, R.E., Pierce, M.I. Studies on the relationship of viral infections to leukemia in mice: The accelerating agent in AKR mice. *Cancer* 1962; 15: 750–8.

75 Shimizu, K., Goldfarb, M., Suard, Y. *et al.* Three human transforming genes are related to the viral ras oncogenes. *Proc. Natl. Acad. Sci. USA* 1983; 80: 2112–16.

76 Strathmann, M.P., Simon, M.I. Gα12 and Gα13 subunits define a fourth class of G protein. *Proc. Natl. Acad Sci USA* 1991; 88: 5582–6.

77 Shih, T.Y., Papageorge, A.G., Stokes, P.E., Weeks, M.O., Scolnick, E.M. Guanine nucleotide-binding and autophosphorylating activities associated with the p21src protein of Harvey murine sarcoma virus. *Nature* 1980; 287: 686–91.

78 Gibbs, J.B., Sigal, I.S., Poe, M., Scolnick, E.M. Intrinsic GTPase activity distinguishes normal and oncogenic ras p21 molecules. *Proc. Natl. Acad. Sci. USA* 1984; 81: 5704–8.

79 Parada, L.F., Tabin, C.J., Shih, C., Weinberg, R.A. Human EJ bladder carcinoma oncogene is homologue of Harvey sarcoma virus ras gene. *Nature* 1982; 297: 474–8.

80 Valencia, A., Chardin, P., Wittinghofer, A., Sander, C. The ras protein family: evolutionary tree and role of conserved amino acids. *Biochemistry* 1991; 30: 4637–48.

81 Broek, D., Toda, T., Michaeli, T. *et al.* The *S. cerevisiae* CDC25 gene product regulates the RAS/adenylate cyclase pathway. *Cell* 1987; 48: 789–99.

82 DeFeo Jones, D., Tatchell, K., Robinson, L.C. *et al.* Mammalian and yeast ras gene products: biological function in their heterologous systems. *Science* 1985; 228: 179–84.

83 Kataoka, T., Powers, S., Cameron, S. *et al.* Functional homology of mammalian and yeast RAS genes. *Cell* 1985; 40: 19–26.

84 Stacey, D.W., Marshall, M.S., Gibbs, J.B., Feig, L.A. Preferential inhibition of the oncogenic form of RasH by mutations in the GAP binding. *Cell* 1991; 64: 625–33.

85 Segal, M., Willumsen, B.M., Levitzki, A. Residues crucial for Ras interaction with GDP-GTP exchangers. *Proc. Natl. Acad. Sci. USA* 1993; 90: 5564–8.

86 Gibbs, J.B. Ras C-terminal processing enzymes: New drug targets? *Cell* 1991; 65: 1–4.

87 Reuveni, H., Geiger, T., Geiger, B., Levitzki, A. Reversal of the Ras-induced transformed phenotype by HR12, a novel ras farnesylation inhibitor, is mediated by the Mek/Erk pathway. *J. Cell. Biol.* 2000; 151: 1179–92.

88 Perucho, M., Goldfarb, M., Shimizu, K. *et al.* Human-tumor-derived cell lines contain common and different transforming genes. *Cell* 1981; 27: 467–76.

89 Seeburg, P.H., Colby, W.W., Capon, D.J., Goeddel, D.V., Levinson, A.D. Biological properties of human c-Ha-ras1 genes mutated at codon 12. *Nature* 1984; 312: 71–5.

90 Feramisco, J.R., Gross, M., Kamata, T., Rosenberg, M., Sweet, R.W. Microinjection of the oncogene form of the human H-ras (T-24) protein results in rapid proliferation of quiescent cells. *Cell* 1984; 38: 109–17.

91 Stacey, D.W., Kung, H.-F. Transformation of NIH 3T3 cells by microinjection of Ha-ras p21 protein. *Nature* 1984; 310: 508–11.

92 Mulcahy, L.S., Smith, L.R., Stacey, D.W. Requirement for ras proto-oncogene function during serum-stimulated growth of NIH 3T3 cells. *Nature* 1985; 313: 241–3.

93 Kung, H.-F., Smith, M.R., Bekesi, E., Manne, V., Stacey, D.W. Reversal of transformed

phenotype by monoclonal antibodies against Ha-ras p21 proteins. *Exp. Cell. Res.* 1986; 162: 363–71.

94 Stacey, D.W., Feig, L.A., Gibbs, J.B. Dominant inhibitory Ras mutants selectively inhibit the activity of either cellular or oncogenic Ras. *Mol. Cell. Biol.* 1991; 11: 4053–64.

95 Bar-Sagi, D., Feramisco, J.R. Microinjection of the ras oncogene protein into PC12 cells induces morphological differentiation. *Cell* 1985; 42: 841–8.

96 Altin, J.G., Wetts, R., Bradshaw, R.A. Microinjection of a p21ras antibody into PC12 cells inhibits neurite outgrowth induced by nerve growth factor and basic fibroblast growth factor. *Growth Factors* 1991; 4: 145–55.

97 Trahey, M., McCormick, F. A cytoplasmic protein stimulates normal N-ras p21 GTPase, but does not affect oncogenic mutants. *Science* 1987; 238: 542–5.

98 Gibbs, J.B., Marshall, M.S., Scolnick, E.M., Dixon, R.A., Vogel, U.S. Modulation of guanine nucleotides bound to Ras in NIH3T3 cells by oncogenes, growth factors, and the GTPase activating protein (GAP). *J. Biol. Chem.* 1990; 265: 20437–42.

99 Nori, M., Vogel, U.S., Gibbs, J.B., Weber, M.J. Inhibition of v-src-induced transformation by a GTPase-activating protein. *Mol. Cell Biol.* 1991; 11: 2812–18.

100 al-Alawi, N., Xu, G., White, R., Clark, R., McCormick F., Feramisco, J.R. Differential regulation of cellular activities by GTPase-activating protein and NF1. *Mol. Cell. Biol.* 1993; 13: 2497–503.

101 Downward, J., Graves, J.D., Warne, P.H., Rayter, S., Cantrell, D.A. Stimulation of p21ras upon T-cell activation. *Nature* 1990; 346: 719–23.

102 Lazarus, A.H., Kawauchi, K., Rapoport, M.J., Delovitch, T.L. Antigen-induced B lymphocyte activation involves the p21ras and ras.GAP signaling pathway. *J. Exp. Med.* 1993; 178: 1765–9.

103 DePaolo, D., Reusch, J.E., Carel, K. *et al.* Functional interactions of phosphatidylinositol 3-kinase with GTPase-activating protein in 3T3-L1 adipocytes. *Mol. Cell Biol.* 1996; 16: 1450–7.

104 Lowy, D.R., Willumsen, B.M. Function and regulation of Ras. *Annu. Rev. Biochem.* 1993; 62: 851–91.

105 Haubruck, H., McCormick, F. Ras p21: Effects and regulation. *Biochim. Biophys. Acta* 1991; 1072: 215–29.

106 Musacchio, A., Gibson, T., Rice, P., Thompson, J., Saraste, M. The PH domain: a common piece in the structural patchwork of signalling proteins. *Trends Biochem. Sci.* 1993; 18: 343–8.

107 Weissbach, L., Settleman, J., Kalady, M.F. *et al.* Identification of a human rasGAP-related protein containing calmodulin-binding motifs. *J. Biol. Chem.* 1994; 269: 20517–21.

108 Martin, G.A., Yataani, A., Clark, R. *et al.* GAP domains responsible for ras p21-dependent inhibition of muscarinic atrial K+ channel currents. *Science* 1992; 255: 192–4.

109 McCormick, F. ras GTPase activating protein: signal transmitter and signal terminator. *Cell* 1989; 13: 5–8.

110 Scheffzek, K., Ahmadian, M.R., Wittinghofer, A. GTPase-activating proteins: Helping hands to complement an active site. *Trends Biochem. Sci.* 1999; 23: 257–62.

111 Toda, T., Uno, I., Ishikawa, T. *et al.* In yeast, RAS proteins are controlling elements of adenylate cyclase. *Cell* 1985; 40: 27–36.

112 Robinson, L.C., Gibbs, J.B., Marshall, M.S., Sigal, I.S., Tatchell, K. CDC25: a component of the RAS-adenylate cyclase pathway in *Saccharomyces cerevisiae*. *Science* 1987; 235: 1218–21.

113 Daniel, J., Becker, J.M., Enari, E., Levitzki, A. The activation of adenylate cyclase by guanyl nucleotides in *Saccharomyces cerevisiae* is controlled by the CDC25 start gene product. *Mol. Cell. Biol.* 1987; 7: 3857–61.

114 Chardin, P., Camonis, J.H., Gale, N.W. *et al.* Human Sos1: a guanine nucleotide exchange factor for Ras that binds to GRB2. *Science* 1993; 260: 1338–43.

115 Fung, B.K. Characterization of transducin from bovine retinal rod outer segments. I. Separation and reconstitution of the subunits. *J. Biol. Chem.* 1983; 258: 10495–502.

116 Marbach, I., Bar-Sinai, A., Minich, M., Levitzki, A. β subunit copurifies with GppNHp-activated adenylyl cyclase. *J. Biol. Chem.* 1990; 265: 9999–10004.

117 Bar-Sinai, A., Marbach, I., Shorr, R.G., Levitzki, A. The GppNHp-activated adenylyl cyclase complex from turkey erythrocyte membranes can be isolated with its βγ subunits. *Eur. J. Biochem.* 1992; 207: 703–8.

118 Hunt, L.T., Dayhoff, M.O. Structural and functional similarities among hormones and active peptides from distantly related eukaryotes. In: Gross, E., Meienhofer J. (Eds) *Peptides: Structure and Biological Function.* Pierce Chemical Co, Rockford, Ill, 1979; 757–60.

119 Hartman, J.L., Northup, J.K. Functional reconstitution in situ of 5-hydroxytryptamine2c (5HT2c) receptors with alphaq and inverse agonism of 5HT2c receptor antagonists. *J. Biol. Chem.* 1996; 271: 22597.

120 Lachance, M., Ethier, N., Wolbring, G., Schnetkamp, P.P.M., Hebert, T.E. Stable association of G proteins with β_2AR is independent of the state of receptor activation Cell. Signal. 1999; 11: 523–33.

121 Brown, P.J., Schonbrunn, A. Affinity purification of a somatostatin receptor-G-protein complex demonstrates specificity in receptor-G-protein coupling. *J. Biol. Chem.* 1993; 268: 6668–76.

122 Law, S.F., Reisine, T. Changes in the association of G protein subunits with the cloned mouse δ opioid receptor on agonist stimulation. *J. Pharm. Exp. Ther.* 1997; 281: 1476–86.

123 Law, S.F., Reisine, T. Agonist binding to rat brain somatostatin receptors alters the interaction of the receptors with guanine nucleotide-binding regulatory proteins. *Mol. Pharm.* 1992; 42: 398–402.

124 Medici, R., Bianchi, E., Di Segni, G., Tocchini-Valentini, G.P. Efficient signal transduction by a chimeric yeast-mammalian G protein α subunit Gpa1-Gsα covalently fused to the yeast receptor Ste2. *EMBO J.* 1997; 16: 7241–9.

125 Milburn, M.V., Tong, L., de Vos, A.M. *et al.* Molecular switch for signal transduction: structural differences between active and inactive forms of protooncogenic ras proteins. *Science* 1990; 247 (4945): 939–45.

126 Prive, G.G., Milburn, M.V., Tong, L. *et al.* X-ray crystal structures of transforming p21 ras mutants suggest a transition-state stabilization mechanism for GTP hydrolysis. *Proc. Natl Acad. Sci. USA* 1992; 89(8): 3649–53.

127 Brunger, A.T., Milburn, M.V., Tong, L. *et al.* Crystal structure of an active form of RAS protein. A complex of a GTP analog and the HRAS p21 catalytic domain. *Proc. Natl Acad. Sci. USA* 1990; 87(12): 4849–53.

128 de Vos, A.M., Tong, L., Milburn, M.V. *et al.* Three-dimensional structure of an oncogene proteincatalytic domain of human c-H-ras p21. *Science* 1988 239: 888–893.

129 Pai, E.F., Krengel, U., Petsko, G.A. *et al.* Refined crystal structure of the triphosphate conformation of H-ras p21 at 1.35 Å resolution: implications for the mechanism of GTP hydrolysis. *EMBO J.* 1990; 9: 2351–9.

5

Effector enzymes coupled to GTP-binding proteins: adenylyl cyclase and phospholipase C

■ Adenylyl cyclase

■ cAMP: the first second messenger

Looking back to the time when most wisdom in mammalian biochemistry was derived from experiments using rat liver (perfusions, slices, homogenates, etc.), one gets a sense of a happy circumstance that led Sutherland to use liver from dog in his investigations of hormonal activation of glycogenolysis.[1,2] During growth and development of rats (6–60 days), the expression of β-adrenergic receptors in the liver declines while that of the α_1-adrenergic receptors increases.[3] Adrenergic stimulation of glycogen breakdown in the liver of adult rats is therefore mediated primarily through the α-receptors, which induce an elevation in the concentration of cytosol Ca^{2+} (see Chapter 7). Using dog liver, Sutherland showed in 1957 that the breakdown of glycogen, in response to adrenaline or glucagon, is consequent on the generation of a heat-stable and dialysable factor.[1] This factor proved to be cyclic AMP (cAMP), the first second messenger to be identified (see page 31).

Although cAMP is most famously the intracellular signal for the breakdown of liver glycogen (and correspondingly for the shut-down of glycogen synthesis), it is a key mediator for many other important receptor responses. We return to these questions in Chapter 9. cAMP is also present in the prokaryotes, both Eubacteria and Archaea, the other fundamental domains of life.[4,5] Although its functions as a regulator in non-nucleated prokaryotes are very different from those of eukaryotes, its generation has signified a response to starvation throughout evolution. cAMP is synthesized in *E. coli* starved of glucose and it induces expression of sugar-metabolizing enzymes such as β-galactosidase. In this case, it

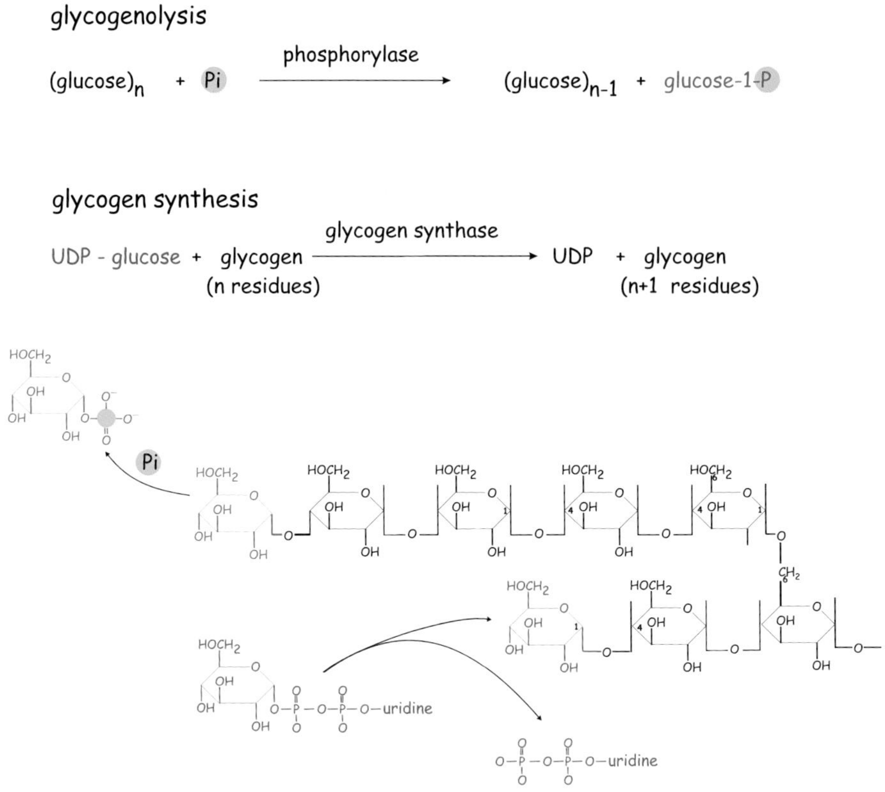

glycogenolysis

$$(glucose)_n \; + \; Pi \xrightarrow{\quad phosphorylase \quad} (glucose)_{n-1} \; + \; glucose\text{-}1\text{-}P$$

glycogen synthesis

$$UDP\text{-}glucose \; + \; glycogen \xrightarrow{\quad glycogen\ synthase \quad} UDP \; + \; glycogen$$
$$\qquad\qquad\qquad (n\ residues) \qquad\qquad\qquad\qquad\qquad (n+1\ residues)$$

Figure 5.1 The basic reactions of glycogen synthesis and breakdown. Both the incoming (glycogen synthesis) and leaving (glycogenolysis) residues are attached to the non-reducing termini of the formed glycogen.

interacts with promoter elements in the vicinity of the RNA polymerase site and is directly involved in the regulation of gene transcription. In yeast (unicellular, eukaryotic), cAMP is a growth signal. The presence of nutrients signals the activation of adenylyl cyclase through the RAS proteins. In starvation conditions, the ratio of GTP- to GDP-bound RAS declines, either because of down-regulation of CDC25 (the guanine nucleotide exchange catalyst) or because of activation of IRA (GTPase activating-protein). As a result, the cellular content of cAMP declines, growth ceases and the cells switch to meiosis and spore formation (see page 97). Under conditions of starvation, cAMP is secreted by the individual cells of the slime mould *Dictyostelium discoides* as a signal for assemblage into a slug.

Although it has generally been hard to discern functions for cAMP in plants,[6] there is now good evidence that it acts as a regulator of ion channels for both K+ and Ca^{2+}. The K+ channels in the guard cells of *Vicia faba* (broad bean) leaves appear to be regulated by a cAMP-dependent kinase (protein kinase A, PKA,[7,8]) and are indirectly sensitive to the actions of cholera and pertussis toxins that modulate the activity of GTP-binding proteins (see below).

■ cAMP is formed from ATP

cAMP (strictly, 3′,5′-cyclic adenosine monophosphate) is synthesized from ATP which is present in all cells at high concentration (5–10 mmol/l). The formation of cAMP is catalysed by adenylyl cyclases and its removal, by conversion to 5′-AMP (adenylate), is catalysed by the action of phosphodiesterases (cAMP-PDE) (Figure 5.2).

■ Adenylyl cyclase and its regulation

The enzyme
Adenylyl cyclase remained an activity without a proper molecular description for about 30 years, until 1990. At first the possibility was entertained that more than one enzyme might be involved in the cyclization reaction. It is the importance of adenylyl cyclase in metabolic regulation, and the simplicity of the reaction it catalyses, that provoked the intense investigation of its properties.

A number of proteins and other agents can activate or inhibit adenylyl cyclase. The known physiological activators and inhibitors include the subunits of the heterotrimeric G-proteins. α_S and $\beta\gamma$-subunits can activate, while α_i, α_o and $\beta\gamma$-subunits can inhibit. Also, Ca^{2+}, through various mechanisms, can activate or inhibit. The plant product forskolin is a potent activator and this poses the interesting question of whether there might exist an endogenous counterpart in animal cells.

A major problem for research in this area was the fact that adenylyl cyclase, an integral membrane protein, generally comprises no more than 0.01–0.001% of total membrane protein. (An exception is the olfactory epithelium in which it comprises more than 0.1%.) The poor solubility of membrane proteins in aqueous media and their dependence upon the phospholipid environment for the maintenance of their tertiary structure, makes them extremely difficult to isolate, purify and reconstitute into a

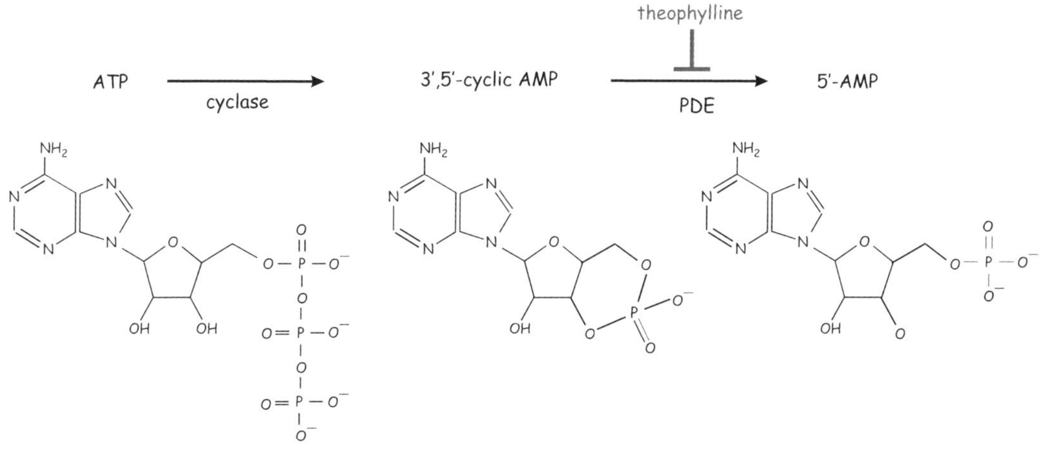

Figure 5.2 **Cyclic AMP is generated from ATP.**

It is simple now to measure the level of cAMP with the use of kits, binding proteins and specific antibodies. Not so simple for Sutherland and his colleagues who measured the product cAMP using a two-stage assay procedure. The first involved incubation together with ATP, phosphorylase b and fractions of liver homogenate (containing phosphorylase kinase and PKA). This was followed up by measurement of the activity of the active product phosphorylase a on its substrate glycogen.

form upon which sensible measurements can be made. Nevertheless the G_s-coupled form of adenylyl cyclase present in many tissues and a brain form of the enzyme have been purified.

From these preparations it has been possible to obtain some limited sequence information and thence a cDNA encoding the full length of type I adenylyl cyclase. The recombinant protein catalyses the conversion of ATP to cAMP and it can be activated by $\alpha_s \cdot$GTP and by Ca^{2+}.calmodulin.

Altogether, nine isoforms (I–IX) of mammalian adenylyl cyclase have been identified and cloned.

▪ Structural organization of adenylyl cyclases

The structure reveals a number of distinct domains (Figure 5.3). Starting from the N-terminus, there is a predominantly hydrophobic domain, M_1, organized as six membrane-spanning α-helices linked on alternate sides of the membrane by short hydrophilic loops. Next, there is an extensive (40 kDa) domain, C_1, then another transmembrane domain M_2, similar to M_1, and finally, at the C-terminus, a second extensive cytoplasmic domain C_2 similar to C_1.[10] Beyond the general conservation of topology, the sequences of all forms of adenylyl cyclase are similar, having 40–60% identity overall. Associated with the C_1 and C_2 domains are the subdomains C_{1a} and C_{2a} that share 50% or even higher similarity. Residues from both C_1 and C_2 contribute to ATP binding and catalysis.

Organized in head-to-tail fashion and having extensive contacts with each other, C_{1a} and C_{2a} are understood to comprise the catalytic centre of the enzyme.[11] The interface between C_{1a} and C_{2a} accommodates two highly conserved nucleotide binding sites, only one of which appears to be crucial for catalysis. The other is the site at which the activating substance forskolin binds. The binding site for α_s is distant (some 3 nm) from the catalytic site and has been localized to a small region at the N-terminus of C_{1a} and to a larger region composed of a negatively charged surface and a hydrophobic groove on C_{2a}. Although the binding sites for the α-subunits of G_i or G_o have not yet been determined, they do not compete with α_s. They too are formed by parts of both the C_{1a} and C_{2a} subdomains, but probably in a location much closer to the catalytic centre. Although the nucleotides, forskolin and α_s can bind to residues on both C_{1a} and C_{2a}, it appears that $\beta\gamma$-subunits potentiate the activity of type II cyclase by interacting only with the C_{2a} subdomain. This causes a conformational change that indirectly enables it to modulate the conformation of the C_{2a} subdomain and so indirectly promotes optimal alignment at the catalytic site. The inhibitory action of $\beta\gamma$-subunits on type I cyclase must occur elsewhere since the sequences are not conserved in this region of the two subtypes.

The C_1 and C_2 domains, shorn of their extensive membrane attachments, are together sufficient to catalyse the conversion of ATP to cAMP.[12,13] Although the activity is very low, the association between them is enhanced and the rate of conversion is increased by more than 100-fold by forskolin in the presence of α_s.[14] These two activators are synergistic, the presence of the one enhancing the affinity of the other, and there is a commensurate increase in reaction rate.

(a)

(b)

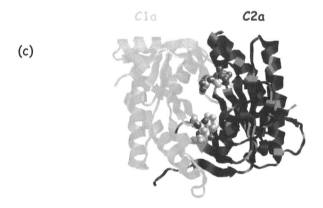

(c)

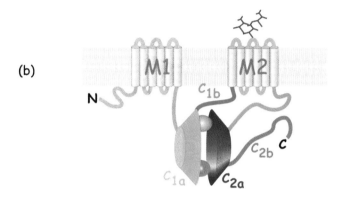

Figure 5.3 Organization of adenylyl cyclases. The figure represents the main structural features of the mammalian adenylyl cyclases: (a) The primary sequence of more than 1000 amino acids is interrupted by a number of specialized domains. (b) General topology and organization of the enzyme. With an even number of membrane spanning α-helices, six each in the domains M_1 and M_2, the N- and C-termini are both exposed on the same (cytosol) side of the plasma membrane. The catalytic unit comprises the two domains C_{1a} and C_{2a} shown as being wrapped over one another so that together they offer binding sites for ATP, forskolin and the α-subunits of G_s and G_i. (c) The molecular structure of the C_{1a} and C_{2a} dimeric cluster containing ATPγS (top) and forskolin (below)

Although the M_1 and M_2 domains clearly anchor the enzyme to the membrane, one would imagine that one or two transmembrane spans would suffice for this purpose and it is reasonable to ask why these enzymes are endowed with twelve. The answer probably lies in their evolutionary ancestry. It is certainly intriguing that the general membrane topology of the mammalian cyclases, particularly the two groups of six membrane-spanning segments, bears a strong resemblance to known pore-forming structures such as the cystic fibrosis transmembrane conductance regulator (CFTR) and the P-glycoproteins. However, no ion-channel-like activity has been detected for any of the mammalian adenylyl cyclases, nor do they exhibit any sequence similarity with known channels.

Sequence similarities certainly extend to the guanylyl cyclases and to some non-mammalian forms of the enzyme expressed in lower organisms, such as the slime mould *D. discoideum*. In general, a common ancestry can be inferred. More distantly, the yeast and bacterial (*E. coli*) enzymes are membrane-attached, but are not true intrinsic membrane proteins. Sequence similarities between these and the mammalian cyclases have been hard to discern.

Regulation of adenylyl cyclase

With the emergence of the eukaryotes, the regulation of cyclase fell under the control of heterotrimeric G-proteins. All isoforms of mammalian adenylyl cyclase are activated by the α-subunit of Gs and all are inhibited by the so-called P-site inhibitors (analogues of adenosine such as 2'-deoxy-3'-AMP). Beyond this, their responses to the many and various inhibitory ligands vary widely. With the exception of type IX, all are activated by forskolin which acts in synergy with α_S. They are influenced, both positively and negatively by the α- and βγ-subunits of other G-proteins, by Ca^{2+} and through phosphorylation by PKA and PKC (see Table 5.1). Although the published data are incomplete, it appears that their evolutionary development has determined their specialized sensitivities towards the various regulatory influences. Broadly, they can be grouped in three main classes:

II, IV and VII which are activated synergistically by α_S- and βγ-subunits;
V and VI which are inhibited by α_i and Ca^{2+};
VIII, I and III which are activated synergistically by α_S together with Ca^{2+}.calmodulin.

Table 5.1 Activation and inhibition of the adenylyl cyclases

isoform	distribution	βγ	α_i	α_o	α_z	PKA	PKC	forskolin	Ca.Calmod	Ca^{2+} (calmod indep)
II	brain, lung	↑	n/e	n/e		↑	↑		n/e	
IV	widely distributed	↑					↑			
VII	widely distributed	↑				↑	↑		n/e	
V	heart > brain	n/e	↓	n/e	↓	↓	↑	↑	n/e	↓
VI	heart, brain > other tissues	n/e	↓	n/e		↓	↓	↑	n/e	↓
VIII	neural tissue							↑	↑	
I	neural tissue	↓	↓	↓	↓			↑	↑	
III	predominantly olfactory	n/e						↑	↑	

60   80   100
amino acid sequence identity

n/e – no effect

■   Regulation by GTP binding proteins

Since the G_i proteins are generally expressed at a level greatly exceeding that of G_s, stimulation of receptors linked to G_i can generate considerable quantities of activated $\beta\gamma$-subunits. It has been proposed that this excess of $\beta\gamma$-subunits should be able to sequester free α_s-subunits, opposing their ability to activate cyclase:

$$\beta\gamma + \alpha_s \rightarrow \alpha_s\beta\gamma$$

For this to operate as an inhibitory mechanism it would be necessary for the subunits to exist as independent entities but, as we have argued, it is unlikely that α_s ever separates from $\beta\gamma$ (see page 95). However, the type I cyclase certainly is inhibited by $\beta\gamma$-subunits, but here the effect is primarily directed through interaction with the enzyme, not by sequestration of α-subunits. On the other hand, $\beta\gamma$-subunits synergize with α_s to activate types II and IV cyclases. As one might expect, there is little, if any, effect of the $\beta\gamma$-subunits in the absence of α_s. The level of $\beta\gamma$-subunits required is much greater than could possibly be derived from G_s alone, but it is fully commensurate with the amounts that could be generated through the activation of G_i or G_o. This implies that two sets of hormone receptors must be activated simultaneously in order to ensure maximal stimulation (Figure 5.4). It is possible that the dimeric structure of many receptors (including the β-adrenergic receptor: see page 61) enables them to operate as coincidence detectors, allowing a powerful response when the two signals are generated, but only a weak or insignificant response to unsynchronized signals. This is the likely explanation for the co-stimulatory effect of noradrenaline and adenosine on the cyclase activity in brain cortical slices[15] mentioned in Chapter 3 (see page 63). The presence of particular subtypes of adenylyl cyclase could then determine distinct patterns of response due to the signals from different classes of receptors. Figure 5.5 presents several possible schemes that explain how the converging pathways might interact to cause activation or inhibition. This could be of particular importance in synapses, allowing them to co-ordinate their responses to incoming stimuli. One of the signals will be transmitted by α_s, the other will be coupled to the more prolific G_i or G_o that would generate the required amounts of $\beta\gamma$-subunits.

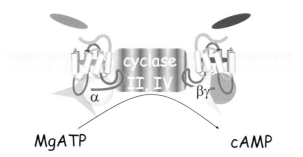

Figure 5.4 Adenylyl cyclase as a coincidence detector, integrating the signals from two G-proteins.

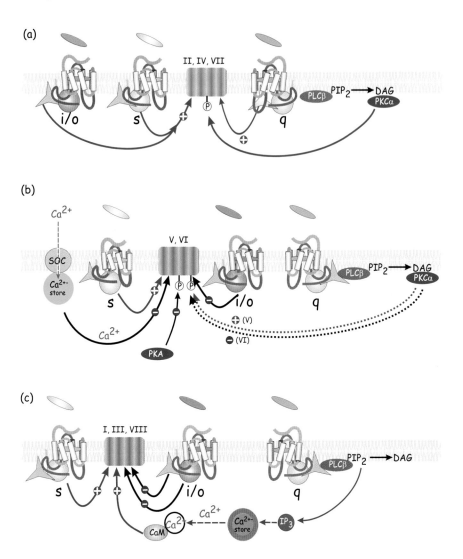

Figure 5.5 Adenylyl cyclase isoforms: patterns of activation and inhibition. The adenylyl cyclases are subject to regulation by multiple signals. These may cooperate or oppose each other. (a) In some instances the activation of types II, IV and VII by α_s together with $\beta\gamma$-subunits occurs with a high degree of synergy so that activation is only achieved when two distinct classes of receptor are activated simultaneously. On the other hand, the phosphorylation of these same enzymes by PKC is likely to be of longer duration so that the enzyme is set up in readiness for stimulation by α_s. (b) Other isoforms, types V and VI, are subject to indirect feedback inhibition by phosphorylation of the enzyme by cyclic AMP-activated PKA. They are also inhibited by Ca^{2+} and by receptors linked through the G_i proteins. (c) Cyclases types I, III and VIII are activated by Ca^{2+} coupled to calmodulin, and inhibited by $\beta\gamma$-subunits. (Note: SOC represents a store-operated Ca^{2+} channel, discussed in Chapter 7.)

So far, with the exception of $\beta_1\gamma_1$ which is located exclusively in the retina, it has been hard to establish any specific functional correlates of the diverse $\beta\gamma$-pairs. Combinations of β_1 or β_2 with any of the four forms of γ (2, 3, 5 or 7), appear to inhibit type 1 cyclase, and to activate type 2

The B-subunits bind with high affinity ($K_d \sim 1$ nmol/l) to the monosialylganglioside GM1, ubiquitously expressed on the surface of all animal cells. (Note: there is also evidence for ADP-ribosylation of GTPases in plant cells by cholera toxin.[29,30]) On gaining entry, the A1 subunit catalyses the ADP-ribosylation reaction. When this reaction was first carried out in the presence of radioactively labelled NAD^+, the label was transferred to a single membrane bound protein, which, when isolated, proved to be the nucleotide-binding component (α-subunit) of G_s.[27,31–33]

As reagents, cholera and pertussis toxins have been instrumental in assigning particular receptors to particular G-proteins in signalling pathways. The substrate arginine (R201) which is modified by cholera toxin is clearly a crucial component of the GTPase catalytic site of α_s. Mutants generated by substituting R201 hydrolyse GTP at a greatly reduced rate (reduced K_{cat}) and are in consequence persistently activated. Acquired (somatic) mutations at this site sometimes occur in non-metastatic endocrine tumours (adenomas). Being non-invasive and restricted to a single clone of cells, they are nonetheless not without dire consequences.

As an example, the physiological stimulus to the secretion of growth hormone from pituitary somatotrophs is provided by pulses of GHRH released from the hypothalamus, especially during periods of deep sleep. This provides both the signal to secrete growth hormone and the trophic stimulus by which these cells maintain their normal number and size. GHRH binds to 7TM receptors on the somatotrophs and the signals are transduced through G_s. In some patients presenting with the symptoms of giantism or acromegaly, it is found that R201 of the α_s in the somatotrophs has been replaced by cysteine or histidine. The consequent and persistent activation of adenylyl cyclase and the elevated level of cAMP causes hypersecretion of growth hormone and gross enlargement (adenoma) of the gland.[34] A similar somatic mutation of α_s in thyroid cells causes excess secretion of thyroid hormone. Mutations in the equivalent position (R179: see Table 4.4, page 88), suppressing the GTPase activity of G_{i2} are associated with adrenocortical and ovarian tumours.[35]

Note that in addition to the ADP-ribosylating toxins, both *Vibrio cholerae* and *Bordetella pertussis* synthesize and release many other cell-penetrating toxins. Interestingly, one of the major virulence factors of *B. pertussis* is a secreted form of adenylyl cyclase which becomes activated in contact with cellular calmodulin.

The word **somatic** refers to any cell of a multicellular organism that does not contribute to the production of gametes.

▪ Phospholipase C

▪ First hints of a signalling role for inositol phospholipids

The first hints of a role for the inositol-containing phospholipids in cell regulation emerged as long ago as 1953.[36] However, some 20 years elapsed before a viable proposal concerning their role was forthcoming. R. H. Michell[37] then pointed to the striking correlation between the activation of receptors that cause an increase in the metabolism of phosphatidylinositol (PI) and activation of processes such as secretion and the contraction of smooth muscle that are dependent on Ca^{2+}. Although a phospholipase C (PLC)-catalysed reaction was inferred, the precise identities of the substrate and the resulting second messengers remained elusive. Then it was found that the rate of depletion of the *poly*phosphoinositides, in particular phosphatidylinositol-4,5-bisphosphate (PI-4,5-P_2) (Figure 5.9) and not

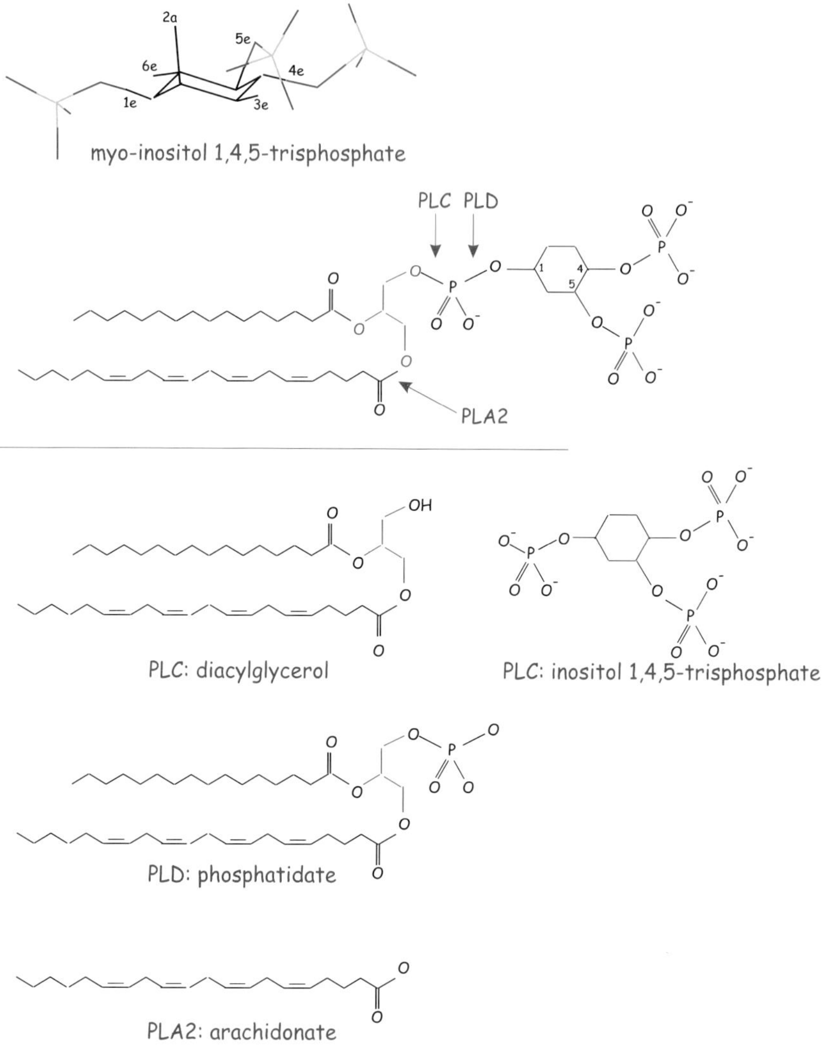

Figure 5.9 **Phospholipases cleave specific phospholipid ester bonds.** The structures of p PI(4,5)P$_2$ (specifically, 1-stearoyl,2-arachidonyl PIP$_2$) and its breakdown products. The glycerol nucleus is picked out in blue. The biological form of inositol, based on cyclohexane, is *myo*-inositol, in which the hydroxy substitutions at positions 1, 3, 4, 5 and 6 are all equatorial (e), while the 2 position is substituted axially (a). Together with diacylglycerol, this is the product of PLC. Phosphatidate is the product of PLD (together with choline, the main reacting phospholipid being phosphatidyl choline). Arachidonate (together with a lysophospholipid) is the product of PLA$_2$.

phosphatidylinositol itself, correlates with the onset of cellular activation.[38] The water-soluble product IP$_3$ resulting from the hydrolysis of PI(4,5)P$_2$, was found to release Ca^{2+} from intracellular storage sites when introduced into permeabilized pancreatic cells.[39] Within a very short time it was shown that this mechanism is ubiquitous and that the immediately accessible Ca^{2+} stores are present in the endoplasmic reticulum which is endowed with receptors for IP$_3$ (see Chapter 7). The hydrophobic product of PI(4,5)P$_2$ breakdown is diacylglycerol (DAG) and this is retained in the membrane bilayer. It causes the activation of protein kinase C (see Figure 7.6, page 153; also Chapter 9).

■ The phospholipase family

Phospholipids possessing a glycerol backbone contain four ester bonds, all of which are susceptible to enzyme-catalysed hydrolysis by specific phospholipases (see Figure 5.10).

Hydrolysis at three of these positions give rise to products that are either second messengers or substrates for enzymes that yield further signalling molecules.

▨ *PLA$_2$ and PLD*

Phospholipase A$_2$ (PLA$_2$) hydrolyses mostly phosphatidylcholine yielding a lysophospholipid and releasing the fatty acid bound to the 2-position, generally arachidonate. Arachidonate is the substrate for the formation of prostaglandins and leukotrienes (potent signalling molecules with autocrine and paracrine effects involved in pathological processes such as inflammation). Phospholipase D (PLD) provides another, less transient, source of DAG by cleaving phosphatidyl choline (the major membrane phospholipid) to form phosphatidate and the water-soluble choline. Removal of the phosphate (by phosphatidate phosphohydrolase) produces the DAG.

▨ *Phospholipase C*

The process by which IP$_3$ is formed in response to agonist stimulation requires the activation of a plasma membrane-associated enzyme of the phospholipase C (PLC) family. Its substrate, PI(4,5)P$_2$ or PIP$_2$, is derived from phosphatidylinositol through the action of lipid kinases (Figure 5.9). In addition to the PLCs, PI(4,5)P$_2$ is also the substrate of the phosphoinositide 3-kinases that cause further phosphorylation, generating the lipid second messenger PI-3,4,5-P$_3$ (see Chapter 13).

■ The isoenzymes of PLC

A number of PLC isoforms have been purified and cloned.

They fall into three main classes, β, γ and δ. Examination of the amino acid sequence, (starting at the N-terminus), reveals that all of the PLCs possess a PH domain (defined in Chapter 18) and then a stretch containing EF-hand motifs (Chapters 8 and 18). All forms of PLC possess a C2 domain in the vicinity of the C-terminus. The catalytic centres, split in two separated subdomains, X and Y, possess high sequence similarity. In the case of PLCγ, the intervening sequence extends to more than 500 amino acids and encompasses two SH2 domains and a single SH3 domain (Figure 5.11; see also Table 18.1, page 393). In spite of their considerable separation in the primary sequences of these enzymes, in the folded structure the X and Y segments come together forming two halves of a single structural unit which comprises the catalytic centre.

The status of a possible α-isoform is uncertain since the cDNA originally associated with this enzyme was later found to code for a protein disulphide isomerase. Isoenzymes also exist within each class, as PLCβ$_{1-4}$, PLCγ$_{1-2}$, PLCδ$_{1-4}$.

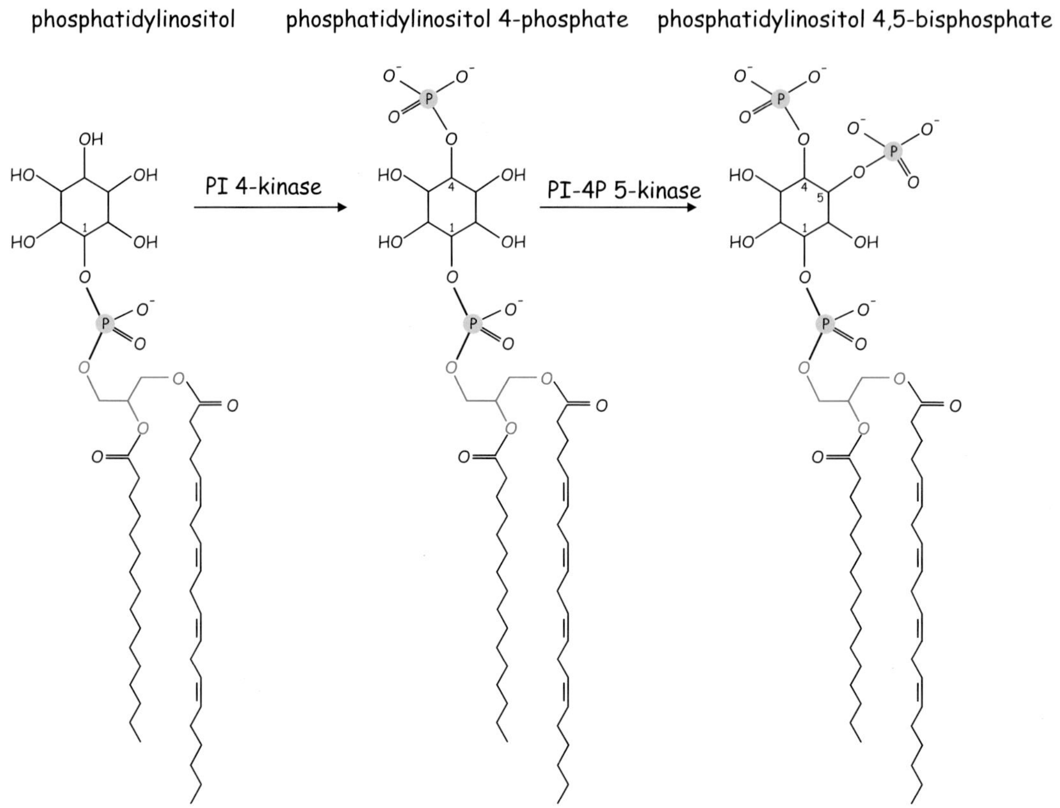

phosphatidylinositol   phosphatidylinositol 4-phosphate   phosphatidylinositol 4,5-bisphosphate

Figure 5.10 Phosphatidyl inositol (PI) and its derivatives PI4-P and PI(4,5)P$_2$.

■ PLCδ: a prototype

The δ-forms of PLC are likely to be ancestral to the other animal phospholipase C isoenzymes. They are of lower molecular weight and possess no domains that are unique to themselves (i.e. that are lacking in either of the other two classes). Only the δ-forms are present in organisms such as the yeasts, *Dictyostelium* and flowering plants. It is has been speculated that PLCδ emerged initially as a Ca^{2+}-dependent generator of DAG, the activator of PKC.[41] In this case, the water-soluble by-product, IP$_3$, would have had no role to play in signalling. Later in evolution, as it too became a second messenger, regulating Ca^{2+}, more complex forms of PLC (β and γ) emerged that are regulated by receptors.

The relevance of the domains that are common to all the known isoforms is unclear. Although EF-hand motifs may bind Ca^{2+} (Chapter 8), the N-terminal EF-hand-containing structure in the PLCs does not appear to be the site through which regulation by Ca^{2+} is expressed. However, it is certainly of importance, possibly in a structural context, since the activity of the enzyme is abolished when about half of this domain is deleted. X-ray diffraction of PLCδ co-crystallized with IP$_3$, which mimics the head

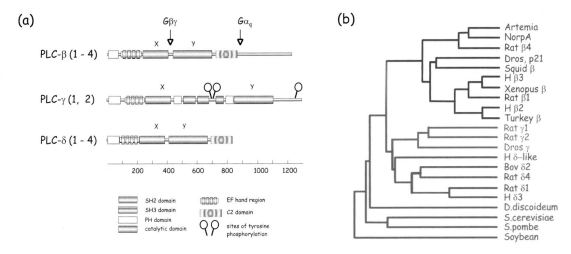

Figure 5.11 Organization of domains in the amino acid sequences of the PLC enzymes: (a) The main structural domains are indicated. In all of the inositide specific PLCs, the catalytic domain is divided in two parts, designated X and Y. PLCδ possesses a minimalist design, having only the PH, C2 and EF-hand domains, all of which are present in the other structures. The long (approximately 500 amino acids) C-terminal extension on PLCβ enables its association with membranes and allows for regulation by α-subunits of the $G_{q/11}$ family of G-proteins. In PLCγ the X and Y components are separated by an extended stretch of more than 500 amino acids and which includes two SH2 and one SH3 domains. These mediate the interaction of the PLCγ enzymes with phosphorylated growth factor receptors and other signalling molecules. The sites of phosphorylation by receptor tyrosine kinases are indicated. (b) Evolutionary relationships of the PI-specific phospholipases C based on sequence comparisons of the conserved X and Y catalytic domains. Adapted from Singer et al.[40]

group of PI(4,5)P$_2$, has revealed that the regulating Ca^{2+} ions are bound at the active site of the enzyme and also to the IP$_3$ molecule.[42] It is likely that the PH domain tethers the enzyme to the membrane by high-affinity attachment to PI(4,5)P$_2$. The C2 domain may then reinforce this attachment in a Ca^{2+} dependent manner, optimizing the orientation so that the active site can insert into the membrane for catalysis (Figure 5.12).

While the β and γ isoforms of PLC are certainly more central to the regulation of cellular activity by receptors, a basis for the understanding of their regulation is provided by knowledge of the simpler δ isoform. It may also offer clues about the operation of other signal transduction enzymes, such as PKC (Chapter 9), cytosolic PLA$_2$ and PI 3-kinase (Chapter 13), that use similar structural modules for reversible association with membranes.

■ Regulation of PLC

There are two ways in which PLC may be activated to hydrolyse PI(4,5)P$_2$. One of these involves G-proteins and the other a protein tyrosine kinase.

▒ G-protein-coupled activation of phospholipase Cβ

The β isoforms of PLC are activated by G-protein-mediated mechanisms involving members of the G$_q$, G$_i$ or G$_o$ families. For proteins of the G$_q$ class it is the α-subunit that conveys the activation signal, interacting at a site

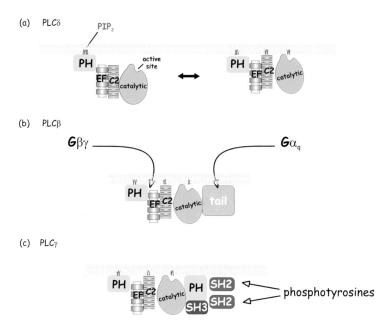

(a) PLCδ

(b) PLCβ

(c) PLCγ

phosphotyrosines

Figure 5.12 Membrane-association of PLC. (a) Initial attachment of PLCδ is through its PH domain. Subsequently, the C2 domain brings the catalytic site into contact with its substrate. (b) PLCβ interacts additionally with Gβγ or Gα$_q$ to initiate catalysis. (c) PLCγ also interacts with phosphotyrosines, through its SH2 domains. Adapted from Katan and Williams.[43]

in the extended C-terminal region of these enzymes. In the case of receptors which communicate through G$_i$ and G$_o$ (for example, α$_2$-adrenergic, M2 and M4 muscarinic) it is the βγ heterodimer that activates PLCβ interacting at a site between the X and Y catalytic subdomains.[44] This may be prevented by treating cells with pertussis toxin which catalyses ADP-ribosylation of the α-subunits of G$_i$ and G$_o$, but not the members of the G$_q$ family. In addition, it is important to remember that the G-proteins, their subunits and the PLC effector enzymes are all membrane-associated. Also, that the substrate, PI(4,5)P$_2$, and one of the products of reaction, DAG, are hydrophobic and are retained as integral components of the membrane phospholipid bilayer. In contrast, the other product, IP$_3$, is water-soluble and is released into the cytosol.

Phospholipase Cγ is activated through interaction with a protein tyrosine kinase

The γ isoenzymes of PLC possess an additional region, absent from the β and δ forms, inserted between the X and Y sequences of the catalytic domain (Figure 5.11). This region contains one SH3 domain and two SH2 domains. PLCγ is present in cells that have receptors that respond to growth factors (e.g. EGF, PDGF and insulin receptors) and also in cells that bear receptors belonging to the immunoglobulin superfamily (e.g. T and B lymphocytes). The SH2 domains on PLCγ enable it to bind to motifs on the intracellular chains of receptors that have become phosphorylated on tyrosine residues following the binding of a stimulatory ligand (see

Chapter 11). This brings PLC into a signalling complex, where it becomes phosphorylated itself by the tyrosine kinase activity of the receptor (or an accessory, non-receptor tyrosine kinase). This signals its activation.

■ References

1 Rall, T.W., Sutherland, E.W., Berthet, J. The relationship of epinephrine and glucagon to liver phosphorylase: Effect of epinephrine and glucagon on the reactivation of phosphorylase in liver homogenates. *J. Biol. Chem.* 1957; 224: 463–75.
2 Sutherland, E.W. Studies on the mechanism of hormone action. *Science* 1972; 177: 401–8.
3 Nakamura, T., Tomomura, A., Kato, S., Noda, C., Ichihara, A. Reciprocal expressions of α1- and α-adrenergic receptors, but constant expression of glucagon receptor by rat hepatocytes during development and primary culture. *J. Biochem. Tokyo* 1984; 96: 127–36.
4 Botsford, J.L., Harman, J.G. cAMP in prokaryotes. *Microbiol. Rev.* 1992; 56: 100–22.
5 Leichtling, B.H., Rickenberg, H.V., Seely, R.J., Fahrney, D.E., Pace, N.R. The occurrence of cAMP in archaebacteria. *Biochem. Biophys. Res. Commun.* 1986; 136: 1078–82.
6 Bolwell, G.P. cAMP, the reluctant messenger in plants. *Trends Biochem. Sci.* 1995; 20: 492–5.
7 Li, W.W., Luan, S., Schreiber, S.L., Assmann, S.M. CyclicAMP stimulates K^+ channel activity in mesophyll cells of *Vicia faba* K. *Plant Physiol.* 1994; 105: 957–61.
8 Wu, W.H., Assmann, S.M. A membrane delimited pathway of G-protein regulation of the guard cell inward K^+ channel. *Proc. Natl. Acad. Sci. USA* 1994; 91: 6310–14.
9 Sutherland, E.W., Rall, T.W., Menon, T. Adenyl cyclase: Distribution, preparation and properties. *J. Biol. Chem.* 1962; 237: 1220–7.
10 Sunahara, R.K., Dessauer, C.W., Gilman, A.G. Complexity and diversity of mammalian adenylyl cyclases. *Annu. Rev. Pharm. Toxicol.* 1996; 36: 461–80.
11 Tang, W.-J., Hurley, J.H. Catalytic mechanism and regulation of mammalian adenylyl cyclase. *Mol. Pharm.* 1998; 54: 231–40.
12 Dessauer, C.W., Gilman, A.G. Purification and characterization of a soluble form of mammalian adenylyl cyclase. *J. Biol. Chem.* 1996; 271: 16967–74.
13 Yan, S.Z., Hahn, D., Huang, Z.H., Tang, W.J. Two cytoplasmic domains of mammalian adenylyl cyclase form a Gs α- and forskolin-activated enzyme in vitro. *J. Biol. Chem.* 1996; 271: 10941–5.
14 Whisnant, R.E., Gilman, A.G., Dessauer, C.W. Interaction of the two cytosolic domains of mammalian adenylyl cyclase. *Proc. Natl. Acad. Sci. USA* 1996; 93: 6621–5.
15 Sattin, A., Rall, T.W., Zanella, J. Regulation of cyclic adenosine 3′,5′-monophosphate levels in guinea-pig cerebral cortex by interaction of α-adrenergic and adenosine receptor activity. *J. Pharm. Exp. Ther.* 1975; 192: 22–32.
16 Sternweis, P.C. The active role of βγ in signal transduction. *Curr. Opin. Cell Biol.* 1994; 6: 198–203.
17 Rodbell, M. Metabolism of isolated fat cells. V. Preparation of ghosts and their properties; adenyl cyclase and other enzymes. *J. Biol. Chem.* 1967; 242: 5744–50.
18 Birnbaumer, L., Pohl, S.L., Rodbell, M. Adenyl cyclase in fat cells. 1. Properties and the effects of adrenocorticotropin and fluoride. *J. Biol. Chem.* 1969; 244: 3468–76.
19 Sternweis, P.C., Gilman, A.G. Aluminum: a requirement for activation of the regulatory component of adenylate cyclase by fluoride. *Proc. Natl. Acad. Sci. USA* 1982; 79: 4888–91.
20 Mittal, R., Ahmadian, M.R., Goody, R.S., Wittinghofer, A. Formation of a transition-state analog of the Ras GTPase reaction by Ras-GDP, tetrafluoroaluminate, and GTPase-activating proteins. *Science* 1996; 273: 115–17.
21 Ahmadian, M.R., Mittal, R., Hall, A., Wittinghofer, A. Aluminum fluoride associates with the small guanine nucleotide binding proteins. *FEBS Lett.* 1997; 408: 315–18.
22 Harwood, J.P., Low, H., Rodbell, M. Stimulatory and inhibitory effects of guanyl nucleotides on fat cell adenylate cyclase. *J. Biol. Chem.* 1973; 248: 6239–45.

23 Harwood, J.P., Rodbell, M. Inhibition by fluoride ion of hormonal activation of fat cell adenylate cyclase. *J. Biol. Chem.* 1973; 248: 4901–4.

24 Laurenza, A., Sutkowski, E.M., Seamon, K.B. Forskolin: a specific stimulator of adenylyl cyclase or a diterpene with multiple sites of action? *Trends Pharm. Sci.* 1989; 10: 442–7.

25 Peterson, J.W., Ochoa, L.G. Role of prostaglandins and cAMP in the secretory effects of cholera toxin. *Science* 1989; 245: 857–9.

26 Pittman, M. The concept of pertussis as a toxin-mediated disease. *Pediatr. Infect. Dis.* 1984; 3: 467–86.

27 Gill, D.M., Meren, R. ADP-ribosylation of membrane proteins catalyzed by cholera toxin: basis of the activation of adenylate cyclase. *Proc. Natl. Acad. Sci. USA* 1978; 75: 3050–4.

28 Cassel, D., Pfeuffer, T. Mechanism of cholera toxin action: covalent modification of the guanyl nucleotide-binding protein of the adenylate cyclase system. *Proc. Natl. Acad. Sci. USA* 1978; 75: 2669–73.

29 Seo, H.S., Kim, H.Y., Jeong, J.Y. *et al.* Molecular cloning and characterization of RGA1 encoding a G protein α subunit from rice (*Oryza sativa* L. IR-36). *Plant Mol. Biol.* 1995; 27: 1119–31.

30 Wu, W.H., Assmann, S.M. A membrane-delimited pathway of G-protein regulation of the guard-cell inward K+ channel. *Proc. Natl. Acad. Sci. USA* 1994; 91: 6310–4.

31 Katada, T., Ui, M. Direct modification of the membrane adenylate cyclase system by islet activating protein due to ADP-ribosylation of a membrane protein. *Proc. Natl. Acad. Sci. USA* 1982; 79: 3129–33.

32 Pfeuffer, T. GTP-binding proteins in membranes and the control of adenylate cyclase activity. *J. Biol. Chem.* 1977; 252: 7224–34.

33 Pfeuffer, T., Helmreich, E.J. Activation of pigeon erythrocyte membrane adenylate cyclase by guanylnucleotide analogues and separation of a nucleotide binding protein. *J. Biol. Chem.* 1975; 250: 867–76.

34 Landis, C.A., Masters, S.B., Spada, A. *et al.* GTPase inhibiting mutations activate the α chain of Gs and stimulate adenylate cyclase in human pituitary tumours. *Nature* 1989; 340: 690–6.

35 Lyons, J., Landis, C.A., Harsh, G. *et al.* Two G protein oncogenes in human endocrine tumors. *Science* 1990; 249 (4969): 655–9.

36 Hokin, M.R., Hokin, L.E. Enzyme secretion and the incorporation of ^{32}P into phospholipides of pancreas slices. *J. Biol. Chem.* 1953; 203: 967–77.

37 Michell, R.H. Inositol phospholipids in cell surface receptor function. *Biochim. Biophys. Acta* 1975; 415: 81–147.

38 Creba, J.A., Downes, C.P., Hawkins, P.T. *et al.* Rapid breakdown of phosphatidylinositol 4-phosphate and phosphatidylinositol 4,5-bisphosphate in rat hepatocytes stimulated by vasopressin and other Ca^{2+}-mobilizing hormones. *Biochem J.* 1983; 212: 733–47.

39 Streb, H., Irvine, R.F., Berridge, M.J., Schultz, I. Release of Ca2+ from a non-mitochondrial intracellular store in pancreatic acinar cells by inositol-1,4,5-trisphosphate. *Nature* 1983; 306: 67–9.

40 Singer, W.D., Brown, H.A., Sternweis, P.C. Regulation of eukaryotic phosphatidylinositol-specific phospholipase C and phospholipase D. *Annu. Rev. Biochem.* 1997; 66: 475–509.

41 Irvine, R. Phospholipid signalling. Taking stock of PI-PLC. *Nature* 1996; 380: 581–3.

42 Essen, L.-O., Perisic, O., Cheung, R., Katan, M., Williams, R.L. Crystal structure of a mammalian phosphoinositide-specific phospholipase Cδ. *Nature* 1996; 380: 595–602.

43 Katan, M., Williams, R.L. Phosphoinositide-specific phospholipase C: structural basis for catalysis and regulatory interactions. *Semin. Cell Dev. Biol.* 1997; 8: 287–96.

44 Kuang, Y., Wu, Y., Smrcka, A., Jiang, H., Wu, D. Identification of a phospholipase C β2 region that interacts with Gβγ. *Proc. Natl. Acad. Sci. USA* 1996; 93: 2964–8.

6

The regulation of visual transduction

Do not the Rays of Light in falling upon the bottom of the Eye excite Vibrations in the Tunica retina? Which Vibrations, being propagated along the solid Fibres of the optick Nerves into the Brain, cause the sense of seeing.

Sir Isaac Newton, *Opticks* (1704)

The visual system provides an exceptional opportunity to investigate the transduction mechanism of a 7TM receptor at the level of a single molecule. The molecule is the photoreceptor pigment rhodopsin. This consists of the protein opsin to which is bound the photosensitive compound 11-*cis*-retinal. The stimulus is light and the second messengers are cyclic GMP and Ca^{2+}.[1]

Sensitivity of photoreceptors

Significant advances in our understanding of molecular mechanisms in biology have, from time to time, emerged from opportunities that allow us to observe unitary events. For example, our appreciation of ion channels has benefited from the observation of the opening and closing of individual channels. In a similar way, our understanding of the first steps of the visual transduction mechanism has followed from the investigation of the interaction of light with single photoreceptor molecules. To detect this it is necessary to illuminate photoreceptors with very low light intensities to establish the minimal conditions for excitation. The first attempts were made in the 19th century, well before it had been realized that light possesses both wave and particle properties.

Only then could estimates of the minimal intensity in terms of quanta could be made. With little knowledge of the basic physiology of the human eye, Hecht, Shlaer and Pirenne in 1942 set about measuring its quantum sensitivity.[2] Using dark-adapted human subjects responding verbally to precisely calibrated single flashes (1 ms duration at 510 nm, close to the optimal wavelength for vision in dim light), they determined that the eye's detection limit corresponds to energies, incident at the cornea, in the range 2–6×10^{-17} J, equivalent to 54–148 photons. To calculate the energy actually absorbed by the retinal receptors, they applied

For an excellent review of visual phototransduction, see Lagnado and Baylor.[1]

Light is defined by its wave properties – frequency (v) and wavelength (λ) where the velocity $c = v \times \lambda$. The intensity of a beam of light is the rate at which energy is delivered per unit area. It is measured in W m^{-2}. Light also possesses particle properties. The quanta are called photons and the energy (in J) of a single photon is given by $E = h \times \nu$, where h is Planck's constant (6.626×10^{-34} J s).

corrections to take account of losses due to reflection (4% at the cornea) and absorption by the ocular medium (50%). Also at least 80% of the light falling on the retina passes through unabsorbed. The estimated number of photons absorbed by the visual pigment was then 5–14. Since this is very small in comparison with the number of rods in the field illuminated, it was concluded that human photoreceptor rod cells are of such sensitivity that the coincidence of single photons impinging simultaneously on five cells is sufficient to strike consciousness in a human being.[2]

> The fact that for the absolute visual threshold the number of quanta is small makes one realize the limitation set on vision by the quantum structure of light. Obviously the amount of energy required to stimulate any eye must be large enough to supply at least one quantum to the photosensitive material. No eye need be so sensitive as this. But it is a tribute to the excellence of natural selection that our own eye comes so remarkably close to the lowest level.[2]

Much later, estimation of the quantum sensitivity of individual photoreceptors was performed electrophysiologically using suction electrodes applied to single rod cells obtained from the toad.[3] This approach avoids the subjectivity of human psychophysical experimentation and many of the assumptions made concerning the proportion of the light signal that actually reaches the photoreceptors. An outward membrane current was recorded during illumination and when dim flashes of light were applied, the current fluctuated in a quantal manner. Similar quantal responses have been observed in photocurrent records from primate rod cells; examples are shown in Figure 6.1. Each unitary event is considered to be the result of an interaction between a single photon and a single pigment molecule.

Although the human eye is able to sense fluxes of just a few photons per second, it can also detect subtle intensity differences under conditions of very bright illumination. This gives it a remarkable dynamic range. Additionally, the human eye can detect transient events and recover rapidly. For example, when exposed to a train of dim flashes, we can distinguish individual events up to a frequency of 24 Hz. The flickering image is an enduring (and endearing) feature of old movies, in which the frame rate is slower than the flicker-fusion frequency of the audience. The faster frame rate of modern cinematography is just sufficient to allow successive images to fuse and give the impression of continuous motion. The flicker-fusion frequency of other species, such as some insects, can exceed 100 Hz.

> And when a Coal of Fire moved nimbly in the circumference of a Circle, makes the whole circumference appear like a Circle of Fire; is it not because the Motions excited in the bottom of the Eye by the Rays of Light are of a lasting nature, and continue till the Coal of Fire in going round returns to its former place? And considering the lastingness of the Motions excited in the bottom of the eye by Light, are they not of a vibrating nature?
>
> Sir Isaac Newton, *Opticks*

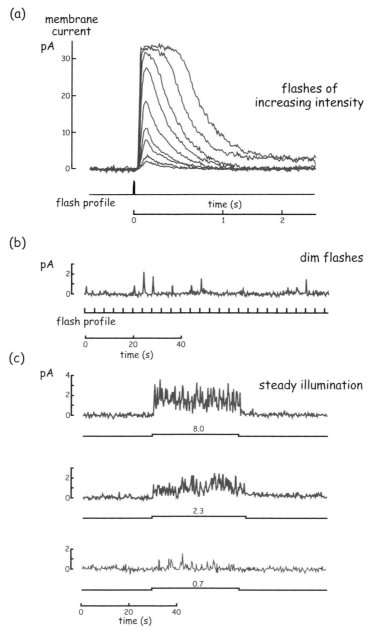

Figure 6.1 Detecting light responses in single rod cells. (a) Responses of a single retinal rod cell to flashes of light. Light flashes evoke transient photocurrents in the membrane of a single rod cell outer segment. The amplitude of the transient outward currents increases with flash intensity up to a saturating level of 34 pA; (wavelength 500 nm, flash duration 11 ms, photon density 1.7–503 photons μm^{-2}, species: monkey, *Macaca fascicularis*). (b) Quantal responses of single retinal rod cells to dim flashes. The membrane current responses of a single rod cell to a train of dim flashes are variable in amplitude. This variability together with the presence of background noise gives the appearance of a sequence of successes and failures. Amplitude histograms (not shown) reveal a quantal response to light. The unitary events have an average amplitude of 0.7 pA and they correspond to the detection of a single photon (photon density 0.6 photons μm^{-2}), other conditions similar to panel (a). (c) Response of a single rod cell to steady illumination. Records of photocurrents evoked by periods of steady illumination correspond to the superposition of random photon responses. The photon flux density is indicated below each trace (photons μm^{-2} s^{-1}). Adapted from Baylor et al.[4]

■ Photoreceptor mechanisms

We know more about signalling through vertebrate photoreceptors than through any of the other 7TM receptors interacting with heterotrimeric GTP-binding proteins. However, a warning is in order. It should not be imagined that what we learn here is a general reflection of G-protein-coupled receptor mechanisms. Almost everything that happens in vertebrate phototransduction is the very converse of the normal sequence of events by which receptors signal the generation of second messengers and then their interactions with effectors. This is a world turned upside down. 11-*cis*-Retinal is the prosthetic group in rhodopsin that absorbs visible light (see Figure 6.3). It is not only the chromophore, it is also the ligand and it is already bound *covalently* to opsin before receipt of the light signal. After photoisomerization it detaches and diffuses into a neighbouring cell. The mechanism of transduction is coupled to membrane *hyper*polarization and a *reduction* rather than *de*polarization and an elevation of the concentration of cytosol Ca^{2+}. The mechanism of down-regulation and preparation of the system for a subsequent bout of illumination continues in this vein, generally the very reverse of what we have learned about hormone receptor systems. By contrast, in the invertebrate photoreceptor system we appear to be on rather more familiar ground. Here, the excitation of rhodopsin is coupled mostly through G_q to the activation of PLC and elevation of cytosol Ca^{2+}.

■ Photoreceptor cells

The photoreceptor cells of the retina (illustrated schematically in Figure 6.2) are of two types. There are the cones which collectively provide colour discrimination (photopic vision), and the rods which are responsible for sensing low levels of light (scotopic vision). The human retina contains about 120 million rods and about 6 million cones. It is common experience that the appearance of colour is lost when objects are viewed in dim light and this is because the image is then detected only by the rods, which are more sensitive than cones but which possess no means of colour discrimination. The mechanism of phototransduction in cones is very similar to that of rods, differing only in terms of colour specificity. There are three types of cone, each containing one of three different pigments. Each of these, like rhodopsin, consists of 11-*cis*-retinal embedded in an opsin molecule. There are three different cone opsins that tune the absorption spectrum of the contained retinal to form either a red, green or blue sensitive photopigment. In the case of rods the tuning is matched to the spectral distribution of dim natural light, with an optimum at around 500 nm. Subsequent discussion will concentrate on rod cells.

■ Rod cells

Photoreceptor cells are highly differentiated epithelial cells in which the light-sensing region is segregated from the main cell body as the outer

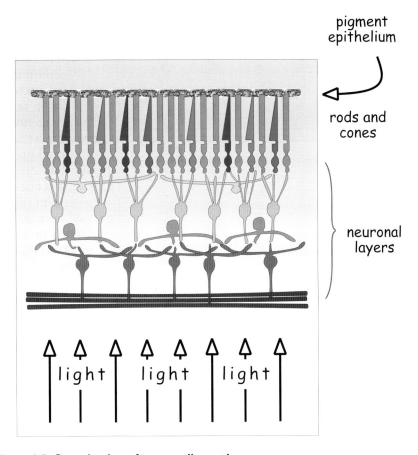

pigment
epithelium

rods and
cones

neuronal
layers

light   light   light

Figure 6.2 **Organization of mammalian retina.**

segment (Figure 6.3). It is a salient (if not another perverse) feature of the
vertebrate visual system that the light enters these cells through the end
of the cell body distal to the photosensitive outer segment. Not only this –
before it enters the photoreceptors, it first traverses a network of blood
vessels and then several layers of neuronal cells. The outer segments of
the photoreceptors contain arrays of 1000–2000 intracellular discs, flattened vesicular structures about 16 nm thick, the membranes of which
contain the photopigment rhodopsin. Each disc is formed by the invagi-
nation of the plasma membrane to produce a structure that eventually
becomes detached, so that the intradiscal space is topologically equivalent to the extracellular space. By weight, 50% of the disc membrane is
protein, mostly rhodopsin, and a single human rod cell contains on average 10^8 molecules of rhodopsin. This protein is organized in the same way
as the conventional 7TM receptors that are found in plasma membranes.
The glycosylated N-terminus projects into the intradiscal space and the
C-terminal domain, containing several potential phosphorylation sites
(Ser/Thr) is exposed to the cytosol. Transducin, the G-protein that transduces the light signal,[5] is also very abundant, though clearly outnumbered

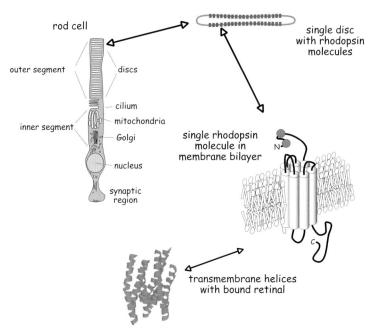

Figure 6.3 Rod photoreceptor cell and the rhodopsin molecule.

by rhodopsin. It comprises about 20% of the total (50% of the soluble) protein and unlike other G-proteins, it is soluble. Photoreceptor cells provide the only instance in which the number of receptors are present in large excess over the number of G-protein molecules to which they couple. The need for such vast amounts of rhodopsin is probably determined by the fact that photons travel in straight lines – unlike normal soluble ligands, they are unable to diffuse in the extracellular medium until they are captured by their favoured receptor.

The chromophore 11-*cis*-retinal (Figure 6.4) is coupled, through its aldehyde group, to the α-amino group of a lysine (K296) in the centre of the G transmembrane segment, as a protonated Schiff base. Although covalent, this linkage is broken subsequent to the photoisomerization to all-*trans*-retinal (Figure 6.4), which is released for reprocessing.

The extremities of the rod outer segments protrude into the layer of cells that form the pigment epithelium. This layer serves a number of purposes. Most significantly in connection with phototransduction, its cells contain the metabolic enzymes that regenerate the active isomer of retinal, 11-*cis*-retinal, from the inactive all-*trans* form following its detachment from the visual pigment in the neighbouring rods. As the name suggests, the cells of this layer contain their own pigment, which is melanin. This absorbs light that has penetrated past the arrays of discs in the rod cells and so prevents it from being scattered back into the photoreceptor layer. A failure to synthesize melanin (associated with the inherited conditions described as albinism) causes problems associated with bright lights and glare.

Beyond these difficulties, in **albinism** the fovea (the central region of the retina responsible for visual acuity) may also fail to develop properly, so that the eye cannot process sharp images well.

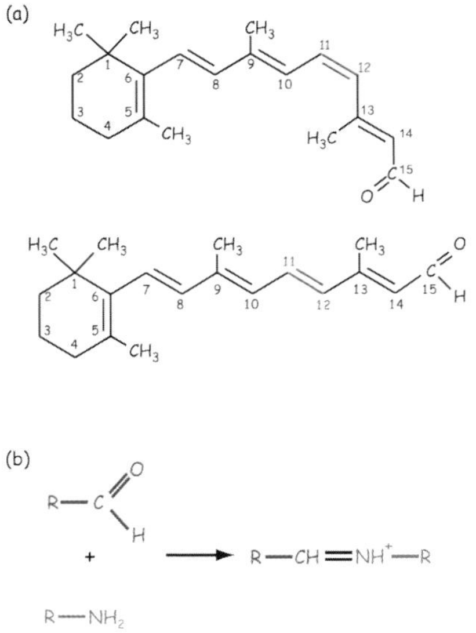

Figure 6.4 **Structures of (a) 11-*cis*-retinal and all-*trans*-retinal. (b) Formation of a protonated Schiff base from an aldehyde and an amine.**

The principal enzymes of the transducing cascade are the G-protein transducin (G_t) and the effector enzyme cyclic GMP phosphodiesterase. The rhodopsin is an integral membrane protein but the transducin and the phosphodiesterase are soluble. The cell body contains the nucleus, mitochondria and the organelles that form the protein synthetic apparatus. This is very active, enabling the photosensory cells to keep pace with the loss of the outer segment discs which have a life of just a few weeks.[6] Over this period, the discs migrate progressively to the distal end of the outer segments where they are shed and phagocytosed by the pigment epithelial cells.[7] The inner ends of the photosensory cells form synapses with intermediate neurons. The neurotransmitter is glutamate.

Dark current and signal amplification

In the absence of illumination, the membrane potential of the photoreceptor cells is about -30 mV. This is much less negative than the potential of excitable nerve and muscle cells and is due to the photoreceptor dark current caused by an influx of Na^+ and Ca^{2+} through non-specific cation channels. The ions are extruded from the cells by ion pumps (Na^+,K^+-ATPase) and exchangers (Na^+/Ca^{2+}) which are located in the plasma membrane of the inner segment of the photoreceptor cells. Thus there is a constant flow of cations into and out of the cell (Figure 6.5).

Cyclic GMP: a second messenger in reverse

In the dark, the cation channels are held in their open configuration by the presence of cyclic GMP. The effect of light is to initiate a series of

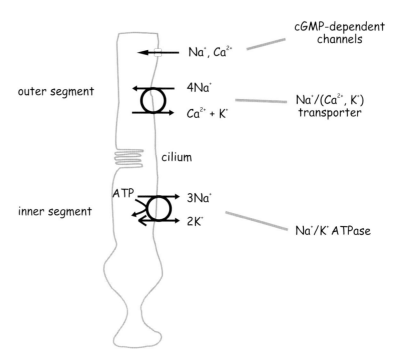

cGMP-dependent channels

Na⁺, Ca²⁺

outer segment

4Na⁺

Ca²⁺ + K⁺

Na⁺/(Ca²⁺, K⁺) transporter

cilium

ATP

inner segment

3Na⁺

2K⁺

Na⁺/K⁺ ATPase

Figure 6.5 Principal cation movements across a rod cell plasma membrane in the dark.

molecular rearrangements in rhodopsin that generate activated rhodopsin (Rh*) which then catalyses the exchange of GTP for GDP on transducin releasing α_T.GTP and $\beta\gamma$-subunits. The α-subunit then activates cyclic GMP phosphodiesterase and cyclic GMP is hydrolysed to 5′-GMP.

In the dark-adapted eye, cyclic GMP is present in the cytosol of rod outer segments at a concentration (40–80 μM), about 300-fold higher than in the nerve cells of brain. There are effectively 10–20 moles of cyclic GMP for each mole of rhodopsin. It binds directly to the cation channels to keep them in the open state. The sequence of events following illumination ensures enormous amplification of the signal such that a single photon, causing a single photoisomerization, can lead during the next second to the hydrolysis of more than 10^5 molecules of cyclic GMP, leading to the closure of about 500 cation channels, so blocking the influx of as many as 10^7 Na⁺ ions.

The cyclic GMP regulated channels are ligand-gated, but they differ from the ion channels operated by neurotransmitters, in that the ligand is applied from the inside of the cell and it is normally in place in the unstimulated eye in the dark. Three molecules of cyclic GMP are required to maintain the channel in its open state, with the result that the activation curve for channel closure is very steep indeed. The consequence of this is that once committed to opening or closing, due to an appropriate change in cyclic GMP concentration, there is very little tendency for indeterminacy. Like an electrical switch, the channel tends to be either open or shut. The 500 or so cation channels in a rod cell that close following the transduction of a single photon represent 3% of the total number that are open in the dark. The resultant hyperpolarization (membrane potential more negative) is about 1 mV and lasts about a second. This is sufficient to depress the rate of neurotransmitter release at the synapse which impinges on the nerve cells that transmit the onward signals.

■ Adaptation: calcium acts as a negative regulator

Given that the incidence of a singe photon can lead to the closure of 3% of the channels open in the dark, it might seem that steady illumination of even modest intensity would cause closure of all of the channels. Indeed, if the response of the system was directly proportional to the *number* of photoreceptor molecules activated, then saturation would be reached at very low light levels, as all the channels become closed. Yet, the human eye is capable of sensing small differences in the intensity of light against high background levels. Clearly, there must be an adaptive mechanism that reduces the amplification, so that there are always some open channels when the background is bright. Overall the eye is responsive over an energy range of about 11 orders of magnitude. Much of this dynamic range can be accounted by the ability of individual photoreceptors to adapt by decreasing their sensitivity. Thus rod cells can adapt over 2 orders of magnitude and cones can adapt even more (over 7–9 orders; see Figure 6.6).

An important component of the adaptive mechanism by which the sensitivity is adjusted against ambient light level is contributed by the cytosol concentration of free Ca^{2+}. As the cation channels close in response to the hydrolysis of cyclic GMP, the concentration of Ca^{2+} declines from its dark level of about 300 nM to 10 nM under strong illumination. As always, the

For example, this response allows us to perceive light–dark contrasts at levels of illumination that range from that of a dark overcast night sky (about 10^{-6} candela m^{-2}) to the brilliance of sunlight reflected from snow fields (about 10^7 candela m^{-2}). The candela is an SI unit of measure of the intensity of **visible** light, called the luminous intensity. One candela is the luminous intensity, in a given direction, of a source that emits monochromatic radiation of frequency 540×10^{12} Hz (green light) and that has a radiant intensity in that direction of 1/683 W per steradian.

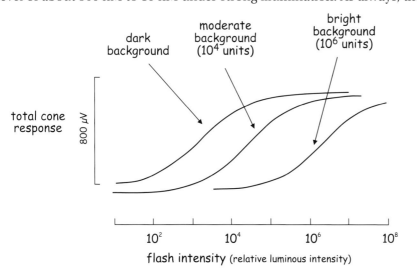

Figure 6.6 The shift in the flash response curve of mammalian cone photoreceptors as the background light level increases. Cone responses (instantaneous illumination plus background) measured with a bipolar electrode. Modified from Valeton and Van Norren.[8]

effects of Ca^{2+}, whether stimulatory (e.g. causing the contraction of muscle) or inhibitory (as in the present case), are mediated through specific Ca^{2+}-binding proteins. The synthesis of cyclic GMP by guanylyl cyclase is inversely and cooperatively regulated by Ca^{2+}. This is mediated through its interaction with guanylyl cyclase activating protein (GCAP), a member of the EF-hand group of Ca^{2+}-binding proteins (see Chapter 8). It is the free form of GCAP which binds and activates guanylyl cyclase so that in the dark, when the level of Ca^{2+} in the cytosol is high, the rate of cyclic GMP synthesis is low (Figure 6.7). This means that cyclic GMP hydrolysed under low light conditions is not so readily replenished and the system is kept at its most sensitive. In strong light, the conversion of GTP to cyclic GMP is accelerated due to the reduction in the concentration of free Ca^{2+}. The free GCAP binds to the cyclase and stimulates the resynthesis of cyclic GMP, which in turn opposes the closure of the membrane channels. The effect is to counteract the long-term effects of the light-activated hydrolysis of cyclic GMP.

Calcium also acts as a negative signal at a number of other points in the chain of events. It controls the influx of cations and restricts the extent of membrane hyperpolarization. Although the reaction rate of the phosphodiesterase reaction appears to be unaffected, the concentration of Ca^{2+}

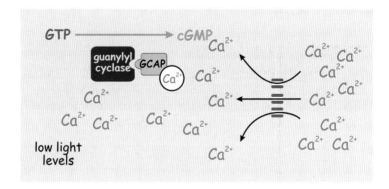

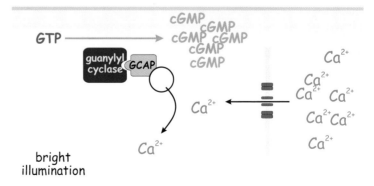

Figure 6.7 Photo-adaptation: a fall in intracellular Ca^{2+} promotes cGMP synthesis, opposing the effect of cGMP phosphodiesterase.

acts to regulate the lifetime of the activated enzyme. In the dark, through its interaction with another EF-hand Ca^{2+}-binding protein, recoverin (see Chapter 8), Ca^{2+} prolongs the signal through inhibition of rhodopsin kinase. The decline in the concentration of intracellular Ca^{2+} under intense illumination is permissive of rhodopsin phosphoryation and this necessarily curtails the lifetime of the activated phosphodiesterase (see Figure 6.11). Other sites of action of Ca^{2+} may include transducin (low Ca^{2+} might accelerate the GTPase reaction and so curtail the activation signal at this level) and the cyclic GMP-regulated ion channels. As the concentration of free Ca^{2+} declines during illumination, the affinity of the channels for cyclic GMP increases, and this must oppose the effects of the fall in the concentration of cyclic GMP and favour the open state.

To summarize, these highly integrated forms of automatic gain control arise in response to the generation of two signals as the membrane cation channels close in response to light. One of these is purely electrical and is conveyed as a pulse of hyperpolarization to the synaptic body, depressing the rate of release of the transmitter, glutamate. The other signal is chemical and results from the reduction in the concentration of cytosol Ca^{2+}. In essence, the light signal to these cells acts as a negative stimulus. In the dark, they are effectively partially depolarized and the rate of transmitter release at the synaptic body is maximal. In the light, they become hyperpolarized and the rate of transmitter release is depressed.

Photo-excitation of rhodopsin

The initial light-induced isomerization of 11-*cis*-retinal to all-*trans*-retinal, which occurs within picoseconds, does not immediately trigger the rhodopsin to catalyse the exchange of guanine nucleotides on transducin. The first rapid event is followed by a series of dark reactions. These have been characterized spectrally as the stepwise shift of the peak wavelength of light absorption of the photoreceptor protein to shorter wavelengths as the ligand and the opsin form a series of intermediates (Figure 6.8).

The early forms, bathorhodospin, lumirhodopsin and metarhodopsin-I are unstable. It is metarhodopsin-II, also called photo-excited rhodopsin (Rh*), that is formed within milliseconds of light absorption and which initiates the transduction cascade by inducing guanine nucleotide exchange on transducin.[9–11] However, in contrast with 'conventional' G-protein transduction mechanisms, the ensuing step takes place without amplification. The effector, cyclic GMP phosphodiesterase, is a heterotetramer having two independent catalytic subunits, α and β, and two inhibitory subunits, γ. The γ-subunits are understood to impede the access of the substrate, cyclic GMP to the catalytic sites.[12] The nucleotide exchange catalysed by Rh* occurs upon G_T.GDP that is bound to the inhibitory γ-subunits on the phosphodiesterase (Figure 6.9). When GDP is replaced by GTP, the α_T.GTP relieves the inhibition. As pointed out in Chapter 4, this is a cooperative process that involves both the conserved Ras-like (rd) domain of the α_T subunit that interacts with the γ subunits

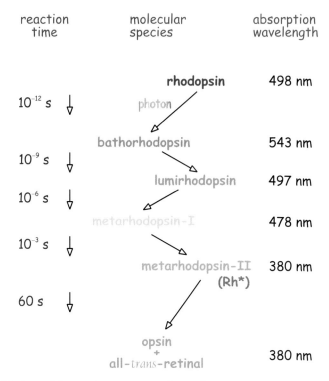

reaction time	molecular species	absorption wavelength

rhodopsin 498 nm

10^{-12} s ↓ photon

bathorhodopsin 543 nm

10^{-9} s ↓

lumirhodopsin 497 nm

10^{-6} s ↓

metarhodopsin-I 478 nm

10^{-3} s ↓

metarhodopsin-II 380 nm
(Rh*)

60 s ↓

opsin
+
all-*trans*-retinal 380 nm

Figure 6.8 Steps following isomerization of retinal.

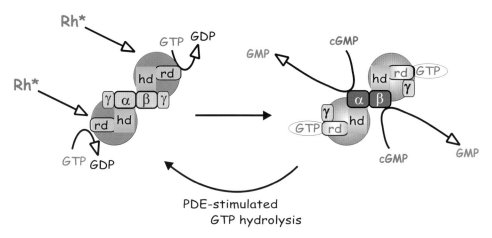

Figure 6.9 Activation of transducin by photo-excited rhodopsin. The activation of cyclic GMP phosphodiesterase by α_T is a cooperative process. In the resting state α_T (GDP) interacts through the helical domain (hd) with the α- and β- subunits. This reduces the affinity of the catalytic units for the inhibitory γ-subunits. Interaction of the Ras.GTP domain (rd) of α_T with the inhibitory γ-subunit of the PDE then acts as the switch, enabling access of the substrate cyclic GMP to the active sites on the α- and β-subunits. The return to the resting state is determined by the hydrolysis of GTP, assisted by the immediate proximity of the phosphodiesterase, acting as a GTPase activating protein (GAP).

and the unique helical domain (hd: see Figure 4.12, page 88) that interacts with the catalytic units. The α_T remains associated with the phosphodiesterase, so preventing it from interacting with further effectors and by its proximity, ensuring rapid deactivation of the complex following hydrolysis of GTP.[13,14]

Switching off the mechanism

We have described how a transient light signal rapidly initiates the phototransduction pathway and we have seen how the transduction of a single photon results in a hyperpolarization that lasts for about a second. However, this does not accord with our perception of transient optical signals which can be much briefer than this; the flicker-fusion frequency of 24 Hz corresponds to an image every 40 ms. This implies that there is a mechanism that shuts down visual transduction as soon as the stimulus is removed. Thus it is important for photoreceptor cells to be able to terminate signals emanating from the activated pigment and also to be able to terminate any signals that are in progress further down the cascade. These terminations therefore involve inhibition of the activities of the photopigment (the receptor), transducin (the transducer) and cyclic GMP phosphodiesterase (the catalytic unit or effector). Cyclic GMP is replenished through activation of guanylyl cyclase.

The conversion of metarhodopsin-I to metarhodopsin-II involves the loss of the proton from the Schiff base attaching the ligand to the protein. After metarhodopsin-II has activated transducin and over the next minute or so, the linkage is hydrolysed to yield all-*trans*-retinal and the colourless apoprotein opsin. The pigment is said to be bleached. Later on, rhodopsin will be regenerated by the binding of 11-*cis*-retinal, regenerated from all-*trans*-retinal in the adjacent cells of the pigment epithelium.[15,16]

Retinal, an inverse agonist?

A potential problem here is that opsin, although it is insensitive to visible light, in the presence of all-*trans*-retinal is capable of catalysing guanine nucleotide exchange on transducin, albeit to a low extent.[17–19] Furthermore, as noted in Chapter 3, the very high levels of rhodopsin in photoreceptor cells must pose the risk that spontaneous activation might give rise to some form of spurious 'dark vision' (see page 61). It is important to bear in mind that the catalytic potency of opsin is very small, only 10^{-6} that of photo-excited rhodopsin.[19] However, with 10^8 molecules of rhodopsin per cell, even a very low level of spontaneous activation would be sufficient to cause the sensation of light perception in total darkness. It is now thought that this is prevented by the ligand, 11-*cis*-retinal, acting as an inverse agonist and stabilizing the inactive conformation of the photoreceptor molecule (Figure 6.10).[17] Indeed, in the cell-free situation, the apoprotein opsin can catalyse guanine nucleotide exchange on the transducin and this can be inhibited by 11-*cis*-retinal. Conversely, all-*trans*-retinal behaves as an activator.[18]

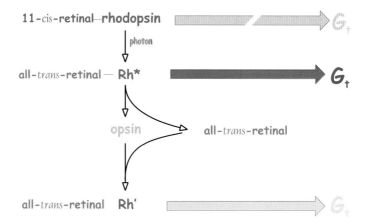

Figure 6.10 Possible four-state model of rhodopsin. Of the four states of the rhodopsin molecule that are shown, both Rh* and Rh′ can activate transducin. Rh′ is a non-covalent complex of opsin and all-*trans*-retinal and it is a weak activator of the G-protein. Rhodopsin itself does not activate G_t because of the presence of 11-*cis*-retinal, acting as an inverse agonist. After Surya *et al.*[17]

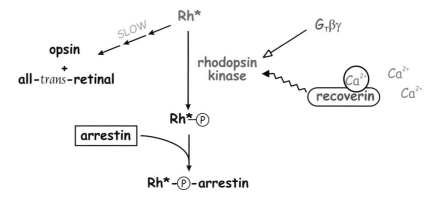

Figure 6.11 Phosphorylation of activated rhodopsin and the binding of arrestin shuts off the signal.

Activation of transducin by the apoprotein is avoided when it is phosphorylated by rhodopsin kinase. This is activated by the transducin βγ-subunits in a manner similar to the action of the receptor kinases (Chapter 4). As a result, the phosphorylated C-terminal domain of rhodopsin acts as a binding site for arrestin, the most abundant protein in the cytosol of the outer segments. This effectively blocks any further interaction with transducin and prevents any signals emanating from the light-insensitive opsin (Figure 6.11).

The rate of decline of the photoresponse is primarily determined by the deactivation of transducin which constitutes the rate–limiting step in the sequence of reactions leading from activated rhodopsin to the cyclic GMP phosphodiesterase.[20] The persistence of the active form of the α-subunits must depend entirely on the rate of hydrolysis of the bound GTP. The hydrolysis rate by isolated $α_T$ subunits is far too slow to account for the physiological rate of recovery, but this is enormously accelerated when it is reconstituted *in vitro* together with the phosphodiesterase, approaching the rate that can be measured in disrupted retinal rod outer segments.[21] In this sense, the phosphodiesterase appears to act not only as the downstream effector of transducin, but also as a GTPase activating protein (GAP) that functions to ensure abrupt signal termination. However, this is not the full story because for knock-out mice lacking RGS-9 (a member of the RGS family of GAP proteins, uniquely expressed in rod outer segments[22]), recovery is not only slow, but it is insensitive to the presence of the phosphodiesterase.[23] From this it appears that the GAP activity actually resides in the RGS protein, which lies in wait until the α_t is in communication with the phosphodiesterase before it pounces. The system has the advantage that the activated form of α_T is likely to persist until it scores an effective hit but, of course, under no circumstances must it be allowed to linger.

■ A note on phototransduction in invertebrates

We have described the basic elements of the signal transduction process as it occurs in the rod cells of vertebrate eyes. The situation in invertebrates is very different, as summarized in Figure 6.12.

In some ways, phototransduction in flies and other spineless creatures appears to follow more familiar pathways. Here, the light-activated rhodopsin is coupled through the G-protein G_q, not transducin. This regulates phospholipase Cβ, producing IP_3 and DAG, and results in the elevation of intracellular Ca^{2+} and the activation of protein kinase C. As with vertebrates, these events initiate both electrical and chemical signals, but here the consequence is the opening, not the closure of plasma membrane ion channels. This causes a very transient depolarization (instead of hyperpolarization) and further elevation of the concentration of cytosol Ca^{2+}. As with vertebrates, Ca^{2+} plays several roles ensuring the deactivation of the photosignal. In addition, PLC is both the target and a regulator (GTPase activator: see Chapter 4) of G_q,[24] and PKC is a negative regulator of the Ca^{2+} channels, Trp and TrpL (Trp-like; see Chapters 7 and 9).

As with vertebrates, the chromophore retinal is an integral component of rhodopsin, but with the difference that it remains attached to the opsin throughout the cycle of activation and recovery. Following illumination by blue light (below 490 nm), the retinal undergoes rapid isomerization to

---

Is not Vision preform'd chiefly by the Vibrations of this medium, excited in the bottom of the Eye by the Rays of Light, and propagated through the solid, pellucid and uniform Capillamenta of the optick Nerves into the place of Sensation? And is not Hearing perform'd by the Vibrations either of this or some other Medium, excited in the auditory Nerves by the Tremors of the Air, and propagated through the solid, pellucid and uniform Capillamenta of those Nerves into the place of Sensation? And so of the Other Senses.

Sir Isaac Newton, *Opticks*

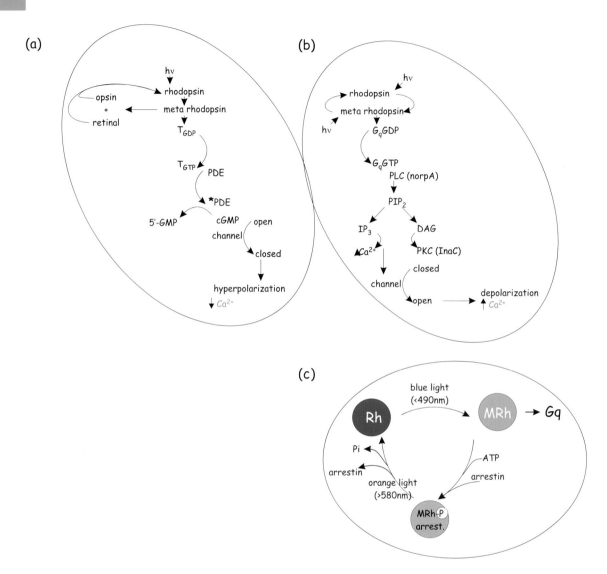

Figure 6.12 Comparing the main features of retinal phototransduction in vertebrates and flies. The main elements of visual phototransduction in vertebrates (a) and in *Drosophila* (b, c) are illustrated. For further detail, see text. The diagram in (c) illustrates the rhodopsin cycle in *Drosophila* in which the chromophore remains attached to the receptor throughout the process of activation and subsequent recovery. As with vertebrates, activation is initiated by a photon of light that causes the conversion of 11-*cis*-retinal to the all-*trans* form. In the invertebrate, however, recovery is effected by a second, longer wavelength pulse of light following phosphorylation of the rhodopsin.

the all-*trans* form and the resulting metarhodopsin catalyses guanine nucleotide exchange on G_q. Unlike vertebrates, the retinal remains firmly attached and the activated metarhodopsin is so stable that it would have a half-life of more than 5 hours.[25,26] This would not be much help to flies except that phosphorylation and then the binding of arrestin, sensitizes the system to a second photon, this one of longer wavelength (580 nm, orange). This triggers the reinstatement of 11-*cis*-retinal.

■ References

1 Lagnado, L., Baylor, D. Signal flow in visual transduction. *Neuron* 2001; 8: 995–1002.

2 Hecht, S., Shlaer, S., Pirenne, M.P. Energy, quanta and vision. *J. Gen. Physiol.* 1942; 25: 819–40.

3 Baylor, D.A., Lamb, T.D., Yau, K.W. Responses of retinal rods to single photons. *J. Physiol. Lond.* 1979; 288: 613–34.

4 Baylor, D.A., Nunn, B.J., Schnapf, J.L. The photocurrent, noise and spectral sensitivity of rods of the monkey *Macaca fascicularis*. *J. Physiol. Lond.* 1984; 357: 575–607.

5 Wheeler, G.L., Matuto, Y., Bitensky, M.W. Light-activated GTPase in vertebrate photoreceptors. *Nature* 1977; 269: 822–4.

6 Young, R.W. The renewal of photoreceptor cell outer segments. *J. Cell. Biol.* 1967; 33: 61–72.

7 Young, R.W. Shedding of discs from rod outer segments in the rhesus monkey. *J. Ultrastruct. Res.* 1971; 34: 190–203.

8 Valeton, J.M., Norren, D. Van. Light adaptation of primate cones: an analysis based on extracellular data. *Vision Res.* 1983; 23: 1539–47.

9 Liebman, P.A., Pugh, E.N. The control of phosphodiesterase in rod disk membranes: kinetics, possible mechanisms and significance for vision. *Vision Res.* 1979; 19: 375–80.

10 Pappone, M.C., Hurley, J.B., Bourne, H.R., Stryer, L. Functional homology between signal-coupling proteins. Cholera toxin inactivates the GTPase activity of transducin. *J. Biol. Chem.* 1982; 257: 10540–3.

11 Fung, B.K., Hurley, J.B., Stryer, L. Flow of information in the light-triggered cyclic nucleotide cascade of vision. *Proc. Natl. Acad. Sci. USA* 1981; 78: 152–6.

12 Granovsky, A.E., Natochin, M., Artemyev, N.O. The gamma subunit of rod cGMP-phosphodiesterase blocks the enzyme catalytic site. *J. Biol. Chem.* 1997; 272: 11686–9.

13 Liu, W., Northup, J.K. The helical domain of a G protein 3α subunit is a regulator of its effector. *Proc. Natl. Acad. Sci. USA* 1998; 95: 12878–83.

14 Liu, W., Clark, W.A., Sharma, P., Northup, J.K. Mechanism of allosteric regulation of the rod cGMP phosphodiesterase activity by the helical domain of transducin 3α subunit. *J. Biol. Chem.* 1998; 273: 34284–92.

15 Flannery, J.G., O'Day, W., Pfeffer, B.A., Horwitz, J., Bok, D. Uptake, processing and release of retinoids by cultured human retinal pigment epithelium. *Exp. Eye Res.* 1990; 51: 717–28.

16 Bridges, C.D., Alvarez, R.A. The visual cycle operates via an isomerase acting on all-trans retinal in the pigment epithelium. *Science* 1987; 236: 1678–80.

17 Surya, A., Stadel, J.M., Knox, B.E. Evidence for multiple, biochemically distinguishable states in the G protein-coupled receptor, rhodopsin. *Trends Pharm. Sci.* 1998; 19: 243–7.

18 Surya, A., Foster, K.W., Knox, B.E. Transducin activation by the bovine opsin apoprotein. *J. Biol. Chem.* 1995; 270: 5024–31.

19 Melia, T.J., Cowan, C.W., Angleson, J.K., Wensel, T.G. A comparison of the efficiency of G protein activation by ligand-free and light-activated forms of rhodopsin. *Biophys. J.* 1997; 73: 3182–91.

20 Sagoo, M.S., Lagnado, L. G-protein deactivation is rate-limiting for shut-off of the phototransduction cascade. *Nature* 1997; 389: 392–5.

21 Arshavsky, V.Y., Dumke, C.L., Zhu, Y. *et al.* Regulation of transducin GTPase activity in bovine rod outer segments. *J. Biol. Chem.* 1994; 269: 19882–7.

22 He, W., Cowan, C.W., Wensel, T.G. RGS9, a GTPase accelerator for phototransduction. *Neuron* 1998; 20: 95–102.

23 Chen, C.K., Burns, M.E., He, W. *et al.* Slowed recovery of rod photoresponse in mice lacking the GTPase accelerating protein RGS9-1. *Nature* 2000; 403: 557–60.

24 Cook, B., Bar-Yaacov, M., Cohen Ben-Ami, H. *et al.* Phospholipase C and termination of G-protein-mediated signalling in vivo. *Nat. Cell Biol.* 2000; 2: 296–301.

25 Scott, K., Zuker, C. Lights out: deactivation of the phototransduction cascade. *Trends Biochem. Sci.* 1997; 22: 350–4.

26 Feiler, R., Bjornson, R., Kirschfeld, K. *et al.* Ectopic expression of ultraviolet-rhodopsins in the blue photoreceptor cells of *Drosophila*: visual physiology and photochemistry of transgenic animals. *J. Neurosci.* 1992; 12: 3862–68

7

Calcium and signal transduction

Calcium is an abundant element. It constitutes 3% of the Earth's crust and it is prominent throughout the biosphere. It is present in fresh and sea water at levels that range from micromolar to millimolar. Ca^{2+} ions and Ca^{2+}-binding-proteins are essential for many biochemical processes in both prokaryotic and eukaryotic cells. In this chapter and in Chapter 8, we shall see that Ca^{2+} has special roles in the signalling mechanisms that regulate the activities of eukaryotic cells and that this depends upon the sensitivity of some proteins to changes in Ca^{2+} concentration.

■ A new second messenger is discovered

A number of the more enduring truths of science have been discovered when a failure of vigilance is coupled with uncommon perspicacity. Sidney Ringer's discovery of a requirement for calcium in a biological process must surely be placed in this category.[1,2]
Henry Dale's account:

Ringer was a physician to University College Hospital, and, in such time as he could spare from his practice, one of the pioneers of pharmacological research in this country. In his early experiments he had found that a solution containing only pure sodium chloride, common salt, in the proportion in which it is present in the serum of frog's blood would keep the beat of the heart in action for only a short time, after which it weakened and soon stopped. And then suddenly the picture changed: apparently the same pure salt solution would now maintain the heart in vigorous activity for many hours. Ringer was puzzled, and thought for a time that the difference must be due to a change in the season of the year – until he discovered what had really happened. Being busy with other duties, he had trusted the preparation of the solutions to his laboratory boy, one Fielder; and as Fielder himself, who I knew as an ageing man, explained to me, he didn't see the point of spending all that time distilling water for Dr Ringer, who wouldn't notice any difference if the salt solution was made up with water straight out of the tap. But, as we have seen, Ringer did notice the difference; and when he discovered what had happened he did not merely become angry and insist on having distilled water for his saline solution, he took full advantage of the opportunity which accident had thus offered him and soon discovered that water from the tap, supplied then to North London by the New River Water Company, contained just the right small proportion of calcium ions to make a physiologically balanced solution with his pure sodium chloride. . . .

Hints of a wider role for calcium soon followed with Locke's demonstration[3] that removal of calcium could block the transmission of impulses at the neuromuscular junction in a frog sartorius muscle preparation. The realization that this requirement for calcium is due to its role in controlling the secretion of a chemical messenger (neurotransmitter) had to wait 50 years.[4,5]

The direct introduction of Ca^{2+} ions into muscle fibres to cause contraction was first reported by Kamada and Kinoshita in 1943.[6] At this time Japan and the USA were at war, and as a result of the breakdown of communication, the credit for this finding has generally been ascribed to the Americans Heilbrunn and Wiercinski, who reported their results 4 years later.[7] Other cations such as Na^+, K^+ and Mg^{2+} were found to be without effect. It had been Heilbrunn's contention, at that time at least, that the effect of calcium was on the general colloid properties of the protoplasm and that the effects of ions on isolated proteins would lack biological relevance.[8] However, Otto Loewi (see Chapter 1) who attended their presentation at the New York Academy of Sciences, was heard to growl '*Kalzium ist alles*'. So it remained, until the biochemical understanding of signalling mechanisms became more developed, taking in cAMP (Chapter 5), GTPases (Chapter 4), tyrosine kinases (Chapter 11), PI 3-kinases (Chapter 13) and much more. Even with his foresight, however, had Loewi lived to the end of the century, he would surely have been astounded by the prominence of calcium in contemporary biology.

In previous chapters we have seen how the idea of second messengers followed from the work of Sutherland and Rall,[9] showing that the generation of cAMP represents the essential link between membrane events and the metabolic process in the signalling of glycogenolysis. If cAMP can be said to be the first second messenger, then Ca^{2+} is certainly the second. Indeed, it is a more ubiquitous messenger than cAMP, regulating a very diverse range of activities, including secretion, muscle contraction, fertilization, gene transcription and cell proliferation.

■ Calcium and evolution

Although prokaryotic cells do not make extensive use of intracellular Ca^{2+} as a signal, its level is kept very low. The evolution of the multicompartment structure of eukaryotic cells from their prokaryotic forerunners is thought to have occurred in stages, involving successive invaginations and internalizations of the cell membrane. Some modern prokaryotes provide evidence for this (Figure 7.1). The internalization of elements of the plasma membrane had the effect of segregating specific functions on to specialized intracellular structures, such as the nuclear envelope, the smooth and rough endoplasmic reticulum, the Golgi membranes, lysosomes, etc. A probable early outcome of this process was the formation of a reticular compartment responsible for protein synthesis – a 'protoendoplasmic' reticulum. This organelle inherited powerful ion translocation mechanisms that, instead of expelling Ca^{2+} from the cell, could now draw it from the cytosol into its lumen, providing an intracellular Ca^{2+} store.

The formation of specialized membranes and organelles imposed a need for eukaryotic cells to elaborate the means of communication between their various internal compartments and the external world. Furthermore, the later evolution of complex metazoan organisms having differentiated cells with individual specialized functions necessitated increasingly complex signalling systems. These would enable the coupling of extracellular signals to intracellular signalling pathways and thence to specific downstream effectors. Calcium is well suited for the role of an intracellular messenger for two main reasons. One is that its immediate proximity provides the opportunity for it to enter the cytosol directly, through membrane channels, either from the extracellular environment or from the internal reticular compartment; the other is provided by calcium's distinctive coordination chemistry.

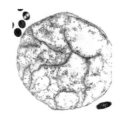

Figure 7.1 *Thiovulum,* **a sulphur-oxidizing bacterium, possesses endoplasmic membrane structures.** This large eubacterium (diameter about 6 μm) possesses an intracellular membrane system possibly including a rudimentary rough endoplasmic reticulum. The black dots inside the cell are probably ribosomes. The small blobs at the top left are 'ordinary' bacteria. From Tom Fenchel, Marine Biological Laboratory Helsingør, Denmark.

■ Distinguishing Ca^{2+} and Mg^{2+}

Although nature seems to have taken advantage of the presence of Ca^{2+}, its suitability for signalling depends upon its ability to form stable complexes with particular biological molecules. The ability of metal ions such as Ca^{2+} and Mg^{2+} to form coordination complexes with anionic or polar ligands can be described by equilibria of the form

$$M + L \overset{K}{\rightleftharpoons} ML$$

in which ML represents the complex of the metal ion M with the ligand L.

The stability (equilibrium) constant K is a measure of the stability of the complex.

All ions in aqueous solution are surrounded by a hydration shell, so in order for coordination to occur, water molecules must be displaced from both the metal and the ligand. The factors that determine the stability therefore include not just the electrostatic attraction between cation and ligand, but also attractive and repulsive interactions between adjacent coordinating ligands. Importantly, water is itself a ligand that stabilizes the ionized state. Not all of the water molecules present in the hydration shell need necessarily be removed to form a complex. The stability of the complex will therefore depend upon the energy of association of M and L, set against the energies of association of M and L separately with water.

Ca^{2+} is remarkably different from Mg^{2+} in these respects.[10] This is principally because Ca^{2+} is larger and has less difficulty accommodating ligand atoms around its surface. The number of atoms that bind to it in a complex (the coordination number) may vary between 6 and 12, and the

The term **ligand** was first coined by chemists to describe an electron-donating group that forms coordination complexes with metal ions.

The **stability constant** K is the reciprocal of the dissociation constant K_D, which represents the concentration of free cation M at which the ligand is half saturated.

geometry of the arrangement around the coordination sphere is rather flexible. Conversely, Mg^{2+} forms complexes in which the ligand atoms are nearly always arranged in an octahedral formation (a coordination number of 6). For small anions that can be accommodated in such a structure, the strength of association with Mg^{2+} is greater than with Ca^{2+}. For bulky and irregularly shaped anions, for which the octahedral requirement cannot be satisfied, the order is reversed.

Chelating agents are synthetic compounds that provide an array or cage of several ligand atoms. Such multidentate ('many-toothed') assemblies, particularly those that form an irregular structure, can be very selective for Ca^{2+} over Mg^{2+}, because of the ability of Ca^{2+} to tolerate an irregular geometry. One of the most specific Ca^{2+} chelators is EGTA (Figure 7.2). This molecule possesses four carboxyl groups and two ether oxygens and exhibits a 10^5-fold selectivity for Ca^{2+} over Mg^{2+}. (Note that Ca^{2+} overwhelmingly prefers oxygen atom donors.) EGTA and similar compounds can be used as Ca^{2+} buffers to control the concentration of free Ca^{2+} in experiments.

> The word **chelate** is derived from Latin *chela*, meaning a claw of a crab or lobster.

■ Free, bound and trapped Ca^{2+}

In the biological world, Ca^{2+} may be considered to exist in three main forms: free, bound and trapped. In vertebrates, calcified tissues such as bones and teeth account for the major proportion of body calcium. Within these tissues calcium is trapped in mineral form (with phosphate) as hydroxyapatite. Apart from its structural function, bone provides a reservoir of slowly exchangeable calcium that can be mobilized, when needed, to maintain a steady extracellular concentration (Table 7.1). The total amount of calcium in extracellular fluid, in comparison with that in mineralized tissues, is very small, and only about half of this exists as the free ion. The remainder is bound mostly to proteins in the extracellular milieu. Within cells, the concentration of free Ca^{2+} inside the cytosolic compartment is four orders of magnitude lower than its concentration in extracellular fluid (Table 7.1). A considerable proportion of intracellular

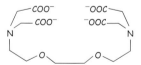

Figure 7.2 **EGTA (ethylene-bis(2-aminoethylether)-*N,N,N´,N´*-tetraacetic acid).**

Table 7.1 Approximate levels of free Ca^{2+} and Mg^{2+}

	$[Ca^{2+}]$		$[Mg^{2+}]$
Plasma, extracellular fluid	1–2 mmol/l	pCa ~3	1 mmol/l
Intracellular cytosolic (resting cells)	50–100 nmol/l	pCa ~7	0.5–1 mmol/l
Intracellular lumenal (ER)	30–300 μmol/l	pCa 5–4	

calcium is also bound to proteins and these exhibit a wide range of affinities. More importantly, many are able to bind Ca^{2+} ions in the presence of a huge excess of Mg^{2+}. This is because the arrangement of the amino acid side chains that bear the coordinating atoms tends to create sites of irregular geometry. There is also structural flexibility. This has important consequences for the 'on–off' kinetics of binding.

In the extracellular environment, free Ca^{2+} and Mg^{2+} are both present at millimolar concentrations (Table 7.1). Low affinity Ca^{2+}-binding-proteins provide a short-term, local buffer, and in vertebrates at least, long-term (homeostatic) control of free extracellular Ca^{2+} is maintained by a hormonal mechanism that utilizes the mineralized tissue reservoirs. Failure to maintain a stable level of extracellular Ca^{2+} can have serious consequences. Excess circulating Ca^{2+} (hypercalcaemia) reduces neuromuscular transmission and causes myocardial dysfunction and lethargy. Hypocalcaemia affects the excitability of membranes and, if left untreated, leads to tetany, seizures and death.

■ Cytosol Ca^{2+} is kept low

Ca^{2+}-binding proteins within prokaryotic and eukaryotic cells fulfil a wide range of functions (see Chapter 8). Some are employed to keep cell Ca^{2+} low. These are membrane pumps and ion exchangers. Others act as intracellular Ca^{2+} buffers. In eukaryotic cells, low Ca^{2+} levels are maintained in the cytosolic compartment by ejecting Ca^{2+} from the cell and also by pumping it into the lumen of the endoplasmic reticulum (ER). The pumps are transmembrane proteins that move Ca^{2+} ions across membranes against their electrochemical potential gradients.

They are called Ca^{2+}-ATPases and they provide primary active transport. Extrusion of Ca^{2+} from the cytosol by this mechanism is supported by secondary active transport through Na^+–Ca^{2+} exchangers: transmembrane proteins that use the inward Na^+ electrochemical gradient across the plasma membrane to transport Ca^{2+} ions out of the cell.

Electrochemical potential gradient The combined effect of the concentration and voltage gradients.

Changes in cytosol Ca^{2+} concentration are also resisted by resident, high-affinity Ca^{2+}-binding-proteins, some of which may be confined to particular locations within the cytosolic compartment. All of them may act as buffers that tend to restrict local changes of Ca^{2+}, although their capacity is not high and the speed at which they take effect will depend on the binding kinetics. Within the ER, low-affinity Ca^{2+}-binding-proteins (K_D ~2 mmol/l) such as calreticulin, can retain up to 20 mol of Ca^{2+} per mol of protein, greatly increasing the capacity of this intracellular store.

Over the long term, the coordination of these mechanisms underpins cellular Ca^{2+} homeostasis. Passive inward leaks are balanced by the action of the Ca^{2+}-ATPases which provide a low capacity extrusion pathway. Their rather high affinity (K_D ~0.2 µmol/l) enables them to maintain low cytosolic Ca^{2+} in resting cells. During cell activation, cytosol Ca^{2+} levels can rise considerably and the pumps and exchangers then operate to restore Ca^{2+} to its resting level. The plasma membrane Na^+–Ca^{2+} exchangers, present in most tissues and of particular importance in heart, are of low affinity

(K_D = 0.5–1 µmol/l), but provide a high-capacity extrusion mechanism which enables them to fulfil this task.

◼ Detecting changes in cytosol Ca²⁺

The early evidence underlying a role for Ca^{2+} in cell activation was based mainly on the experience that some tissue responses are suppressed when Ca^{2+} is not provided, much as Locke and Ringer had demonstrated over 100 years ago. Others are seemingly unaffected because, as we now realize, the cells can call upon their own intracellular stores. Confirmation of a direct role for Ca^{2+} in cell activation required more direct means of controlling and sensing changes in its level *in situ*.

◼ Using Ca²⁺ ionophores to impose a rise in Ca²⁺

The Ca^{2+} ionophores are lipid-soluble, membrane permeant ion carriers that are specific for Ca^{2+}. They were originally isolated from cultures of *Streptomyces* and include the antibiotics A23187 (524 Da) and ionomycin (709 Da). They may be used to convey Ca^{2+} ions into cells and they have been applied to numerous preparations, particularly in the field of secretion. Among the observed effects were responses similar to those that follow biological stimulation, but without the involvement of receptor-associated mechanisms. This revealed a widespread involvement of Ca^{2+} in activation mechanisms, establishing it as a legitimate second messenger alongside cAMP. However, ionophores are blunt weapons and although they certainly enable the introduction of Ca^{2+}, seemingly to induce biological responses, it is hard to exert control over the concentrations achieved.

◼ Sensing changes in Ca²⁺ concentration

The modern approach relies on the principle of introducing an exogenous Ca^{2+}-sensing compound into the cytosolic compartments of living cells. This was first attempted in 1928, by injecting the dye alizarin sulphonate into an amoeba. Apparent evidence of a local increase in Ca^{2+} was obtained when a visible precipitate formed at the site of pseudopod formation.[11] Unfortunately, it later emerged that alizarin is not specific for Ca^{2+}. More unfortunately, the development of practical Ca^{2+} indicators suitable for use with mammalian cells took a further 50 years. In the meantime, intracellular calcium levels were mostly investigated by measuring transmembrane fluxes of the radioisotope ^{45}Ca. It is possible to measure changes in total cell Ca^{2+} by this means, but impossible to determine how levels have changed within particular intracellular compartments, such as the cytosol. To make matters worse, the bulk of cell calcium is retained within organelles and there is very little in the cytosolic pool.

◼ Ca²⁺-sensitive photoproteins

Ca²⁺-sensitive photoproteins, extracted in their active form from marine invertebrates such as the

Among the more colourful fauna of the oceans are luminescent jellyfish from which the Ca^{2+}-activated, light-emitting-protein aequorin can be extracted.[12] Aequorin was the first effective indicator for intracellular Ca^{2+}. Its molecular mass is 22 kDa and it cannot cross membranes, so its use was initially limited to larger cells into which it could easily be microinjected. Once in the cytosol, an increase in free Ca^{2+} concentration causes the protein to release a pulse of light. Analysis of the time course of the emission can be used to follow changes in Ca^{2+} concentration (up to 100 μM). In important early work, aequorin was used with preparations such as squid giant axon[13] and permeabilized skeletal muscle.[14] These studies provided the first direct observations of so-called 'Ca2+ transients', the biphasic rise and fall of cytosol Ca^{2+} following a stimulus.

Fluorescent Ca^{2+} indicators

A major advance in Ca^{2+} measurement came with the design by Roger Tsien of the first of a series of synthetic Ca^{2+}-indicators. The fluorescence of these compounds is sensitive to Ca^{2+} concentration in the range 50 nmol/l to 1 μmol/l.[15] The first indicator to be developed was called Quin2.[16] Its structure (Figure 7.3) is based on a Ca^{2+} chelator that resembles EGTA. Like EGTA, Quin2 has high affinity for Ca^{2+} and low affinity for Mg^{2+}. Because it is an anionic species, it cannot cross membranes and a trick is needed to deliver it to the cytosolic compartment without resorting to microinjection. This problem affects all Ca^{2+} indicators of this class and is overcome when they are provided to cells as uncharged ester derivatives. In this form they can diffuse freely across the plasma membrane. Upon entry, endogenous esterase activity releases the acidic form of the indicator, which then accumulates in the cytosol of the intact cells. Importantly, Quin2 and its successors paved the way towards the monitoring and mapping of intracellular Ca^{2+} levels at high temporal and spatial resolution.

For Quin2 and many of the other indicator dyes, the intensity of the fluorescent emission increases with Ca^{2+} concentration over a limited range, providing an uncalibrated estimate of the level of free cytosolic Ca^{2+}. Among the second generation of indicator dyes, Fura2 (Figure 7.4) exhibits wavelength shifts in its excitation spectrum when Ca^{2+} is bound. For this indicator the ratio of the fluorescence emission at two selected excitation wavelengths can provide a calibrated estimate of cytosol Ca^{2+} concentration.

The major significance of these developments was that it became possible, at last, to monitor intracellular Ca^{2+} in almost every kind of cell, large or small, simply by recording fluorescence. Initially, measurements were

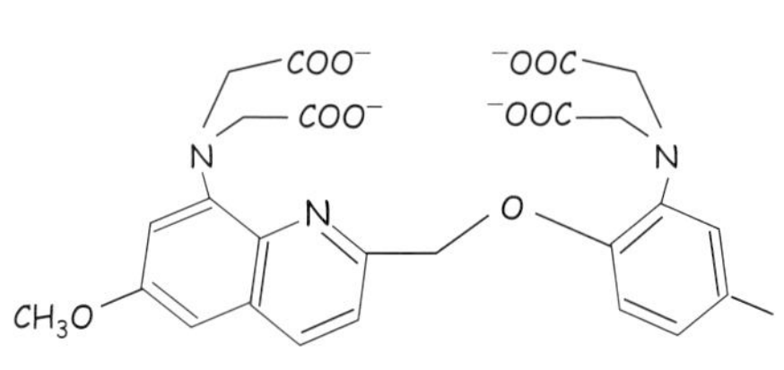

cnidarians *Aequorea*[12] and *Obelia*,[13] undergo an internal molecular rearrangement activated by Ca^{2+}. The conformational change oxidizes a luminophoric prosthetic group, coelenterazine, emitting a flash of blue light.

Recently, **aequorin**, expressed in mammalian cells, has enjoyed a revival. Recombinant DNA, incorporated in an expression vector, can be used to transfect cells so that they express their own calcium reporter. Targeting this to specific intracellular compartments and adjusting its Ca^{2+} affinity has enabled the detection of Ca^{2+} changes in particular organelles, for example mitochondria.[17]

Figure 7.3 **Quin2.**

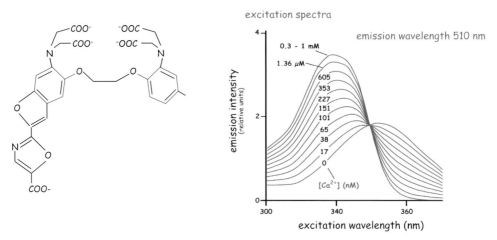

Figure 7.4 **Molecular structure and fluorescence excitation of Fura2.** The graph shows how the excitation spectrum of Fura2 varies with Ca²⁺ concentration. For example, if excitation occurs at 340 nmol/l (near the UV region), the intensity of emission (green fluorescence, measured here at 510 nmol/l) will increase with Ca²⁺ as indicated.

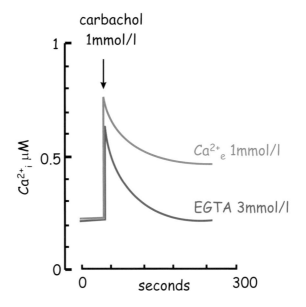

Figure 7.5 **Changes in the concentration of cytosol Ca²⁺: effect of extracellular Ca²⁺.** The traces show elevations of Ca^{2+} in parotid acinar cells loaded with Fura2. Carbachol (1 mmol/l) was added at the time indicated by the arrow to cells incubated in a medium containing Ca^{2+} (1 mmol/l) or without added Ca^{2+} salts, but with the chelator EGTA to remove residual Ca^{2+} ions ($[Ca^{2+}] < 1$ nmol/l). Adapted from Nowak et al.[20]

performed on suspensions of isolated cells undergoing activation in a fluorimeter cuvette. In this situation, the recorded emission represents the sum of contributions from each cell in the light beam. Very soon, studies on a wide variety of electrically non-excitable cells revealed characteristic response patterns as illustrated in Figure 7.5. Application of a stimulatory ligand typically produces a biphasic change, in which the concentration rises from the resting level (50–100 nmol/l) towards the micromolar range and then falls back to a level somewhat higher than the resting concentration. The transient phase of the signal is not affected when Ca^{2+} is removed from the external solution, so that its source must be Ca^{2+} mobilized from internal stores. Its duration varies from seconds to minutes, depending on the type of cell. The residual component of the signal, on the other hand, is eliminated by removal of extracellular Ca^{2+} prior to stimulation. This must therefore require entry of Ca^{2+} into the cell across the plasma membrane (Figure 7.5).

The problem with measurements (of any type) on cell suspensions is that the signal represents the summation of individual, unsynchronized contributions. We shall see that when Ca^{2+} indicator dyes are used with single cells, the time courses appear very different (for example see panel (a) of Figure 7.11).

Mechanisms that elevate cytosol Ca^{2+} concentration

The evolution in eukaryotic cells of enzymes that are sensitive to changes in Ca^{2+} concentration was accompanied by the development of switch-able membrane channels, which allow the ion to flow down its concentration gradient into the cytosolic compartment upon demand. The resultant rise in Ca^{2+} concentration activates Ca^{2+}-sensitive enzymes that initiate a wide range of effects. These are discussed in Chapter 8. Although Ca^{2+}-dependent processes take place in both prokaryotes and eukaryotes,

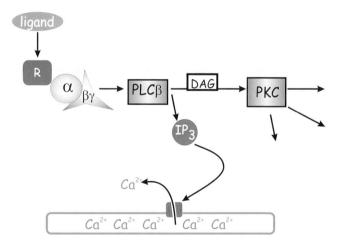

Figure 7.6 Activation of phospholipase C and the mobilization of Ca^{2+} from intracellular stores. Calcium mobilizing receptors are coupled through G_q or G_i to PLCβ (for details, see Chapter 5) which hydrolyses PI(4,5)P$_2$ releasing two products. DAG is retained in the membrane and activates protein kinase C. The water-soluble IP$_3$ binds to ligand gated channels present in the membranes of intracellular Ca^{2+} stores in the endoplasmic reticulum, releasing Ca^{2+} into the cytosol.

it is only in eukaryotic cells that extensive use of Ca^{2+} as an intracellular signal has evolved.

In the following sections, the mechanisms that elevate cytosol Ca^{2+} will be described. In short, these involve its entry into the cell from the exterior and its release from Ca^{2+}-storing organelles. Mobilization of intracellular Ca^{2+} most commonly follows a ligand-receptor interaction at the cell surface that leads to the activation of phospholipase C. This key intracellular enzyme is described in Chapter 5. It hydrolyses the phospholipid, phosphatidylinositol 4,5-bisphosphate $(PI(4,5)P_2)$ to produce two second messengers, one of which is inositol trisphosphate (IP_3) which triggers release of Ca^{2+} from the intracellular stores into the cytosol. When the stores become depleted, they are replenished from the extracellular Ca^{2+} pool by a mechanism that involves the opening of plasma membrane channels.

In electrically excitable cells, Ca^{2+} elevation following a depolarizing stimulus is generally more rapid and it is achieved by a different mechanism. This increase is initiated by the opening of voltage-dependent Ca^{2+}-channels in the plasma membrane. Consequent entry of Ca^{2+} produces high concentrations in the immediate vicinity. This localized increase may itself then cause release of further quantities of the ion from intracellular stores in a regenerative manner. A more detailed description of these processes is now presented.

■ Two sources of Ca^{2+}

▨ *Ca^{2+} release from intracellular stores, IP_3 and ryanodine receptors*

Ca^{2+} release into the cytosolic compartment from the ER occurs through two families of structurally related ion channels, inositol trisphosphate receptors (IP_3Rs) and ryanodine receptors (RyRs).

IP_3Rs are virtually universal, whereas RyRs, which were first identified in the sarcoplasmic reticulum (SR) of skeletal and cardiac muscle, are most evident in excitable cells.

Three RyR isoforms are known. Type 1 RyRs occur in skeletal muscle, where they interact directly with the voltage-sensing subunits of the Ca^{2+} channel complex. This requires the close apposition of the sarcoplasmic reticulum and the plasma membrane (see page 182). In contrast, the type 2 RyRs that exist in cardiac muscle and some nerve cells, do not interact with plasma membrane Ca^{2+} channels in such a direct manner. A third isoform, type 3, is present in brain, muscle and some non-excitable cells.

RyRs are very large. The chains of the type 1 RyR consist of some 5000 amino acids organized in a complicated tetrameric arrangement (Figure 7.7), only 20% of which is associated with channel activity. The remainder forms a prominent structure that extends into the cytoplasm and possesses binding sites for a number of physiological modulators. These include Ca^{2+}, ATP, calmodulin (described in Chapter 8) and FKBP12. (This 12 kDa protein binds the immunosuppressive drug FK506 and is known as an immunophilin; see Figure 17.13, page 389.) It is not clear how these

Ryanodine is a plant alkaloid that affects the release of Ca^{2+} from the sarcoplasmic reticulum in muscle. It binds to ryanodine receptors with high affinity. At low concentrations the channel open probability is increased, at high concentrations (micromolar) it decreases.

Sarcoplasmic reticulum The smooth endoplasmic reticulum of muscle cells.

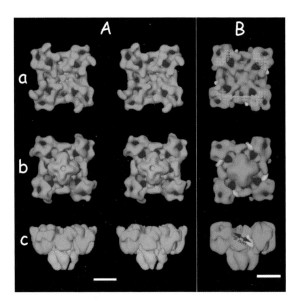

Figure 7.7 Three-dimensional structure of the ryanodine receptor obtained by cryoelectron microscopy: (A) Three-dimensional reconstruction of the ryanodine receptor from skeletal muscle. Stereoscopic views of (a), the surface that faces the T-tubule; (b) that facing the lumen of the SR; (c) a side view. The transmembrane part is pink and the cytoplasmic region is green. Instructions for viewing stereo images can be found on page 396. (B) A depiction of the receptor (cyan) indicating the locations of calmodulin (yellow) and FKBP12 (magenta). The orange shading indicates regions likely to interact with dihydropyridine receptors. The scale bar is 10 nm. Adapted from Malenka *et al.*[21]

factors are coordinated to regulate channel activity, but FKBP12 is an integral part of the structure and it may interact with the protein phosphatase calcineurin. The most important regulator of the channel is Ca^{2+} itself, and this is discussed below.

IP$_3$Rs are also tetramers and each of the component subunits possesses six transmembrane segments (Figure 7.8). There are three isoforms of the basic subunit, types 1 to 3. For all of these, both the N- and C-termini lie within the cytosol and the N-terminal chain, which bears the IP$_3$ binding site, is particularly large (over 2000 amino acids for type 1 IP$_3$Rs). This cytosolic domain, like that of the RyRs, possesses a number of modulatory sites and Ca^{2+} can also bind to the lumenal (ER) domain of the protein. Type 1 IP$_3$Rs are present at high levels in cerebellar tissue, but apart from the CNS, nearly all mammalian cells express more than one isotype. A complicating factor in the study of the characteristics of the different forms has been their propensity to assemble as hetero- as well as homotetramers.

Both IP$_3$Rs and RyRs are sensitive to local Ca^{2+} concentration, so that as cytosol Ca^{2+} increases, it induces further release from the stores. This is known as Ca^{2+}-induced Ca^{2+}-release or CICR. As it proceeds it causes positive feedback, and this underlies the regenerative character of many Ca^{2+} signals. In a sense, the names of these channels are misleading. Both are Ca^{2+}-induced Ca^{2+}-release channels, that are modulated by other factors.

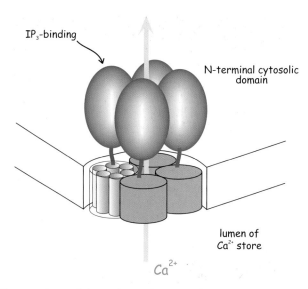

IP$_3$-binding

N-terminal cytosolic domain

lumen of Ca^{2+} store

Ca^{2+}

Figure 7.8 **IP$_3$ receptor calcium channel.**

The opening of the IP$_3$R channels is modulated by IP$_3$, the binding of which increases the sensitivity of the channels to the surrounding Ca^{2+} concentration, so that Ca^{2+}-induced Ca^{2+} release commences at resting cytosol Ca^{2+} levels. To prevent a runaway situation, which would lead ultimately to complete depletion of the stores, the Ca^{2+}-sensitivity of both IP$_3$Rs and RyRs then decreases as the concentration of Ca^{2+} rises, i.e. the channels begin to close again. The overall effect is that the dependence of channel open probability upon Ca^{2+} concentration is bell-shaped (see Figure 7.15). The manner in which IP$_3$ modulates this relationship is discussed below.

Efforts to elucidate the physiological characteristics that distinguish the three types of IP$_3$R have produced somewhat contradictory data, due to their relative inaccessibility (in the ER), sensitivity to the environment and heterogeneity (possible heterotetramers). The most direct measurements, involving electrophysiological studies of IP$_3$R channels expressed in native membranes (albeit nuclear, rather than ER), indicate that all three isoforms have very similar gating and ion permeation characteristics.[19] Any observed differences in their behaviour in intact cells must then be due to differences in the way they are modulated by intracellular factors (or perhaps to differences in their location). However, the modulatory mechanisms that cause the closure of IP$_3$R and indeed RyR channels at high Ca^{2+} levels are not well understood. Both channel types are phosphorylated by Ca^{2+}/calmodulin-dependent kinase II (see Chapter 8) and both are dephosphorylated by the phosphatase calcineurin. (Like the RyRs, IP$_3$Rs are also associated with the immunophilin FKB12.) There is evidence that different IP$_3$R isoforms are affected differently by Ca^{2+}, calmodulin and phosphorylation and such mechanisms could modify the temporal and spatial characteristics of Ca^{2+} signals.

Finally, the release of Ca^{2+} from intracellular stores may be even more complicated than all this. The notion that cells contain only one type of store has been challenged by a number of findings, including the discovery that there are other potential messenger molecules that can release Ca^{2+}. These include the nucleotides cyclic ADP-ribose and nicotinic acid adenine dinucleotide phosphate. These are discussed below.

Ca^{2+} influx through plasma membrane channels

Voltage-operated channels

In the context of activation there are a number of different pathways by which Ca^{2+} ions can enter cells from the exterior. Nerve and muscle cells possess a variety of cation channels that are sensitive to changes in the membrane potential. Some of these conduct Ca^{2+} and have roles in the initiation of action potentials, the control of excitability and the activation of exocytosis. Such voltage-operated calcium channels (VOCCs) open in response to a membrane depolarization and allow Ca^{2+} entry. They are transducers that convert an electrical signal into a second messenger (Ca^{2+}) signal. VOCCs are distinguished electrophysiologically and pharmacologically into classes termed L, N, P/Q, R and T-type and they are tightly regulated. L-type channels are primarily modulated by phosphorylation and N, P/Q and R-type channels are modulated by G-proteins.

L-type Ca^{2+} channels are expressed by many vertebrate excitable cells and they are the major VOCCs of skeletal muscle. They are a target for 1,4-dihydropyridine drugs (such as nifedipine), which act as channel blockers. The part of the channel complex that binds the drug is termed the dihydropyridine receptor (DHPR), although this term is often used to denote the whole channel. In vertebrate skeletal muscle, plasma membrane DHPRs and RyRs on the SR are in close apposition. Possible points of contact are shown in Figure 7.7. This occurs at the triad regions, where the terminal cisternae of the SR almost contact the T-tubules (see Figure 8.8, page 183). Interaction between these proteins initiates the Ca^{2+}-induced Ca^{2+}-release that underlies excitation-contraction coupling (see page 182).

VOCCs also play important roles at synapses. In the presynaptic neurone, they are situated close to the sites at which neurotransmitter-containing vesicles await the stimulus to undergo exocytosis. Membrane depolarization causes the entry of Ca^{2+} ions that interact locally with (relatively) low affinity Ca^{2+}-binding proteins (see page 181). VOCCs are also found in certain endocrine cells, such as pituitary cells and pancreatic β cells, where elevated intracellular Ca^{2+} is also a trigger for secretion. Common characteristics of Ca^{2+} entry through VOCCs are the speed, brevity and intensity of the observed Ca^{2+} transients. The rather low affinity of the sensor(s) is matched to the high concentrations of Ca^{2+} reached (up to 10 μmol/l or higher) and the limited spatial volume in which it is confined. Rapid dissipation may then follow, ensuring that the subsequent effects are local and transient.

■ Receptor-operated channels

In some neuronal cells, non-specific ion channels in the plasma membrane may open as a consequence of activation and allow Ca^{2+} to enter. Glutamate receptors sensitive to NMDA are important post-synaptic ion channels, which mediate excitatory transmission at central synapses.

In order for these channels to open, two conditions must be satisfied: they must bind glutamate and the membrane in which they reside must already be depolarized in order to remove a blocking Mg^{2+} ion.[20] This may be achieved by the simultaneous effects of two different neurotransmitters, one of which depolarizes the cell in preparation for the action of glutamate itself. Alternatively, a sustained release of glutamate from the pre-synaptic cell could provide both signals, by first activating AMPA receptors to depolarize the post-synaptic cell. The ability of NMDA receptors to act as coincidence detectors, coupled with other processes, gives them a role in the long-term potentiation of synaptic signalling.[21,22]

■ Store-operated channels

The rapid transmission of signals between nerves and the fast responses of skeletal muscles, all of which depend on signalling through VOCCs, are essential for survival. For other types of tissue, there is often less urgency. In hormone-secreting cells, the changes in Ca^{2+} concentration may take the form of a series of oscillations that continue over a period of minutes (see Figure 7.12). In cells stimulated by growth factors or cytokines, there may be a need for a protracted period (hours) of Ca^{2+} elevation in order to ensure full commitment to a proliferative response.[23] Such demands present a problem since repetitive or sustained elevation of cytosol Ca^{2+} must lead to depletion of the Ca^{2+} stores. This is overcome by allowing extracellular Ca^{2+} to enter through plasma membrane cation channels, called store-operated channels (SOCs). This allows the stores to be replenished and has been termed capacitative Ca^{2+} entry.[24,25]

The mechanism that senses store depletion and couples it to the opening of plasma membrane channels is not well understood. There appears to be a close or even a direct interaction between the IP_3R proteins in the ER and the SOCs, resembling the coupling between RyR1s and VOCCs[26] as illustrated in Figure 7.9.

Evidence for such interactions has been obtained from studies of the photoreceptor cells of the fruit fly *Drosophila melanogaster*. Adaptation to bright illumination requires the entry of Ca^{2+} ions to overcome store depletion.[29,30] In these cells, elements of the ER, close to the plasma membrane, possess a Ca^{2+} channel encoded by the *trp* gene (transient receptor potential). A presumably related channel has been characterized electrophysiologically in rodent mast cells, where an inward current is activated within about 10 s following store depletion. The channel is specific for Ca^{2+} and the current is termed I_{CRAC} (calcium release-activated calcium current).[31] Ca^{2+} entry through I_{CRAC} channels is required for the activation of mast cells by antigens reactive at the receptors for immunoglobulin E (Chapter 12).

NMDA   *N*-methyl-D-aspartic acid, a glutamate analogue that activates a sub-class of glutamate receptors.

Mast cells are associated with the immune system. They secrete inflammatory mediators in response to an antigen challenge. If the antigen is also an allergen they can present a particular problem for allergic individuals. See page 288.

(a)

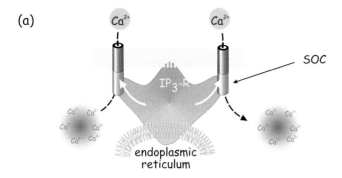

(b)

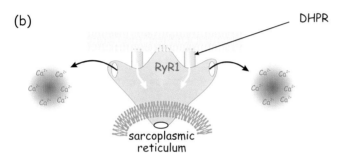

Figure 7.9 Coupling of ryanodine and IP$_3$ receptors to plasma membrane channels. Model suggesting that conformational changes in the IP$_3$Rs (a) and RyRs (b) may be induced by direct interaction with plasma membrane channel proteins: store-operated channels (SOCs) in (a) and dihydropyridine receptors (DHPRs), in (b). Adapted from Hardie and Minke.[29]

Finally, mammalian homologues of *Drosophila trp* have been identified including the human form, Htrp. One isoform, Htrp3, is a non-specific cation channel that interacts directly with IP$_3$Rs.

Store-operated channel opening requires that the receptors should be occupied by IP$_3$ and that the stores should also be depleted.[32] How depletion is sensed remains unknown. A possibility is that it is detected by the IP$_3$Rs, through their lumenal Ca^{2+} binding sites, and then signalled by a conformational change.

Elevation of Ca^{2+} by cyclic ADP-ribose, NAADP and sphingolipid metabolites

IP$_3$ has a clear and firmly established role as the major intracellular Ca^{2+}-mobilizing messenger. There is evidence, however, of other messenger molecules that can also bring about the release of Ca^{2+} from stores. Two of these are nucleotides, cyclic ADP-ribose (cADPr) and nicotinic acid adenine dinucleotide phosphate (NAADP). Their structures are shown in Figure 7.10. Remarkably, each of these compounds is formed by the same enzyme, ADP-ribosyl cyclase, acting upon two different substrates, NAD$^+$ or NADP$^+$ respectively.

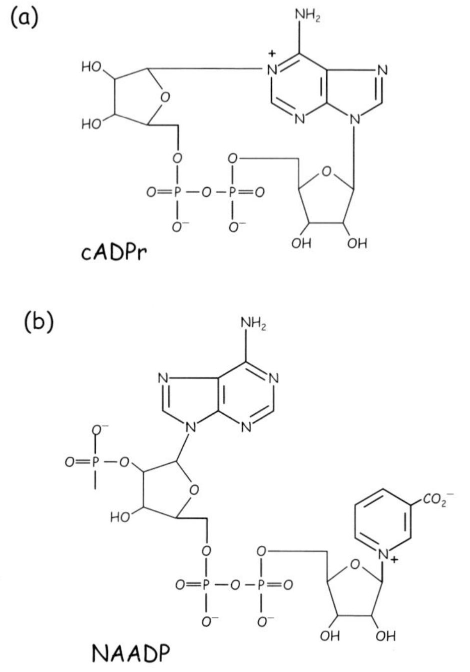

(a)

cADPr

(b)

NAADP

Figure 7.10 (a) Cyclic ADP-ribose (b) NAADP.

■ Cyclic ADP-ribose

Experimentally, cADPr introduced into cells at concentrations in the region of 10–50 nmol/l has been found to modulate Ca^{2+}-induced Ca^{2+}-release from a ryanodine-sensitive pool. Evidence for a biological role for cADPr exists for a number of mammalian cell types, but is strongest in sea urchin eggs, where it is an endogenous Ca^{2+} regulator, activated through a mechanism involving cGMP.[33] Its effects are mediated by types 2 and 3 RyR channels (but not type 1) and it is likely that the cADPr/RyR interaction is not direct, but requires an accessory protein. In the sea urchin egg, this appears to be calmodulin.

In a number of different types of mammalian cell, there is evidence that the level of cADPr rises in response to an extracellular stimulus.[34,35] For example, stimulation of T lymphocytes through their T-cell receptor complexes (see Chapter 12), activates ADP-ribosyl cyclase to produce a sustained elevation of the level of cADPr and a long-lasting Ca^{2+} signal (a series of oscillations and waves).[36] Sustained Ca^{2+} elevations may be necessary for lymphocytes to become committed to a full proliferative response.

■ Nicotinic acid adenine dinucleotide phosphate

We have seen that, in sea urchin eggs, the formation of cADPr from NAD^+ is mediated by an increase in the level of cytosolic cGMP. This may occur when guanylyl cyclase is activated by nitric oxide (see Chapter 8). The effect of cGMP is to activate a cytosolic form of ADP-ribosyl cyclase to produce cADPr. On the other hand, an elevation of the more familiar cyclic nucleotide, cAMP, causes a membrane-bound form of the enzyme to act upon $NADP^+$ to produce NAADP instead. This nucleotide has recently also been found to bring about the release of Ca^{2+} from intracellular stores and it does so at concentrations in the range 10–50 nmol/l (similar to cADPr, but compare with IP_3, Figure 7.15).

The activity of NAADP has been studied in mammalian cells as well as in sea urchin eggs but, unlike cADPr, NAADP does not modulate Ca^{2+}-induced Ca^{2+} release directly and it may have its own receptor. Importantly, NAADP also differs from cADPr in that it is self-desensitizing. Application of subthreshold levels to sea urchin egg microsomes abrogates a further response. It would seem therefore that its formation physiologically could be an initiating event, leading to the activation of Ca^{2+}-induced Ca^{2+} release at neighbouring RyRs or IP_3Rs.

In some cells, NAADP-activated Ca^{2+} release is insensitive to inhibitors (heparin and ryanodine respectively) of release induced by IP_3 or cADPr. Moreover, the stores of Ca^{2+} accessed by NAADP (and cADPr) can be spatially distinct from those accessed by IP_3. The nucleotide-based messengers tend to interact, at least initially, with cortical stores, while the IP_3-sensitive stores are more widespread and may permeate the entire cytoplasm. A general picture emerges in which cells may use different Ca^{2+}-mobilizing messengers to release Ca^{2+} from different reserves, to produce and orchestrate a variety of responses.

Sphingolipid metabolites

Another mechanism of Ca^{2+} mobilization and influx has been proposed that is quite independent of the products of phospholipase C and of the nucleotide messengers above. High affinity receptors for the constant region of certain immunoglobulins (FcεRI in mast cells, FcγRI in macrophages) initiate their signalling through recruited tyrosine kinases (see Chapter 12). A consequence is the activation of phospholipase D to produce phosphatidate and this, in turn, activates sphingosine kinase. Its substrate, sphingosine, is a product of the breakdown of sphingomyelin and it is converted to sphingosine-1-phosphate (Figure 7.11). It has been proposed that this phosphorylated lipid acts as a Ca^{2+}-mobilizing second messenger in a number of cell types. There has, however, been some confusion and controversy in this matter, in part because sphingosine-1-phosphate can also bind to G-protein-coupled receptors that communicate with PLC.[26]

Sphingosine kinase also has a role in the activation of Ca^{2+} influx by store depletion. This is through the effect of sphingosine, which can influence Ca^{2+} influx by inhibiting the opening of I_{CRAC} channels (see above).[37] Activation of sphingosine kinase can remove this inhibition and allow Ca^{2+} entry. This would augment the overall Ca^{2+} signal and also help to replenish the stores.

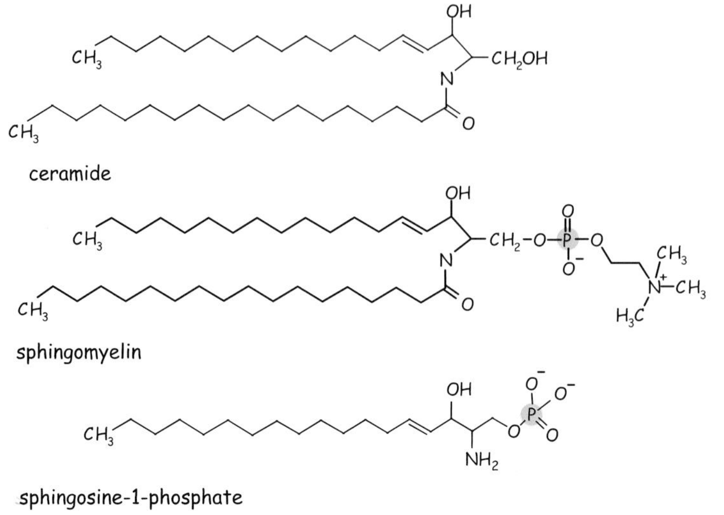

Figure 7.11 Ceramide, sphingomyelin and sphingosine-1-phosphate. Note that the acyl chains of these sphingolipids tend to be fully saturated.

Although new Ca^{2+}-mobilizing messengers are emerging, IP_3 is still the most widespread and important.

The pattern of cytosol Ca^{2+} changes in single cells

Temporal aspects

Ca^{2+}-sensing fluorescent dyes have been widely used to monitor the time-course of Ca^{2+} concentration changes in single cells under the fluorescence microscope. Following activation, the signals may appear as trains of sharp spikes or as a series of smooth oscillations (Figure 7.12). The periodicity, frequency and amplitude tend to vary with the strength of the stimulus.[38] Low concentrations, or brief exposure to an activating ligand may elicit short episodes of oscillations. These then die away and the concentration of Ca^{2+} reverts to the basal level. Stronger stimulation may evoke longer trains of waves that can merge to give a more persistent elevation.

Resolving the spatial detail

Although ions are very mobile compared with molecular solutes, the spatial distribution of Ca^{2+} within the cytosol during activation is often far from uniform. As a consequence, the interpretation of fluorescence intensity recordings is hampered by the absence of corresponding spatial information. This is a particular problem with excitable cells. Here, the speed of response is important and processes such as neurotransmitter release at nerve terminals are highly localized events. A more detailed picture can be obtained by observing single cells with a fluorescence microscope. Commonly, digital images of Ca^{2+} indicator fluorescence are obtained using a confocal laser microscope.

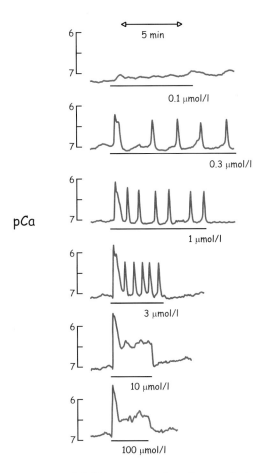

Figure 7.12 Time course of Ca²⁺ concentration changes induced by histamine in a single vascular endothelial cell. The effect of different concentrations of histamine on [Ca²] in a single human endothelial cell in the presence of 1 mmol/l Ca²⁺. The bars indicate the presence of histamine. Adapted from Jacob et al.[38]

To interpret these data in terms of intracellular Ca²⁺ levels, images are usually presented as intensity maps, using a pseudocolour display in which 'warmer' colours conventionally indicate higher Ca²⁺ levels. For example, Figure 7.13 shows the localized Ca²⁺ increases that occur in different types of cell.

■ Miniature calcium release events and global cellular signals

Digital imaging techniques have added considerable detail to our knowledge of transient Ca²⁺ signals. In excitable cells, the initial increments of Ca²⁺ tend to be confined to regions, called microdomains, that lie close to the locations where action takes place, for example in the vicinity of the triad structures in skeletal muscle or near to exocytotic release sites in neurones. Under strong stimulation, the localized changes in Ca²⁺ merge to generate a global increase, that permeates the whole cytosolic

By eliminating the out-of-focus contributions to the image from regions of the specimen lying above and below the image plane, **confocal microscopy** provides a picture of a selected slice through the specimen. The resolution and quality of such images are higher than those of a standard fluorescence microscope. Moreover, a set of such two-dimensional images, obtained by moving the plane of focus successively through the specimen, can be combined to give a three-dimensional image of cell fluorescence.

(a)

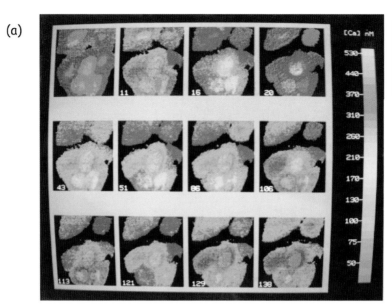

(b)

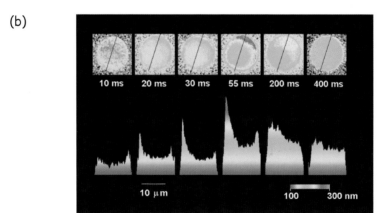

Figure 7.13 **Pseudocolour images of Ca²⁺ levels in cells activated by extracellular ligands and by plasma membrane depolarization:** (a) Ca²⁺ changes visualized in a confluent layer of human endothelial cells following stimulation with histamine (1 μmol/l). The calibration, on the right, indicates intracellular Ca²⁺ concentration. The map on the extreme right shows the individual cell boundaries. The time, in seconds, after histamine addition is indicated in each frame. There is a complex pattern of spatial and temporal changes in Ca²⁺ level within each cell. (Dye: Fura2.) Courtesy of Ron Jacob.[38] (b) Time course of Ca²⁺ changes in a single chromaffin cell, following membrane depolarization. The cell is voltage clamped in the whole cell configuration (patch pipette not shown). Six fluorescence images of a cell were collected at fixed times after a series of depolarizations, using pulsed laser imaging. The colour-coded images in the top panel show the Ca²⁺ concentration rising mostly in peripheral regions of the cell, especially on one edge (55 ms) and then falling. The bisecting line in each image indicates the region sampled to give the cross-sections shown beneath. (Dye: Rhod2.) Courtesy of Julio Hernandez and Jonathan Monck.[39]

compartment. In non-excitable cells the response tends to take the form either of propagating waves or global changes in Ca^{2+} concentration.

When these changes are visualized, a variety of optical phenomena may be observed. These range from punctate flashes, lasting tens of milliseconds (muscle cells), to broad waves that spread within seconds across non-excitable cells. Examples are shown in Figure 7.14. Waves of Ca^{2+} can also pass from cell to cell through gap junctions. In both electrically excitable and non-excitable cells, cytoplasm that is permeated by elements of the ER or SR, may form an 'excitable medium'[40] in terms of its ability to release Ca^{2+}. That is, once initiated, the process of Ca^{2+}-induced Ca^{2+}-release can produce an expanding, regenerative response.

■ Ca^{2+} signals in electrically excitable cells

The generation of transient Ca^{2+} signals in skeletal muscle cells is thought to commence with localized Ca^{2+} release at individual complexes of dihydropyridine receptors with type 1 ryanodine receptors (DHPR-RyR1 complexes, 4 DHPRs and 1 RyR1). The regions of high cytosol Ca^{2+} then become more extensive as RyRs that are not coupled to DHPRs are

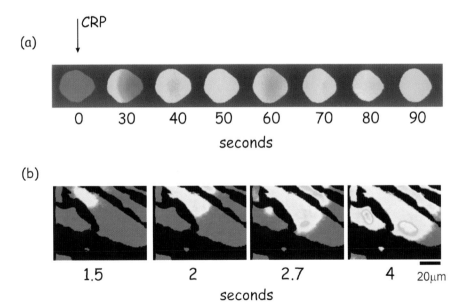

Figure 7.14 **Examples of Ca^{2+} waves in a megakaryocyte and in astrocytes in response to stimulation by soluble ligands:** (a) Time courses of Ca^{2+} levels in cells preloaded with a Ca^{2+}-sensitive fluorescent dye. Warmer colours indicate higher Ca^{2+} concentrations. (b) Activation of a single megakaryocyte in response to a collagen peptide (CRP). The increase in Ca^{2+} peaks at 50 s. This is followed by a sustained increase lasting up to 5 min (not shown). The extracellular Ca^{2+} concentration was 200 μM. The cell is approximately 40 μm in diameter. From Melford et al.[41] (b) Ca^{2+} waves in single astrocytes stimulated with ATP (20 μmol/l) which activates purinergic receptors. From Michael Duchen.[42]

activated. The Ca^{2+}-sensitivity of RyRs, and also of (type 1) IP_3Rs, is bell-shaped (Figure 7.15), so that more SR Ca^{2+} channels open as the concentration rises above the resting level. They then close again when it approaches micromolar levels. The closure prevents the stores from emptying completely and it also sets a limit to the magnitude of the Ca^{2+} signal.

Whether a membrane depolarization evokes a brief local transient (a 'spark') or a global rise depends on the strength of the stimulus. In resting skeletal muscle cells, spontaneous sparks are observed and these may represent the opening of single type 1 RyR channels. The DHPR-RyR1 complex could then be considered an 'elementary unit' of Ca^{2+} release. The frequency and size of the sparks is increased by plasma membrane depolarization as further units are recruited, leading ultimately to a global rise in Ca^{2+} concentration.[27]

The situation in cardiac muscle is somewhat different. Here, the RyRs (type 2) are not directly coupled to DHPRs. Instead, the elementary unit appears to be a loosely associated combination of a DHPR with approximately four RyR2s. The high Ca^{2+}-conductance and longer open-times of the RyR2 channels, synchronized by the plasma membrane depolarization, are then sufficient to cause a global Ca^{2+} elevation that effects rapid contraction throughout the cell. The initiation of contraction in striated muscle is considered in Chapter 8 (page 182).

■   Calcium signals in non-excitable cells

The Ca^{2+} release process in electrically non-excitable cells is generally slower than its counterpart in nerve and striated muscle. In part, this is because the IP_3R channels release Ca^{2+} into the cytosol only after they have been sensitized by IP_3. Also, the build-up of Ca^{2+} is slower because there is no equivalent of the depolarization signal that synchronizes Ca^{2+} entry and its secondary release from the internal stores of excitable cells.

Elementary or unitary Ca^{2+} release events are less conspicuous in non-excitable cells. It appears that a 'unit of release' may consist of one or more IP_3Rs clustered together. Imaging of Ca^{2+} transients reveals phenomena that range from tiny 'blips' (possibly release from single IP_3Rs), through more extensive 'puffs' (IP_3R clusters), up to full-scale regenerative calcium waves, as neighbouring channels are successively recruited. These waves can propagate across the entire cytoplasm of a cell in a few seconds (Figure 7.14).

■ Localization of intracellular second messengers

Unlike other second messengers, the spatial and temporal fluctuations of Ca^{2+} are amenable to measurement and they give the hint that similar localizations may underlie other signalling processes. The following chapter describes how some of these signals are translated into action.

■ Graded responses and Ca^{2+}-induced Ca^{2+} release

Many non-excitable cells exhibit responses that are graded according to the level of stimulation. Because cellular activation mechanisms characteristically involve amplifying enzyme cascades, the most sensitive control points will lie early in the pathway, at the receptor or the second messenger level. In processes mediated by Ca^{2+}, a low stimulus concentration can initiate release from intracellular stores in limited regions, generating isolated puffs or incomplete waves. As the concentration of the stimulus is increased, release occurs more extensively and ultimately the wave of Ca^{2+} may pervade the entire cytosol, maximizing the cellular response.

The problem is that a mechanism that relies solely on Ca^{2+}-induced Ca^{2+}-release should produce a maximal response, regardless of the degree of stimulation. A possible reason why this does not occur is because each elementary unit of Ca^{2+}-induced Ca^{2+}-release provides a quantal amount of Ca^{2+}. A graded response might then result from successive recruitment of such units, each in an all-or-none fashion, as the local IP$_3$ concentration rises.[43] However, a closer look at the IP$_3$-dependence of the Ca^{2+} sensitivity of IP$_3$Rs suggests an alternative explanation. For example, the Ca^{2+} sensitivity of type 1 IP$_3$Rs expressed in *Xenopus* eggs has been measured at different concentrations of IP$_3$.[44] In this setting, the relationship between the open probability of the release channel and Ca^{2+} concentration is, as elsewhere, bell-shaped. At low levels of IP$_3$, the probability peaks at pCa 7 (100 nmol/l, green line in Figure 7.15). However, as the level of IP$_3$ is increased (blue and red lines), the descending limb of the curve moves to the right and at the same time the maximum open probability increases. Thus the amount of Ca^{2+} released into the cytosol

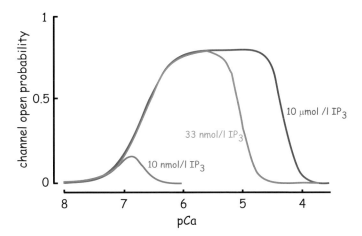

Figure 7.15 IP$_3$ and Ca^{2+}-dependence of Type 1 IP$_3$ receptor channel open probability. This graph shows the open probability of IP$_3$ receptors in the outer membrane of nuclei of *X. laevis* oocytes. Adapted from Mak *et al.*[44]

through a single channel can increase with IP$_3$ concentration and this may provide a basis for a graded response.

■ References

1   Ringer, S. Regarding the action of the hydrate of soda, hydrate of ammonia, and hydrate of potash on the ventricle of the frog's heart. *J. Physiol. Lond.* 1883: 3; 195–202.
2   Ringer, S. A further contribution regarding the influence of the different constituents of the blood on the contraction of the heart. *J. Physiol. Lond.* 1884: 4; 29–42.
3   Locke, F.S. Notiz über den Einflüss, physiologischer Kochsalzläsung auf die Errebarkeit von Muskel und Nerv. *Zent. Physiol.* 1894; 8: 166–7.
4   Mann, P.J.G., Tennenbaum, M., Quastel, J.H. Acetylcholine metabolism in the central nervous system: The effects of potassium and other cations on acetylcholine liberation. *Biochem. J.* 1939; 33: 822–35.
5   Harvey, A.M., MacIntosh, F.C. Calcium and synaptic transmission in a sympathetic ganglion. *J. Physiol. Lond.* 1940; 97: 408–16.
6   Kamada, T., Kinoshita, H. Disturbances initiated from naked surface of muscle protoplasm. *Jap. J. Physiol.* 1943; 10: 469–93.
7   Heilbrunn, L.V., Wiercinski, F.J. The action of various cations on muscle protoplasm. *J. Cell Comp. Physiol.* 1947; 29: 15–32.
8   Heilbrunn, L.V. The action of calcium on muscle protoplasm. *Physiol. Zool.* 1940; 13: 88–94.
9   Sutherland, E.W. Studies on the mechanism of hormone action. *Science* 1972; 177: 401–8.
10 Levine, B.A., Williams, R.J.P. The chemistry of calcium ion and its biological relevance. In: Anghilieri, L.J., Tuffet-Anghilieri, A.-M. (Eds) *The Role of Calcium in Biological Systems* vol. 1. CRC Press, Boca Raton, Fla., 1982; 3–26.
11 Pollack, H. Micrurgical studies in cell physiology: Calcium ions in living protoplasm. *J. Gen. Physiol.* 1928; 11: 539–45.
12 Shimomura, O., Johnson, F.H. Properties of the bioluminescent protein aequorin. *Biochemistry* 1969; 8: 3991–7.
13 Llinas, R., Blinks, J.R., Nicholson, C. Calcium transient in presynaptic terminal of squid giant synapse: detection with aequorin. *Science* 1972; 176: 1127–9.
14 Ashley, C.C. Aequorin-monitored calcium transients in single *Maia* muscle fibres. *J. Physiol. Lond.* 1969; 203: 32P–33P.
15 Tsien, R.Y., Pozzan, T., Rink, T.J. Calcium homeostasis in intact lymphocytes: cytoplasmic free calcium monitored with a new, intracellularly trapped fluorescent indicator. *J. Cell Biol.* 1982; 94: 325–34.
16 Tsien, R.Y. New calcium indicators and buffers with high selectivity against magnesium and protons: design, synthesis, and properties of prototype structures. *Biochemistry* 1980; 19: 2396–404.
17 Merritt, J.E., Rink, T.J. Rapid increases in cytosolic free calcium in response to muscarinic stimulation of rat parotid acinar cells. *J. Biol. Chem.* 1987; 262: 4958–60.
18 Samsó, M., Wagenknecht, T. Contributions of electron microscopy and single particle techniques to the determination of the ryanodine receptor three-dimensional structure. *J. Struct. Biol.* 1998; 121: 172–80.
19 Mak, D.O., McBride, S., Raghuram, V. *et al.* Single-channel properties in endoplasmic reticulum membrane of recombinant type 3 inositol trisphosphate receptor. *J. Gen. Physiol.* 2000; 115: 241–56.
20 Nowak, L., Bregestovski, P., Ascher, P., Herbet, A., Prochiantz, A. Magnesium gates glutamate-activated channels in mouse central neurones. *Nature* 1984; 307: 462–5.
21 Malenka, R.C., Kauer, J.A., Perkel, D.J. *et al.* An essential role for postsynaptic calmodulin and protein kinase activity in long-term potentiation. *Nature* 1989; 340: 554–7.
22 Bliss, T.V.P., Collingridge, G.L. A synaptic model of memory: Long-term potentiation in the hippocampus. *Nature* 1993; 361: 31–9.
23 Wacholtz, M.C., Lipsky, P.E. Anti-CD3-stimulated Ca^{2+} signal in individual human

peripheral T cells. Activation correlates with a sustained increase in intracellular Ca²⁺. *J. Immunol.* 1993; 150: 5338–49.

24 Putney, J.W.J. Capacitative calcium entry revisited. *Cell Calcium* 1990; 11: 611–24.

25 Putney, J.W.J. A model for receptor-regulated calcium entry. *Cell Calcium* 1986; 7: 1–2.

26 Schneider, M.F., Chandler, W.K. Voltage dependent charge movement of skeletal muscle: a possible step in excitation-contraction coupling. *Nature* 1973; 242: 244–6.

27 Berridge, M.J. Elementary and global aspects of calcium signalling. *J. Physiol.* 1997; 499: 291–306.

28 Meissner, G., Lu, X. Dihydropyridine receptor-ryanodine receptor interactions in skeletal muscle excitation-contraction coupling. *Biosci. Rep.* 1995; 15: 399–408.

29 Hardie, R.C., Minke, B. Calcium-dependent inactivation of light-sensitive channels in *Drosophila* photoreceptors. *J. Gen. Physiol.* 1994; 103: 409–27.

30 Hardie, R.C., Minke, B. The trp gene is essential for a light-activated Ca²⁺ channel in *Drosophila* photoreceptors. *Neuron* 1992; 8: 643–51.

31 Hoth, M., Penner, R. Calcium release-activated calcium current in rat mast cells. *J. Physiol.* 1993; 465: 359–86.

32 Kiselyov, K., Xu, K., Mozhayeva, G. *et al.* Functional interaction between InsP3 receptors and store-operated Htrp3 channels. *Nature* 1999; 396: 478–82.

33 Galione, A., Lee, H.C., Busa, W.B. Ca²⁺-induced Ca²⁺ release in sea urchin egg homogenates: modulation by cyclic ADP-ribose. *Science* 1991: 253; 1143–6.

34 Okamoto, H. The CD38-cyclic ADP-ribose signaling system in insulin secretion. *Mol. Cell. Biochem.* 1999; 193: 115–18.

35 Kato, I., Yamamoto, Y., Fujimura, M. *et al.* CD38 disruption impairs glucose-induced increases in cyclic ADP-ribose. [Ca²⁺]i, and insulin secretion. *J. Biol. Chem.* 1999; 274: 1869–72.

36 Guse, A.H., da-Silva, C. P., Berg, I. *et al.* Regulation of calcium signalling in T lymphocytes by the second messenger cyclic ADP-ribose. *Nature* 1999; 398: 70–3.

37 Mathes, C., Fleig, A., Penner, R. Calcium release-activated calcium current (ICRAC) is a direct target for sphingosine. *J. Biol. Chem.* 1998; 273: 25020–30.

38 Jacob, R., Merritt, J.E., Hallam, T.J., Rink, T.J. Repetitive spikes in cytoplasmic calcium evoked by histamine in human endothelial cells. *Nature* 1988; 335: 40–5.

39 Monck, J.R., Robinson, I.M., Escobar, A.L., Vergara, J.L., Hernandez, J. M. Pulsed laser imaging of rapid Ca²⁺ gradients in excitable cells. *Biophys. J.* 1994; 67: 505–14.

40 Lechleiter, J.D., Clapham, D.E. Molecular mechanisms of intracellular calcium excitability in *X. laevis* oocytes. *Cell* 1992; 69: 283–94.

41 Melford, S.K., Briddon, S.J., Turner, M., Tybulewicz, V.L., Watson, S.P. Syk and Fyn are required by mouse megakaryocytes for the rise in intracellular calcium induced by a collagen-related peptide. *J. Biol. Chem.* 1997; 272: 27539–42.

42 Boitier, E., Rea, R., Duchen, M.R. Mitochondria exert a negative feedback on the propagation of intracellular Ca²⁺ waves in rat cortical astrocytes. *J. Cell Biol.* 1999; 145: 795–808.

43 Bootman, M.D., Cheek, T.R., Moreton, R.B., Bennett, D.L., Berridge, M.J. Smoothly graded Ca²⁺ release from inositol 1,4,5-trisphosphate-sensitive Ca²⁺ stores. *J. Biol. Chem.* 1994; 269: 24783–91.

44 Mak, D.O., McBride, S., Foskett, J.K. Inositol 1,4,5-tris-phosphate activation of inositol tris-phosphate receptor Ca²⁺ channel by ligand tuning of Ca²⁺ inhibition. *Proc. Natl. Acad. Sci. USA* 1998; 95: 15821–52.

8

Calcium signalling

◼ Calcium binding by proteins

The evidence of the previous chapter shows that Ca^{2+} is an important second messenger in signal transduction. Now we ask how its elevation within the cytosol can activate downstream pathways. Clearly, the first step must be the binding of Ca^{2+} to specific proteins which then carry the message forwards. However, the mere fact that a protein binds Ca^{2+} does not mean that it has to be involved in signalling. Proteins with the highest affinities will always be saturated even under resting conditions and an elevation in the concentration of Ca^{2+} cannot affect them. Those with dissociation constants above 0.1 µM will certainly bind more Ca^{2+} as its concentration is elevated within the physiological range, but again, it does not necessarily follow that they will take part in signalling. Instead, their relative on and off rates may be important in shaping Ca^{2+} transients and their overall buffering capacity may help prevent excursions to very high concentrations. More interestingly, there are important signalling proteins that bind Ca^{2+} at regulatory sites and that can be activated by local or global increases in Ca^{2+} level.

◼ Polypeptide modules that bind Ca^{2+}

Common Ca^{2+}-binding sites on proteins include the EF-hand motif and the C2 domain. The structures of these are described in Chapter 18. The EF-hand motifs occur in pairs and a single motif may bind a single Ca^{2+} ion. However, their Ca^{2+}-dissociation constants vary among the proteins in which they are situated, lying anywhere between 10^{-7} and 10^{-5} mol/l. The Ca^{2+}-dissociation constants of C2 domains also vary widely (10^{-6}–10^{-3} mol/l). C2 domains and EF-hands with low Ca^{2+} affinity may represent components of systems that sense large elevations of Ca^{2+} or which, through the course of evolution, have lost their ability to detect physiological changes in Ca^{2+} concentration, for example the EF-hands of PLC (see Chapter 5). Inspection of the sequence databases reveals that close to 100 known proteins possess C2 domains (Table 8.1).

C2 domains and EF-hands are not the only sites at which Ca^{2+} ions may bind. For example, the annexins are a family of Ca^{2+}-dependent, phospholipid-binding proteins that are located on the inner surface of the plasma membrane and are associated with the cytoskeleton.[1–3] Ca^{2+} is

Table 8.1 Examples of intracellular Ca^{2+}-binding proteins of vertebrate origin

Protein	Type	Ca^{2+}-binding domains	Probable function	Special location
Calsequestrin			Ca^{2+} buffering	SR in muscle
Calreticulin			Ca^{2+} buffering	ER
Parvalbumin		EF-hand	Cytosolic Ca^{2+} buffering	Muscle/nerve
Ca^{2+}-ATPase	CaM dependent		Pumps Ca^{2+} out of cell	Plasma membrane
Ca^{2+}-ATPase (SERCA)[a]			Pumps Ca^{2+} into stores	ER membrane
Calmodulin (CaM)		EF-hand	Multipurpose Ca^{2+}-sensing mediator	
Troponin C		EF-hand	Ca^{2+}-sensing mediator of contraction	Striated muscle
Ca^{2+}/calmodulin kinase II	CaM is a regulatory subunit		Multipurpose signalling	
Myosin light chain kinase	Ca^{2+}/CaM dependent		Phosphorylates myosin II	Smooth muscle
Adenylyl cyclase (I, III, VIII)	Ca^{2+}/CaM dependent		Makes cyclic AMP	
Cyclic nucleotide phosphodiesterase (1A-C)	Ca^{2+}/CaM dependent		Breaks down cyclic AMP	
Phosphorylase b kinase	CaM is a regulatory subunit		Phosphorylates glycogen phosphorylase	Skeletal muscle
Recoverin	Ca^{2+}-myristoyl switch	EF-hand	Ca^{2+}-sensing mediator	Photoreceptor cells
Calpain		EF-hand	Protease	
α-Actinin		EF-hand	Cytoskeleton	
Gelsolin			Actin severing and capping	
Synaptotagmin	Putative Ca^{2+} sensor	C2	Signalling	Secretory cells
Calcineurin (protein phosphatase 2B)	Ca^{2+}/CaM dependent	EF-hand	Signalling, e.g. transcription	
Protein kinase C (α, β1, β2, γ)		C2	Signalling	
Phospholipase C (all isoforms)		EF-hand, C2	Signalling	
Diacylglycerol kinase		EF-hand		
Nitric oxide synthase	CaM is a regulatory subunit		Production of NO for signalling	

[a] sarco(endo)plasmic reticulum Ca^{2+} ATPase.

bound at loops in a structure termed the endonexin fold. Other Ca^{2+}-binding proteins that lack C2 domains and EF-hands may be found among the wide range of channels and ATPases that conduct Ca^{2+} ions across membranes.

◼ Effects of elevated calcium

Ca^{2+} can activate a wide range of Ca^{2+}-sensitive regulatory enzymes; a selection of these effectors and transducers is listed in Table 8.1. The waves and oscillations of Ca^{2+} in a stimulated cell may activate specific downstream targets that respond within a particular range of Ca^{2+} concentrations or that require periodic stimulation. For instance, transcription factors that control gene expression and that are activated through calcineurin (see below), respond not only at particular Ca^{2+} concentrations, but also to particular oscillation frequencies.[4,5]

Because of the temporal fluctuations, the target proteins must be able to sense and respond to short-term or local increases in the concentration of Ca^{2+} before it subsides back to the resting level. For signalling mechanisms in which Ca^{2+} rises and falls rapidly, it is not only the stability constant of the binding that is important, but also the rate of the 'on' and 'off' reaction. The forward reaction requires the displacement of water molecules from the cation and these are removed one at a time. In general, if a multidentate coordinating ligand is to associate with or dissociate from a cation rapidly, then the ligand framework needs to be flexible. The evolution of such sites in proteins has resulted in Ca^{2+}-activated regulatory enzymes that can bind and respond very rapidly to changes in Ca^{2+} concentration. It is conceivable then, that temporally coded signals may be timed so that a key set of Ca^{2+}-binding effectors with low effective off-rates keep their bound Ca^{2+} during a down-swing in its concentration, while those with high off-rates may lose it.

Calmodulin and troponin C

Calmodulin

Calmodulin and its isoform troponin C of skeletal muscle, represent the major Ca^{2+}-sensing proteins in animal cells. Calmodulin itself is a highly conserved protein, present at significant levels in all eukaryotic cells. In mammalian brain it comprises about 1% of the total protein content. While the effects of cAMP are mostly conveyed through the regulatory subunit of protein kinase A (PKA), Ca^{2+}, through its interaction with calmodulin, can cause the activation of more than 100 different enzymes. PKA has multiple substrates, but the individual Ca-calmodulin activated enzymes tend to have a more restricted specificity and are not necessarily themselves Ca^{2+}-sensitive (Figure 8.1). Either way the outcome is similar: each second messenger is linked to multiple targets.

Calmodulin first came to light as the companion of cyclic nucleotide phosphodiesterase.[6,7] It is an acidic protein of modest size (17 kDa), consisting of a single, predominantly helical polypeptide chain with four Ca^{2+}-binding EF-hands, two at each end (see Figure 8.3). The affinities of the individual Ca^{2+} binding sites are in the range 10^{-6}–10^{-5} mol/l and adjacent sites bind Ca^{2+} with positive cooperativity so that the attachment of the first Ca^{2+} ion enhances the affinity of its neighbour. This has the effect of making the protein sensitive to small changes in the concentration of Ca^{2+} within the signalling range. Ca^{2+}-calmodulin itself has no intrinsic catalytic activity. Its actions depend on its close association with a target enzyme, in some instances, as in phosphorylase kinase and calcineurin, acting as a permanent component of a multisubunit complex. In the absence of bound Ca^{2+} the central helix is shielded by the terminal helices and the protein is unable to interact with its targets (Figure 8.2). Binding of Ca^{2+} to the four sites induces a conformational change causing the terminal regions to expose hydrophobic surfaces and also exposing the central α-helical segment (Figure 8.3).[8] Ca^{2+}-bound calmodulin binds to its targets with high affinity (K_D ~10^{-9} mol/l). To form the bound state, the central residues of the link region unwind from their α-helical arrangement to form a hinge that allows the molecule to bend and wrap itself

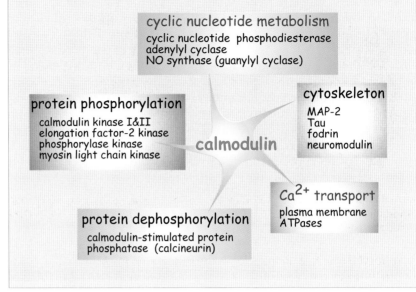

Figure 8.1 **Multiple signal transduction pathways initiated by calmodulin.** Calmodulin bound to Ca²⁺ interacts and activates many enzymes, opening up a wide range of possible cellular responses. Abbreviations: MAP-2, microtubule associated protein-2; NO, nitric oxide; Tau, tubulin assembly unit.

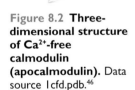

Figure 8.2 **Three-dimensional structure of Ca²⁺-free calmodulin (apocalmodulin).** Data source 1cfd.pdb.[46]

Figure 8.3 **Three-dimensional structure of calmodulin with bound Ca²⁺ ions.** Data source 1cll.pdb.[45]

around the target protein. The N- and C-terminal regions approach each other and by their hydrophobic surfaces bind to it, rather like two hands holding a rope. This encourages an α-helical arrangement of the target sequence so that it that occupies the centre of a hydrophobic tunnel (Figure 8.4). The consequence of this interaction is a conformational change in the target protein. This state persists only as long as the Ca²⁺ concentration remains high. When it falls, the bound Ca²⁺ dissociates and calmodulin is quickly released, inactivating the target. However, at least one important target protein is an exception to this rule. This is CaM-kinase II, which can remain in an active state once has been activated by calmodulin (see below). Selected calmodulin-binding proteins are now described.

Troponin C

Troponin C is effectively an isoform of calmodulin. It is present in striated muscle where it regulates the interaction between actin and myosin. Like calmodulin it possesses two pairs of Ca²⁺-binding EF-hands located at opposite ends of a peptide chain. The affinities of these sites for Ca²⁺ lie between 10^{-5} and 10^{-7} mol/l. Its function is described below (page 182).

Ca²⁺/calmodulin-dependent kinases

Within the class of enzymes controlled by calmodulin are the Ca²⁺/calmodulin-dependent kinases (CaM-kinases). These enzymes include

Figure 8.4 Structure of calmodulin bound to a peptide corresponding to the calmodulin-binding domain of smooth muscle myosin light chain kinase. Data source 2bbm.pdb.[47]

phosphorylase kinase, myosin light chain kinase (MLCK), which activates smooth muscle contraction, and the CaM-kinases I–IV. Each of these interacts with calmodulin to convert a Ca^{2+} signal into a phosphorylation signal.

Multifunctional Ca^{2+}/calmodulin activated protein kinases

On the basis of *in vitro* phosphorylation assays, these enzymes have been termed multifunctional, suggesting that each member can phosphorylate a range of different physiological targets. Whether this multifunctionality, observed *in vitro*, extends to signalling events *in vivo* has been questioned. However, the most prominent member of the family, CaM kinase II, is certainly a true broad-spectrum serine/threonine kinase. It is widely expressed, with particularly high levels in brain. On activation, it undergoes autophosphorylation and this is why it remains active (or 'autonomous'), even after the Ca^{2+} signal has been withdrawn. Eventually the action of phosphatases terminates the process. This 'latching' behaviour is thought to underlie a role for CaM kinase II in learning and memory.

Phosphorylase kinase and glycogen synthase kinase

Phosphorylase kinase was the first protein kinase to be discovered.[9] It is controlled by protein kinase A under β-adrenergic stimulation and it is also activated through the effect of Ca^{2+} on calmodulin (Figure 8.5). Indeed, calmodulin itself comprises the δ-subunit of phosphorylase kinase.[10] To reinforce the consequent glycogenolysis, glycogen synthesis is simultaneously inhibited through the action of Ca^{2+} and calmodulin on glycogen synthase kinase, which phosphorylates and inactivates glycogen synthase (Figure 8.5). These themes will be developed further in Chapters 9 and 17, where we shall see that phosphorylase kinase is also activated by phosphorylation mediated by the second messenger cAMP.

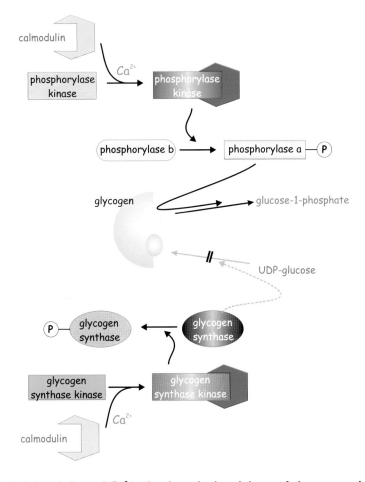

Figure 8.5 Calmodulin and Ca²⁺ stimulate the breakdown of glycogen and inhibit its synthesis in skeletal muscle.

■ Other Ca^{2+}-calmodulin dependent enzymes

▨ *Ca^{2+}-calmodulin-sensitive adenylyl cyclase and phosphodiesterase*

Two contrasting themes are apparent in the expression of the signalling pathways initiated by Ca^{2+} and cAMP. These two second messengers may either operate towards opposite outcomes or they may complement each other and act towards a similar goal. The activation of phosphorylase kinase through either PKA or Ca^{2+} is an example of such convergence. Ca^{2+}- and cAMP-mediated pathways may also be coordinated through Ca^{2+}-calmodulin dependent isoforms of adenylyl cyclase (such as type I, a predominant isoform in brain: see Chapter 5). Thus a rise in Ca^{2+} can strongly promote the formation of cAMP. Following this, cyclic nucleotide phosphodiesterase, which breaks down cAMP to 5′-AMP (and cyclic GMP to 5′-GMP), is also activated by Ca^{2+}-calmodulin. Therefore in brain, Ca^{2+} entering a nerve cell may first act to generate cAMP and then, as cytosolic calmodulin-dependent phosphodiesterase becomes activated, break it down. The overall effect is a brief pulse of cAMP.

Calcineurin

Calcineurin is a calmodulin-dependent Ser/Thr phosphatase. It exists as a heterodimer in mammalian cells and it has a number of targets (Figure 8.6). The catalytic subunit, calcineurin A, is also familiar as protein phosphatase 2B (see Chapter 17). The regulatory subunit, calcineurin B, possesses four EF-hands and binds Ca^{2+} with high affinity, but Ca^{2+}-calmodulin is still required as an additional regulatory subunit for the activation of phosphatase activity.

The immunosuppressive drugs cyclosporin and FK506, which are used to prevent organ rejection after transplant surgery, bind to cytosolic immunophilins (cyclophilin and FKBP12 respectively) and inhibit the action of calcineurin. In this way, they prevent the activation of T lymphocytes as discussed in Chapter 17. Note, however, that immunophilins are not restricted to cells of the immune system. They are abundant in other cells as components of ryanodine and IP_3 receptor complexes (see Chapter 7). In nerves, they also influence neuronal growth and repair.[11] In this context, their actions are also calmodulin-dependent, but they do not require elevated Ca^{2+} nor is calcineurin implicated in their function.

Nitric oxide synthase

One of the more surprising discoveries of the last decade has been that nitric oxide (NO), familiar as a noxious gas, has an important role as an *inter*cellular messenger. Nitric oxide was first perceived as endothelium-derived relaxing factor (EDRF) in the vascular system.[12] It is formed by the oxidation of L-arginine by the haem protein, nitric oxide synthase (NOS). There are three members of the NOS family: the membrane-bound endothelial enzyme called eNOS (or NOS III), a soluble enzyme first

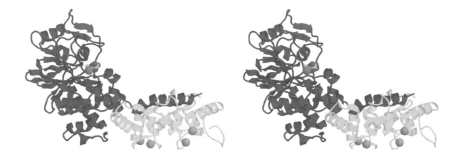

Figure 8.6 Stereo diagram of the structure of the calcineurin heterodimer. The regulatory subunit, calcineurin B (cyan chains), resembles calmodulin with EF-hand binding sites for four Ca^{2+} ions (grey spheres). The catalytic subunit, calcineurin A or protein phosphatase 2B (blue chains) contains a Zn^{2+} (lilac) and an Fe^{3+} ion (purple) at the active site, where the target phosphate is bound. A molecule of calmodulin also forms part of the complex, but is not shown here. Instructions for viewing stereo-images can be found on page 396. Data source: 1tc0.pdb.[48]

characterized in brain called nNOS (or NOS I) and in macrophages a cytokine-inducible form called iNOS (or NOS II).[13] nNOS is most apparent in nerve and skeletal muscle, though not restricted to these cell types. eNOS is also found in a wide variety of cells. Both eNOS and nNOS are Ca^{2+}/calmodulin-dependent enzymes and produce NO in response to a rise in cytosol Ca^{2+}. By contrast, the transcriptionally regulated iNOS binds calmodulin at resting Ca^{2+} levels and is constitutively active. It has high output and, for example in macrophages, it can remain activated for many hours, maximizing the cytotoxic damage that these cells can inflict.

NO is by no means a classical messenger molecule. It diffuses rapidly in solution and, because it is short-lived *in vivo*, it has mostly short range (paracrine) effects.[14] It can readily cross cell membranes to reach neighbouring cells and these can be activated without the need for plasma membrane receptors. A wide range of cells are affected by NO. For example, as mentioned above, it is a potent smooth muscle relaxant, not only in the vasculature, but also in the bronchioles, gut and genito-urinary tract. In intestinal tissue, nNOS in varicosities of myenteric neurones is activated by an influx of Ca^{2+}. The NO formed diffuses into neighbouring smooth muscle cells, where at nanomolar concentrations it activates a soluble guanylyl cyclase (also a haemprotein) that forms cyclic GMP.[15,16] This then activates cGMP-dependent kinase I (G kinase) which brings about relaxation by interacting with the mechanisms that regulate the cytosol Ca^{2+}, essentially keeping it low.[17] In the cardiovascular system, eNOS in endothelial cells responds to Ca^{2+} in a similar way, to relax vascular smooth muscle and reduce blood pressure.[18,19] (Note: as well as stimulating vasodilatory responses, NO has a homeostatic role in the regulation of vascular tone.)

In the central nervous system, nNOS is tethered close to NMDA-type glutamate receptors (mentioned in Chapters 3 and 7) so that it can respond to the transient, but intense increases in Ca^{2+} concentration that occur in the vicinity of the open channels.[20] The NO that is formed has the potential of acting as either an anterograde or retrograde messenger.[21] Because of its diffusibility, it may also influence other cells in the vicinity such as glial cells. All this may have implications for synaptic plasticity (the modulation of synaptic potency).[22]

An important downstream consequence of NOS activation is the effect of NO on cellular Ca^{2+} homeostasis.[23] Many of the effects of NO are mediated by cGMP and the consequent activation of G kinase. This can phosphorylate and inactivate PLC and IP_3 receptors,[24] and modulate SOCs.[25] It also inhibits Ca^{2+} release from intracellular stores in a variety of cells[26] (but not endothelial cells or liver). Finally, in sea urchin eggs the generation of cyclic ADP-ribose (see Chapter 7), an activator of ryanodine receptors in these cells, is cGMP-dependent.[27] How these actions are coordinated in the control of Ca^{2+} concentration is not fully understood.

■ Calcium-dependent enzymes that are not regulated by calmodulin

Calmodulin is a key Ca^{2+} sensor and mediator of intracellular enzymes that are not in themselves Ca^{2+}-binding proteins, but there are also many enzymes that can respond to changes in Ca^{2+} directly. To fulfil this function, they must possess high-affinity binding sites and these may be in the form of EF-hand motifs, C2 domains or other structures. Yet other cellular Ca^{2+}-binding proteins serve as Ca^{2+} buffers within the cytosol (high-affinity binding) and within the lumen of the endoplasmic reticulum (low-affinity binding) where they increase its storage capacity. We consider some specific examples.

Calpain

Calpain is a member of a widely distributed family of cytosolic, Ca^{2+}-activated cysteine proteases, possessing EF-hand sites.[28] It cleaves a wide range of intracellular proteins modifying their functions, often destructively. It operates in cell death pathways and is involved in neurodegeneration and apoptosis. It also degrades cytoskeletal and other proteins in the vicinity of the plasma membrane and is thought to be involved in processes where remodelling of the cytoskeleton takes place.

Although selective, the effects of this neutral protease are mostly irreversible, which is perhaps one reason why cells do not permit Ca^{2+} levels to remain high for a prolonged period. To make additionally sure, calpain is also held in check by calpastatin.[28] This not only inhibits its activity, but also prevents it from binding to membranes.

Synaptotagmin

In neurones and in many types of endocrine cell, the release of a neurotransmitter or hormone by exocytosis is activated by an increase in intracellular Ca^{2+}.

This event is mediated by a Ca^{2+}-sensitive protein (or proteins). Among the numerous proteins that are evident on the surface of secretory vesicles and secretory granules is synaptotagmin.[29,30] This is a highly conserved transmembrane protein having two C2 domains in its cytosolic chain and it is thought to be the Ca^{2+} sensor for exocytosis. Indeed it has been shown to bind Ca^{2+} ions at physiological concentrations. This is discussed further below.

Exocytosis The fusion of the membrane of a secretory vesicle with the cell plasma membrane, allowing the vesicle contents to be released to the exterior, without affecting the integrity of the plasma membrane.

DAG kinase

The hydrophobic second messenger diacylglycerol (DAG) formed by the phospholipases PLC and (indirectly by) PLD (see Chapter 5) is the activator of protein kinase C (PKC, see Chapter 9). It is short-lived because it is rapidly phosphorylated to form phosphatidate by diacylglycerol kinase (DGK). This reaction is part of the cycle that regenerates phosphatidyl-inositol and since it removes DAG, it acts to terminate the activation of PKC (Figure 8.7). Of the eight known isoforms of DGK, α, β and γ possess EF-hand motifs and are Ca^{2+}-dependent.[31]

Recoverin

Recoverin is a Ca^{2+} sensor important in vision. It takes its name because it promotes recovery of the dark state. As described in Chapter 6, Ca^{2+}-bound recoverin inhibits rhodopsin kinase and so regulates the phosphorylation

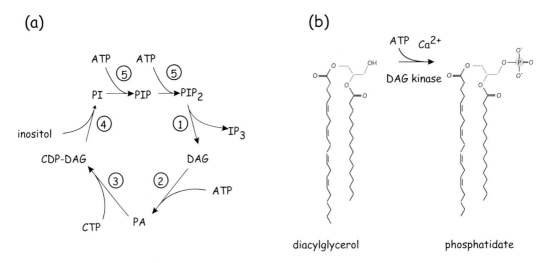

Figure 8.7 The reactions of the inositol lipid cycle: diacylglycerol kinase terminates PKC activation: (a) The enzymes are as follows: 1, phospholipase C; 2, diacylglycerol kinase; 3, CDP-diacylglycerol synthase; 4, phosphatidylinositol synthase; 5, phosphatidylinositol kinases. (b) The structures of diacylglycerol and phosphatidate.

state of the vertebrate photoreceptor, rhodopsin. Recoverin operates as a Ca^{2+}-myristoyl switch.[32] Of the four EF-hands, only two bind Ca^{2+} and the presence of a myristoyl group at the N-terminus promotes cooperative Ca^{2+} binding at the two sites (apparent K_D = 17 µmol/l). A model for the operation of the switch proposes that recoverin can exist in two states, one in which the hydrophobic myristoyl group is sequestered within the protein and the other in which it is extruded.[33,34] Ca^{2+} binds ~10000 times more tightly to the latter form and the exposure of the myristoyl group enables it to bind to the retinal disc membrane, where it extends the life-time of photo-excited rhodopsin (see Chapter 6).

Cytoskeletal proteins

Cytoskeletal proteins are responsible for the maintenance of cell shape and for motile functions. Almost every form of cellular activation is either preceded, accompanied or followed by a rearrangement of at least part of the cytoskeleton. Such a change can be an essential or even defining component of the cellular response. In non-muscle cells, the cytoskeletal arrangement of microfilaments (F-actin) is controlled by a large array of proteins, some of which bind actin.[35] Some of these, for example α-actinin and gelsolin, are sensitive to the concentration of cytosol Ca^{2+}. These proteins have various effects on the cytoskeleton. α-actinin is a cross-linker, whereas gelsolin is a Ca^{2+}-regulated actin-severing (and capping) protein. However, the cross-linking action of α-actinin (which has two EF-hands) is inhibited by Ca^{2+} and its consequent removal from F-actin allows access for gelsolin.

Paradigms of calcium signalling

Triggering neurotransmitter secretion

In the animal kingdom, in a world of predators and prey, the speedy transmission of neural signals is crucial. The exchange of information between nerve cells (and between nerve and muscle) involves the release, by exocytosis, of neurotransmitter substances at synapses. Ca^{2+} plays a critical role in mediating this process. Highly localized voltage-sensitive channels in the plasma membrane of the presynaptic cell admit Ca^{2+} directly into the active zone. Here secretory vesicles containing neurotransmitter are located, primed to undergo exocytosis. Secretion is triggered rapidly. After channel opening, it can occur in less than a millisecond. Within the presynaptic cell, the sites of exocytosis lie on average only 50 nm from the Ca^{2+} channels and the concentration of Ca^{2+} in this confined region may rise as high as 0.1 mmol/l, but cytosolic Ca^{2+} buffers and transport mechanisms prevent a widespread Ca^{2+} increase, ensuring that its action is local and transient. The resting level is resumed within a few tens of milliseconds.

The mechanism by which the brief elevation of Ca^{2+} leads to the fusion of the membrane of the synaptic vesicle with the plasma membrane is incompletely understood, but it clearly involves one or more Ca^{2+}-sensing proteins, having rather low affinity coupled with fast binding kinetics. The speed of binding is demonstrated by the inability of the chelator EGTA to inhibit secretion, when injected into the presynaptic cell to prevent a rise in Ca^{2+} concentration.[36] EGTA has a relatively slow forward rate of binding, presumably less than that of the intracellular Ca^{2+} sensor. The identity of this protein (or proteins) has not been finally established, but a candidate for the task is synaptotagmin I, a transmembrane protein of synaptic vesicles. Its single cytosolic chain consists of two consecutive C2 domains, C2A and C2B, linked by a short sequence. The C2A domain binds to acidic phospholipids in a Ca^{2+}-dependent manner[37] and this may help the protein to become tethered to the plasma membrane (see Chapter 18, page 402). The C2B domain tends to self-associate when Ca^{2+} is elevated, so that synaptotagmin may form oligomeric structures.

To add to this complexity, synaptotagmin associates with components of the so-called SNARE complex of proteins in Ca^{2+}-dependent secretory cells (reviewed by Brunger[38]). This intricate array of vesicle, plasma membrane and cytosolic proteins forms tightly bound complexes that, it is thought, may draw the vesicle and plasma membranes together to enable them to fuse to form the pore that initiates exocytosis. Self-associating molecules of synaptotagmin might contribute to this scaffold (or they may even act independently). At any rate, it is only when Ca^{2+} binds to synaptotagmin (synaptotagmin I in nerve cells), that fusion of the two membranes to form the initial fusion pore that precedes exocytosis occurs. In support of this, the C2A domain of synaptotagmin also binds to syntaxin and the C2B domain to SNAP25 (both plasma membrane SNARE proteins) and, interestingly, the phospholipids that bind most strongly to both C2 domains are the polyphosphoinositides, PI-4,5-P_2 and PI-3,4,5-P_3. All this provides the elements of a system for linking protein chains to membrane surfaces and to each other in a Ca^{2+}-dependent manner, but how it works and if the mechanism is universal is far from clear. To complicate matters, there are at least 13 isoforms of synaptotagmin, many of which are not Ca^{2+} sensitive. Furthermore, there are types of secretory cell that can be triggered to undergo exocytosis in the effective absence of Ca^{2+}.

■ Initiation of contraction in striated muscle

The contraction of all types of muscle depends on an increase in intracellular Ca^{2+}. In vertebrate striated muscle, rapid mechanisms have evolved that are mediated by transient increases in cytosol Ca^{2+}. The operating units are complexes of three different membrane proteins, two on the plasma membrane and one on the membrane of the sarcoplasmic reticulum (SR). At the skeletal muscle plasma membrane (sometimes called the sarcolemma) nicotinic acetylcholine receptors are closely associated with the Ca^{2+} channels that are voltage-gated. The third protein in the complex is a ryanodine receptor (RyR). This is resident on the membrane of the SR which is very close to the plasma membrane in the vicinity of the T-tubules, in characteristic structures called junctional triads. The ryanodine receptors are visible in electron micrographs of triads as so-called 'junctional feet' (Figure 8.8).

When the released acetylcholine binds to nicotinic receptors on the muscle endplate, the integral, non-selective cation channels open, the membrane depolarizes and the change in potential causes voltage-operated Ca^{2+} channels (VOCCs) to open. These admit Ca^{2+} into the immediate vicinity of the RyR and, although this may bring about the release of Ca^{2+} from the SR by Ca^{2+}-induced Ca^{2+} release (CICR), there seems to be a direct interaction between the type 1 RyRs and part of the neighbouring VOCC, the dihydropyridine receptor (DHPR). The result is substantial Ca^{2+} release from the SR (figure 7.9) and a steep rise in cytosolic Ca^{2+} concentration in the close vicinity of the contractile machinery. This tight localization ensures a rapid response and has the further advantage that the resting state can be rapidly resumed, since only a small quantity of Ca^{2+} ions have to be removed. Heart muscle possesses type 2 RyRs and lack direct coupling between VOCCs and RyRs, so that Ca^{2+}-induced Ca^{2+} release is the predominant mechanism of Ca^{2+} mobilization.

Within each skeletal muscle cell, the contractile machinery of actin and myosin filaments is controlled by the proteins tropomyosin and troponin. The attachment of myosin heads to actin, which causes contraction, is prevented by the presence of threads of tropomyosin organized on the surface of the actin filaments. Troponin, a complex of three subunits, I, T and C, is distributed at intervals along the actin filaments and it acts to mediate the effects of Ca^{2+}. When elevated, Ca^{2+} binds to troponin C, initiating conformational changes in the troponin complex so that the inhibition of the myosin ATPase activity is lifted. This allows contraction to occur. As the concentration of Ca^{2+} returns to the resting level, inhibition by tropomyosin is re-established and contraction ceases. Troponin C,

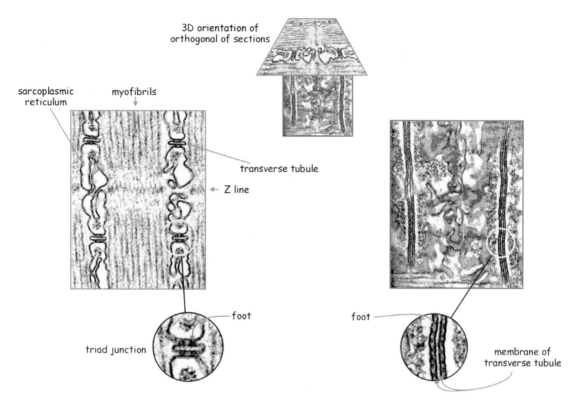

Figure 8.8 **Triad junctions and junctional feet.** Electron micrographs of longitudinal sections of striated swimbladder muscle from the toad fish (*Opsanus tau*). Orthogonal sections cut perpendicular and parallel to the T-tubule axis are shown and the junctional 'feet', which are ryanodine receptors, are indicated. Micrographs courtesy of Clara Franzini-Armstrong.[39]

which has a structure that closely resembles calmodulin, possesses four EF-hand Ca^{2+} binding sites, two in the C-terminal 'structural' lobe, that are of high affinity ($K_D \sim 10^{-7}$ mol/l), and two in the N-terminal or 'regulatory' lobe of low affinity ($K_D \sim 10^{-5}$ mol/l). (The C-terminal sites can bind Ca^{2+} or Mg^{2+}.) Figure 8.9 shows the structure of troponin C with two Ca^{2+} ions bound and in association with a fragment of troponin I (residues 1–47). When the concentration of Ca^{2+} rises, it binds to the low affinity N-terminal sites and a conformational change occurs that strengthens the association with troponin C_{1-47}. It also interacts with a regulatory segment of troponin I (96–127) which leads to the lifting of the inhibition of contraction by tropomyosin.[40]

The speed with which these events occur can be very fast. For example, the muscles that surround the swimbladder of the toadfish can twitch at frequencies in excess of 100 Hz, enabling it to emit a characteristic 'boat-whistle' sound. Humming birds flap their wings at about 80 times per second, but this can rise to 200 in courtship flight. A failure to clear Ca^{2+} between successive stimuli (depolarizations) will cause a steady contraction (tetanus). The achievement of such high frequency contractions

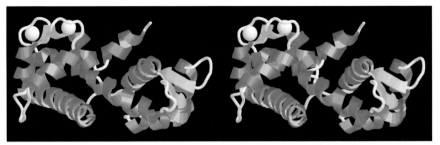

Figure 8.9 Troponin C interacts directly with troponin I. Stereoscopic view of the three-dimensional structure of troponin C interacting with a fragment (1–47) of troponin I (purple helix). Although structurally similar to calmodulin, the two C-terminal Ca^{2+} binding sites of troponin C are occupied (grey spheres) at resting Ca^{2+} concentrations and the molecule adopts a compact structure that forms hydrophobic contacts with troponin I_{1-47}. When the Ca^{2+} binding sites in the N-terminal region become occupied, further interactions take place with troponin I_{1-47} and also with troponin I_{96-127} (not shown).

without fusion requires both fast generation and the fast removal of Ca^{2+}. The transience of the individual Ca^{2+} signals then calls for particularly rapid on *and* off rates of Ca^{2+} binding to troponin C.

■ Smooth muscle contraction

Compared with skeletal and cardiac muscle, smooth muscle contraction is a slow process. The enormous variability that exists between different smooth muscle cells makes it difficult to draw general conclusions concerning their calcium signalling mechanisms. They use both depolarization-dependent and agonist-dependent mechanisms.[41] Some smooth muscle cells in arteries and veins, display a mechanism similar to that in the heart where entry of calcium through VOCCs is further amplified by Ca^{2+}-induced Ca^{2+} release from internal stores.[42–44] In contrast to cardiac cells, however, these smooth muscle cells operate a low gain mechanism so that the entry of external calcium contributes a larger proportion of the global calcium signal. In addition to this depolarization-dependent mechanism, many smooth muscle cells also employ an agonist-dependent process which uses IP_3 to release calcium from internal stores.[41] Calcium signals are generated by both IP_3Rs and RyRs and there are indications that these two release channels might cooperate with each other in some smooth muscle cells. In smooth muscle there is no troponin. The contraction of actomyosin is achieved through the phosphorylation of the P chain of myosin by myosin light chain kinase (MLCK) and this is activated through an interaction with calmodulin and elevated Ca^{2+}. Note: calmodulin-dependent kinase also phosphorylates MLCK in skeletal muscle but here it does not cause contraction. Instead it appears to affect the force of contraction.

■ Adrenergic control of contraction in the heart

Interaction of signalling pathways in cardiac myocytes

The mammalian fight or flight reaction includes stimulation of the heart through β-adrenergic receptors. Catecholamines acting at β_1 and β_2 receptors bring about an elevation of cAMP and activation of PKA to phosphorylate VOCCs. This leads to greater Ca^{2+} entry and a consequent increase in the force of contraction during systole. Relaxation (diastole) is equally important and efficient clearance of the accumulated Ca^{2+} in the cytosol requires efficient pumping by the Ca^{2+}-ATPases.

Another protein that is phosphorylated by PKA following the stimulation of β-receptors is the SR membrane protein phospholamban. In its phosphorylated form its inhibition of the SR Ca^{2+}-ATPase is lifted. This has the effect of increasing the rate of relaxation by enhancing the rate of removal of Ca^{2+} from the cytosol. Failure to remove Ca^{2+} efficiently, for example as a result of defective Ca^{2+}-ATPases or associated proteins such as phospholamban, can lead to heart failure.

Systole Contraction of the ventricles and the expulsion of blood into the arteries.
Diastole Relaxation and filling of the ventricles.

References

1 Tzima, E., Trotter, P.J., Orchard, M.A., Walker, J.H. Annexin V binds to the actin-based cytoskeleton at the plasma membrane of activated platelets. *Exp. Cell. Res.* 1999; 251: 185–93.

2 Pollard, T.D., Almo, S., Quirk, S., Vinson, V., Lattman, E.E. Structure of actin binding proteins: insights about function at atomic resolution. *Annu. Rev. Cell Biol.* 1994; 10: 207–49.

3 Thiel, C., Osborn, M., Gerke, V. The tight association of the tyrosine kinase substrate annexin II with the submembranous cytoskeleton depends on intact p11- and Ca$(^{2+})$-binding sites. *J. Cell Sci.* 1992; 103: 733–42.

4 Dolmetsch, R.E., Xu, K., Lewis, R.S. Calcium oscillations increase the efficiency and specificity of gene expression. *Nature* 1998; 392: 933–6.

5 Li, W.-H., Llopis, J., Whitney, M., Zlokarnik, G., Tsien, R.Y. Cell-permeant caged InsP3 ester shows that Ca^{2+} spike frequency can optimize gene expression. *Nature* 1998; 392: 936–41.

6 Cheung, W.Y. Cyclic 3',5'-nucleotide phosphodiesterase. Demonstration of an activator. *Biochem. Biophys. Res. Commun.* 1970; 38: 533–8.

7 Kakiuchi, S., Yamazaki, R. Calcium dependent phosphodiesterase activity and its activating factor (PAF) from brain studies on cyclic 3',5'-nucleotide phosphodiesterase. *Biochem. Biophys. Res. Commun.* 1970; 41: 1104–10.

8 Babu, Y.S., Sack, J.S., Greenhough, T.J. *et al.* Three-dimensional structure of calmodulin. *Nature* 1985; 315: 37–40.

9 Cori, G.T., Cori, C.F. The enzymatic conversion of phosphorylase a to b. *J. Biol. Chem.* 1945; 158: 321–32.

10 Cohen, P., Burchell, A., Foulkes, J.G., Cohen, P.T. Identification of the Ca^{2+}-dependent modulator protein as the fourth subunit of rabbit skeletal muscle phosphorylase kinase. *FEBS Lett.* 1978; 15: 287–93.

11 Snyder, S.H., Sabatini, D.M., Lai, M.M. *et al.* Neural actions of immunophilin ligands. *Trends Pharm. Sci.* 1998; 19: 21–6.

12 Palmer, R.M., Ferrige, A.G., Moncada, S. Nitric oxide release accounts for the biological activity of endothelium-derived relaxing factor. *Nature* 1987; 327: 524–6.

13 Andrew, P.J., Mayer, B. Enzymatic function of nitric oxide synthases. *Cardiovasc. Res.* 1999; 43: 521–31.

14 Balligand, J.L., Cannon, P.J. Nitric oxide synthases and cardiac muscle. Autocrine and paracrine influences. *Arterioscler. Thromb. Vasc. Biol.* 1997; 17: 1846–58.

15   Rodgers, K.R. Heme-based sensors in biological systems. *Curr. Opin. Chem. Biol.* 1999; 3: 158–67.

16   Foster, D.C., Wedel, B.J., Robinson, S.W., Garbers, D.L. Mechanisms of regulation and functions of guanylyl cyclases. *Rev. Physiol. Biochem. Pharm.* 1999; 135: 1–39.

17   Lucas, K.A., Pitari, G.M., Kazerounian, S. *et al.* Guanylyl cyclases and signaling by cyclic GMP. *Pharm. Rev.* 2000; 52: 375–414.

18   Huang, P.L., Huang, Z., Mashimo, H. *et al.* Hypertension in mice lacking the gene for endothelial nitric oxide synthase. *Nature* 1995; 377: 239–42.

19   Gyurko, R., Kuhlencordt, P., Fishman, M.C., Huang, P.L. Modulation of mouse cardiac function in vivo by eNOS and ANP. *Am. J. Physiol.* 2000; 278: H971–81.

20   Duchen, M.R. Mitochondria and calcium: from cell signalling to cell death. *J. Physiol.* 2000; 529: 57–68.

21   Park, J.H., Straub, V.A., O'Shea, M. Anterograde signaling by nitric oxide: characterization and in vitro reconstitution of an identified nitrergic synapse. *J. Neurosci.* 1998; 18: 5463–76.

22   Gudi, T., Hong, G.K., Vaandrager, A.B., Lohmann, S.M., Pilz, R.B. Nitric oxide and cGMP regulate gene expression in neuronal and glial cells by activating type II cGMP-dependent protein kinase. *FASEB J.* 1999; 13: 2143–52.

23   Martinez-Serrano, A., Borner, C., Pereira, R., Villalba, M., Satrustegui, J. Modulation of presynaptic calcium homeostasis by nitric oxide. *Cell Calcium* 1996; 20: 293–302.

24   Rooney, T.A., Joseph, S.K., Queen, C., Thomas, A.P. Cyclic GMP induces oscillatory calcium signals in rat hepatocytes. *J. Biol. Chem.* 1996; 271: 19817–25.

25   Xu, X., Star, R.A., Tortorici, G., Muallem, S. Depletion of intracellular Ca^{2+} stores activates nitric-oxide synthase to generate cGMP and regulate Ca^{2+} influx. *J. Biol. Chem.* 1994; 269: 12645–53.

26   Clementi, E. Role of nitric oxide and its intracellular signalling pathways in the control of Ca^{2+} homeostasis. *Biochem. Pharm.* 1998; 55: 713–18.

27   Galione, A., White, A., Willmott, N. *et al.* cGMP mobilizes intracellular Ca^{2+} in sea urchin eggs by stimulating cyclic ADP-ribose synthesis. *Nature* 1993; 365: 456–9.

28   Kawasaki, H., Kawashima, S. Regulation of the calpain-calpastatin system by membranes. *Mol. Membr. Biol.* 1996; 13: 217–24.

29   Geppert, M., Goda, Y., Hammer, R.E. *et al.* Synaptotagmin I: a major Ca^{2+} sensor for transmitter release at a central synapse. *Cell* 1994; 79: 717–27.

30   Li, C., Davletov, A.B., Sudhof, T.C. Distinct Ca^{2+} and Sr^{2+} binding properties of synaptotagmins. Definition of candidate Ca^{2+} sensors for the fast and slow components of neurotransmitter release. *J. Biol. Chem.* 1995; 270: 24898–902.

31   Sakane, F., Kanoh, H. Molecules in focus: diacylglycerol kinase. *Int. J. Biochem. Cell. Biol.* 1997; 29: 1139–43.

32   Ames, J.B., Porumb, T., Tanaka, T., Ikura, M., Stryer, L. Amino-terminal myristoylation induces cooperative calcium binding to recoverin. *J. Biol. Chem.* 1995; 270: 4526–33.

33   Zozulya, S., Stryer, L. Calcium-myristoyl protein switch. *Proc. Natl. Acad. Sci. USA* 1992; 89: 11569–73.

34   Ames, J.B., Tanaka, T., Stryer, L., Ikura, M. Portrait of a myristoyl switch protein. *Curr. Opin. Struct. Biol.* 1996; 6: 432–8.

35   Holt, M.R., Koffer, A. Cell motility: proline-rich proteins promote protrusions. *Trends Biochem. Sci.* 2001; 11: 38–46.

36   Adler, E.M., Augustine, G.J., Duffy, S.N., Charlton, M.P. Alien intracellular calcium chelators attenuate neurotransmitter release at the squid giant synapse. *Neuroscience* 1991; 11: 1496–507.

37   Brose, N., Petrenko, A.G., Sudhof, T.C., Jahn, R. Synaptotagmin: a calcium sensor on the synaptic vesicle surface. *Science* 1992; 256: 1021–5.

38   Brunger, A.T. Structural insights into the molecular mechanism of Ca^{2+}-dependent exocytosis. *Curr. Opin. Neurobiol.* 2000; 10: 293–302.

39   Franzini-Armstrong, C., Nunzi, G. Junctional feet and particles in the triads of a fast-twitch muscle fibre. *J. Muscle Res. Cell Motil.* 1983; 4: 233–52.

40   Vassylyev, D.G., Takeda, S., Wakatsuki, S., Maeda, K., Maeda, Y. Crystal structure of troponin C in complex with troponin I fragment at 2.3-Å resolution. *Proc. Natl. Acad. Sci. USA* 1998; 95: 4847–52.

41 Somlyo, A.P., Somlyo, A.V. Signal transduction and regulation in smooth muscle. *Nature* 1994; 372: 231–6.

42 Ganitkevich, V.Y., Isenberg, G. Contribution of Ca^{2+}-induced Ca^{2+} release to the $[Ca^{2+}]_i$ transients in myocytes from guinea-pig urinary bladder. *J. Physiol. Lond.* 1992; 458: 119–37.

43 Ya, Ganitkevich, V., Isenberg, G. Efficacy of peak Ca2+ currents (I Ca) as trigger of sarcoplasmic reticulum Ca2+ release in myocytes from the guinea-pig coronary artery. *J. Physiol. Lond.* 1995; 484: 287–306.

44 Gregoire, G., Loirand, G., Pacaud, P. Ca^{2+} and Sr^{2+} entry induced Ca^{2+} release from the intracellular Ca^{2+} store in smooth muscle cells of rat portal vein. *J. Physiol. Lond.* 1993; 472: 483–500.

45 Chattopadhyaya, R., Meador, W.E., Means, A.R., Quiocho, F.A. Calmodulin structure refined at 1.7 Angstrom resolution. *J. Mol. Biol.* 1992; 228: 1177–92.

46 Kuboniwa, H., Tjandra, N., Grzesiek, S. *et al.* Solution structure of calcium-free calmodulin. *Nat. Struct. Biol.* 1995; 2: 768–76.

47 Ikura, M., Clore, G.M., Gronenborn, A.M. *et al.* Solution structure of a calmodulin-target peptide complex by multidimensional NMR. *Science* 1992; 256: 632–8.

48 Griffith, J.P., Kim, J.L., Kim, E.E. *et al.* X-ray structure of calcineurin inhibited by the immunophilin-immunosuppressant FKBP12-FK506 complex. *Cell* 1995; 82: 507–22.

9

Phosphorylation and dephosphorylation: Protein kinases A and C

▪ Protein phosphorylation as a switch in cellular functioning

The first indications that protein phosphorylation might be an important regulator of enzyme activities emerged at a time when the regulatory roles of non-covalent interactions of various substances (substrates, co-factors, end products, etc.) were already well established. In particular, the allosteric influence of end products, interacting with regulatory subunits of multi-subunit enzymes, provides precise control over metabolic pathways made up of many steps. The regulation of the activation state of phosphorylase (see below) by 5′-AMP (positive) and glucose-6-phosphate (negative) is a pertinent example. Although neither of these metabolites takes part in the reaction catalysed by phosphorylase, the presence of G-6-P indicates metabolic sufficiency whereas 5′-AMP indicates hunger. Both can be described as allosteric effectors. Then it emerged that the addition of a single phosphate group to phosphorylase *b* also switches the enzyme between inactive and active states (Figure 9.1).[1] Thus, in addition to allosteric influences, covalent modification by phosphorylation (and dephosphorylation) also affects enzyme activity. Although these two regulatory processes operate through similar conformational changes, phosphorylation is primarily a response to extracellular influences expressed through hormone receptors, while allosteric effectors allow the system to respond to intracellular conditions.

The importance of phosphorylation is underlined by the fact that the genome of the budding yeast *Saccharomyces pombe* specifies more than 120 different protein kinases.[3] And this is only yeast! It is predicted that the human genome will specify more than 1000. One in three cytoplasmic proteins contains covalently bound phosphate. Phosphorylation modifies proteins by the addition of negatively charged groups to serines, threonines and, less commonly, tyrosines. These neutral hydroxy amino acid residues are typically exposed on surfaces and often in the interfaces

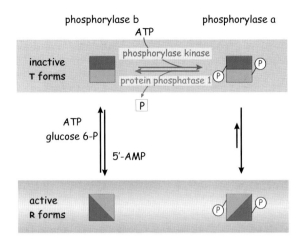

Figure 9.1 Regulation of phosphorylase through phosphorylation and by the action of allosteric effectors. The figure represents the catalytically inactive (T) and active (R) conformations of the enzyme. The phosphorylated form (phosphorylase *a*) favours the relaxed conformation but, except under conditions of starvation (signalled by the presence of 5′-AMP), phosphorylase *b* tends to the inactive T conformation. In general, the activity of phosphorylase is determined by its phosphorylation status. The letters T and R, indicating tense and relaxed, refer to the alternative quarternary structures of allosteric enzymes.[2] This terminology was originally applied in the analysis of oxygen binding to the constrained low-affinity deoxyhaemoglobin and the more open high-affinity oxyhaemoglobin (having lost eight salt bridges).

between the subunits of regulatory proteins. Phosphorylation alters the chemical properties of proteins substantially. As a result, a protein may then recognize, bind, activate, deactivate, phosphorylate or dephosphorylate its substrate. Phosphorylation can switch enzymes on and it can switch them off.

Things can go badly wrong when the balance of phosphorylation and dephosphorylation is disturbed. An example is the explosive diarrhoea due to the inhibition of protein phosphatases, by toxins present in infected shellfish (okadaic acid). The reverse is grim as, for example, the consequences of uncontrolled stripping of phosphate groups, which occurs in cells affected by the virulence factor of *Yersinia pestis* (the agent of bubonic plague, still a major killer in some parts of the world). This factor is a protein tyrosine phosphatase.[4,5] The prospects are quite as bad when tyrosine kinases become constitutively activated (oncogenic mutations, Chapters 11 and 12). On a happier note, however, a number of important therapeutic procedures, especially in the fields of tumour suppression, immunosuppression and inflammation, are based on drugs which either promote or suppress phosphorylation reactions. For example, cyclosporin, used as an immunosuppressive agent, interacts with calcineurin, a subunit of which is protein phosphatase 2B (see Chapter 17).

cAMP and the amplification of signals

Most, but not all of the signals conveyed by cAMP are mediated through phosphorylation reactions catalysed by the protein kinases A (PKA). Following stimulation, for example, of β-adrenergic receptors on liver cells by adrenaline, at concentrations around 10^{-10} mol/l, the intracellular concentration of cAMP increases to about 1 μmol/l. As a rule of thumb, such 'typical' receptors are present at a density of about 10^5 per cell and, in general, only a small proportion of these need to be occupied or activated to maximize the response (see Chapter 2). This small signal can activate a great deal more of the enzyme glycogen phosphorylase (100 μmol/l) and this liberates enormous quantities of glucose into the bloodstream.

Intermediate amplification of the signal provided by cAMP is provided through two rounds of protein phosphorylation. The first of these is catalysed by the cAMP-dependent protein kinase (protein kinase A, PKA). This phosphorylates and activates phosphorylase kinase, which in turn phosphorylates and activates glycogen phosphorylase (Figure 9.2). During exercise or under conditions of starvation, the elevated level of 5′-AMP acts as an allosteric signal for the activation of phosphorylase *b* regardless of its state of phosphorylation. Thus, phosphorylation of the enzyme and its activated state are not synonymous. However, it is fair to say that phosphorylase *b* is generally inactive and that phosphorylase *a* is generally active. On the other hand, if there is no need to break down glycogen, then why do so? Under resting conditions, the catalytic activity of phosphorylase *b* is suppressed by the presence of glucose-6-phosphate and ATP (which competes at the binding site for 5′-AMP) (Figure 9.1). In effect, an elevated level

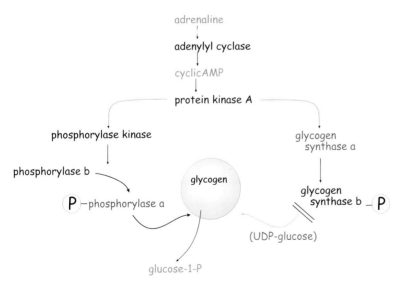

Figure 9.2 Control of glycogen breakdown: the main plan. The figure illustrates the effect of a receptor activating PKA and the phosphorylation of phosphorylase (active) and glycogen synthase (inactive).

In general, the concentration of ATP is unaffected even after prolonged periods of acute exercise or starvation but the level of 5'-AMP is elevated as the cell attempts to maintain its level of ATP by transphosphorylation of ADP in the reaction catalysed by myokinase: 2ADP = ATP + AMP.

of 5 '-AMP indicates metabolic insufficiency. Phosphorylase *a* is insensitive to allosteric regulation by 5'-AMP, ATP or glucose-6-phosphate.

The net result of all this is the transduction of a very weak first signal (low receptor occupancy by hormone) into a large kinetic signal (rapid glycogenolysis). However, although activated target enzymes can process their substrates rapidly (in our example the release of glucose-1-phosphate from liver glycogen), the penalty for all this stage-by-stage complexity, control and amplification is a delay in the onset of the response. The really fast systems, such as the signalling for neurotransmitter release or the contraction of skeletal muscle, are not regulated in this way. In skeletal muscle, the phosphorylase kinase is activated by elevation of cytosol Ca^{2+} which is also the signal that initiates contraction (see Figure 8.5). The effect of phosphorylation by PKA is not only to maximize the rate of conversion of phosphorylase *b* to phosphorylase *a*, but also to increase its affinity for Ca^{2+}. This ensures that muscle phosphorylase kinase is operative even under the ambient conditions of low intracellular Ca^{2+} concentrations. In terms of the fright, fight and flight response, adrenaline readies the muscle for maximal activity even before it receives its signal to contract (acetylcholine). The cAMP signal also acts to inhibit dephosphorylation of phosphorylase and hence its deactivation. We return to this topic in Chapter 17.

Protein kinase A

Protein kinase A is a broad-spectrum kinase. The catalytic subunit is directed at serine or threonine residues embedded in a sequence of amino acids, RRxS/Tx (x is variable) and this is widespread. What distinguishes different forms of PKA are the tissue-specific regulatory subunits with which they are associated.

In the absence of cAMP, PKA is inactive and exists as a stable tetramer (k_D ~ 0.2 nmol/l), R_2C_2, composed of two regulatory subunits (R) and two catalytic subunits (C)[6] (Figure 9.3). The tetramer is stabilized through an interaction between the catalytic sites and a pseudosubstrate sequence on the regulatory subunits which closely resembles the substrate phosphorylation consensus sequence.[7] For type I regulatory subunits, isolated from skeletal muscle, this is a true pseudosubstrate motif, in which the serine is replaced by glycine or alanine, RRxG/Ax, neither of which can be phosphorylated. In type II isoforms (from cardiac muscle) the serine is retained and the regulator is itself a substrate and undergoes phosphorylation.[8]

Autoinhibition by such motifs is a common feature of the serine/threonine protein kinases. The regulatory subunits, tightly and stably attached to each other, R_2, both possess two similar (35% identity) binding sites for cAMP. The binding of cAMP is cooperative, so that occupation of the first site by cAMP enhances the affinity of the vacant sites. With the binding sites occupied, the stability of the R_2C_2 complex declines by a factor in the region of 10^4–10^5, so liberating the catalytic units that can then phosphorylate their target substrates.[9]

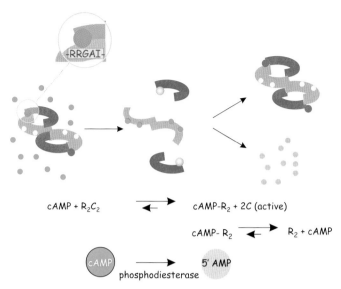

$$cAMP + R_2C_2 \rightleftharpoons cAMP\text{-}R_2 + 2C \text{ (active)}$$

$$cAMP\text{-}R_2 \rightleftharpoons R_2 + cAMP$$

$$cAMP \xrightarrow{\text{phosphodiesterase}} 5'\,AMP$$

Figure 9.3 Activation of protein kinase A by cAMP. The holoenzyme is composed of two regulatory components (green) that each bind two molecules of cAMP and two catalytic units (red). The components are linked between the catalytic sites and a 'pseudo-substrate' sequence on the regulatory units. This has the sequence –RRGAI– which closely resembles the consensus sequence for phosphorylation (in which the place of the alanine is taken by the substrate serine). On binding four molecules of cAMP the catalytic units are released and their catalytic sites exposed for interaction with real as opposed to 'pseudo' substrates. The affinity of the regulatory subunits declines allowing the bound cAMP to dissociate. It is converted to 5'-AMP by a phosphodiesterase.

The concentration of PKA in mammalian cells is so high, in the range 0.2–2 µmol/l,[10] that one might expect that little or no cAMP would ever be free in the cytosol and accessible to the phosphodiesterases which catalyse its hydrolysis. However, following the dissociation of the inactive (R_2C_2) PKA complexes, the cooperative nature of the binding of cAMP to the regulatory subunits is lost and, with it, the high-affinity state.[9] As a result, the cAMP dissociates and becomes subject to conversion to 5'-AMP by phosphodiesterase (Figure 9.3). Furthermore, the activity of PDE may be enhanced through phosphorylation by PKA, so ensuring abrupt termination of the signal.[11,12]

The specific actions of protein kinase A and other phosphorylating enzymes are controlled by targeting of the regulatory subunits. Specific anchoring proteins are present in particular locations such as the Golgi membranes, centrioles (microtubule organizing centres), cytoskeleton, etc.[6,9,13] With four isoforms of the regulatory subunits that can form heterodimers (as in RαRβ), and three isoforms of the catalytic subunit, it is possible that up to 24 different forms of the holoenzyme could exist, although to be sure, not all isoforms are expressed in all tissues.

Protein kinase A and the regulation of transcription

Activation of the CREB transcription factor

The action of PKA in the control of glycogen metabolism is an example of short-term regulation, as between one meal and the next. PKA also acts to regulate events in the longer term by switching on the transcription of specific genes. For this purpose, the signals have to be conveyed into the nucleus. The first identified example of such long-term control by PKA was the expression of the hypothalamic peptide somatostatin.

Other well-established examples are the stimulation by catecholamines of the synthesis of the RNA coding for the β_2-adrenergic receptor and by glycoprotein hormones that generate receptors for TSH and LH. The phosphorylation substrate is the transcription factor <u>c</u>AMP response element <u>b</u>inding protein (CREB, see Figure 9.4). As a result of phosphorylation by PKA, the CREB dimer interacts with DNA at the <u>c</u>AMP response element (CRE). This is an eight base-pair palindromic sequence (TGACGTCA) generally located within 100 nucleotides of the TATA box.

However, transcriptional partners or co-activators are necessary for subsequent activation of gene transcription. Such factors are <u>C</u>REB

Somatostatin, a tetradecapeptide, is best known as an inhibitor of growth hormone secretion but it appears to inhibit the release of other pituitary hormones and also insulin from pancreatic β-cells. In view of this, it has been suggested that its name should be changed to *panhibin*.

Palindrome A phrase that reads the same backwards or forwards. In the present case it refers to the sequence ACTGCAGT on the corresponding strand of DNA.

In eukaryotic cells, the **TATA box** is an important component of the promoter site that binds DNA-dependent RNA polymerase and determines where transcription commences.

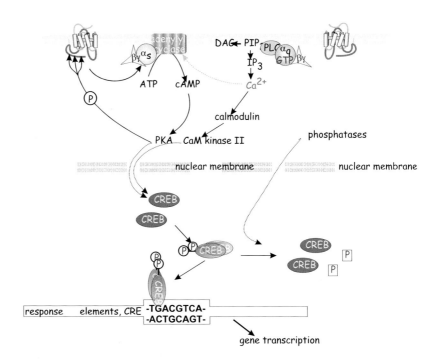

Figure 9.4 Pathways leading to activation of CREB-mediated gene expression. The figure illustrates the basic cytosolic and nuclear events leading to the activation of the cAMP response elements (CRE). Both PKA and (Ca^{2+} activated) CaM kinase can diffuse into the nucleus and phosphorylate CREB at the same serine residue. The phosphorylation promotes CREB dimer formation and, in this form, the active transcription factor binds specifically at the CRE resulting in gene transcription. Dephosphorylation of CREB leads to attenuation of the response.

binding protein (CBP) and p300, both large proteins that only interact with the phosphorylated form of CREB and direct the transcription factors to the transcriptional machinery situated at the TATA box.[14]

Not surprisingly, the activation of CREB can also occur as a consequence of hormonal stimuli that cause an elevation in the cytosol concentration of Ca^{2+} acting through a Ca^{2+} sensitive (type I) adenylyl cyclase.[15] Less expected was the finding that, in some cells, Ca^{2+} can activate CREB directly as a result of phosphorylation by CaM kinases at the same serine-133 residue.[16,17] These effects of Ca^{2+} may have important consequences, in the medium and long term, in cells which also react abruptly to transient elevations in the concentration of cytosol Ca^{2+}. Clearly, for cells that must be able to secrete proteins repeatedly and 'on demand' (such as pancreatic endocrine (islet) and exocrine (acinar) cells), it is essential to restore the complement of secretory proteins after the event. The Ca^{2+} that provides the stimulus to secretion may thus also provide the stimulus to specific protein synthesis. In pancreatic α cells, Ca^{2+} entering through voltage-sensitive channels acts both as the stimulus to the secretion of glucagon and, through the action of CaM kinase II, as a stimulus to transcription of the glucagon gene.[18] The CREB is also understood to regulate many of the long-term effects of stimulus-induced plasticity at synapses and to underlie the process of long-term potentiation.[17,19]

In addition to phosphorylation, the effect of CREB on cellular responses may be also be regulated through the expression of CREB itself. In testicular Sertoli cells, follicle stimulating hormone (FSH) acting through cAMP not only induces phosphorylation to activate the CREB, but also results in the accumulation of CREB-specific mRNA.[20] As with other hormones of pituitary origin, FSH is released in a pulsatile manner under the command of the hypothalamus. This observation therefore suggests that the resulting pulses of cAMP could prime the cells to become more sensitive to subsequent stimulation by the hormone.

Long-term potentiation
A form of synaptic plasticity that is thought to be involved in learning and memory and in the formation of synaptic connections in the CNS.

■ Attenuation of the cAMP response elements by dephosphorylation

Over the 15–20 minutes following hormonal activation, the catalytic units of PKA diffuse into the nucleus to cause phosphorylation of CREB[21] and the initiation of transcription. Then, over 4–6 hours (the attenuation phase), transcription of the target genes gradually declines. This is probably due to dephosphorylation of CREB, since it can be prolonged by application of phosphatase inhibitors (such as okadaic acid[22]). Furthermore, fibroblasts over-expressing phosphatase inhibitor-1, which specifically inhibits protein phosphatase-1 (PP1), manifest an enhanced transcriptional response to CRE.[23] In hepatic cells manipulation of the activity of the levels of PP2A causes dephosphorylation of CREB.[24] It appears that the magnitude of CRE-induced responses is regulated, at least in part, through dephosphorylation, but that different phosphatases are involved in different cells.

As mentioned above, specificity of PP1 is determined by its association with regulatory subunits, which target it to particular cellular locations. In the nucleus, PP1 is attached to the chromatin by its association with NIPP, the endogenous nuclear inhibitor of PP1.[25] Phosphorylation of NIPP by PKA drastically reduces its affinity for PP1 and so cAMP may act to down-regulate its own signals through the activation of the phosphatase (see Chapter 17).

As will become apparent in the later chapters of this book, the binding of a dimerized transcription factor complex to a palindromic enhancer element on DNA is a recurring feature in the regulation of specific gene expression.

■ Protein kinase A and the activation of ERK

Most of the actions of the β_2-adrenergic receptor are transduced through G_s and mediated by phosphorylation catalysed by PKA. However, following phosphorylation of the receptor by PKA (heterologous desensitization: see Figure 4.10), its specificity alters and it shifts its attention from G_s to G_i. The mechanism that uncouples the receptor from its normal transducer also enables it to couple with G_i. This opens up a whole new range of possibilities, not least because of the much greater quantities of G_i proteins expressed in most cells, and hence the much greater availability of $\beta\gamma$-subunits. An important end result of this switching is the involvement

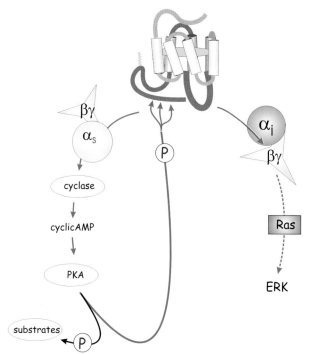

Figure 9.5 **All change trains.** Down-regulated receptors are not necessarily down-regulated, merely looking the other way. Phosphorylation of the β_2-adrenergic receptor by PKA diverts its attention from G_s to G_i. The resulting activation of $\beta\gamma$-subunits initiates a pathway of reactions leading to the activation of ERK. Adapted from Daaka *et al.*[26]

of Ras leading to activation of extracellular signal regulated protein kinase (ERK) and the generation of nuclear signals (Figure 9.5). Further details are given in Chapter 11.

This line of communication is interrupted in cells which have been treated with pertussis toxin, a characteristic of the involvement of G_i proteins, not G_s. It can also be prevented by over-expression of a PH-domain-containing peptide derived from the C-terminus of β-adrenergic receptor kinase (βARK: see Chapter 4) which has the effect of sequestering the βγ-subunits. All this effectively places cAMP and PKA upstream of the receptor, conditioning it to communicate with G_i, and then Ras conventionally on the pathway leading to the activation of ERK (Chapter 11).

Actions of cAMP not mediated by PKA

Regulation of ion channels by cyclic nucleotides

Important among the actions of cAMP that are not mediated through PKA is the regulation of a gated cation channel, a component of the signalling mechanism in olfactory cells.[27–30]

Here, it is understood that the interaction of odorants with receptors coupled to G_{olf} (an heterotrimeric G-protein homologous to G_s) leads to activation of adenylyl cyclase and the elevation of the concentration of cAMP. This causes the ion channels to open with consequent depolarization of the cell membrane. Unlike the retinal cyclic GMP regulated channel (see Chapter 6), which is relatively unresponsive to cAMP, the olfactory channel is equally responsive to both nucleotides. Also, unlike the ion channel of photoreceptor cells, the olfactory channel closes in response to cyclic nucleotides, causing the cells to hyperpolarize. Cloning of cyclic nucleotide dependent channels has shown that they are closely related and probably share common ancestry with the superfamily of voltage regulated ion channels.

The **cAMP gated K⁺ channel** is composed of four similar protein chains that are assembled together so that the membrane-crossing domains (generally five or six membrane spans) combine to form a central pore. In the voltage-gated Na^+ and Ca^{2+} channels the subunits are linked as a single protein chain.

Epac, a guanine nucleotide exchange factor directly activated by cAMP

Rap1 is a monomeric GTPase that appears to be involved in a number of cellular processes including the activation of platelets, cell proliferation, differentiation and morphogenesis.[31] It first came to notice as a regulator that suppresses oncogenic effects of Ras (Figure 9.6). Unlike Ras, Rap1 is confined to organellar membranes. This may explain how, despite the two proteins sharing some downstream effectors, their effects can be antagonistic. Depending on the cell type, several ligands, through interaction with their receptors and following the generation of second messengers such as Ca^{2+}, diacylglycerol and cAMP, are able to activate Rap1. Important however is the observation that this works equally well in cells that express a mutant form of PKA that has a low affinity for cAMP.

The activation of Rap1 is mediated through the direct action of cAMP with a guanine nucleotide exchanger, Epac (exchange protein directly activated by cAMP).[32] This was identified by screening sequence databases for proteins having homology with other GEFs for Ras and Rap1 and

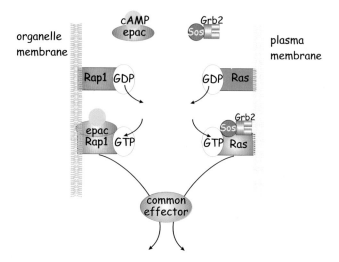

Figure 9.6 Activation of Rap1 by Epac, a guanine nucleotide exchange factor activated by cAMP. Rap1 is a monomeric GTPase that is confined to organellar membranes and that counteracts the oncogenic effects of Ras. Its activation is an example of a process mediated by cAMP without the mediation of PKA.

also with proteins having binding sites for cAMP. The nucleotide binding site is similar to those present in the regulatory subunits of PKA and the cyclic nucleotide regulated ion channels. Using purified Epac it was shown that cAMP is required to initiate the release of GDP from Rap1, whilst the interaction of a mutant form of Epac, lacking the cyclic nucleotide binding site, is constitutively active. It can release GDP from Rap1 in the absence of cAMP.

As with other monomeric GTPases, the return to the GDP-bound state is promoted through interaction with a GTPase activating protein. The specific Rap1GAP is itself regulated by the heterotrimer Go in its resting state.[33] Thus, α_0GDP can sequester Rap1GAP and, in this way, regulate the various activity of Rap1.

In these experiments, an analogue of **GDP**, labelled with a fluorescent group sensitive to its environment (attachment to the protein binding site) was used to allow continuous kinetic monitoring of the release reaction.

Protein kinase C

Discovery of a phosphorylating activity independent of cAMP

The enzymes that we call protein kinase C (PKC) became apparent during the late 1970s as a cyclic nucleotide-independent protein kinase present in bovine cerebellum.[34] As first described, this activity appeared to be the product of limited proteolysis by a calcium-dependent protease,[35] but shortly afterwards it was found to be activated by Ca^{2+} (50 µmol/l) and phospholipids with no need of proteolysis.[36] More significantly, the activity of this protein kinase could also be stimulated by the addition of a small quantity of diacylglycerol (DAG, product of the phospholipase C [PLC] reaction), together with phospholipids (generally phosphatidylserine) and Ca^{2+}, now at concentrations not far above the physiological range

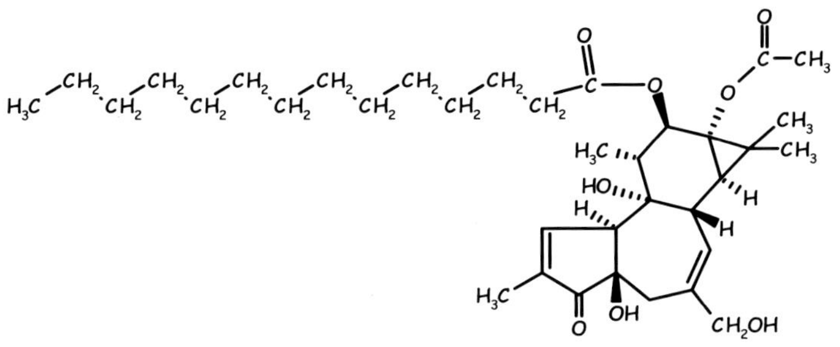

Figure 9.7 **Phorbol myristate acetate (PMA).**

(2–6 μmol/l).[37] The new enzyme became a focus of interest when it was found to be the predominant intracellular target for the active principles of croton oil, long recognized as skin irritants having tumour-promoting capacities.[38]

Such phorbol esters[39] activate PKC, substituting for the physiological diacylglycerol.[40] Phorbol myristate acetate (PMA, also known as 12-*O*-tetradecanoylphorbol-13-acetate, TPA), has since been applied to every cell and system imaginable. It is apparent that PKC is involved in an enormous number of cellular processes. These range from tumour formation, host defence, embryological development, pain perception, neurite outgrowth and the development of long-term memory. In this chapter we consider the roles of protein kinase in tumour formation and host defence.

The protein kinase C family

Molecular cloning reveals the mammalian PKCs to be a family of 12 distinct members[41,42] not all of which are activated by phorbol esters.[43] They can be subdivided into three subfamilies (see Table 9.1, also Figure 9.8).

In addition there are more distantly related enzymes such as PKC-related kinase (PRK, also referred to as PKN).[44] The various isoforms are also referred to as 'isozymes' but this term is better reserved for an enzyme such as lactate dehydrogenase, since the isoforms of PKC have separate and characteristic substrates.

The subfamilies of PKC are classified on the basis of sequence similarities and their modes of activation. The conventional PKCs are all activated by phospholipid, in particular phosphatidylserine, DAG and Ca^{2+}. The novel PKCs require phospholipid and diacylglycerol but not Ca^{2+}. This

Croton oil A poisonous viscous liquid obtained from the seeds of a small Asiatic tree, *Croton tiglium*, of the spurge family (Euphorbiaceae). The tree is native to India and the Indonesian archipelago. The oil is pale yellow to brown and transparent, with an acrid persistent taste and disagreeable odour. It is a violent irritant. The use of croton oil as a drug apparently originated in China. It was introduced to the West by the Dutch in the 16th century, and the 19th-century physician (and pioneer in the field of signal transduction) Sidney Ringer describes its use as a purgative and its topical application in the treatment of ringworm. It is now considered too dangerous for medicinal use.

Table 9.1 PKC isoforms

Subfamily	Isoforms	Requirements for activation
Conventional (cPKC)	α, β1, β2 and γ	DAG, PS, Ca^{2+}
Novel (nPKC)	δ, ε, η and θ	DAG, PS
Atypical (aPKC)	λ, ι, ξ and μ (PKD)	PS

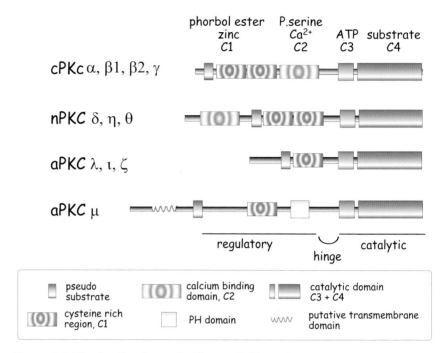

Figure 9.8 The family of protein kinases C. Mammalian PKC comprises a family of 12 distinct members subdivided into 3 subfamilies: conventional (c), novel (n) and atypical (a). They are identified as PKC by the conserved domains C1, C2, C3 and C4. The protein is made up of regulatory and catalytic domains, connected by the hinge region.

may be explained by the different location of their Ca^{2+}-binding domain. The atypical PKCs respond neither to DAG nor to Ca^{2+} but they still require phospholipids. All the PKCs have in common a catalytic domain, comprising two highly conserved subdomains, C3 and C4. They all contain a cysteine-rich region, C1, and, in addition, the conventional and novel PKCs contain the C2 domains that bind Ca^{2+}. They are all characterized by the presence of a pseudosubstrate sequence that plays a role in maintaining the kinase inactive in the absence of a stimulus. The atypical PKCμ has a transmembrane domain and a pleckstrin homology domain (PH domain).

Phosphorylation by the PKCs occurs only at serine and threonine residues in the close vicinity of arginine residues situated in the consensus sequence, RxxS/TxRx.[45] This is present in many proteins (for a detailed analysis see Nishikawa *et al.*[46]) and, as a result, PKCs can be regarded as broad-specificity protein kinases though there are some differences in substrate recognition between the subfamilies. Thus, all forms of PKC (with the exception of PKCζ) are capable of phosphorylating MARCKS and GAP-43, while ribonucleoprotein A1 (hnR A1) is only efficiently phosphorylated by PKCζ. The different expression of the various PKCs in the tissues may contribute to particular tissue- or cell-specific responses to hormones, growth factors, cytokines or neurotransmitters.[47]

MARCKS <u>M</u>yristoylated <u>a</u>lanine-<u>r</u>ich <u>C</u>-protein <u>k</u>inase <u>s</u>ubstrate.
GAP-43 Neuronal <u>g</u>rowth cone <u>a</u>ssociated <u>p</u>rotein (43 kDa).

Genes coding for PKC (*tpa-1* and *pkc*) are present in *C. elegans*,[48,49] and three isozymes, InaC and dPKC2–3, have been found in *Drosophila*.[50] Genetic screening of these organisms has supplied insights into the functioning of PKC, in particular with respect to the role of assembling proteins in the formation of large signalling complexes (see below).

InaC denotes a locus defined from a screen for mutations in *Drosophila* visual transduction. When mutated the fly displays inactivation no-after potential (see below).

▨ Structural domains and activation of protein kinase C

▨ The C1–C4 regions

The deduced amino acid sequences reveal four reasonably conserved functional domains, C1–C4, having similarities to those present in other signalling proteins (see Chapter 18)[41,51] (Figure 9.9). Proceeding from the N-terminus, C1 and C2 constitute the regulatory domains and then C3 and C4 together constitute the catalytic domain characteristic of all kinases (see Chapter 18). C1 contains a cysteine–histidine-rich motif ('zinc finger') that coordinates two Zn atoms and this forms the binding site for DAG and phorbol esters. This is duplicated in most isoforms. C1 domains, occurring singly or doubly, are present in a wide range of proteins, some of which bind phorbol esters ('typical') whereas others do not ('atypical'). The proteins that contain a typical C1 domain are predicted to be regulated by DAG or phorbol esters.[52] Examples of phorbol ester binding proteins lacking kinase activity are the chimaerins (a family of Rac-GTPase-activating proteins), RasGRP (a Ras exchange factor), and Unc-13/Munc-13 (a family of proteins involved in exocytosis) (for review, see Kazanietz[53]). The C2 domain binds negatively charged phospholipid, such as phosphatidyl serine, and in some isoforms a Ca^{2+}-binding site responsible for the Ca^{2+}-dependence of lipid binding is also present.

The regulatory (C1 and C2) and the catalytic (C3 and C4) domains are linked by a hinge region. When the enzyme is membrane bound the hinge is vulnerable to proteolytic enzymes such as trypsin. The fragment

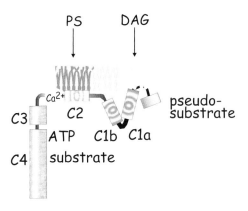

Figure 9.9 PKC domains. The conserved domains, C1–C4, are functional modules. C1 binds to DAG or phorbol ester, C2 is involved in the attachment to phospholipid, which is enforced by the binding of Ca^{2+}, and C3+C4 constitute the catalytic domain.

(protein kinase m), containing the kinase domain, necessarily detached from the membrane, is constitutively active.

■ Pseudosubstrate

In the segment immediately N-terminal to the C1 domain is a stretch of amino acids (19–36 in PKCα) which constitutes the autoinhibitory pseudosubstrate. Its sequence resembles the consensus phosphorylation sites present in target proteins that are phosphorylated by PKC. However, in the pseudosubstrate the serine is replaced by alanine and is consequently not amenable to phosphorylation.[54] In the absence of a stimulus, the catalytic domain binds to the pseudosubstrate, causing the enzyme to fold about the hinge linking C2 and C3, resulting in a suppression of kinase activity.

■ Activation

Activation of PKC requires phosphorylation of the catalytic domain in the activation loop (see Chapter 18) and the detachment of the pseudosubstrate domain from the active site.[55] The following description applies to conventional and novel PKCs but not the atypical isoforms.

PKC is synthesized as an inactive non-phosphorylated precursor present in a detergent-insoluble fraction (probably associated with the cytoskeleton). At this stage the catalytic site is accessible to ATP but the enzyme is catalytically incompetent.[56] In the further processing of PKC, the activation loop, near the sequence APE in the catalytic domain, becomes phosphorylated (Figure 9.10), possibly through the action of PDK1. This is a phospholipid-dependent protein kinase, also involved in the activation of protein kinase B (see also Table 13.1).[57] Replacement of the activation-loop threonine by a neutral residue results in an enzyme that cannot be activated. In contrast, replacement with glutamate, intended to mimic the effect of phosphate, results in an enzyme that can be activated. There are a number of problems regarding the regulation of phosphorylation in the activation loop. For instance, regardless of the presence or absence of agonists known to stimulate PDK1, at least 50% of cPKCβ2 is normally present in its phosphorylated form.[58] For nPKCδ the situation is different because the protein is not normally phosphorylated and the phosphorylation is clearly induced by activation of PDK1.[59] The phosphorylation in the activation loop is followed by two autophosphorylations at the C-terminus of the catalytic domain (sites 1 and 2) (see Table 9.2). aPKCζ and λ lack the second phosphorylation site and have substituted a glutamate residue (see Table 13.1, page 311). aPKCμ lacks both these phosphorylation sites (see page 404). For review see Parekh *et al.*;[60] further information is available at the protein kinase resource at http://www.sdsc.edu/Kinases.

The mature form of PKC detaches from the detergent-insoluble fraction and it is now catalytically competent, though still inactive due to the apposition of the pseudosubstrate with the catalytic domain. The

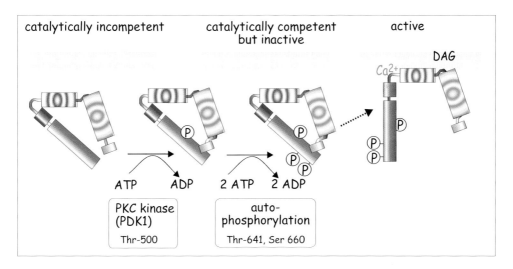

Figure 9.10 Activation of PKC occurs in four steps. As an example, PKCβ2 is synthesized as a 74 kDa precursor, its catalytic site is accessible to substrate but inactive. Three different phosphorylations follow, rendering the enzyme catalytically competent but still inactive. Substrate access is prevented by the attachment of the pseudosubstrate to the catalytic site. On binding DAG and elevation of Ca^{2+}, the enzyme binds firmly to the membrane, the pseudosubstrate detaches from the catalytic site and the system becomes both competent and accessible.

Table 9.2 Phosphorylation sites that render PKC isoforms catalytically competent

Isoform	In activation loop		In C-terminus site 1		In C-terminus site 2	
hPKCα	GVTTRTFCGTPDYIAPE	T497	RGQPVLTPPDQLVI	T638	QSDFEFGSYVNPQ	S657
HPKCβ1	GVTTKTFCGTPDYIAPE	T500	RQPVELTPTDKLFI	T642	QNEFAGFSYTNPE	S661
HPKCβ2	GVTTKTPCGTPDYIAPE	T500	RHPPVLTPPDQEVI	T641	QSEFEGFSFVNSE	S660
hPKCγ	GTTTRTFCGTPDYIAPE	T514	RAAPAVTPPDRLVL	T655	QADFQGFTYVNPD	T674
hPKCδ	ESRASTFCGTPDYIAPE	T507	NEKARLSYSDKNLI	S645	QSAFAGFSFVNPK	S664
hPKCι	GVTTTTFCGTPDYIAPE	T566	REEPVLTLVDEAIV	T710	QEEFKGFSYFGED	S729
hPKCη	GVTTATFCGTPDYIAPE	T510	KEEPVLTPIDEGHL	T650	QDEFRNFSYVSPE	S672
hPKCθ	DAKTNTFCGTPDYIAPE	T538	NEKPRLSPADRALI	S676	QNMFRNFSFMNPG	S695
hPKCζ	GDTTSTFCGTPNYIAPE	T410	SEPVQLTPDDEDAI	T552	QSEFEGFEYINPL	E579
hPKCλ	GDTTSTFCGTPNYIAPE	T411	NEPVQLTPDDDDIV	T563	QSEFEGFEYINPL	E582
hPKCμ	KSFRRSVVGTPAYLAPE	S742	Absent		Absent	

The first phosphorylation occurs in the activation loop of the catalytic domain, near the APE sequence. This is not an inter- or autophosphorylation but involves another type of protein kinase. The phosphorylations in the C-terminus of the catalytic domain are autophosphorylations. Note the substitution of the site 2 serines by a glutamate (E) in PKCζ and PKCμ. Phosphorylation targets are indicated in red.
Adapted from Toker.[61]

stimulus-mediated generation of DAG effectively plugs a hydrophilic site in the C1 domain, making the surface more hydrophobic, and this allows C1 to bury into the membrane. The DAG also reduces the Ca^{2+} requirement for the binding of the C2 domain to phospholipids hence increasing the strength of the interaction. Lastly, it brings about a conformational change in the catalytic domain, separating the catalytic and pseudo-substrate domains (Figure 9.9). The protein kinase is now fully active.

■ Multiple sources of DAG and other lipids to activate PKC

In view of the requirement for DAG, the regulation of PKC is clearly linked to the activity of PLC (Figure 9.11; see Chapter 5).[62] The hydrophobicity and small size of DAG enable it to diffuse laterally in membranes with some rapidity. It is also short-lived, either broken down by DAG lipase (to form glycerol and fatty acids) or converted to phosphatidate by the action of DAG kinase (see Figure 8.7, page 180). Virtually all ligands, growth factors, hormones or neurotransmitters, promote the production of DAG (and IP_3) in one way or another, and this means that PKC is implicated in a large number of cellular responses. Sustained activation of PKC requires the simultaneous presence of elevated levels of DAG and high-frequency Ca^{2+}-spikes.[63]

Several other lipid second messengers and mediators either potentiate the effect of DAG and Ca^{2+} or activate the PKCs directly (Figure 9.11). In particular, the 3-phosphorylated inositol lipids $PI(3,4)P_2$ and $PI(3,4,5)P_3$, products of the phosphoinositide 3-kinases (see Chapter 13) activate both the novel (δ, ε and θ) and the atypical (ζ) PKCs in a phosphatidyl serine environment.[64] Unsaturated fatty acids (most notably arachidonic acid, lysophosphatidic acid (LPA) and lyso-phosphatidylcholine (lysoPC)) can also enhance the activity of PKC. These lipids potentiate the effects of DAG at basal Ca^{2+} concentrations (100 nmol/l range).[65] Breakdown products of sphingolipids, such as sphingosine and lysosphingolipids, inhibit the conventional PKCs most likely by masking their interaction with phosphatidyl serine.[66]

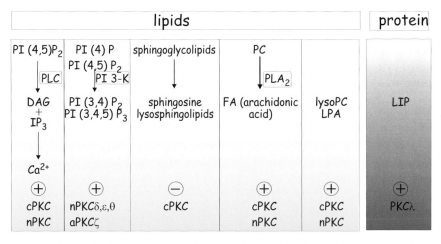

Figure 9.11 Multiple lipid sources to activate PKC. Different lipid sources and different phospholipases and kinases activate different isoforms of PKC. Uniquely, the atypical PKCλ can be activated by a protein, LIP.

The atypical PKCs are not activated by DAG or phorbol esters but are activated by other lipid products, such as PI(3,4,5)P$_3$, lysophosphatidic acid and ceramide. Uniquely, PKCλ can be activated by a protein, lambda-PKC interaction protein (LIP).[67]

Differential localization of PKC isoforms

The existence of numerous isoforms, but the lack of individual substrate specificities, raises the question of whether the various PKCs might have redundant roles. Immunocytochemical analysis reveals, however, that particular isoforms are present in different cells and subcellular compartments and bind to unique protein complexes. They may have specialized functions. The first indication was provided by a study of inductive signalling in *Xenopus laevis* embryos. Here, dorsal ectoderm is more competent to develop as neural tissue than the ventral ectoderm. This difference persists even when an artificial stimulus such as phorbol ester is applied. PKCα is preferentially expressed in dorsal ectoderm, whereas PKCβ is uniformly distributed.[68] Over-expression of PKCα in ventral ectoderm annuls the difference in competence between the two tissues, the ventral ectoderm now being fully competent to form neural tissue.[69] In fibroblasts induced to over-express various isoforms of PKC, the majority tend to be diffusely distributed throughout the cytoplasm with just a fraction of PKCβ and θ attached to the membrane and concentrated in the Golgi apparatus. However, within minutes of the addition of phorbol ester there occurs an extensive redistribution. PKCα and ε concentrates at the cell margin, PKC-α accumulates in the endoplasmic reticulum, PKCβ2 associates with actin-rich microfilaments of the cytoskeleton, PKCγ in Golgi organelles, and PKCε attaches to the nuclear membrane.[70]

PKC anchoring proteins, STICKs, PICKs and RACKs

There are several proteins that bind and target PKCs to specific subcellular sites. The STICKs (substrates that interact with C-kinase, including vinculin, talin, MARCKS, α- and γ-adducin and the annexins) link the actin cytoskeleton with the plasma membrane (see Table 9.3). They bind phosphatidyl serine and they act as substrates for PKC. This implies that PKC plays a role in the regulation of membrane-cytoskeletal interactions. None of these is selective between different isozymes. In the case of MARCKS and γ-adducin, phosphorylation disrupts both the interaction with the PKC and with the phospholipids, so releasing the protein into the cytosol.[71]

There are other PKC-binding proteins, the RACKs (receptors for activated C-kinase) and the PICKs (proteins interacting with C-kinase), that are not substrates but nonetheless show specificity among the various isoforms. They are thought to anchor activated PKC to particular membrane domains in the vicinity of appropriate substrate proteins. RACK 1 links PKCβ with the intracellular domain of the receptors for interleukin-5 (ref 72) and β-COP (β-cytosolic coat protein, another RACK), interacts with PKC-ε, linking it to the Golgi membrane.[73] InaD (a PICK) orchestrates the association of the proteins of the retinal visual transduction pathway in *Drosophila* (see below).

The primary function of all these anchoring proteins is to position individual PKCs in the appropriate location to respond to specific receptor-mediated activating signals. A second role is to bring them into close contact with their substrates. Given the promiscuity of PKC in test-tube protein kinase A assays, these anchoring proteins may act to prevent inappropriate phosphorylation events. A similar theme emerges with the serine-threonine protein phosphatases (see Chapter 17).

■　A matter of life or death: the role of PKC signalling complexes in the evasive action to the fly-swat

Anyone who has tried unsuccessfully to swat a fly must know that the speed of their visual response is faster by far than ours. More than this, their sensitivity and adaptational capacity far outstrips any vertebrate photosensors. The phototransduction cascade in *Drosophila* (see Figure 6.12, page 141) reveals how anchoring proteins operate in signal trans-

Table 9.3　PKC binding proteins

Binding protein/substrate	Cellular location	Known functions
STICKs		
Vinculin/talin	Focal contacts	Adhesion to extracellular matrix
Annexins I and II	–	–
MARCKS	Vesicles, plasma membrane	Vesicle trafficking, secretion, cell spreading
Desmoyokin, AHNAK	Desmosomes/nucleus	Cell–cell binding
α-Adducin	Cortical cytoskeleton	Cell polarity, actin capping
γ-Adducin	Cortical cytoskeleton	Interaction of actin with spectrin
STICK72 and gravin	Plasma membrane	–
STICK34	Cytoskeleton, caveolae	–
GAP43 (B-50)	Nerve growth cone	Neurite outgrowth, neurotransmitter release
P47^{phox}	Neutrophil cytoplasm	Activation NADPH oxidase
AKAP79	Synaptic densities	Neurotransmitter release
PAR-3	CNS	Role in cell polarity
PICKs		
InaD (*Drosophila*)	Photoreceptor	Links PKC to PLC
ASIP	Tight junction epithelial cells	–
PICK1	Perinuclear	–
RACKs		
RACK1 (PKCβ)	Interleukin-5 receptor	–
β-COP (PKCε)	Golgi	Protein traffic

Adapted from Jaken and Parker.[74]

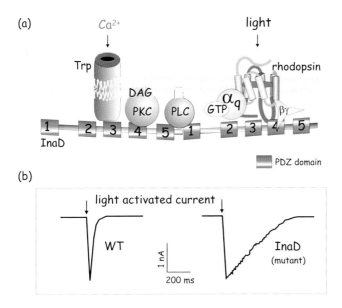

Figure 9.12 Anchoring proteins: (a) The formation of a signalling circuit in *Drosophila* phototransduction. Two copies of InaD gather together a number of signalling molecules to form a signalling circuit. The events are as follows (see also Figure 6.12): Light activates rhodopsin, that in turn activates Gq-α, which acts on phospholipase C (PLC), causing the generation of DAG. This opens a Ca^{2+} channel, Trp, causing membrane depolarization. This rapid event is followed by a PKC-mediated deactivation of Trp and re-polarization of the membrane potential. (b) A loss of InaD causes a retardation of the return to the resting state and hence a loss of visual resolution. Adapted from Xu *et al.*[75]

duction through PKC (Figure 9.12, upper panel). Photoactivation of retinal rhodopsin leads to the opening of the light-sensitive cation influx channels Trp and TrpL (trp: transient receptor potential, homologues of human store operated Ca^{2+}-channels). This rapid response results in a very transient depolarization of the photoreceptor cells. Most of the proteins that operate in the phototransduction cascade associate as a single supramolecular signalling complex. The key component in this complex is InaD having multiple PDZ domains. InaD together with InaC, a PKC, emerged in genetic screens for mutations in *Drosophila* visual transduction.

InaD, present in the microvillar compartment of the photoreceptor (the rhabdomere: see Figure 11.9, page 267) is closely associated with several proteins including a β-type PLC (gene product of norpA, 'no receptor potential'), the light-sensitive cation influx channels (Trp and TrpL), InaC, the *Drosophila* variant of PKC and the photosensor rhodopsin (product of the *ninaE* gene).[75] The binding occurs through the five PDZ domains of InaD (Figure 9.12), PDZ-3 interacting with the Ca^{2+}-channel, PDZ-4 with rhodopsin or PKC and PDZ-5/PDZ-1 with PLC. At least two InaDs are required to combine all the signal transduction components involved, linking the rhodopsin to the ion-channel. This supramolecular complex ensures both the rapid opening of the Ca^{2+} channel and then its almost instantaneous desensitization. Disruption of any one of the interacting components alters the characteristic photo-response (Figure 9.12). In InaD mutants, the various components of the signalling complex, such as the Ca^{2+}-channel, PLC and PKC are no longer exclusively located in the rhabdomeres though rhodopsin and the Ca^{2+}-channel (intrinsic membrane proteins) of course remain. Photoreceptor cells in flies having mutations at either InaC or InaD desensitize sluggishly, remaining depolarized (hence ina, inactivation no-after potential). Such flies have problems distinguishing light of different intensities and flash duration.[76]

The macromolecular assembly predicates against random collisions of individual components of the signalling pathway and ensures high precision.

■ PKC and cell transformation

Long before the association with PKC became apparent, it was evident that the phorbol esters can act as co-carcinogens, enhancing the tumorigenic activity of other substances such as urethane.[38,77] In some cell cultures they increase the number of mitoses[78] but in the absence of a tumour promoter they do not induce the formation of fully transformed cells. One might think that PKC should be implicated in all this, maybe acting to phosphorylate a transcription factor, in a manner similar to PKA that works through the CREB protein, to induce specific gene expression and the formation of tumours. However, the matter remains unclear and PKC has failed to qualify as a true oncogene.

With the aim of understanding the mechanisms underlying tumour promotion by phorbol esters, two independent experimental strategies have been applied. By searching for transcriptional control elements that mediate the phorbol ester-induced alterations in gene expression it should be possible to work backwards, identifying first the transcription factor(s) that bind these elements and then the signal transduction pathway that regulates their activation.[79] The alternative has been to over-express various isoforms of PKC and to study changes in cell phenotype.[80]

■ The search for transcription factors that mediate phorbol ester effects

Analysis of the promoter regions of a number of genes (for instance, collagenase, metallothionein IIA and stromelysin) induced by phorbol ester (PMA or TPA) revealed a conserved 7 base-pair palindromic motif (TGACTCA). This TPA-responsive element (TRE) is recognized by AP-1.

It was understood that AP-1 is at the receiving end of a complex pathway that transmits the effects of phorbol ester tumour promoters from the plasma membrane to the transcriptional machinery, possibly involving PKC[81]. AP-1 is a complex of two proteins, c-Fos and c-Jun, both known oncogenes, linked by a protein–protein interaction motif known as a leucine zipper (Figure 9.13).[82] These transcription factors are similar in structure to CREB.[83,84]

Defining the upstream signal transduction pathway that regulates AP-1

PMA or TPA? Phorbol myristate acetate and 12-O-tetradecanoylphorbol-13-acetate are synonymous. Unfortunately, the pharmacologists generally prefer PMA but the molecular biologists prefer TPA. Hence we have TPA-inducible genes, the TRE, etc.

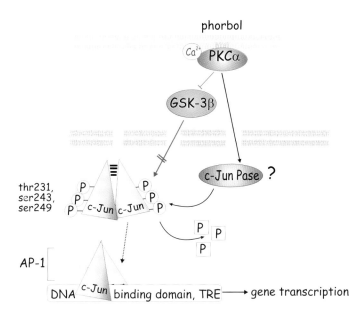

thr231,
ser243,
ser249

AP-1

DNA c-Jun binding domain, TRE ⟶ gene transcription

AP-1 encompasses a group of dimeric transcription factor complexes composed of Jun, Fos or ATF subunits that bind to either the TRE (Jun–Jun or Jun–Fos) or the CRE (Jun–ATF). The oncogenic variants of these transcription factors have increased half-lives and show enhanced transcriptional activity as a consequence of partial deletions.[83,84] CREB can also bind to the TRE[85] and so inhibit activation through c-Jun.[86]

Figure 9.13 PKC and the activation of the TRE. Activation of PKCα by TPA (phorbol ester) results in phosphorylation and inactivation of glycogen synthase kinase-3β. Inactivation of GSK-3β allows dephosphorylation of c-Jun by an unidentified phosphatase and the c-Jun dimer becomes competent to bind to the TRE and induce gene transcription.

activity proved far from straightforward. It was found that activation of PKC causes the *de*phosphorylation of c-Jun just in the basic region where it binds DNA.[87] Phosphorylation of this segment can also be achieved (in the test tube) by glycogen synthase kinase-3β (GSK) and so it was postulated that PKC stimulates the binding of c-Jun DNA through the inhibition of GSK-3β which would result in the dephosphorylation of the basic region. Consistent with this idea is that activation of PKC (α, β1, β2 and γ) causes phosphorylation and thus de-activation of GSK-3β.[87,88] However, a molecular interaction between GSK-3β and c-Jun has not been demonstrated, nor is it clear which phosphatase strips the phosphate residues from c-Jun. That this cannot be the whole story became clear from the finding that phosphorylation at the N-terminus is also crucial for both transcriptional activity and cell transformation by c-Jun.[89,90]

The discovery of a Jun N-terminal protein kinase, JNK-1, that phosphorylates c-Jun through interaction with a specific kinase docking site,[91,92] drew the field away from PKC and focused attention on the serum response element (SRE) and the newly emerging family of mitogen activated protein kinases (MAP kinases) (Figure 9.14; see Chapter 11). In addition to its role in regulating serum-mediated expression of c-*fos*,[93] the SRE is also involved in the cellular response to phorbol ester.[94] The SRE binds two transcription factors: the serum response factor (SRF) and the TCF, p62[TCF] (Elk). Growth factors present in serum (see Chapter 10) regulate the transcriptional activity through phosphorylation of p62[TCF] (ref. 95) a mode of activation that also applies for phorbol ester (Figure 9.14).[96]

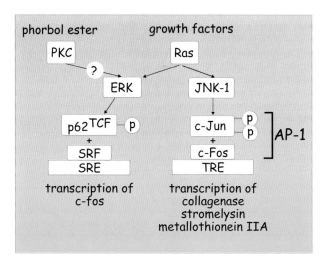

Figure 9.14 TPA causes activation of SRE. PKC and growth factors were initially thought to activate distinct signal transduction pathways resulting in the activation of TRE and SRE respectively. This notion ended when it was realized that TPA also activates the SRE and that growth factors can activate the TRE through activation of Jun N-terminal kinase, JNK. The question remains, how is PKC involved in all this?

As the signal transduction pathway emanating from growth factor receptors that activate the SRE was gradually resolved, and found to involve Ras and members of the MAP kinase family (see Chapter 11),[95,97] the role of PKC remained obscure. PKCα was found to activate the Ras-activated kinase c-Raf (Chapter 4) that cooperate in the transformation of NIH3T3 fibroblasts.[98] In rat embryo fibroblasts, activation of Raf-1 is also essential for the transforming effect of PKCε.[99] Raf-1 is involved in growth factor receptor mediated signalling leading to the activation of ERK (one of the mitogen activated protein kinases, MAP kinases: see Chapter 11). Since all growth factors induce the generation of DAG and hence activate PKC, it follows that PKC reinforces the Ras-initiated growth factor signal at the level of Raf-1. However, this does not necessarily result in enhanced cell proliferation. More recent findings, using kinase-dead and constitutively activated mutants, confirm that several PKC isoforms can activate members of the MAP kinase family, in some cases leading to the activation of both ERK and JNK. This dual signal reintegrates at the level of phosphorylation of p62[Tcf] (Figure 9.15).[96,100] Activation of JNK could equally result in the phosphorylation of c-Jun, resulting in the activation of AP-1 at a TRE site.

Collectively, all this suggests that PKC acts primarily as a modulator of the Ras signal transduction pathways that emanate from growth factor receptors. The commitment, either to promote or to suppress activity, is determined at the level of the MAP kinases.

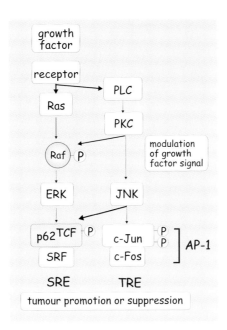

Figure 9.15 **PKC causes activation of the SRE through activation of Raf and JNK.** Activation of PLC-γ by growth factor receptors results in activation of PKC. From here two substrates have been identified, Raf and JNK, both components of the Ras-controlled MAP kinase pathway. Phosphorylated and activated Raf and JNK enhance the phosphorylation of p62^TCF, thereby enforcing the Ras signal. Moreover, JNK activates the TRE through phosphorylation of c-Jun.

■ Over-expression of PKC isoforms and cellular transformation

Over-expression of PKCβ1 in rat embryo fibroblasts (R6 cells) induces the characteristic transformed cell phenotype (see 'Cancer and transformation', page 246). These cells have a reduced requirement for growth factors and generate an autocrine mitogenic factor that possibly induces their own transformation.[101] However, when injected into nude mice, R6-cells over-expressing PKCβ1 induce tumours with a lower frequency and longer latency period than cells transformed by the Ha-Ras oncogene (see Chapter 4, also Chapter 11 for the role of Ras in growth factor signalling). This should not be entirely unexpected in view of the idea that phorbol esters act as tumour promoters or co-carcinogens, not carcinogens in their own right. Co-expression of Ha-ras and PKCβ1 certainly results in greatly enhanced transformation.[102] In contrast, for R6 cells over-expressing PKC-α there is no tendency for transformation.[101] Instead, the growth is retarded and these cells achieve lower saturation densities than normal. In glioma cells, PKCα plays a growth-promoting role by suppressing the expression of the cell cycle inhibitor p21^WAF1/CIP1. (ref. 103) PKC-ζ has tumour suppressor qualities actually reverting v-Raf transformation of NIH-3T3 cells,[104] and PKC-δ inhibits proliferation of vascular smooth muscle cells by suppressing the expression of the cyclins D and E (components of the cell cycle machinery that prepare the cell for DNA synthesis: see 'The cell cycle', page 232) (Table 9.4).[105] It appears that, depending

Table 9.4 Phenotypic changes after over-expression of PKC isoforms

PKCα	PKCα	PKCβ1	PKCδ	PKCε	PKCζ
Rat colonic epithelial cells	Glioma cells	Rat embryo Fibroblasts (R6)	Vascular smooth muscle cells	Rat fibroblasts and colonic epithelial cells	v-raf transformed NIH-3T3
Tumour suppression	Tumour promotion	Tumour promotion	Tumour suppression	Tumour promotion	Tumour suppression
Slower growth Low densities	Suppression of p21$^{WAF/CIP}$	Growth in nude mice Anchorage independence Synergy with Ha-ras	Slower growth Reduced expression of cyclin D and E	Growth in nude mice Anchorage independence Synergy with ras	Reversion of transformed phenotype

on the cells and on the circumstances, different isoforms of PKC when over-expressed, can either induce or suppress the formation of the transformed cell phenotype.

Unfortunately, it has not been possible to draw any strong generalizations from these experiments. The tumour-promoting tendency of phorbol esters cannot be readily explained by a simple mechanism involving the activation of PKC. In addition, because prolonged treatment with phorbol esters has the effect of down-regulating the expression of conventional and novel PKCs (a phenomenon often used to dissect their role in signal transduction), it is possible in some instances that the proliferative response may be due to absence of PKC rather than its activation. Lastly, there are many other proteins that possess C1 domains similar to those present in the PKCs that may also be regulated by phorbol esters. One of these is Ras-GRP, a guanine nucleotide exchange factor for Ras, that promotes malignant transformation in fibroblasts when induced by phorbol ester.[106] Lastly, few natural tumours have been found that express PKC mutations or express PKC aberrantly. An example is presented in a subpopulation of human thyroid tumours that express a mutated PKCα.[107] Among other characteristics, this has a selective loss of substrate recognition and an altered subcellular localization.[108] How this explains aberrant growth regulation is not known.

▪ PKC and inflammation

▪ Phorbol ester and activation of endothelial cells and leukocytes

Phorbol ester (TPA) is a skin irritant and inflammatory agent. It induces prompt remodelling of the vasculature, resulting in oedema formation,[109] and this response is enhanced by the induced secretion of histamine from mast cells and blood-borne basophils (Figure 9.16).[110,111] It activates adhesion molecules present on the surface of leukocytes causing them to bind and then to migrate through the vascular endothelial layer into the tissues underlying the skin[112] (see Chapter 15). In the tissues, TPA potentiates antigen-mediated stimulation of T-lymphocytes, enhancing the produc-

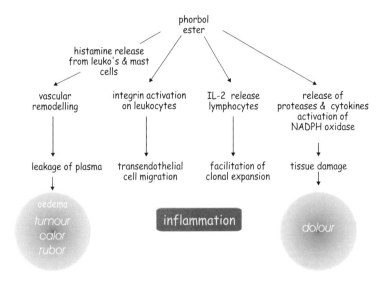

Figure 9.16 Induction of an inflammatory response by phorbol ester. Effects of phorbol on the vasculature and leukocytes and on release of mediators are illustrated. These events result in tumour (swelling), calor (heat), rubor (redness) and dolour (pain), four well-recognized characteristics of the inflammatory response.

tion of the cytokine interleukin-2 (IL-2) which is essential for the induction of clonal expansion (see Chapter 12).[113] This effect is mediated through activation of PKCθ and the transcription factor complex AP-1.[114,115] TPA also causes degranulation of neutrophils, releasing pro-inflammatory cytokines and matrix proteases, and activation of the respiratory burst.[116] This metabolic response is vital for first line host defence and is also implicated in the tissue destruction associated with chronic inflammatory diseases.

Neutrophils (polymorphonuclear neutrophilic granulocytes) are dedicated phagocytic cells and they constitute the first line of defence against invading micro-organisms. Their action can be dissected into three distinct processes:

1   direct migration towards micro-organisms (chemotaxis), followed by

2   engulfment of the micro-organism (phagocytosis) and finally

3   its killing (Figure 9.17).

The intracellular killing mechanism is based on two independent but mutually supportive mechanisms, the generation of toxic oxygen metabolites and the release of microbicidal proteins from the storage granules into the phagosome.[117] The generation of toxic oxygen metabolites, O_2^- and H_2O_2, is accompanied by a 20–30-fold increase in oxygen consumption, called the respiratory burst.[118] Its importance is made apparent by the finding that individuals having a genetic defect in any one of the components regulating the respiratory burst chain manifest recurrent bacterial infections due a reduced ability to kill invading microbes.[119]

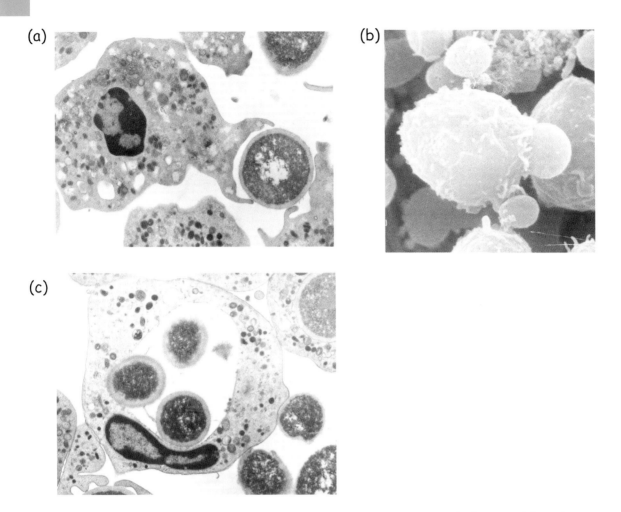

Figure 9.17 Killing of microbes by neutrophils. (a) Transmission electron micrograph of a neutrophil attaching to a yeast particle coated with IgG (immunoglobulin-treated zymosan). Note the formation of a protrusion which, using the Fc-IgG receptor to direct the movement, gradually engulfs the particle. This contact also results in the activation of the respiratory burst with the formation of toxic oxygen metabolites. (b) Scanning electron micrograph of the same event. (c) Transmission electron micrograph of a neutrophil containing several IgG-coated yeast particles in a large phagosome after 15 minutes of incubation. From Dirk Roos and Wim Leene, Amsterdam.

This condition is called chronic granulomatous disease (CGD), because of the formation of fibroblast-encapsulated granulomas, packed with neutrophils that harbour the microbes.[120,121] The enzyme responsible for the respiratory burst, NADPH : O_2 oxido-reductase (NADPH oxidase) catalyses the reaction

$$\dot{N}ADPH \rightarrow NADP^+ + H^+ + 2e^-$$

$$2O_2 + 2e^- \rightarrow 2O_2^-$$

It transfers two electrons from NADPH (oxidation) to molecular oxygen (reduction), generating $2O_2^-$. From reconstitution studies with recomb-

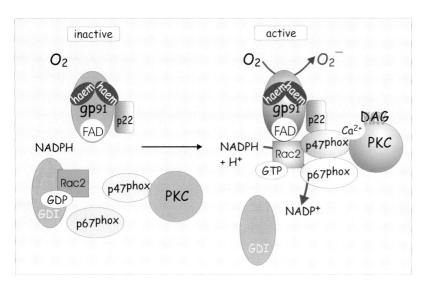

Figure 9.18 Assembly of the NADPH oxidase components upon activation of neutrophilic granulocytes. The membrane components act as docking sites for the cytosolic components. Rac is activated, detaches from GDI and exchanges GDP for GTP. Activated Rac interacts with p67[phox] and together they dock on to the gp91 subunit of cytochrome *b*. PKC is bound to p47[phox] and this enables its localization to the p22 subunit of cytochrome *b*.

inant proteins it became apparent that the NADPH oxidase is composed of a number of separate components (Figure 9.18):[122]

- p47[phox] a cytosolic component
- p67[phox] a cytosolic component
- Rac2   a monomeric GTP binding protein of the Rho family
- flavocytochrome b558 an $\alpha 1\beta 1$ transmembrane protein-dimer (gp91[phox]/p22[phox]).

Loss of function mutations (or partial deletions) have been observed in all of these components, all resulting in immunodeficiencies.[123]

phox Phagocyte oxidase.
gp Glycoprotein.
b558 A cytochrome-*b* having a haem absorption peak at 558 nm.

■ The role of PKC in the regulation of the respiratory burst

Phosphorylation of p47[phox] by PKC
Bacterial infection elicits activation of the serum complement cascade, generating the complement derived peptides C3 and C3bi which bind to the microbial surface. In addition, specific antibodies (immunoglobulin G, IgG) recognize and bind to microbial antigens. The binding (opsonization) of the complement peptides and IgG to the neutrophil Mac-1 and Fcγ-receptors activates phagocytosis and the respiratory burst. In addition the microbe may shed lipopolysaccharides (LPS) or small fragments of peptidoglycan and these act as priming agents, making the neutrophil more responsive to the opsonized particles. Occupation of the Mac-1 and the Fcγ-receptors activates PLCβ, resulting in the generation of DAG and

IP$_3$ (Figure 9.19).[124] Simultaneously, the predominant PKC isoforms, βI and βII, δ, and ζ redistribute from the cytosol to the plasma membrane.[125] This is followed by phosphorylation of p47^{phox} and the assembly of the other components of the NADPH oxidase. p67^{Phox} is recruited to the plasma membrane through activation of the GTPase Rac by the engaged Fcγ-receptors.[126] In the resting state, Rac is associated with GDI, that inhibits GDP dissociation and keeps it in the cytosol. p47^{phox} associates with the NADPH oxidase by interaction of its SH3 domain with the proline-rich region in p22^{phox}. The assembly effectively enables the transport of electrons from NADPH to O$_2$ (Figure 9.18).

PKC is an important regulator of the respiratory burst in human neutrophils, in particular, because of its role in the multiple (seven) phosphorylations of p47^{phox}. The respiratory burst activity can be induced directly by phorbol ester and in this case the level of phosphorylation of p47^{phox} matches the level of respiratory burst activity. Conversely, compounds that inhibit PKC prevent the respiratory burst.[127,128] It is also reduced by 50% in neutrophils obtained from mice that fail to express PKCβ(I+II),

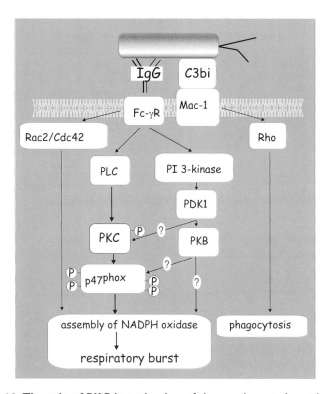

Figure 9.19 **The role of PKC in activation of the respiratory burst in neutrophilic granulocytes.** Activation of serum complement and binding of antibodies by microbial antigens results in occupation of both Fc-γR and Mac-1. Fc-γR initiates activation of PLC, PI 3-kinase and Rac2/Cdc42. Rac determines the activity of the NADPH oxidase. PKC phosphorylates p47^{phox} and this permits the activation of the respiratory burst. PI 3-kinase is also vital to respiratory burst activity but the downstream components are not yet resolved.

normally located around the phagosome during the process of bacterial uptake and killing.[128] PKCβ(I+II) is also associated with p47^{phox} (see Table 9.3) and this is necessary for the phosphorylation of further substrates. In individuals having the autosomal form of CGD due to the deletion of p47^{phox} the PKC necessarily loses its way.[129] Finally, addition of fully phosphorylated p47^{phox} to reconstituted NADPH oxidase enables electron transport to occur; the non-phosphorylated form does not.[130]

Here, however, is where the story runs into the sand. Surprisingly, the only serine residue that is vital to respiratory burst activity, Ser-379, is not phosphorylated by PKC.[131] Serine to alanine mutations of any of the other PKC phosphorylation sites still allow full respiratory burst activity in reconstituted NADPH oxidase systems. Secondly, stimulation of neutrophils with formylmethionyl peptides (see Chapter 15) causes phosphorylation of p47^{phox} to an extent that does not match respiratory burst activity.[127] Thirdly, the respiratory activity induced by addition of highly phosphorylated p47^{phox} to the other components of NADPH oxidase, is but a fraction of that which can be maximally obtained by the addition of anionic detergents (in the presence of non-phosphorylated p47^{phox}).[130,132]

In conclusion, PKC and p47^{phox} play important roles in the regulation of respiratory burst activity but the precise way in which they interact remains far from clear.

The respiratory burst can be reconstituted *in vitro* by addition of p47^{phox} and p67^{phox} to a membrane fraction of neutrophils, together with GTP-γ-S, NADPH with arachidonic acid or the detergent sodium dodecyl sulphate.

References

1 Krebs, E.G., Fischer, E.H. The phosphorylase *b* to a converting enzyme of rabbit skeletal muscle. *Biochim. Biophys. Acta* 1956; 20: 150–7.

2 Monod, J., Wyman, J., Changeux, J.P. On the nature of allosteric transitions: A plausible model. *J. Mol. Biol.* 1965; 12: 88–118.

3 Hunter, T., Plowman, G.D. The protein kinases of budding yeast: Six score and more. *Trends Biochem. Sci.* 1997; 22: 18–22.

4 Bliska, J.B., Guan, K.L., Dixon, J.E., Falkow, S. Tyrosine phosphate hydrolysis of host proteins by an essential *Yersinia* virulence determinant. *Proc. Natl. Acad. Sci. USA* 1991; 88: 1187–91.

5 Guan, K.L., Dixon, J.E. Protein tyrosine phosphatase activity of an essential virulence determinant in *Yersinia*. *Science* 1990; 249: 553–6.

6 Francis, S.H., Corbin, J.D. Structure and function of cyclic nucleotide-dependent protein kinases. *Annu. Rev. Physiol.* 1994; 56: 237–72.

7 Soderling, T.R. Protein kinases and phosphatases: regulation by autoinhibitory domains. *Biotechnol. Appl. Biochem.* 1993; 18: 185–200.

8 Erlichman, J., Rosenfeld, R., Rosen, O.M. Phosphorylation of a cyclic adenosine 3′:5′-monophosphate-dependent protein kinase from bovine cardiac muscle. *J. Biol. Chem.* 1974; 249: 5000–3.

9 Doskeland, S.O., Maronde, E., Gjertsen, B.T. The genetic subtypes of cAMP-dependent protein kinase – functionally different or redundant? *Biochim. Biophys. Acta* 1993; 1178: 249–58.

10 Shabb, J.B., Corbin, J.D. Cyclic nucleotide-binding domains in proteins having diverse functions. *J. Biol. Chem.* 1992; 267: 5723–6.

11 Gettys, T.W., Vine, A.J., Simonds, M.F., Corbin, J.D. Activation of the particulate low Km phosphodiesterase of adipocytes by addition of cAMP-dependent protein kinase. *J. Biol. Chem.* 1988; 263: 10359–63.

12 Gettys, T.W., Blackmore, P.F., Redmon, J.B., Beebe, S.J., Corbin, J.D. Short-term feedback regulation of cAMP by accelerated degradation in rat tissues. *J. Biol. Chem.* 1987; 262: 333–9.

13 Hubbard, M.J., Cohen, P. On target with a new mechanism for the regulation of protein phosphorylation. *Trends Biochem. Sci.* 1993; 18: 172–7.

14 Cesare, D. De, Fimiaa, G.M., Sassone-Corsi, P. Signalling routes to CREM and CREB: plasticity in transcriptional activation. *Trends Biochem. Sci.* 1999; 24: 281–5.

15 Impey, S., Wayman, G., Wu, Z., Storm, D.R. Type I adenylyl cyclase functions as a coincidence detector for control of cyclic AMP response element-mediated transcription: synergistic regulation of transcription by Ca^{2+} and isoproterenol. *Mol. Cell Biol.* 1994; 14: 8272–81.

16 Hanson, P.I., Schulman, H. Neuronal Ca^{2+}/calmodulin-dependent protein kinases. *Annu. Rev. Biochem.* 1992; 61: 559–601.

17 Hu, S.C., Chrivia, J., Ghosh, A. Regulation of CBP-mediated transcription by neuronal calcium signaling. *Neuron* 1999; 22: 799–808.

18 Schwaninger, M., Lux, G., Blume, R. *et al.* Membrane depolarization and calcium influx induce glucagon gene transcription in pancreatic islet cells through the cyclic AMP-responsive element. *J. Biol. Chem.* 1993; 268: 5168–77.

19 Bito, H., Deisseroth, K., Tsien, R.W. CREB phosphorylation and dephosphorylation: a Ca^{2+}- and stimulus duration-dependent switch for hippocampal gene expression. *Cell* 1996; 87: 1203–14.

20 Walker, W.H., Fucci, L., Habener, J.F. Expression of the gene encoding transcription factor cyclic adenosine 3′,5′-monophosphate (cAMP) response element-binding protein (CREB): regulation by follicle-stimulating hormone-induced cAMP signaling in primary rat Sertoli cells. *Endocrinology* 1995; 136: 3534–45.

21 Harootunian, A.T., Adams, S.R., Wen, W. *et al.* Movement of the free catalytic subunit of cAMP-dependent protein kinase into and out of the nucleus can be explained by diffusion. *Mol. Biol. Cell* 1993; 4: 993–1002.

22 Hagiwara, M., Alberts, A., Brindle, P. *et al.* Transcriptional attenuation following cAMP induction requires PP-1-mediated dephosphorylation of CREB. *Cell* 1992; 70: 105–13.

23 Alberts, A.S., Montminy, M., Shenolikar, S., Feramisco, J.R. Expression of a peptide inhibitor of protein phosphatase 1 increases phosphorylation and activity of CREB in NIH 3T3 fibroblasts. *Mol. Cell Biol.* 1994; 14: 4398–407.

24 Wadzinski, B.E., Wheat, W.H., Jaspers, S., *et al.* Nuclear protein phosphatase 2A dephosphorylates protein kinase A-phosphorylated CREB and regulates CREB transcriptional stimulation. *Mol. Cell Biol.* 1993; 13: 2822–34.

25 Van Eynde, A., Wera, S., Beullens, M., Torrekens, S. *et al.* Molecular cloning of NIPP-1, a nuclear inhibitor of protein phosphatase-1, reveals homology with polypeptides involved in RNA processing. *J. Biol. Chem.* 1995; 270: 28068–74.

26 Daaka, Y., Luttrell, L.M., Lefkowitz, R.J. Switching of the coupling of the α2-adrenergic receptor to different G proteins by protein kinase A. *Nature* 1999; 390: 88–91.

27 Nakamura, T., Gold, G.H. A cyclic nucleotide-gated conductance in olfactory receptor cilia. *Nature* 1987; 325: 442–4.

28 Liman, E.R., Buck, L.B. A second subunit of the olfactory cyclic nucleotide-gated channel confers high sensitivity to cAMP. *Neuron* 1994; 13: 611–21.

29 Bradley, J., Li, J., Davidson, N., Lester, H.A., Zinn, K. Heteromeric olfactory cyclic nucleotide-gated channels: a subunit that confers increased sensitivity to cAMP. *Proc. Natl. Acad. Sci. USA* 1994; 91: 8890–4.

30 Chen, T.Y., Yau, K.W. Direct modulation by $Ca^{(2+)}$-calmodulin of cyclic nucleotide-activated channel of rat olfactory receptor neurons. *Nature* 1994; 368: 545–8.

31 Zwartkruis, F.J., Bos, J.L. Ras and Rap1: two highly related small GTPases with distinct function. *Exp. Cell Res.* 2000; 1253: 157–65.

32 de Rooij, J., Zwartkruis, F.J.T., Verheijen, M.H.G. *et al.* Epac is a Rap1 guanine-nculeotide-exchange factor directly activated by cyclic AMP. *Nature* 1998; 396: 474–7.

33 Jordan, J.D., Carey, K.D., Stork, P.J., Iyengar, R. Modulation of rap activity by direct interaction of Gα(o) with Rap1 GTPase-activating protein. *J. Biol. Chem.* 1999; 274: 21507–10.

34 Takai, Y., Kishimoto, A., Inoue, M., Nishizuka, Y. Studies on a cyclic nucleotide-independent protein kinase and its proenzyme in mammalian tissues. I. Purification and characterization of an active enzyme from bovine cerebellum. *J. Biol. Chem.* 1977; 252: 7603–9.

35 Inoue, M., Kishimoto, A., Takai, Y., Nishizuka, Y. Studies on a cyclic nucleotide-independent protein kinase and its proenzyme in mammalian issues. II. Proenzyme and its activation by calcium-dependent protease from rat brain. *J. Biol. Chem.* 1977; 252: 7610–16.

36 Takai, Y., Kishimoto, A., Iwasa, Y. *et al.* Calcium-dependent activation of a multifunctional protein kinase by membrane phospholipids. *J. Biol. Chem.* 1979; 254: 3692–5.

37 Kishimoto, A., Takai, Y., Mori, T., Kikkawa, U., Nishizuka, Y. Activation of calcium and phospholipid-dependent protein kinase by diacylglycerol, its possible relation to phosphatidylinositol turnover. *J. Biol. Chem.* 1980; 255: 2273–6.

38 Duuren, B.L. Van, Orris, L., Arroyo, E. Tumour-enhancing activity of the active principles of *Croton tiglium* L. *Nature* 1963; 200: 1115–16.

39 Crombie, L., Games, M.L., Pointer, D.J. Chemistry and structure of phorbol, the diterpene parent of the co-carcinogens of croton oil. *J. Chem. Soc. (Perkin)* 1968; 1: 1362.

40 Castagna, M., Takai, Y., Kaibuchi, K. *et al.* Direct activation of calcium-activated, phospholipid-dependent protein kinase by tumor-promoting phorbol esters. *J. Biol. Chem.* 1982; 257: 7847–51.

41 Parker, P.J., Coussens, L., Totty, N. *et al.* The complete primary structure of protein kinase C – the major phorbol ester receptor. *Science* 1986; 233: 853–9.

42 Coussens, L., Parker, P.J., Rhee, L. *et al.* Multiple, distinct forms of bovine and human protein kinase C suggest diversity in cellular signaling pathways. *Science* 1986; 233: 859–66.

43 Nishizuka, Y. Protein kinase C and lipid signaling for sustained cellular responses. *FASEB J.* 1995; 7: 484–96.

44 Dekker, L.V., Palmer, R.H., Parker, P.J. The protein kinase C and protein kinase C related gene families. *Curr. Opin. Struct. Biol.* 1995; 5: 396–402.

45 Klemp, B.E., Pearson, R.B. Protein kinase recognition sequence motifs. *Trends Biochem. Sci.* 1990; 15: 342–6.

46 Nishikawa, K., Toker, A., Johannes, F.J., Songyang, Z., Cantley, L.C. Determination of the specific substrate sequence motifs of protein kinase C isozymes. *J. Biol. Chem.* 1997; 272: 952–60.

47 Ohno, S., Kawasaki, H., Imajoh, S. *et al.* Tissue-specific expression of three distinct types of rabbit protein kinase C. *Nature* 1987; 325: 161–6.

48 Tabuse, Y., Nishiwaki, K., Miwa, J. Mutations in a protein kinase C homolog confer phorbol ester resistance on *Caenorhabditis elegans*. *Science* 1989; 243: 1713–16.

49 Wu, S.L., Staudinger, J., Olson, E.N., Rubin, C.S. Structure, expression, and properties of an atypical protein kinase C (PKC3) from *Caenorhabditis elegans*. PKC3 is required for the normal progression of embryogenesis and viability of the organism. *J. Biol. Chem.* 1998; 273: 1130–43.

50 Rosenthal, A., Rhee, L., Yadegari, R. *et al.* Structure and nucleotide sequence of a *Drosophila melanogaster* protein kinase C gene. *EMBO J.* 1987; 6: 433–41.

51 Newton, A.C. Protein kinase C. Seeing two domains. *Curr. Biol.* 1995; 5: 973–6.

52 Hurley, J.H., Newton, A.C., Parker, P.J., Blumberg, P.M., Nishizuka, Y. Taxonomy and function of C1 protein kinase C homology domains. *Protein Sci.* 1997; 2: 477–80.

53 Kazanietz, M.G. Eyes wide shut: protein kinase C isozymes are not the only receptors for the phorbol ester tumor promoters. *Mol. Carcinog.* 2000; 28: 5–11.

54 House, C., Kemp, B.E. Protein kinase C contains a pseudosubstrate prototope in its regulatory domain. *Science* 1987; 238: 1726–8.

55 Newton, A.C. Regulation of protein kinase C. *Curr. Opin. Cell Biol.* 1997; 9: 161–7.

56 Dutil, E.M., Newton, A.C. Dual role of pseudosubstrate in the coordinated regulation of protein kinase C by phosphorylation and diacylglycerol. *J. Biol. Chem.* 2000; 275: 10697–70.

57 Le Good, J.A., Ziegler, W.H., Parekh, D.B. *et al.* Protein kinase C isotypes controlled by phosphoinositide 3-kinase through the protein kinase PDK1. *Science* 2000; 281: 2042–5.

58 Keranen, L.M., Dutil, E.M., Newton, A.C. Protein kinase C is regulated in vivo by three functionally distinct phosphorylations. *Curr. Biol.* 1995; 5: 1394–403.

59 Le Good, J.A., Ziegler, W.H., Parekh, D.B. *et al.* Protein kinase C isotypes controlled by

phosphoinositide 3-kinase through the protein kinase PDK1. *Science* 1998; 281: 2042–2045.

60 Parekh, D.B., Ziegler, W., Parker, P.J. Multiple pathways control protein kinase C phosphorylation. *EMBO J.* 2000; 19: 496–503.

61 Toker, A. Signaling through protein kinase C. *Front. BioSci.* 1998; 3: 1134–47.

62 Berridge, M.J., Irvine, R. Inositol trisphosphate, a novel second messenger in cellular signal transduction. *Nature* 2000; 312: 315–21.

63 Oancea, E., Meyer, T. Protein kinase C as a molecular machine for decoding calcium and diacylglycerol signals. *Cell* 1998; 95: 307–18.

64 Toker, A., Meyer, M., Reddy, K.K. *et al.* Activation of protein kinase C family members by the novel polyphosphoinositides PtdIns-3,4-P2 and PtdIns-3,4,5-P3. *J. Biol. Chem.* 1994; 269: 32358–67.

65 Sando, J.J., Chertihin, O.I. Activation of protein kinase C by lysophosphatidic acid: dependence on composition of phospholipid vesicles. *Biochem. J.* 1996; 317: 583–8.

66 Hannun, Y.A., Bell, R.M. Functions of sphingolipids and sphingolipid breakdown products in cellular regulation. *Science* 1989; 243: 500–7.

67 Diaz-Meco, M.T., Municio, M.M., Sanchez, P., Lozano, J., Moscat, J. λ-interacting protein, a novel protein that specifically interacts with the zinc finger domain of the atypical protein kinase C isotype λ/ι and stimulates its kinase activity in vitro and in vivo. *Mol. Cell Biol.* 1996; 16: 105–14.

68 Otte, A.P., Kramer, I.M., Durston, A.J. Protein kinase C and regulation of the local competence of *Xenopus* ectoderm. *Science* 1991; 251: 570–3.

69 Otte, A.P., Moon, R.T. Protein kinase C isozymes have distinct roles in neural induction and competence in *Xenopus. Cell* 1992; 68: 1021–9.

70 Goodnight, J.A., Mischak, H., Kolch, W., Mushinski, J.F. Immunocytochemical localization of eight protein kinase C isozymes overexpressed in NIH 3T3 fibroblasts. Isoform-specific association with microfilaments, Golgi, endoplasmic reticulum, and nuclear and cell membranes. *J. Biol. Chem.* 1995; 270: 9991–10001.

71 Dong, L., Chapline, C., Mousseau, B. *et al.* 35H, a sequence isolated as a protein kinase C binding protein, is a novel member of the adducin family. *J. Biol. Chem.* 1995; 270: 25534–40.

72 Geijsen, N., Spaargaren, M., Raaijmakers, J.A. *et al.* Association of RACK1 and PKCβ with the common β-chain of the IL-5/IL-3/GM-CSF receptor. *Oncogene* 1999; 18: 5126–30.

73 Csukai, M., Chen, C.H., Matteis, M.A. De, Mochly-Rosen, D. The coatomer protein β-COP, a selective binding protein (RACK) for protein kinase Cε. *J. Biol. Chem.* 1997; 272: 29200–6.

74 Jaken, S., Parker, P.J. Protein kinase C binding partners. *Bioessays* 2000; 22: 245–54.

75 Xu, X.Z., Choudhury, A., Montell, C. Coordination of an array of signaling proteins through homo- and heteromeric interactions between PDZ domains and target proteins. *J. Cell. Biol.* 1998; 142: 545–55.

76 Smith, D.P., Ranganathan, R., Hardy, R.W. *et al.* Photoreceptor deactivation and retinal degeneration mediated by a photoreceptor-specific protein kinase C. *Science* 1991; 254: 1478–84.

77 Pound, A.W., Bell, J.R. The influence of croton oil stimulation on tumour initiation by urethane in mice. *Br. J. Cancer* 1962; 16: 690–5.

78 Sivak, A., Duuren, B.L. Van. Phenotypic expression of transformation: induction in cell culture by a phorbol ester. *Science* 1967; 157: 1443–4.

79 Imbra, R.J., Karin, M. Phorbol ester induces the transcriptional stimulatory activity of the SV40 enhancer. *Nature* 1986; 323: 555–8.

80 Housey, G.M., Johnson, M.D., Hsiao, W.L. *et al.* Overproduction of protein kinase C causes disordered growth control in rat fibroblasts. *Cell* 1998; 52: 343–5.

81 Angel, P., Imagawa, M., Chiu, R. *et al.* Phorbol ester-inducible genes contain a common cis element recognized by a TPA-modulated trans-acting factor. *Cell* 1987; 49: 729–39.

82 Sassone-Corsi, P., Ransone, L.J., Lamph, W.W., Verma, I.M. Direct interaction between fos and jun nuclear oncoproteins: role of the 'leucine zipper' domain. *Nature* 1988; 336: 692–5.

83 Karin, M. Zg. Liu, Zandi, E. AP-1 function and regulation. Cell 1997; 59: 709–17.

84 Bohmann, D., Tjian, R. Biochemical analysis of transcriptional activation by Jun: differential activity of c- and v-Jun. *Cell* 1989; 59: 709–17.

85 Kramer, I.M., Koornneef, I., de Laat, S.W., van den Eijnden-van Raaij, A.J. TGF-β1 induces phosphorylation of the cyclic AMP responsive element binding protein in ML-CCl64 cells. *EMBO J.* 1991; 10: 1083–9.

86 Masquilier, D., Sassone, C.P. Transcriptional cross-talk: nuclear factors CREM and CREB bind to AP-1 sites and inhibit activation by Jun. *J. Biol. Chem.* 1992; 267: 22460–66.

87 Boyle, W.J., Smeal, T., Defize, L.H. *et al.* Activation of protein kinase C decreases phosphorylation of c-Jun at sites that negatively regulate its DNA-binding activity. *Cell* 1991; 64: 573–84.

88 Goode, N., Hughes, K., Woodgett, J.R., Parker, P.J. Differential regulation of glycogen synthase kinase-3 beta by protein kinase C isotypes. *J. Biol. Chem.* 1992; 267: 16878–82.

89 Smeal, T., Binetruy, B., Mercola, D.A., Birrer, M., Karin, M. Oncogenic and transcriptional cooperation with Ha-Ras requires phosphorylation of c-Jun on serines 63 and 73. *Nature* 2000; 354: 494–6.

90 Behrens, A., Jochum, W., Sibilia, M., Wagner, E.F. Oncogenic transformation by ras and fos is mediated by c-Jun N-terminal phosphorylation. *Oncogene* 2000; 19: 2657–63.

91 Derijard, B., Hibi, M., Wu, I.H. *et al.* JNK1: a protein kinase stimulated by UV light and Ha-Ras that binds and phosphorylates the c-Jun activation domain. *Cell* 1994; 76: 1025–37.

92 Kallunki, T., Deng, T., Hibi, M. and M. Karin, c-Jun can recruit JNK to phosphorylate dimerization partners via specific docking interactions. *Cell* 1996; 87: 929–39.

93 Treisman, R. Transient accumulation of c-fos RNA following serum stimulation requires a conserved 5′ element and c-fos 3′ sequences. *Cell* 1985; 42: 889–902.

94 Buscher, M., Rahmsdorf, H.J., Litfin, M., Karin, M., Herrlich, P. Activation of the c-fos gene by UV and phorbol ester: different signal transduction pathways converge to the same enhancer element. *Oncogene* 1988; 3: 301–11.

95 Marais, R., Wynne, J., Treisman, R. The SRF accessory protein Elk-1 contains a growth factor-regulated transcriptional activation domain. *Cell* 1993; 73: 381–93.

96 Whitmarsh, A.J., Shore, P., Sharrocks, A.D., Davis, R.J. Integration of MAP kinase signal transduction pathways at the serum response element. *Science* 1995; 269: 403–7.

97 Egan, S.E., Weinberg, R.A. The pathway to signal achievement. *Nature* 1993; 365: 781–3.

98 Kolch, W., Heidecker, G., Kochs, G. *et al.* Protein kinase Cα activates RAF-1 by direct phosphorylation. *Nature* 1993; 364: 249–52.

99 Cacace, A.M., Ueffing, M., Philipp, A. *et al.* PKC epsilon functions as an oncogene by enhancing activation of the Raf kinase. *Oncogene* 1996; 13: 2517–26.

100 Schonwasser, D.C., Marais, R.M., Marshall, C.J., Parker, P.J. Activation of the mitogen-activated protein kinase/extracellular signal-regulated kinase pathway by conventional, novel, and atypical protein kinase C isotypes. *Mol. Cell Biol.* 1998; 18: 790–8.

101 Borner, C., Ueffing, M., Jaken, S., Parker, P.J., Weinstein, I.B. Two closely related isoforms of protein kinase C produce reciprocal effects on the growth of rat fibroblasts. Possible molecular mechanisms. *J. Biol. Chem.* 1995; 270: 78–86.

102 Hsiao, W.L., Housey, G.M., Johnson, M.D., Weinstein, I.B. Cells that overproduce protein kinase C are more susceptible to transformation by an activated H-ras oncogene. *Mol. Cell Biol.* 1989; 9: 2641–7.

103 Besson, A., Yong, V.W. Involvement of p21(Waf1/Cip1) in protein kinase C alpha-induced cell cycle progression. *Mol. Cell Biol.* 2000; 20: 4580–90.

104 Kieser, A., Sietz, T., Adler, H.S. *et al.* Protein kinase Cα reverts v-raf transformation of NIH-3T3 cells. *Genes Dev.* 1966; 10: 1455–66.

105 Fukumoto, S., Nishizawa, Y., Hosoi, M. *et al.* Protein kinase C delta inhibits the proliferation of vascular smooth muscle cells by suppressing G1 cyclin expression. *J. Biol. Chem.* 1997; 272: 13816–22.

106 Tognon, C.E., Kirk, H.E., Passmore, L.A. *et al.* Regulation of RasGRP via a phorbol ester-responsive C1 domain. *Mol. Cell Biol.* 1998; 18: 6995–7008.

107 Prevostel, C., Alvaro, V., Boisvilliers, F. de *et al.* The natural protein kinase C alpha mutant is present in human thyroid neoplasms. *Oncogene* 1995; 11: 669–74.

108 Prevostel, C., Alvaro, V., Vallentin, A. *et al.* Selective loss of substrate recognition induced by the tumour-associated D294G point mutation in protein kinase Cα. *Biochem. J.* 1998; 334: 393–7.

109 Janoff, A., Klassen, A., Troll, W. Local vascular changes induced by the cocarcinogen, phorbol myristate acetate. *Cancer Res.* 2000; 30: 2568–71.

110 Katakami, Y., Kaibuchi, K., Sawamura, M., Takai, Y., Nishizuka, Y. Synergistic action of protein kinase C and calcium for histamine release from rat peritoneal mast cells. *Biochem. Biophys. Res. Commun.* 1984; 121: 573–8.

111 Schleimer, R.P., Gillespie, E., Lichtenstein, L.M. Release of histamine from human leukocytes stimulated with the tumor-promoting phorbol diesters. I. Characterization of the response. *J. Immunol.* 1981; 126: 570–4.

112 Kavanaugh, A.F., Lightfoot, E., Lipsky, P.E., Oppenheimer-Marks, N. Role of CD11/CD18 in adhesion and transendothelial migration of T cells. Analysis utilizing CD18-deficient T cell clones. *J. Immunol.* 1991; 146: 4149–56.

113 Rosenstreich, D.L., Mizel, S.B. Signal requirements for T lymphocyte activation. I. Replacement of macrophage function with phorbol myristic acetate. *J. Immunol.* 1979; 123: 1749–54.

114 Werlen, G., Jacinto, E., Xia, Y., Karin, M. Calcineurin preferentially synergizes with PKC-theta to activate JNK and IL-2 promoter in T lymphocytes. *EMBO J.* 1998; 17: 3101–11.

115 Sun, Z., Arendt, C.W., Ellmeier, W. *et al.* PKC0 is required for TCR-induced NF-κB activation in mature but not immature T lymphocytes. *Nature* 2000; 23: 402–7.

116 Repine, J.E., White, J.G., Clawson, C.C., Holmes, B.M. The influence of phorbol myristate acetate on oxygen consumption by polymorphonuclear leukocytes. *J. Lab. Clin. Med.* 1974; 83: 911–20.

117 Klebanoff, S.J., Clark, R.A. *The Neutrophil: function and clinical disorders.* Elsevier Biomedical Press, Amsterdam, 1987; 1–3.

118 Sbarra, A.J., Karnovsky, M.L. The biochemical basis of phagocytosis. I. Metabolic changes during the ingestion of particles by polymorphonuclear leukocytes. *J. Biol. Chem.* 1995; 234: 1355–62.

119 Holmes, B., Page, A.R., Good, R.A. Studies of the metabolic activity of leukocytes from patients with a genetic abnormality of phagocytic function. *J. Clin. Invest.* 1967; 46: 1422–32.

120 Baehner, R.L. The CGD leukocyte – a model of impaired phagocytic function. *N. Engl. J. Med.* 1969; 280: 1355.

121 Tauber, A.J., Borregaard, N., Simons, E., Wright, J. Chronic granulomatous disease: a syndrome of phagocyte deficiencies. *Medicine (Baltimore)* 1983; 62: 268–309.

122 Abo, A., Boyhan, A., West, I., Thrasher, A.J., Segal, A.W. Reconstitution of neutrophil NADPH oxidase activity in the cell-free system by four components: p67-phox, p47-phox, p21rac1, and cytochrome b-245. *J. Biol. Chem.* 1992; 267: 16767–70.

123 Ambruso, D.R., Knall, C., Abell, A.N. *et al.* Human neutrophil immunodeficiency syndrome is associated with an inhibitory Rac2 mutation. *Proc. Natl. Acad. Sci. USA* 2000; 97: 4654–9.

124 Garcia Gil, M., Alonso, F., Alvarez Chiva, V., Sanchez Crespo, M., Mato, J.M. Phospholipid turnover during phagocytosis in human polymorphonuclear leucocytes. *Biochem. J.* 1982; 206: 67–72.

125 Sergeant, S., McPhail, L.C. Opsonized zymosan stimulates the redistribution of protein kinase C isoforms in human neutrophils. *J. Immunol.* 1997; 159: 2877–85.

126 Caron, E., Hall, A. Identification of two distinct mechanisms of phagocytosis controlled by different Rho GTPases. *Science* 1998; 282: 1721.

127 Kramer, I.M., van der Bend, R.L., Tool, A.T. *et al.* 1-O-hexadecyl-2-O-methylglycerol, a novel inhibitor of protein kinase C, inhibits the respiratory burst in human neutrophils. *J. Biol. Chem.* 1989; 264: 5876–84.

128 Dekker, L.V., Leitges, M., Altschuler, G. *et al.* Protein kinase C-β contributes to NADPH oxidase activation in neutrophils. *Biochem. J.* 2000; 347: 285–9.

129 Reeves, E.P., Dekker, L.V., Forbes, L.V. *et al.* Direct interaction between p47phox and

protein kinase C: evidence for targeting of protein kinase C by p47phox in neutrophils. *Biochem. J.* 1999; 344: 859–66.

130 Lopes, L.R., Hoyal, C.R., Knaus, U.G., Babior, B.M. Activation of the leukocyte NADPH oxidase by protein kinase C in a partially recombinant cell-free system. *J. Biol. Chem.* 1999; 274: 15533–7.

131 Faust, L.R., el Benna, J., Babior, B.M. and Chanock, S.J. The phosphorylation targets of p47phox, a subunit of the respiratory burst oxidase. Functions of the individual target serines as evaluated by site-directed mutagenesis. *J. Clin. Invest.* 1995; 96: 1499–505.

132 Bromberg, Y., Pick, E. Activation of NADPH-dependent superoxide production in a cell-free system by sodium dodecyl sulfate. *J. Biol. Chem.* 1985; 260: 13539–45.

Growth factors: Setting the framework

The trails that led to the discovery of the growth factors (and related messengers) are very different from those that revealed the first hormones and neurotransmitters. For a start, people knew, more or less, what to look for and where to go looking and, in general, the tale is somewhat less romantic and less fraught with angst and vehement disagreements.

However, what began with a simple search for factors that would sustain living cells in laboratory conditions has expanded into a plethora of subject areas which are subjects in their own right. These include inflammation, wound healing, immune surveillance, development and carcinogenesis. To confront this bewildering prospect we first set out some details of how it evolved initially so that the reader is aware of how the major questions developed and have been confronted. A difficulty arises from the convergence of different disciplines, each bringing with it the baggage of its favoured nomenclature. We return to this matter at the end of this chapter.

For the ultimate in aggressive rivalry and claims to primacy in hormone discovery, coupled to barely restrained personality conflict, see Nicholas Wade's accounts of the discovery by Roger Guillemin and Andrew Schally of thyrotropin releasing factor and other hypothalamic peptides.[1–3]

Viruses and tumours

The first report of a tumour linked to a virus appeared in 1908, when Ellermann and Bang obtained a filterable agent from a chicken leukaemia and were able to make six passages of it, from fowl to fowl, producing the same disease each time.[4] Their report was generally disregarded. Leukaemia was not considered to be a tumour, though as should have been evident from the work of Aldred Warthin (Figure 10.1), first reporting in 1904,[5,6] a link between leukaemias and tumours, and hence cell growth and proliferation, had already been established:

These conditions are comparable to malignant tumors. The formation of metastases, the infiltrative and destructive growth, the failure of inoculations and transplantations etc., all favor the view that they are neoplasms, and present the same problems as do the malignant tumors.

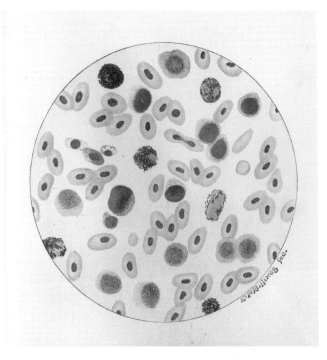

Figure 10.1 Stained blood smear from a leukaemic fowl.
In December 1905, there came into my hands a Buff Cochin Bantam hen showing signs of illness in the way of indisposition to move about and a general weakness of a progressive character. No symptoms of ordinary fowl diseases were present . . . Examination of blood smears showed, however, a great increase of white cells of the large lymphocyte type . . . A diagnosis of leukemia was therefore made . . . A great variety of staining methods were used, including the most recent methods for the staining of spirochetes and protozoan parasites . . . No evidence of the existence of any infective agent could be obtained.[6]
Note that avian red blood cells (yellow stain) are nucleated.

Francis Peyton Rous (1879–1970) shared the 1966 Nobel Prize for Physiology or Medicine for his discovery of tumour-inducing viruses.

Sarcoma A form of cancer that arises in the supportive tissues such as bone, cartilage, fat or muscle.
Carcinoma A malignant new growth that arises from epithelium, found in skin or, more commonly, the lining of body organs, for example: breast, prostate, lung, stomach or bowel. Carcinomas tend to infiltrate into adjacent tissue and spread (metastasize) to distant organs, for example: to bone, liver, lung or the brain.

More than 20 years were to pass before the leukaemias were eventually recognized as being tumours or neoplastic diseases.

In 1910, Francis Peyton Rous described a chicken tumour, identified as a sarcoma, that could be propagated by transplanting its cells, these then multiplying in their new hosts giving rise to tumours of the same sort.[7] The cells yielded a virus, now known as Rous sarcoma virus (RSV), from which, in 1980, the first protein tyrosine kinase, v-Src, was isolated.[8,9] Writing at the end of his career in 1967, Rous[10] gives an apt description of how things were done:

Those were primitive times in the raising of chickens. They were sold in a New York market not far from the institute, and many individual breeders brought their stock there. Every week F. S. Jones, a gifted veterinarian attached to my laboratory, went to it, not only to buy living chickens with lumps which might be tumors, but any that seemed sickly and had a pale comb, as perhaps having leukemia. Thus within less than four years we got more than 60 spontaneous tumors of various sorts . . .

■ The discovery of NGF . . . and EGF

Among the fractions that I assayed *in vitro* the following day, there was one containing snake venom. Having not been told which of the fractions had been specially treated, I was completely stunned by the stupendous halo radiating from the ganglia. I called Stan in without telling him what I had seen. He looked through the microscope's eyepieces, lifted his head, cleaned his glasses which had fogged up, and looked again. 'Rita', he murmured, 'I'm afraid we've just used up all the good luck we're entitled to. From now on, we can only count on ourselves'. . . . Events were to prove him wrong.

Nerve growth factor (NGF) may perhaps be regarded as the first identified growth factor, but there were many early clues hinting at their existence. The embryologist Hans Spemann (Nobel Prize, 1935) had described the eponymous 'organizer' that directs the creation of the antero-posterior axis in the gastrula stage in the development of the amphibian embryo. This work had clearly indicated that soluble factors made by the embryonic cells must be instrumental in regulating cell proliferation and differentiation.[13]

Rita Levi-Montalcini's affair with growth factors had its origin at that time, when, dismissed from her university post by the Nazi racial edicts, she determined to continue on alone. She had been alerted to the work of Spemann by an article written by one of his protégés, Viktor Hamburger. He described the concept of the inductive reaction of certain tissues on others during early development. In particular, he cited the effect of ablating the embryo limb buds of chicks upon the reduction in the volume of the motor column and the spinal ganglia responsible for the innervation of the limbs. The idea was that the failure of the cells to differentiate and to develop is due to the absence of an inductive factor normally released by the innervated tissues. She aimed to understand how the excision of non-innervated tissue could affect differentiation and subsequent development.

Confined to working in her bedroom, she made use of just the most basic materials and instruments needed for histological investigation. In her examination of this problem, Levi-Montalcini found that nerve cell differentiation proceeds quite normally in the embryos with excised limbs, but that a degenerative process (what we would now call apoptosis: see Apoptosis, page 335) commences as soon as the cells emerge from the cord and ganglia appear at the stump of the amputated limb.[14,15] It appeared to her that the failure to develop was best explained by the absence of a trophic factor.

In 1946, she sailed for the USA in the company of Renato Dulbecco (see below), a friend from her student days. She was destined for St Louis, he for Bloomington. Intending to pay a brief visit of a few weeks to the laboratory of Viktor Hamburger, she remained there for 30 years. The work that led eventually to the discovery of NGF had its origins in the observation by a late student of Hamburger's, Elmer Beuker. He had had reported that fragments of an actively growing mouse tumour, grafted on to chick

Stanley Cohen and Rita Levi-Montalcini, awarded the Nobel Prize, 1986, "for their discoveries of growth factors"

The quotations in this section are taken from *In Praise of Imperfection, My life and work* by Rita Levi-Montalcini[11]). A facsimile collection of Levi-Montalcini's major papers, with annotations by the author, is presented in reference [12].

embryos, caused a great ramification of nerve fibres into the mass of tumour cells.[16] Seeking advice and encouragement, he suggested that the tumour generated conditions favourable for the differentiation of the nerve cells which was reflected in the increased volume of the ganglia. Repeating the experiment, Levi-Montalcini describes the extraordinary spectacle of seeing bundles of nerve fibres passing between the tumour cells like rivulets of water flowing steadily over a bed of stones. In no case did they make any connection with the cells, as is the rule when fibres innervate normal embryonic or adult tissue. Later she describes how the sympathetic fibres invaded the embryonal viscera, even entering into lumen of the venous, but not the arterial blood vessels so that the smaller veins were quite obstructed.

> The penetration of the nerve fibers into the veins, furthermore, suggested to me that this still unknown humoral substance might be exerting a neurotropic effect, or what is known as a chemotactic directing force, one that causes nerve fibers to grow in a particular direction. . . . Among these, I guessed, was undoubtedly also the humoral growth factor that the cells produced. This hypothesis would explain this most atypical finding of sympathetic fibers gaining access inside the veins. . . . Now, with the hindsight of the nearly forty years gone by since those moments of keenest excitement – it appears that the new field of research that was opening up before my eyes was, in reality, much vaster than I could possibly have imagined.

It was clearly necessary to develop an in-vitro assay. These were early days in the field of cell culture and it took the best part of 6 months to develop a practical method that could be used as the basis for measuring the biological activity of fractions in protein purification. Only then was the point reached when a biochemical approach could usefully be applied in the pursuit of NGF, the name by which the factor became known shortly after.

> 'Rita', Stan said one day, 'you and I are good, but together, we are wonderful'.

After a year's intense work, they had narrowed down the factor as a nucleoprotein, though Stanley Cohen suspected that the nucleic acid component was likely to be a contaminant. On the advice of Arthur Kornberg, he applied an extract of snake venom as a source of nuclease activity with the aim of removing nucleic acids, present as an impurity in their material. To their great surprise this yielded a preparation that enhanced neuronal growth still further. It emerged that the snake venom alone was active.[17] Making the connection between venom and saliva, they tested mouse salivary glands and found that this too is an excellent source of NGF activity. (If they had been able to purchase purified nuclease enzyme, the course of the discovery must surely have been prolonged.)

Later, Cohen discovered a new phenomenon that 'was destined to become a magic wand that opened a whole new horizon to biological studies'. A contaminating factor was present that caused precocious

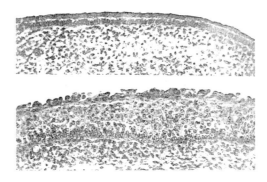

Figure 10.2 **An early demonstration of the effects of EGF.** Thickening of the epidermal layer in skin explants from chick embryos after 3 days in culture.[18]

growth in epidermis (Figure 10.2). It was later discovered that it also had a powerful proliferative effect on connective tissues and it became clear that there is a link between the mechanisms that control normal proliferation and neoplastic growth.

Since, under culture conditions, the stimulus to proliferate could not involve systemic or hormonal influences, Cohen called the new protein epidermal growth factor (EGF).[18,19] Later, it was shown that mouse EGF enhances DNA synthesis in cultured human fibroblasts. It was also found that EGF is similar to uragastrone, a peptide that had been isolated from human urine and recognized by pharmacologists because of its ability to inhibit gastric acid secretion.[20] Out of 53 residues in the amino acid sequence, 37 share a common location and the two polypeptides have similar effects on both gastric acid secretion and the growth of epidermal cells.

All this now paved the way for a more molecular approach using isolated cells. It was found that rat kidney cells, transformed with the Kirsten sarcoma virus (see Chapter 4) fail to bind EGF. This down-regulation, which is due to internalization of the receptor, is caused by the elevated expression and release of EGF by the cells themselves (an autocrine mechanism of feedback regulation). The possibility that internalization might be a necessary step initially found many adherents, but it became apparent that the transduction mechanism emanates from events at the plasma membrane. In particular, ligand binding directly induces phosphorylation (on tyrosine residues) of a membrane protein, later shown to be the EGF receptor (EGF-R) itself.[21] This was an important breakthrough because tyrosine kinase activity had already been associated with virus-induced sarcomas (the v-*src* gene product). Thus, a firm link was established between neoplasia and the physiological regulation of cellular growth.

Further evidence for the role of growth factors in tumour generation came with the revelation that the avian erythroblastosis virus oncogene, v-*erb-B*, codes a product having similarities with the EGF receptor.

Indeed, it became apparent that the transformation derives from inappropriate acquisition from the (cellular) c-*erb-B* gene, of a truncated receptor, lacking the binding site for EGF and which is constitutively activated.[22]

The development of this field has generated much excitement but also frustration. It has been exciting because it has yielded a good understanding of cell transformation and growth factors, and frustrating because it has become clear that cancer cells do not readily lend themselves as specific targets for drugs. The main impetus behind these studies was that non-mammalian genes might be the cause of disease. The hope was to discover targets which might be exploited to kill tumour cells selectively, for example the products of the viral genes. We now realize that these are initially hijacked from the mammalian genome itself, then inaccurately transcribed by sloppy DNA or RNA polymerases in the virus which then offers them back to the host upon infection. Apart from this, the large proportion of human tumours are of non-viral origin, arising as a consequence of tumour-promoting substances, radiation, etc.

■ Platelet derived growth factor

In 1912 Alexis Carrel[23,24] reported a number of experiments having the purpose:

> to determine the conditions under which the active life of a tissue outside that of the organism could be prolonged indefinitely.

When he tried to maintain tissues for a few days in a simple buffered salts solution, the cells lost their growth capacity and then their viability. It was supposed that the senility and death of the cultures was due to the accumulation of catabolic substances and exhaustion of essential nutrients. Continuous and more rapid growth was achieved by supplementing the solution with diluted plasma, and then, from time to time, submerging the tissue in serum for a few hours. Interestingly, the notion that the cells might require specific factors present in the serum never appears to have crossed Carrel's mind. In conclusion, he wrote that

> fragments of connective tissue have been kept *in vitro* in a condition of active life for more than two months. As a few cultures are now eighty-five days old and are growing very actively, it is probable that, if no accident occurs, the life of these cultures will continue for a long time (Figure 10.3).

About 40 years later, Temin and Dulbecco, working independently, set out to define the precise requirements for cell culture with respect to amino acids, vitamins, salts (together, Carrel's 'nutrients') and, importantly, growth factors.

Their ambition to culture cells arose from their interest in the role of viruses in cell transformation and tumour formation. Thus an important landmark was the finding that the requirement for serum is drastically reduced in cells infected with tumour viruses. They proposed that trans-

Alexis Carrel was awarded the Nobel Prize 1912 'in recognition of his work on vascular suture and the transplantation of blood vessels and organs'. He tarnished his reputation by his association with the eugenics movement, calling for the establishment of institutions equipped with 'appropriate gases' in order to eliminate the insane. He gave enthusiastic support to the Vichy government during World War II and after the liberation of France he was charged for collaboration with the Nazi occupiers. He died before his case came to trial.

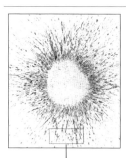
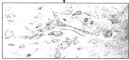

Figure 10.3 A 50-day-old culture of connective tissue. 'Active tissue is encircling a piece of old plasma.' From Carrel, ref.23.

formation might occur as a result of the enhanced capacity of tumour cells to respond to the proliferation signals present in the serum.

An important turning point was the discovery that serum (the soluble component of *clotted* blood) supports growth and proliferation. Plasma merely allows survival and, over a period of about 2 days, cells become quiescent (arrested at the G_0 stage of the cell cycle). Serum contains the products of activated platelets suggesting that they might have a necessary role in the provision of growth factors. In 1974, Russell Ross showed that factors extracted from platelets can induce quiescent smooth muscle cells to synthesize DNA,[25] and in the same year, Kohler and Lipton obtained a similar result with mouse 3T3 fibroblasts. Platelet derived growth factor (PDGF), derived from blood platelets could propel the quiescent cells into the cell cycle S-phase.

With the purification of PDGF from human platelets,[26] the question of how it acts could be faced. PDGF exists as a disulphide-linked dimer, so that binding automatically causes the crosslinking of two receptors. This constitutes the signal for activation.[27] Subsequently, it was found that the oncogene of the simian sarcoma virus (v-*sis*) is homologous to the gene coding for PDGF.[28–30] Here was another clear link between a growth factor and a tumour virus. This time however, the signal to uncontrolled cell proliferation is due to excessive production of growth factor, rather than expression of a constitutively activated receptor. Furthermore, v-*sis* causes cell transformation in primates.[31]

The mechanisms by which an oncogene might cause a tumour were becoming clearer. Here is another example of a mammalian gene, surreptitiously borrowed by a virus, then mutated or mutilated. On return to the host by infection, it causes cell transformation and tumour formation.

Transforming growth factors

The transforming growth factors TGFα and TGFβ were originally isolated from the conditioned medium of a virally infected mammalian fibroblast cell line, 3T3. These are proteins that can bring about transformation of phenotype. The discovery of the TGFs followed some years after the first reports and descriptions of EGF and it derives particularly from the research undertaken in the laboratory of George Todaro.[32]

TGFα, although quite distinct from EGF with respect to amino acid sequence, binds to the EGF receptor and signals cells in a similar fashion (see Chapters 11 and 16). A related factor isolated from these tumour cells, TGFβ, which does not compete with the binding of TGFα or EGF, can nevertheless induce transformation when provided with either of these two factors. Importantly, TGFβ is a normal cellular product and the finding of high quantities in blood platelets and its release during blood coagulation established a clear link with PDGF.[33]

In screening the transforming effect of TGFβ in numerous tumours, there was an unexpected finding. Depending on the cells and the conditions, TGFβ can either promote or suppress cell growth and transformation. It cooperates with TGFα and EGF to cause cell transformation. On

Howard Temin and **Renato Dulbecco** shared the Nobel prize in 1975 for their discoveries concerning the interaction between tumour viruses and the genetic material of the cell.

the other hand, it inhibits colony formation in cells derived from human tumours. It appears that its effects are a function of the total set of growth factors and their receptors that are operational at a given time.[34] In addition, TGFβ plays a number of key roles in the process of tissue remodelling and wound healing.[35] It induces the production of fibronectin and collagen and thus regulates the deposition of the cell matrix, itself a key determinant of cell growth (see Chapter 14).

Problems with nomenclature

As must be evident, nomenclature in this area is arbitrary, to say the least. Some growth factors were named after the cells from which they were first isolated, others from the cells which they stimulated, yet others from the principle action that they appeared to perform.[36] In immunology we hear of interleukins and colony stimulating factors. These direct the maturation and proliferation of white blood cells. In virology, we have the interferons that 'interfere' with viral infection and in cancer research we have tumour necrosis factor (and its relatives) that can influence the growth of solid tumours. In each discipline it seemed that these factors functioned mainly in the category in which they first came to light. Of course, we now know that some growth factors have actions totally unrelated to growth. For instance, PDGF, released from platelets at sites of tissue damage,[37] not only supports the growth of fibroblasts, smooth muscle cells and glial cells, but also acts to regulate the distribution and migration of vascular smooth muscle cells and fibroblasts in wound healing.[38]

To add further complexity to an already complex situation, the conditions in which the cells are studied can determine the cellular response, for instance the presence of other factors, other cells, attachment to substrates. A good example of this is TGFβ. As its name implies, this emerged as a factor enhancing cell transformation, but we now recognize that it can inhibit cell proliferation[39] and that it is a very potent chemotactic factor for neutrophils[40] and fibroblasts.[41] It has been proposed that a common name for these factors should be cytokines.[42]

Cytokines Soluble (glyco)proteins, non-immunoglobulin in nature, released by living cells of the host, acting non-enzymatically in picomolar to nanomolar concentrations to regulate host cell function.

While offering no clues to their various actions, this definition represents a move towards coordinating our understanding of their roles as first messengers. The unity of the cytokines is a concept as important as that of the hormones, defined by Bayliss and Starling nearly 90 years ago (see Chapter 1). However, while the pharmacology of the hormones has been so extensively (some might say exhaustively) investigated, for the protein (growth) factors, this area remains relatively unexplored. Attention here is firmly concentrated on intracellular events.

Essay: The cell cycle

Stages of the cell cycle and random transitions

The cell cycle is the set of cellular events that occur in sequence, taking a cell from division to division (mitosis). Progression through the cell cycle determines the rate of proliferation. Growth, repair and maintenance of organisms all rely on the regulation of the cycle: defects in regulation lie at the root of a number of diseases including cancer.[43]

The cell cycle is conventionally divided into four phases (Figure 10.4):

G_1 (gap phase 1)	DNA integrity, cell size, presence of nutrient and growth factors are checked in preparation for synthesis
S (synthesis)	replication of DNA
G_2 (gap phase 2)	integrity of replication checked
M (mitosis)	cell division

As first described, the cell cycle seemingly comprised two stages, mitosis (visible under the microscope) and interphase. With the realization that DNA is the carrier of genetic information, the cycle gained an S phase, during which DNA synthesis occurs. This was understood to be preceded and then followed by the *gap* phases, G_1 and G_2. However, the word *gap* is a misnomer. Now we recognize that both G_1 and G_2 represent important checkpoints and are as busy as the other phases of the cycle. Indeed, growth factors have their main activity on cells in G_1, when their

For an excellent introduction to this field, see *The Cell Cycle*, by Andrew Murray and Tim Hunt.[43]

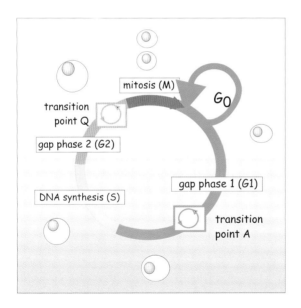

Figure 10.4 A general plan of the cell cycle. The cell cycle is described as a sequence of phases. The defining events of M phase, chromosome segregation and mitosis, can be visualized under the microscope. In most mammalian cells dividing under normal circumstances, this takes about 1 hour. The phases G_1, S and G_2 comprise the interphase. DNA replication is confined to the S phase, a stage that takes 6 hours. During G_1 and G_2, the so-called gap phases, the cell carries out many tests (checkpoints) that determine the commitment to proceed to S or M. The transition points A and Q represent points of hesitation; progress is halted for an undefined period and this has the effect of introducing a stochastic element into the duration of the cell cycle. When the conditions of growth are unfavourable, for instance inhibition due to cell contact or a lack of growth factors, the cell cycle arrests in G_0.

presence is required for a number of hours. A sufficient stimulus drives the cycle to the G_1 checkpoint (formerly 'start' or 'restriction' point), at which the cells become committed to a round of replication.[44] Here, the cells consider the state of their DNA, their mass and the provision of nutrients. On emergence, they are committed to continue until they achieve the next G_1. From here on, removal of growth factors is without effect. If defects in DNA replication are recognized, the cycle can be arrested at the second checkpoint in G_2. It can be arrested again at metaphase by the spindle checkpoint that prevents the segregation of chromosomes in mitosis if they fail to attach to the spindle or if the spindle is damaged. In the absence of a sufficient stimulus, the duration of G_1 extends indefinitely and the cells become quiescent, entering a phase indicated as G_0. Cells that are insensitive to growth factors are said to be senescent.

Although the term *cell cycle* seemingly indicates an ordered, even predictable, series of events, random processes also make contributions. There are two transition points, A and Q, that occur during G_1 and G_2 respectively. At these points progress is halted for an undefined period of time. Emergence from these transition points is a stochastic process and this accounts for variation in cell cycle duration, even of sibling cells under the same conditions (Figure 10.4).[45]

■ Molecules that drive the cell cycle

The procession of events culminating in cell division is largely explained by the sequential expression of the cyclins (Figure 10.5). These proteins act not only to initiate the successive phases of the cell cycle, but also in the termination of the preceding phases. In this way they determine its forward impetus.[46,47] The cyclins exert their actions through association with and activation of cyclin-dependent protein kinases (CDKs). The general idea is that the cyclins are expressed cyclically (although this does not apply to cyclin D) while their partner CDKs are always present.

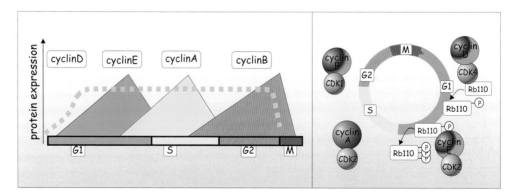

Figure 10.5 Driving the cell cycle. Progress through the cycle is largely controlled by the sequential generation of cyclins that associate with cyclin dependent-protein kinases, the CDKs. Cyclins D and E associate with CDK4 and CDK2 respectively and drive the cell through the G_1 phase. They prepare the cell for DNA replication, a process that requires phosphorylation of Rb110 (retinoblastoma protein). Cyclin A associates with CDK2 and drives the cell through the S phase and cyclin B associates with CDK1 and determines the time of onset of mitosis.

G_1 phase

The main function of the cyclins that operate in G_1 is to inactivate members of the pocket-protein family (Rb110, p107 and p130) (Figure 10.6). These then liberate the transcription factors E2F (six members, E2F1–6). Inactivation of the pocket-proteins can be achieved through phosphorylation and, in the case of p107, by the binding of cyclin E/CDK2.[48] E2F is bound to DNA in association with another transcription factor DP-1. The liberated E2F/DP-1 transcription factors induce expression of cyclin E and cyclin A, necessary for the passage through G_1, and they regulate the transcription of genes implicated in DNA replication. They are therefore essential for the onset of the S phase.

The commencement of the cell cycle coincides with production and stabilization of cyclin D.[49] The D-type cyclins, (three isoforms D1, D2, D3) bind to CDK4 or CDK6. They are activated by phosphorylation in their catalytic loops by the CDK-activating kinase, CAK.

CAK itself is composed of a cyclin and cyclin-dependent protein kinase, cyclinH/CDK7. The D-type cyclins act to integrate the cell cycle machinery with extracellular signals (Figure 10.7). Ras-mediated (see Chapter 11) and focal adhesion-mediated (see Chapter 13) signalling pathways both induce the expression of D-type cyclins, prevent their breakdown, mediate their association with CDK4 or CDK6 and control their translocation

E2F E2 promoter binding factor, a protein isolated from Hela cells that mediates the transcriptional stimulation of the viral E2 promoter.

Adenovirus-E transforming factors such as E1A bind to E2F and prevent the inhibitory action of Rb110. The cells are driven to enter S phase even in the absence of appropriate levels of extracellular stimuli. In this way the virus can cause tumours.

DP DRTF1-associated polypeptide-1.
DRTF Differentiation-regulated transcription factor (the equivalent of E2F).

The activation of kinases by phosphorylation at sites in the catalytic loop is discussed in Chapter 18.

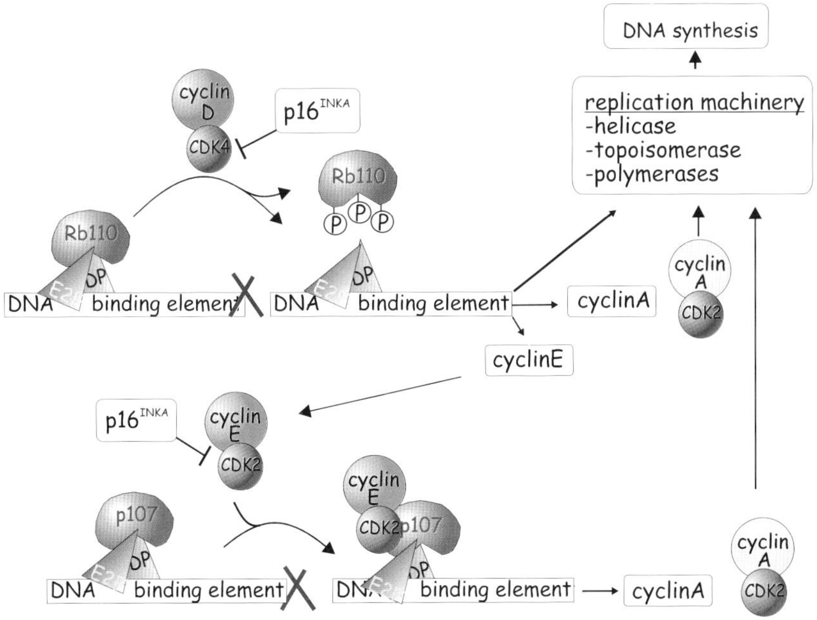

Figure 10.6 Inactivation of pocket-proteins allows entry into S-phase.
Inactivation of Rb110 prepares the system for DNA replication. Cyclin D/CDK4 phosphorylates Rb110, thereby liberating the transcription factor E2F, and cyclin E/CDK2 associates with Rb107 to obtain a functionally active E2F. The transcription factor E2F associates with DP-1 and together they drive expression of cyclin E, cyclin A and genes having products that are necessary for the replication of DNA.

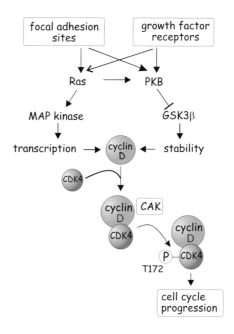

Figure 10.7 Control of cyclin D expression and activity. The D-type cyclins act to integrate extracellular signals with the cell cycle machinery. Signals mediated by Ras and PKB enhance both the transcription rate of cyclin D and its lifetime. The accumulating cyclinD/CDK4 complex is activated through phosphorylation of CDK4 by CDK-activating kinase (CAK). The fully active complex now drives the progression through the G_1 phase of the cell cycle.

into the nucleus.[50–52] Over-expression of cyclin D1 curtails the duration of G_1 but does not prevent the onset of quiescence that normally occurs in the absence of growth factors.[53] The cyclin D/CDK4 (or 6)-activated complex phosphorylates and inactivates Rb110 and p107 and so liberates E2F (Figure 10.6). Following the early induction of D-type cyclins in G_1, they remain elevated, only diminishing at the time of mitosis.

Expression of cyclin E follows activation of cyclin D/CDK4 (or 6) and it associates with CDK2 (Figure 10.6). The cyclin E/CDK2 complex also phosphorylates Rb110 or associates with p107 and drives the cell through the second part of the G_1 phase. It has been proposed that functional inactivation of Rb110 is only achieved after sequential phosphorylation by the complexes containing cyclin D and cyclin E.[54] Both cyclin D/CDK4 (or 6) and cyclin E/CDK2 are necessary for the induction of cyclin A.

S phase

The S phase is dominated by the action of cyclin A in association with CDK2. The cyclin A/CDK2 complex most probably acts in parallel with members of the E2F family to regulate expression of genes necessary for DNA replication (Figure 10.6). Their presence is indispensable throughout S phase, which indicates that they not only play a role in activation of gene expression but also in the activation, through phosphorylation, of the components that are involved in the replication of DNA. This process takes about 6 hours to complete. CyclinA/CDK2 is implicated in the expression of several genes including

- dihydrofolate reductase, needed for the synthesis of purines (dATP, dGTP)
- thymidilate synthase, necessary for the synthesis of pyrimidines (the conversion of dUMP to dTMP)
- DNA helicase, which opens the double helix
- topo-isomerase, which breaks open one of two strands of DNA to facilitate the role of DNA helicase
- DNA polymerase, which replicates the two strands of DNA
- histones H1, H2, H4, needed for the assembly of the newly synthesized DNA into nucleosomes.

G_2 phase

Finally, there is expression of cyclin B (two isotypes, B1 and B2). Its complex with CDK1 (also known as Cdc2, the complex known as maturation promoting factor, MPF) prepares the cell for mitosis (Figure 10.8).

Progress through G2 is tightly regulated by the phosphorylation of CDK1. Its activation requires phosphorylation in the catalytic loop, again catalysed by CAK and further phosphorylation by kinase Myt1 (Thr-14) and Wee1 (Tyr-15), then causes inactivation.

MPF first became apparent as a factor that prepares the immature oöocyte of *Xenopus laevis* for cleavage (meiosis).

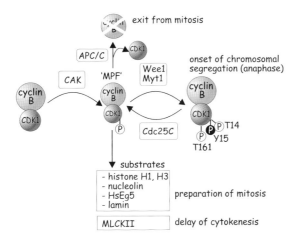

Figure 10.8 Cyclin B and preparation of mitosis. The complex cyclinB/CDK1 (also known as MPF) prepares the cell for mitosis. Its activity is positively regulated by phosphorylation of Thr-161 by CAK, and negatively regulated by phosphorylation at Thr-14 by Myt1 and Tyr-15 by Wee1. Kept in an active state by the dual-specific phosphatase Cdc25 during G_2, the complex promotes condensation of DNA, disassembly of the nucleus and the rearrangement of the tubulin cytoskeleton but also prevents premature cytokinesis. Inactivation by Wee1 and Myt1 initiates the onset of chromosomal segregation and its abrupt destruction terminates the mitotic phase.

Myt1 Membrane associated tyrosine-threonine-specific cdc2-inhibitory kinase.
Wee Referring to the tiny yeast cells obtained with a mutant of wee1 kinase.

This can be counteracted by the dual specificity phosphatase Cdc25. Of the three isoforms of this phosphatase, Cdc25B and Cdc25C are implicated in the G2/M transition (Cdc25A plays a role in the G_1 phase).[55,56] During progression through G_2 and until the metaphase in mitosis, the phosphatase activity of Cdc25 prevails, to maintain the active state of CDK1.

Mitosis

M phase encompasses the two key events of nuclear and cytoplasmic division (mitosis and cytokinesis). The activity of the cyclinB/CDK1 complex is required for condensation of DNA (resulting in the formation of chromatids), depolymerization of lamin (disassembly of the nuclear envelope) and the rearrangement of the tubulin cytoskeleton (formation of kinetochore- and polar microtubules).

The substrates of cyclinB/CDK1 include:

- histones H1 and H3, involved in chromosome condensation
- nucleolin, involved in decondensation of chromatin
- lamin, causing depolymerization of the lamin nuclear-cytoskeleton
- HsEg5, a kinesin-related motor protein associated with centrosomes and the spindle apparatus
- the regulatory light chain of myosin II (LC20), thereby inhibiting actin-dependent myosin ATPase; this may delay segregation of the condensed chromosomes until anaphase [57,58]
- the anaphase promoting complex/cyclosome (APC/C). Phosphorylation of APC/C permits the subsequent activation by one of its subunits, Cdc20 eventually resulting in the degradation of cyclin B at telophase.[59]

At the onset of anaphase the protein kinases Myt1 and Wee1 inactivate the cyclinB/CDK1 complex. Simultaneously there is a peak in the activity of the ubiquitin ligase-APC/C-complex (see point 5 below). The mechanism that activates this complex is still hazy but two regulatory events can be discerned. The first of these is the *de novo* synthesis of two of its subunits Cdc20 and Cdh1 (Figure 10.9b). The second is the phosphorylation of the core components of APC/C itself and one of its subunits, Cdh1, through mitotic kinases (though not all of these have been identified).[60]

In the first instance the APC/C–Cdc20 complex causes the destruction of Pds1.[61] It is believed that it also initiates the destruction of cyclin B. Degradation of Pds1 is a key event in the metaphase-to-anaphase transition because it enables the segregation of the chromatids. Pds1 is localized at the kinetochore, the site of microtubule-chromatid attachment, and it regulates the activity of a complex of proteins, collectively named cohesin. This holds the sister chromatids together (Figure 10.10). Once Pds1 is degraded, the cohesin complex dissipates providing the opportunity for the kinetochore-microtubules to wrench the sister chromatids apart.[62,63]

Pds1 precocious dissociation of sister chromatids-1, mutants that exhibit non-viability after transient exposure to nocodazole and precocious disassociation of sister chromatids.

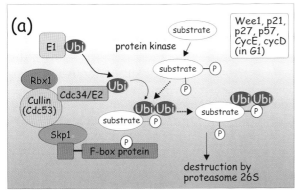

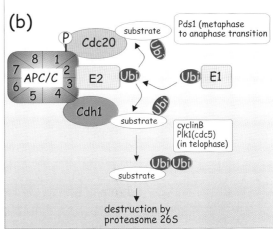

Figure 10.9 Ubiquitin. Proteins are marked for destruction by attachment of ubiquitin at a lysine residue in their 'destruction box' and are degraded by the 26S proteasome. Regulation occurs through two enzyme complexes, SCF and APC/C, each comprising several subunits and each having its own selected substrates.
(a) In the case of SCF (complex of Skp1, cullin and F-box protein), protein phosphorylation at specific residues constitutes the targeting signal. The phosphorylated protein binds the F-box protein and ubiquitination, catalysed by the E1 and E2 subunits, follows. This ubiquitin ligase complex mainly destroys those components of the cell cycle that regulate progression through the G_1 phase. (b) In the case of APC/C, its phosphorylation and the expression of the Cdc20 or Cdh1 subunit constitutes the targeting signal. These subunits bind different substrates that are subsequently ubiquitinated by the E1 and E2 subunits. This ubiquitin ligase complex has an important role in the ordered destruction of proteins regulating progression through G_2 and M phase.

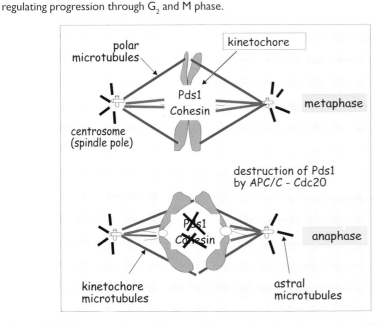

Figure 10.10 Segregation of chromatids requires the destruction of Pds1.
Sister chromatids are held together during metaphase by a complex of proteins. This includes cohesin which is localized at the kinetochore, the site of attachment of tubulin with the chromatids. Pds1 stabilizes the cohesin complex. The activation of APC/C-Cdc20 causes ubiquitination of Pds1 followed by its destruction. This in turn causes the dissociation of the cohesin complex so that the sister chromatids are now available for transfer to the opposing centrosomes by the kinetochore microtubules.

Cohesin The complex comprises proteins Smc1 and 3 and Scc1 and 3. The destruction of Pds1 results in the destruction of Scc1 and this causes the disruption of the cohesin complex. **Scc1** sister chromatids cohesion-1. **Smc1** Structural maintenance of chromosomes.

Plk1 Polo-like kinase, from polo locus in *Drosophila*, where chromosomes appear to be randomly oriented with respect to at least one of the spindle poles.

When this is completed, at telophase, the M phase terminates with the appearance of a second ubiquitin ligase complex, APC/C-Cdh1. This time cyclin B and Plk1 are targeted and their degradation signals the exit from mitosis (Figure 10.9).

Total cyclin B destruction is necessary for the formation of the actin purse-string which is placed perpendicular to the mitotic spindle and which divides the cytoplasm into two compartments (cytokenesis). It is also necessary for the disassembly of the mitotic spindle and to allow cells to re-replicate their genome in the next cell cycle. The logic behind the sequential activation of the two different APC/C-complexes is to ascertain that the sister chromatids are well and truly separated before exiting mitosis. Moreover, the total elimination of cyclin B only at the very end of the cell cycle, when cytokenesis is inevitable, protects the cell against polyploidy. Provided a sufficient stimulatory signal is present, the cells now commence the next cycle with the activation of cyclin D/CDK4 (or 6).

■ Policing the drivers of the cell cycle

Thus far the description of the cell cycle has been rather straightforward, relating the events initiated by growth factors and adhesion molecules and which follow in a well-ordered sequence. Within the strict context of signal transduction this suffices; however, the cycle is of course much more complicated than this. First of all, not all 'growth factors' promote growth and some (depending on the cells, the conditions, etc.) can better be described as differentiation factors. The high level of complexity also originates from the fundamental principle that the integrity of the genetic code must be maintained from generation to generation. Errors must not be passed on to progeny. Therefore an extensive communication circuit has evolved that links the state of the DNA with the checkpoints of the cell cycle. Anomalies are immediately detected and progression through the cycle can be slowed or, to make quite sure, the cell can be handed a loaded gun (to die, by the mechanism of apoptosis).

In the next section we deal with a number of feedback mechanisms that constitute the check-points of the phases S, G_2 and M. Although this cannot be described strictly as signal transduction, the role of signal transduction in cell transformation can only be understood in the context of the complex regulation of the cell cycle. The unrestrained signalling events initiated by a truncated growth factor receptor or constitutively activated Ras only have their disastrous consequences if 'permissive' changes have also occurred in the control of the cell cycle (see also 'Cancer and transformation', page 246).

■ CDK inhibitors

The activity of the cyclin-dependent kinases is regulated not only by their phosphorylation state and by expression of the cyclins but also by the presence of cyclin-specific kinase inhibitors (CKIs). These can be subdivided into two families, INK4 and KIP/CIP.

Members of the INK4 family (p15^{INK4B}, p16^{INK4A}, p18^{INK4C} and p19^{INK4D}) bind to CDK4 and CDK6, preventing their association with cyclin D.[64,65] They inhibit progression through the cycle only in the presence of a functional Rb110. From this it appears that they act to prevent its phosphorylation through inhibition of CDK4 and CDK6 (Figure 10.11).[66] The INK4 inhibitors are appropriately described as tumour suppressors. P16^{INK4A} is frequently mutated or deleted in tumours and p16$^{-/-}$ knock-out mice are highly susceptible to carcinogens. Cells derived from these animals proliferate rapidly and attain high densities in monolayer cultures.[67] The onset of growth inhibition of epithelial cells or keratinocytes by TGFβ1 (see Chapter 14) correlates with an increase in the expression of p15^{INK4B}.[68]

The KIP/CIP inhibitors (p21$^{WAF1/CIP1}$, p27^{KIP1} and p57^{KIP2}) function as heterotrimeric complexes with CDKs and cyclins (Figure 10.11).[69] They inhibit the activity of most of the cyclin/CDK complexes but have their greatest affinity for cyclin E/CDK2. For cyclin D/CDK4 and cyclin E/CDK2 the binding of p27^{KIP1} prevents phosphorylation so preventing activation of the cyclin-dependent kinase. p21$^{CIP1/WAF1}$ also interacts with the transcriptional machinery through binding to proliferating cell nuclear antigen (PCNA), a DNA polymerase implicated in DNA replication and repair.[70,71] Because of its interaction with all CDKs and its capacity to halt DNA replication, this particular inhibitor can arrest or slow down the cell cycle in G$_1$, S and G$_2$.

The discovery of the KIP/CIP family of cell cycle inhibitors arose from

INK Inhibitor of cyclin dependent kinase-4.
KIP Kinase inhibitory protein.
CIP CDK interacting protein.
WAF Wild-type p53 associated factor.
Sdi SDI1-fibroblast inhibitor.

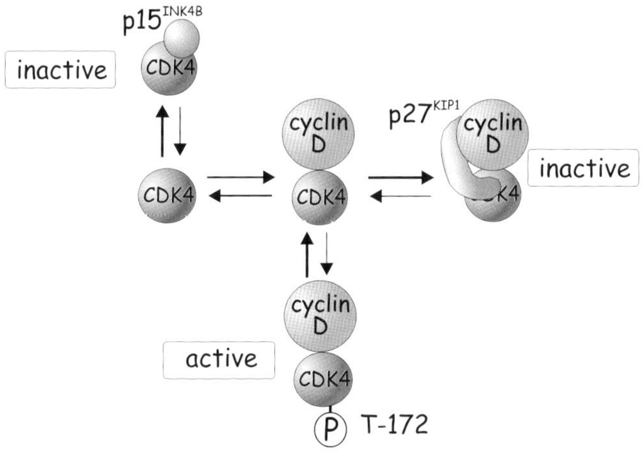

Figure 10.11 **Inhibition of cyclin-dependent kinases.** The CDKs can be locked in an inactive state through the binding of cell cycle inhibitors. These can be broadly divided into two families: (a) Those that bind to the CDK itself and prevent association with the cyclin. These are members of the INK family, for instance p15^{INK4B} binding to CDK4. (b) Those that bind to the complex of CDK and its kinase and prevent activation by CAK. These are members of the KIP family, for instance p27^{KIP1} binding to CDK4/cyclinD. In sufficient amounts, these inhibitors will draw the equilibrium towards the inactive state and bring the cell cycle to a halt.

four independent lines of research. A search for proteins implicated in the repression of DNA synthesis in senescent SD11 fibroblasts yielded SD inhibitor-1 (SD1). A search for proteins induced by the tumour suppressor p53 yielded WAF-1.[72] A search for proteins interacting with CDKs yielded CIP1. We now know that these are one and the same and that p21[CIP1/WAF1] is a transcriptional target of p53 that interacts with CDKs to slow down the cell cycle (although p53 can achieve this in other ways). Finally, the p27 kinase-inhibitory protein KIP-1 emerged from cells treated with TGFβ1.[73] It has become clear that other members of the KIP/CIP family can also be induced by TGFβ1, the choice depending on cell type, so that not all cells are responsive. The expression of p27[KIP1] is elevated in cells undergoing terminal differentiation[74] or slowing down division in response to cell–cell contact.[75] While mutations in p21[WAF/CIP] are not associated with human cancer,[76] abnormally low levels of p27[KIP1] frequently are.[77]

Members of the INK4 and KIP/CIP families compete with each other and with cyclins for binding to cyclin-dependent protein kinases. The general idea is that the amount of active cyclin-dependent protein kinase (i.e. the total amount of kinase coupled to a cyclin, minus the amount of kinase coupled to an inhibitor), determines the rate of progression through the cell cycle (Figure 10.11).

Rb and p53; tumour suppressor proteins

The Rb protein was first identified as the product of a gene that is deleted in patients with retinoblastoma, a tumour originating in the retina.[78] As mentioned above, the Rb protein is a member of the family of nuclear pocket-proteins that bind and inactivate the E2F transcription factor and prevent cells from entering S phase. Inactivation of the Rb gene was later found to occur in numerous other tumours. In the absence of Rb, cells enter into S phase more readily and they do not require the normal array of extracellular growth signals conveyed upon the expression and activation of cyclinD/cdk4 in order to proceed.

p53 was identified as a protein that binds to simian virus 40 (SV40) and to adenovirus proteins implicated in cell transformation.[79] It is expressed throughout the cell cycle but, because it is rapidly degraded, its level is normally kept low. It is a nuclear protein having a number of functions. As a transcription factor it is a regulator of cell cycle progression and apoptosis. Genomic alterations result in its phosphorylation, protecting it against degradation, and causing its activation. The nuclear accumulation of the active transcription factor induces expression of the cell cycle inhibitor p21[waf/cip], which can arrest cell cycle progression at any stage. It can also induce apoptosis through the expression of the pro-apoptotic Bax proteins and by causing elevated expression of a binding protein (IGF-BP3) that blocks the survival signal of IGF-1.

How a cell decides whether to commit suicide or to slow down the cell cycle and repair its DNA remains unknown. As an exonuclease, p53 directly participates in the repair of damaged DNA. This it does without an apparent need for activation, as would be required for its transcrip-

Retinoblastoma affects one in every 15000 live births and is the third most common cancer affecting children. The tumours originate in the retina. The majority (90%) of patients have no family history of the disease. In 40% of the cases, the abnormality is present in every cell of the body including the eye and in 60% of cases, the abnormality is only found in the eye.

p53, called the guardian of the genome, is a transcription factor that is called into play when there is damage to DNA. Its activation causes cell cycle inhibition, allowing repair or apoptosis, to prevent mutations being carried forward into progeny.

Bax Blc-2 associated protein partner with six exons.

tional activity. p53 probably exerts this task as a component of a multi-protein complex comprising other enzymes involved in proof-reading and DNA repair, such as the tumour suppressors BRCA1, hMSH2 and DNA polymerases α and β.[80]

Because of its importance in controlling the integrity of DNA, p53 has been described as the 'guardian of the genome'.[81] Viral oncogenes, such as the large T-antigen of SV40 virus, bind to p53 resulting in its inactivation. Others, such as the E6 protein of papilloma virus, act to accelerate its degradation. In many tumours, p53 is highly expressed but mutated, and in consequence it fails to perform its role as an exonuclease. This results in genetic instability, uncontrolled mismatches and translocations, thus enhancing the frequency of somatic mutations and ultimately resulting in cancer.

Control of DNA integrity

Genes implicated in DNA repair have been studied through mutation analysis of *Saccharomyces pombe*. When exposed to UV or ionizing radiation, pombe normally arrests in G_2 to allow repair of its DNA. Radiation sensitive (rad) mutants were selected that failed to arrest and then die as a consequence of the DNA disarray. A human homologue of one of the affected Rad genes, hRad9, is involved in the control of DNA integrity. It is part of a protein complex having exonuclease activity that associates with DNA (similar to p53) and most likely acts to verify its integrity and 'report' damage to the cell cycle machinery (Figure 10.12).[82–84] hRad 9 communicates with ATM, a protein that has the molecular characteristics of a lipid kinase, but which actually phosphorylates protein. ATM in turn activates the cell cycle checkpoint protein kinases-1 and -2 (Chk-1 and Chk-2).[85]

From here two pathways follow (Figure 10.12). Activation of Chk1 results in the inactivation of the phosphatase Cdc25. As a consequence CDK1 becomes phosphorylated on Tyr-15, by Wee1, and this renders the kinase inactive resulting in a G_2 block (see Figure 10.8).

Activation of Chk2 has two consequences:

- Activation of the tyrosine protein kinase Wee1 causes phosphorylation of CDK1 on Tyr-15 resulting in its inactivation followed by a block in G_2.

- Inactivation of the polo-like protein kinase PLK1, implicated in the formation of centrosomes, the tubulin organization centres essential for the formation of the mitotic spindle. The chromosomes fail to segregate. Inactivation of PLK1 also causes inhibition of APC/C, responsible for ubiquitin-mediated protein degradation (Figure 10.9). Consequently PDS1, an inhibitor of chromosome segregation, is not broken down and again the chromosomes fail to segregate (Figure 10.10). Cyclin B is not broken down and the cells cannot exit anaphase (a block in late mitosis).

In addition, it appears that in mammalian cells, hRad9 induces apoptosis. When hRad9 is activated as a consequence of DNA damage, it binds

IGF-BP3 Insulin-like growth factor binding protein-3, hinders binding of IGF to its receptor.
BRCA1 Breast cancer susceptibility gene-1.
hMSH2 *Escherichia coli* MutS and *Streptococcus pneumoniae* HexA protein homologous, DNA mismatch repair proteins.

Mutation of **p53** is characteristic of many colon cancers in which it causes changes in their microsatellite DNA blocs (short stretches of DNA of very simple, repeating base sequence). The effect is also referred to as 'microsatellite instability' (MIN).

ATM was discovered as a mutated gene implicated in the condition **ataxia telangiectasia**, manifested by a defective immune system and a predisposition to cancer development. This is most likely due to a failure to control the quality of the DNA at the onset of mitosis and a consequent accumulation of mutations.

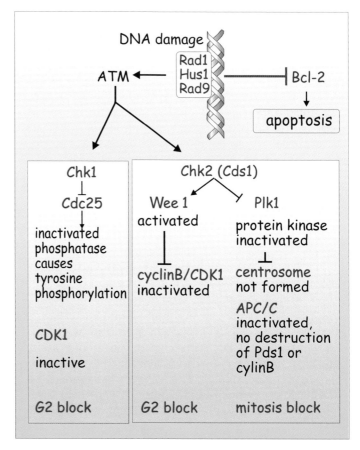

Figure 10.12 Components of the G$_2$ checkpoint that come into action when DNA is incorrectly replicated or damaged. The exonuclease complex Rad1, Hus1 and Rad9 detects anomalies in DNA and this causes the activation of ATM through interaction with Rad9. ATM activates Chk1 and Chk2 and from there three pathways result in either a G$_2$ block, through inactivation of Cdc25 or activation of Wee1, or a mitosis block, through inhibition of the formation of the centrosome (mitotic spindle) and inhibition of APC/C.

and inactivates the apoptosis inhibitors BCL-2 and BCL-xL (see 'Apoptosis', page 335). This makes the cells much more sensitive to pro-apoptotic signals and may eventually cause cell death.[86]

Control of chromatid segregation

In order to separate, the condensed chromosomes (chromatids) first attach to the tubulin filaments as a structure called the kinetochore. This attachment is mediated by the cohesin complex. Their functioning is controlled by yet another complex consisting of Mad1, Mad2, Bub1 and Bub3.

The general idea is that in case a chromatid remains unattached, and is therefore unable to participate in the segregation, Mad2 detaches and moves away from the kinetochore. It will bind to the CDC20-APC/C complex (Figure 10.9) and inhibit its activity (Figure 10.13). This occurs by

Mad and **Bub** These names are derived from genetic analysis of yeasts, selected because, after mutagenesis, they either failed to arrest in

inhibiting the binding of target proteins to CDC20. As a consequence, the segregation inhibitory protein PDS1 is not degraded and remains attached to the kinetochore, thus preventing the dissociation of the cohesin complex.[87] The two sister chromatids remain firmly attached and the cells are blocked in metaphase.

In most colorectal cancers, and probably in many other cancer types, there is a chromosomal instability (CIN) leading to an abnormal chromosome number (aneuploidy). This is consistently associated with the loss of function of a mitotic checkpoint and in some cancers it is associated with mutational inactivation of the Bub1 gene. Cells progress through mitosis even though not all of the chromatids are attached to tubulin filaments and, as a result, the chromosomes become unevenly distributed between the daughter cells.[88]

The CSF and APC ubiquitination machinery

Thus far we have discussed the role of cyclins, cyclin inhibitors, protein kinases, protein phosphatases and transcription factors. An extra element of complexity is added through their regulation by proteolysis. Proteolysis is well suited to serve as a regulatory switch for unidirectional (irreversible) processes. This principle is clearly evident in the organization of the cell cycle, where initiation of DNA replication, chromosome segregation, and exit from mitosis are all triggered by the destruction of

metaphase in the presence of a tubulin disrupting drug (in case of *S. cerevisiae*, budding uninhibited by benomyl), or show no mitotic arrest in absence of proper alignment of chromatids (in case of *S. pombe*, mitotic arrest deficient).

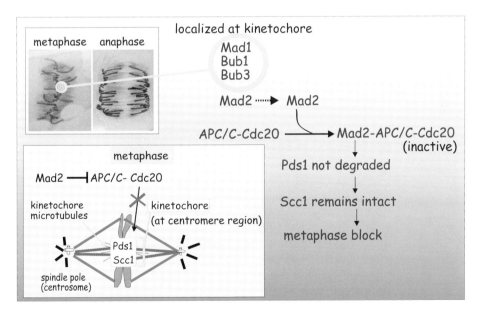

Figure 10.13 Components of the metaphase checkpoint. Defective attachment of chromatids to the kinetochore-microtubules is reported by Mad2 through detachment from the kinetochore complex and association with the APC/C-CDC20 ubiquitin ligase complex. Substrate no longer has access to the CDC20-subunit and destruction of cyclin B or Pds1 does not occur; the cell cycle stagnates in metaphase. Transmission microscope image illustrates the metaphase to anaphase transition of the higher plant *Haemanthus* (*Scadoxus*). Microtubules are stained red and chromosomes are counterstained with toluidine blue. Courtesy of Andrew Bajer, Department of Biology, University of Oregon.

key regulatory proteins. This occurs through the machinery of the 26S proteasome that destroys proteins ligated with ubiquitin. Ubiquitin, a 73-residue polypeptide is attached to a lysine in the 'destruction box' motif of the targeted protein. The process is carried out by the sequential action of three enzymes, the ubiquitin-activating enzyme E1, a ubiquitin-carrier protein E2 and a ubiquitin-protein ligase E3. The E3 component exists in two forms, the SCF-and the APC/C-complex (Figure 10.9).

The SCF complex, so called because of the presence of Skp1, Cullin (Cdc53) and the F-box (CDC4) protein, mainly acts in the G_1 and S phases of the cell cycle. For ubiquitination to occur, the substrate must first be phosphorylated at specific residues, allowing recognition by the F-box protein. The SCF complex adds several ubiquitin groups to the substrate that then detaches, to become a target for the proteasome 26S. Several substrates are targeted for destruction in G_1 and S, including Wee1 (protein tyrosine protein kinase that inactivates CDK1), cyclin E (to prevent cells restarting a G1 without having passed through S phase), $p21^{WAF/CIP}$, $p27^{KIP}$ and $p57^{KIP}$ (three inhibitors of CDK1 that have to be removed in order to pass through G_1).

The APC/C complex acts mainly in the G_2 and S phases (Figure 10.8). Recognition and enzyme activation occurs through the regulatory sub-units Cdc20 and Cdh1 and substrates attached to these subunits will be ubiquitinated (Figure 10.9). It is believed that destruction is controlled largely through the regulation of expression of these subunits, although phosphorylation of components of the core enzyme also plays an important role. For instance, in early mitosis, expression of Cdc20 rapidly increases in order to destroy Pds1, which normally inhibits segregation of the chromatids (and thus blocks the metaphase to anaphase transition). Later in mitosis, levels of Cdh1 rise rapidly causing the destruction of cyclin B and Plk1,so allowing the cells to complete mitosis and commence a new cycle.

■ Essay: Cancer and transformation

■ The essence of cancer

The essence of cancer is that cells run out of control. They no longer respond correctly to the demands and influences of the environment in which they exist. Cancer in a particular tissue manifests itself by an increased cell mass which we describe as a tumour, meaning a swelling. The acquisition of cell mass is a consequence of new growth (neoplasm). In some cancers the primary tumour gives rise to the dissemination of cells that invade other tissues to form secondary tumours (metastasis). In addition, cancer cells tend to be poorly differentiated. As a consequence, they can lose their capacity to carry out the normal functions of the tissue from which they are derived, or they may take on an entirely inappropriate function (ectopic tumour). This means they may either fail to produce important factors or produce wrong factors in overwhelming excess. Under the microscope, cancerous cells are typically characterized by an enlarged nucleus with a very large nucleolus, less cytoplasm and generally

SCF Skp1, Cullin (Cdc53) and F-box protein-containing complex.
APC/C Anaphase promoting complex/cyclosome, promotes the metaphase to anaphase transition.

■ Definitions

From the *Shorter Oxford English Dictionary* (third edition, 1944, with corrections 1977):

Cancer (kæ·nsəɹ), *sb.* ME. [L. *cancer (cancrum)* crab, also gangrene. OE. *cancer, cancor,* helped by Norman Fr. *cancre,* gave ME. CANKER. The L. form was re-introduced later for techn. use.] **1.** A crab. (Now *Zool.*) **1562.** **b.** *Med.* An eight-tailed bandage 1753. **2.** *Astron.* **a.** The Zodiacal constellation lying between Gemini and Leo. **b.** The fourth of the twelve signs of the Zodiac (♋), beginning at the summer solstitial point, which the sun enters on the 21st of June ME. **3.** *Pathol.* A malignant growth or tumour, that tends to spread and to reproduce itself; it corrodes the part concerned, and generally ends in death. See also CANKER. 1601. Also *fig.* †**4.** A plant: perh. *cancer-wort* –1609.

2. *Tropic of C.*: the northern Tropic, forming a tangent to the ecliptic at the first point of C. **3.** C. is decidedly a hereditary disease ROBERTS. *fig.* Sloth is a C., eating up..Time KEN. Comb. (in sense 3) **c.-root,** *Conopholis (Orobanche) americana* and *Epiphegus virginiana*; **-wort,** *Linaria spuria* and *L. Elatine*; also the genus *Veronica.*

Canker (kæ·ŋkəɹ), *sb.* OE. [a. ONF. *cancre* :—L. *cancrum* crab, also gangrene. See CHANCRE.] **1.** An eating, spreading sore or ulcer; a gangrene. Used as = CANCER till *c* 1700. Now *spec.* A gangrenous affection of the mouth, with fetid sloughing ulcers; *canker of the mouth,* or *water c.* **b.** *Farriery.* A disease of a horse's foot, with a fetid discharge from the frog. **2.** Rust. Now *dial.* 1533. **3.** A disease of plants, *esp.* fruit trees, attended by decay of the bark and tissues 1555. **4.** A canker-worm ME. **5.** The dog-rose *(Rosa canina).* Now *local.* 1582. **6.** *fig.* Anything that frets, corrodes, corrupts, or consumes slowly and secretly 1564.

1. No cankar fretteth flesh so sore 1559. **4.** Cankers in the muske rose buds SHAKS. **5.** 1 *Hen. IV,* ɪ. iii. 176. **6.** Enuie which is the c. of Honour BACON.

Comb.: **c.-berry,** the fruit of the dog-rose; also the plant *Solanum bahamense*; **-bloom,** the blossom of the dog-rose; **-blossom,** a canker (sense 4); also *fig.*; **-rash,** a form of scarlet fever in which the throat is ulcerated; **-rose,** (*a.*) the Dog-rose; (*b.*) the wild poppy *(Papaver Rhæas).*

from the *On-line Medical Dictionary* (http://www.graylab.ac.uk/omd):

Cancer. The first historical description of this condition was in relation to breast carcinoma. This is now a general term for more than 100 diseases that are characterized by uncontrolled, abnormal growth of cells. Cancer cells can spread locally or through the bloodstream and lymphatic system to other parts of the body.

altered morphology. Cancer cells are often more rounded than their normal counterparts and there is generally less contact between them and their neighbours.

Most tumours develop in stages, going from benign to malignant. This is the result of a sequence of mutations that gradually enhance the sensitivity of the cells to growth factor signals, reduce the requirement of cell–cell and cell–matrix contact and render cells insensitive to the signals that determine programmed cell death (see 'Apoptosis', page 335). This change in phenotype is also called transformation and so cancerous cells are often called transformed cells. The classification *benign tumour* means that there is an increase in the number of cells (hyperplasia, there may be a lump), but that the normal functions and morphology are retained. Importantly, there is no infiltration of other tissues, particularly the lymph nodes draining the areas in the vicinity of the tumour. By contrast, malignancy refers to the loss or perversion of normal physiological function, altered morphology and infiltration of other tissues, including metastasis.

Alterations dictating malignancy

Malignancy is a consequence of interplay between the transformed cell and its environment. In order for cells to proliferate in excess and to disseminate, they acquire new functions and they also induce the collaboration of the surrounding tissue.[89] For instance, transformed cells release angiogenic factors that induce the formation of new vasculature.[90] Surrounding tissue also provides activators of metalloproteinases that cause the matrix degradation necessary for tissue invasion.[91]

Fully transformed cells possess one or more of the following characteristics:

- They may not require an exogenous growth signal. In culture conditions, this means that the provision of serum is no longer required. It also means that the cells no longer have to adhere to an extracellular matrix; they can grow in soft agar. They also grow well in athymic (nude) mice.

- They may be insensitive to growth-inhibitory signals. For cells in culture, this means that proliferation of cultured cells continues beyond the point of formation of a tight monolayer. It also means that cells have become insensitive to the cell cycle checkpoints in G_1, G_2 and mitosis.

- They can evade programmed cell death (apoptosis). This means that these cells have an in-built rescue mechanism.

- They have limitless replicative potential. Normal cells can undergo about 50 divisions. This is because progressive erosion of the telomeres (several thousand repeats of 6 base-pair sequence elements) results in exposure of chromosomal ends. When the telomeres are exhausted, chromosome fusion ensues, causing nuclear chaos, cell senescence and cell death. By contrast, tumour cells can continue to divide without limit.

- They can induce angiogenesis. During their development, tumours undergo a so-called angiogenic switch enabling them to induce the formation of new vasculature. This provides them with nutrients and oxygen and allows them to disseminate. Of course, dissemination is only possible when they have also acquired the capacity to survive when detached from the extracellular matrix (see Chapter 14).

- They may invade other tissues (metastasis). This is the ultimate consequence of the preceding alterations. The cells have acquired the capacity to detach without dying, to destroy extracellular matrix in order to migrate through tissue, to be carried by the lymph or bloodstream, to attach and then colonize sites where they would not normally survive due to the absence of specific adhesion and growth factors.

■ Genetic alterations at the basis of malignancy

Cell transformation in cancer is the consequence of mutations of several different types, some inherited, some due to environmental factors, others due to ageing. Loss-of-function or gain-of-function mutations have been found in genes that code for the following.

- Molecules involved in the signalling of growth factors. These are either growth factors themselves (c-Sis, PDGF), their receptors (ErbB, EGF-R), or their downstream signalling components (Ras) (see Chapter 11).

- Molecules involved in the control of the cell cycle (so-called tumour suppressors). Most important are the mutations in the retinoblastoma protein (Rb),[92] the p53 protein or the failure to express p15^{INK4A} and p15^{INK4B}, inhibitors of the cyclin-dependent protein kinases.[93]

- Adhesion molecules. These are mainly implicated in cadherin-mediated cell–cell adhesion (cadherin, β-catenin, APC) and integrins (see Chapter 14).

- Molecules involved in the rescue from apoptosis. Important here are Bcl2 or Bcl-Xl, intracellular inhibitors of apoptosis (see 'Apoptosis', page 335), and the survival factors IGF-1 or IL-3.[94]

- Molecules involved in the signal transduction pathway that regulates cell survival (PTEN) (see Chapter 17).

- Molecules implicated in telomere maintenance. Transformed cells up-regulate the expression of the catalytic subunit of the telomerase reverse transcriptase (hTERT).[95]

- Angiogenic factors or their inhibitors: for instance, VEGF, FGF or thrombospondin-1.[90]

Cells will normally die rather than allow somatic (non-inherited) alterations to their genetic code to persist. The natural rate of somatic mutation (due, for instance, to bombardment by cosmic rays, natural environmental carcinogens, etc.) is low and it is unlikely that a tumour would ever develop over the decades of a human life. Yet, in western Europe and North America, one in three people will develop a cancerous growth. Certain mutations create a form of genomic instability that allows an enhanced and more generalized mutation frequency. The loss of function of the tumour suppressors p53 and Rb or other molecules involved in cell cycle control may be crucial here, the cells becoming too eager to replicate, failing to halt at the appropriate checkpoints and failing to repair their DNA.[96] From here on things go from bad to worse.

■ Constructing cancer in a dish

Cells possess mechanisms which protect the organism against the deleterious effects of their possible transformation. The responses that form this line of defence are set out in Table 10.1.

To understand the mechanism of transformation, it is necessary to establish its minimal requirements. How may this be achieved? Occasionally it has been possible to select rare, spontaneously arising immortalized cells. Alternatively, cells can be transformed by applying chemical and physical agents or a virus. These can be said to represent shot-gun approaches from which little can be learned. However, in certain human cells the imposition of just three mutations are sufficient to generate a transformed phenotype.[97]

- High telomerase activity to maintain telomere size and allow unlimited replication. Achieved by inserting the gene (hTERT) that codes for the catalytic subunit of the enzyme.

- An anti-tumour suppressor activity in the form of a viral oncoprotein (SV40 large-transforming antigen, Large T). This disables the function of p53 and Rb, preventing arrest of the cell cycle in G_1, and enables the cells to evade apoptosis.

- A proliferative signal provided by a constitutively active form of Ras (V12G).

The sequence in which these mutations are applied is important.[98] The enhanced activity of telomerase is rather easily accepted by the cells, whereas an excessive growth factor signal is not. The genetic construction of a transformed human cell line was achieved by inducing telomerase

Unlike human cells, rodent cells express telomerase activity and as a result, there is no erosion of the telomeres during ageing. Because of this, the loss of telomeres does not represent a barrier to transformation in rodents and only two oncogenic mutations are necessary to cause transformation.

Table 10.1 Defences against transformation

Contingency	Consequence
Loss of contact with the extracellular matrix	Apoptosis
Damage to DNA	Arrest in G_1
Defects in replication	Arrest in G_2
Failure of DNA repair	Apoptosis
Excessive proliferation signals	Senescence or apoptosis
Loss of telomeres (after about 50 cell divisions)	Senescence and apoptosis

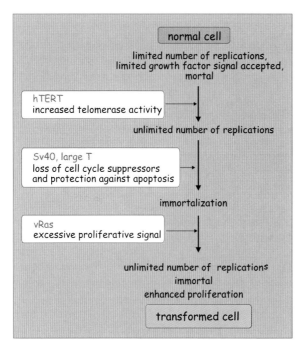

Figure 10.14 **Construction of a transformed cell line.**

activity first, then large-transforming antigen and lastly Ras (see Figure 10.14). With just these three oncogenes at hand the cells continue to proliferate and to grow well in soft agar or in the immunodeficient nude mice.

■ References

1 Wade, N. Guillemin and Schally: the years in the wilderness. *Science* 1978; 200: 279–82.
2 Wade, N. Guillemin and Schally: the three-lap race to Stockholm. *Science* 1978; 200: 411–15.
3 Wade, N. Guillemin and Schally: A race spurred by rivalry. *Science* 1978; 200: 510–13.
4 Ellermann, V., Bang, O. Experimentelle leukämie bei hünern. *Zentr. Bakteriol. Parasitenk.* 1908; 4: 595–609.
5 Warthin, A.S. The neoplasm theory of leukemia, with a report of a case supporting this view. *Trans. Assoc. Am. Phys.* 1904; 19: 421.
6 Warthin, A.S. Leukemia of the common fowl. *J. Infect. Dis.* 1907; 4: 369–80.
7 Rous, P. A transmissable avian neoplasm (sarcoma of the common fowl). *J. Exp. Med.* 1910; 12: 696–705.
8 Erikson, R.L., Purchio, A.F., Erikson, E., Collett, M.S., Brugge, J.S. Molecular events in cells transformed by Rous Sarcoma virus. *J. Cell Biol.* 1980; 87: 319–25.
9 Hunter, T., Sefton, B.M. Transforming gene product of Rous sarcoma virus phosphorylates tyrosine. *Proc. Natl. Acad. Sci. USA* 1980; 77: 1311–15.
10 Rous, P. Comment. *Proc. Natl. Acad. Sci. USA* 1967; 58: 843–5.
11 Levi-Montalcini, R. *In Praise of Imperfection, My life and work* (translation by L. Attardi of *Elogio dell'imperfezione*). Basic Books, New York, 1988.

12 Levi-Montalcini, R. *The Saga of the Nerve Growth Factor: preliminary studies, discovery, further development.* World Scientific Publishing, Singapore, 1997.

13 Hamburger, V. Hans Spemann and the organizer concept. *Experientia* 1969; 25: 1121–5.

14 Hamburger, V., Levi-Montalcini, R. Proliferation, differentiation and degeneration in the spinal ganglia of the chick embryo under normal and experimental conditions. *J. Exp. Zool.* 1949; 111: 457–502.

15 Levi-Montalcini, R. The origin and development of the visceral system in the spinal cord of the chick embryo. *J. Morphol.* 1950; 86: 253–83.

16 Bueker, E.D. Implantation of tumors in the hind limb field of the embryonic chick and the developmental response of the lumbosacral nervous system. *Anat. Rec.* 1948; 102: 369–89.

17 Cohen, S., Levi-Montalcini, R. A nerve-growth stimulating factor isolated from snake venom. *Proc. Natl. Acad. Sci. USA* 1958; 42: 571–4.

18 Cohen, S. The stimulation of epidermal proliferation by a specific protein (EGF). *Dev. Biol.* 1964; 12: 394–407.

19 Cohen, S. Nobel lecture. Epidermal growth factor. *Biosci. Rep.* 1986; 6: 1017–28.

20 Gregory, H. Isolation and structure of urogastrone and its relationship to epidermal growth factor. *Nature* 1975; 257: 325–7.

21 Ushiro, H., Cohen, S. Identification of phosphotyrosine as a product of epidermal growth factor-activated protein kinase in A-431 cell membranes. *J. Biol. Chem.* 1980; 255: 8363–5.

22 Downward, J., Yarden, Y., Mayes, E. *et al.* Close similarity of epidermal growth factor receptor and v-erb-B oncogene protein sequences. *Nature* 1984; 307: 521–7.

23 Carrel, A. On the permanent life of tissues outside of the organism. *J. Exp. Med.* 1912; 15: 516–28.

24 Ross, R., Vogel, A. The platelet-derived growth factor. *Cell* 1978; 14: 203–10.

25 Ross, R., Glomset, J., Kariya, B., Harker, L. A platelet-dependent serum factor that stimulates the proliferation of arterial smooth muscle cells in vitro. *Proc. Natl. Acad. Sci. USA* 1974; 71: 1207–10.

26 Antoniades, H.N., Scher, C.D., Stiles, C.D. Purification of human platelet-derived growth factor. *Proc. Natl. Acad. Sci. USA* 1979; 76: 1809–13.

27 Cooper, J.A., Bowen, P.D., Raines, E., Ross, R., Hunter, T. Similar effects of platelet-derived growth factor and epidermal growth factor on the phosphorylation of tyrosine in cellular proteins. *Cell* 1982; 31: 263–73.

28 Doolittle, R.F., Hunkapiller, M.W., Hood, L.E. *et al.* Simian sarcoma virus onc gene, v-sis, is derived from the gene (or genes) encoding a platelet-derived growth factor. *Science* 1983; 221: 275–7.

29 Waterfield, M.D., Scrace, G.T., Whittle, N. *et al.* Platelet-derived growth factor is structurally related to the putative transforming protein p28sis of simian sarcoma virus. *Nature* 1983; 304: 35–9.

30 Robbins, K.C., Antoniades, H.N., Devare, S.G., Hunkapiller, M.W., Aaronson, S.A. Structural and immunological similarities between simian sarcoma virus gene product(s) and human platelet-derived growth factor. *Nature* 1983; 305: 605–8.

31 Theilen, G.H., Gould, D., Fowler, M., Dungworth, D.L. C-type virus in tumor tissue of a woolly monkey (*Lagothrix* spp.) with fibrosarcoma. *J. Natl. Cancer Inst.* 1971; 47: 881–9.

32 Larco, J.E. De, Todaro, G.J. Growth factors from murine sarcoma virus-transformed cells. *Proc. Natl. Acad. Sci. USA* 1978; 75: 4001–5.

33 Childs, C.B., Proper, J.A., Tucker, R.F., Moses, H.L. Serum contains a platelet-derived transforming growth factor. *Proc. Natl. Acad. Sci. USA* 1982; 79: 5312–16.

34 Roberts, A.B., Anzano, M.A., Wakefield, L.M. *et al.* Type β transforming growth factor: a bifunctional regulator of cellular growth. *Proc. Natl. Acad. Sci. USA* 1985; 82: 119–23.

35 Sporn, M.B., Roberts, A.B., Wakefield, L.M., de-Crombrugghe, B. Some recent advances in the chemistry and biology of transforming growth factor-β. *J. Cell Biol.* 1987; 105: 1039–45.

36 Roberts, A.B., Lamb, L.C., Newton, D.L. *et al.* Transforming growth factors: isolation of polypeptides from virally and chemically transformed cells by acid/ethanol extraction. *Proc. Natl. Acad. Sci. USA* 1980; 77: 3494–8.

37 Kaplan, D.R., Chao, F.C., Stiles, C.D., Antoniades, H.N., Scher, C.D. Platelet α granules

contain a growth factor for fibroblasts. *Blood* 1979; 53: 1043–52.

38 Grotendorst, G.R., Chang, T., Seppa, H.E., Kleinman, H.K., Martin, G.R. Platelet-derived growth factor is a chemoattractant for vascular smooth muscle cells. *J. Cell. Physiol.* 1982; 113: 261–6.

39 Cone, J.L., Brown, D.R., DeLarco, J.E. An improved method of purification of transforming growth factor, type β from platelets. *Anal. Biochem.* 1988; 168: 71–4.

40 Haines, K.A., Kolasinski, S.L., Cronstein, B.N. *et al.* Chemoattraction of neutrophils by substance P and transforming growth factor-β 1 is inadequately explained by current models of lipid remodeling. *J. Immunol.* 1993; 151: 1491–9.

41 Postlethwaite, A.E., Keski Oja, J., Moses, H.L., Kang, A.H. Stimulation of the chemotactic migration of human fibroblasts by transforming growth factor β. *J. Exp. Med.* 1987; 165: 251–6.

42 Nathan, C., Sporn, M. Cytokines in context. *J. Cell Biol.* 1991; 113: 981–6.

43 Murray, A., Hunt, T. *The Cell Cycle: an introduction.* Oxford University Press, Oxford, 1993.

44 Pardee, A.B. A restriction point for control of normal animal cell proliferation. *Proc. Natl. Acad. Sci. USA* 1974; 71: 1286–90.

45 Brooks, R.F., Bennett, D.C., Smith, J.A. Mammalian cell cycles need two random transitions. *Cell* 1980; 19: 493–504.

46 Koepp, D.M., Harper, J.W., Elledge, S.J. How the cyclin became a cyclin: regulated proteolysis in the cell cycle. *Cell* 1999; 97: 431–4.

47 Amon, A., Tyers, M., Futcher, B., Nasmyth, K. Mechanisms that help the yeast cell cycle clock tick: G2 cyclins transcriptionally activate G2 cyclins and repress G1 cyclins. *Cell* 1993; 74: 993–1007.

48 Lavia, P., Jansen-Durr, P. E2F target genes and cell-cycle checkpoint control. *Bioessays* 1999; 21: 221–30.

49 Sherr, C.J. D-type cyclins. *Trends Biochem. Sci.* 1995; 20: 187–90.

50 Gille, H., Downward, J. Multiple ras effector pathways contribute to G(1) cell cycle progression. *J. Biol. Chem.* 1999; 274: 22033–40.

51 Diehl, J.A., Cheng, M., Roussel, M.F., Sherr, C.J. Glycogen synthase kinase-3β regulates cyclin D1 proteolysis and subcellular localization. *Genes Dev.* 1998; 12: 3499–511.

52 Sherr, C.J., Roberts, J.M. CDK inhibitors: positive and negative regulators of G1-phase progression. *Genes Dev.* 1999; 13: 1501–12.

53 Matsushime, H., Quelle, D.E., Shurtleff, S.A. *et al.* D-type cyclin-dependent kinase activity in mammalian cells. *Mol. Cell Biol.* 1994; 14: 2066–76.

54 Lundberg, A.S., Weinberg, R.A. Functional inactivation of the retinoblastoma protein requires sequential modification by at least two distinct cyclin-cdk complexes. *Mol. Cell Biol.* 1998; 18: 753–61.

55 Coleman, T.R., Dunphy, W.G. Cdc2 regulatory factors. *Curr. Opin. Cell Biol.* 1994; 6: 877–82.

56 Dunphy, W.G. The decision to enter mitosis. *Trends Cell Biol.* 1994; 4: 202–7.

57 Nigg, E.A. Targets of cyclin-dependent protein kinases. *Curr. Opin. Cell Biol.* 1993; 5: 187–93.

58 Nigg, E.A., Blangy, A., Lane, H.A. Dynamic changes in nuclear architecture during mitosis: on the role of protein phosphorylation in spindle assembly and chromosome segregation. *Exp. Cell Res.* 1996; 229: 174–80.

59 Rudner, A.D., Murray, A.W. Phosphorylation by Cdc28 activates the Cdc20-dependent activity of the anaphase-promoting complex. *J. Cell Biol.* 2000; 149: 1377–90.

60 Kramer, E.R., Scheuringer, N., Podtelejnikov, A.V., Mann, M., Peters, J.M. Mitotic regulation of the APC activator proteins CDC20 and CDH1. *Mol. Biol. Cell* 2000; 11: 1555–69.

61 Visintin, R., Prinz, S., Amon, A. CDC20 and CDH1: a family of substrate-specific activators of APC-dependent proteolysis. *Science* 1997; 278: 460–3.

62 Michaelis, C., Ciosk, R., Nasmyth, K. Cohesins: chromosomal proteins that prevent premature separation of sister chromatids. *Cell* 1997; 91: 35–45.

63 Uhlmann, F., Lottspeich, F., Nasmyth, K. Sister-chromatid separation at anaphase onset is promoted by cleavage of the cohesin subunit Scc1. *Nature* 1999; 400: 37–42.

64 Serrano, M., Hannon, G.J., Beach, D. A new regulatory motif in cell-cycle control causing specific inhibition of cyclin D/CDK4. *Nature* 1993; 366: 704–7.

65 Serrano, M., Hannon, G.J., Beach, D. Recent advances on cyclins, CDKs and CDK inhibitors. *Trends Cell Biol.* 1997; 7: 95–8.

66 Guan, K.L., Jenkins, C.W., Li, Y. *et al.* Growth suppression by p18, a p16INK4/MTS1- and p14INK4B/MTS2-related CDK6 inhibitor, correlates with wild-type pRb function. *Genes Dev.* 1994; 8: 2939–52.

67 Serrano, M., Lee, H., Chin, L. *et al.* Role of the INK4a locus in tumor suppression and cell mortality. *Cell* 1996; 85: 27–37.

68 Reynisdottir, I., Polyak, K., Iavarone, A., Massagué, J. Kip/Cip and Ink4 Cdk inhibitors cooperate to induce cell cycle arrest in response to TGF-β. *Genes Dev.* 1995; 9: 1831–45.

69 Russo, A.A., Jeffrey, P.D., Patten, A.K., Massagué, J., Pavletich, N.P. Crystal structure of the p27Kip1 cyclin-dependent-kinase inhibitor bound to the cyclin A-Cdk2 complex. *Nature* 1996; 382: 325–31.

70 Waga, S., Hannon, G.J., Beach, D., Stillman, B. The p21 inhibitor of cyclin-dependent kinases controls DNA replication by interaction with PCNA. *Nature* 1994; 369: 574–8.

71 Nakanishi, M., Robetorye, R.S., Adami, G.R., Pereira-Smith, O.M., Smith, J.R. Identification of the active region of the DNA synthesis inhibitory gene p21Sdi1/CIP1/WAF1. *EMBO J.* 1995; 14: 555–63.

72 el-Deiry, W. S., Velculescu V.S., Tokino, T. *et al.* WAF1, a potential mediator of p53 tumor suppression. *Cell* 1993; 75: 817–25.

73 Polyak, K., Kato, J.Y., Solomon, M.J. *et al.* p27Kip1, a cyclin-Cdk inhibitor, links transforming growth factor-β and contact inhibition to cell cycle arrest. *Genes Dev.* 1994; 8: 9–22.

74 Durand, B., Fero, M.L., Roberts, J.M., Raff, M.C. p27Kip1 alters the response of cells to mitogen and is part of a cell-intrinsic timer that arrests the cell cycle and initiates differentiation. *Curr. Biol.* 1998; 8: 431–40.

75 Martin-Castellanos, C., Moreno, S. Recent advances on cyclins, CDK's and CDK inhibitors. *Trends Cell Biol.* 1997; 7: 95–8.

76 Shiohara, M., Koike, K., Komiyama, A., Koeffler, H.P. p21WAF1 mutations and human malignancies. *Leuk. Lymphoma.* 1997; 26: 35–41.

77 Lloyd, R.V., Erickson, L.A., Jin, L. *et al.* p27kip1: a multifunctional cyclin-dependent kinase inhibitor with prognostic significance in human cancers. *Am. J. Pathol.* 1999; 154: 313–23.

78 Lee, W.H., Shew, J.Y., Hong, F.D. *et al.* The retinoblastoma susceptibility gene encodes a nuclear phosphoprotein associated with DNA binding activity. *Nature* 1987; 329: 642–5.

79 Crawford, L.V., Pim, D.C., Gurney, E.G., Goodfellow, P., Taylor-Papadimitriou, J. Detection of a common feature in several human tumor cell lines – a 53,000-dalton protein. *Proc. Natl. Acad. Sci. USA* 1981; 78: 41–5.

80 Albrechtsen, N., Dornreiter, I., Grosse, F. *et al.* Maintenance of genomic integrity by p53: complementary roles for activated and non-activated p53. *Oncogene* 1999; 18: 7706–17.

81 Lane, D.P. Cancer. p53, guardian of the genome. *Nature* 1992; 358: 15–16.

82 Volkmer, E., Karnitz, L.M. Human homologs of *Schizosaccharomyces pombe* rad1, hus1, and rad9 form a DNA damage-responsive protein complex. *J. Biol. Chem.* 1999; 274: 567–70.

83 Bessho, T., Sancar, A. Human DNA damage checkpoint protein hRAD9 is a 3′ to 5′ exonuclease. *J. Biol. Chem.* 2000; 275: 7451–4.

84 St Onge, R.P., Udell, C.M., Casselman, R., Davey, S. The human G2 checkpoint control protein hRAD9 is a nuclear phosphoprotein that forms complexes with hRAD1 and hHUS1. *Mol. Biol. Cell* 1999; 10: 1985–95.

85 Sanchez, Y., Bachant, J., Wang, H. *et al.* Control of the DNA damage checkpoint by chk1 and rad53 protein kinases through distinct mechanisms. *Science* 1999; 286: 1166–71.

86 Komatsu, K., Miyashita, T., Hang, H. *et al.* Human homologue of *S. pombe* Rad9 interacts with BCL-2/BCL-xL and promotes apoptosis. *Nat. Cell Biol.* 2000; 2: 1–6.

87 Brady, D.M., Hardwick, K.G. Complex formation between Mad1p, Bub1p and Bub3p is crucial for spindle checkpoint function. *Curr. Biol.* 2000; 10: 675–8.

88 Cahill, D.P., Lengauer, C., Yu, J. *et al.* Mutations of mitotic checkpoint genes in human

cancers. *Nature* 1998; 392: 300–3.

89 Hanahan, D., Weinberg, R.A. The hallmarks of cancer. *Cell* 2000; 100: 57–70.

90 Hanahan, D., Folkman, J. Patterns and emerging mechanisms of the angiogenic switch during tumorigenesis. *Cell* 1996; 86: 353–64.

91 Johnsen, M., Lund, L.R., Romer, J., Almholt, K., Dano, K. Cancer invasion and tissue remodeling: common themes in proteolytic matrix degradation. *Curr. Opin. Cell Biol.* 1998; 10: 667–71.

92 Harbour, J.W., Dean, D.C. Rb function in cell-cycle regulation and apoptosis. *Nat. Cell Biol.* 2000; 2: 65–7.

93 Serrano, M. The tumor suppressor protein p16INK4a. *Exp. Cell Res.* 1997; 237: 7–13.

94 Evan, G., Littlewood, T. A matter of life and cell death. *Science* 1998; 281: 1317–22.

95 Prescott, J.C., Blackburn, E.H. Telomerase: Dr Jekyll or Mr Hyde? *Curr. Opin. Genet. Dev.* 1999; 9: 368–73.

96 Murphy, K.L., Rosen, J.M. Mutant p53 and genomic instability in a transgenic mouse model of breast cancer. *Oncogene* 2000; 19: 1045–51.

97 Hahn, W.C., Counter, C.M., Lundberg, A.S. *et al.* Creation of human tumour cells with defined genetic elements. *Nature* 1999; 400: 464–8.

98 Weitzman, J.B., Yaniv, M. Rebuilding the road to cancer. *Nature* 1999; 400: 401–2.

Signalling pathways operated by receptor protein tyrosine kinases

The tyrosine kinase family

It is estimated that the human genome will reveal more than 400 tyrosine kinases, a single family of proteins, possibly accounting for as much as 1% of the genomic DNA. Although they are absent from yeasts and protozoans,[1] molecules related to the receptors for EGF and for insulin, having integral catalytic domains, have been identified in marine sponges. It has been suggested that the insulin receptor-like molecules evolved before the Cambrian explosion and contributed to the rapid appearance of the higher metazoan phyla.[2] With respect to the transduction of signals from cell surface receptors, there are two main classes of protein tyrosine kinases (PTKs). Here we consider those that exist as integral domains of transmembrane receptors. In the following chapter we discuss the non-receptor PTKs that are present in the cytosol or are plasma membrane-associated and that can be recruited to receptors.

Spotting phosphotyrosine

Here is another turning point in science that owed as much to chance as to the application of a well-prepared mind. This involved Tony Hunter and the discovery of tyrosine phosphorylation of proteins associated with malignant transformation. He was interested in identifying the transforming antigens of the tumour-causing polyoma virus, of which the main component is the so-called middle T-antigen. A report that the *src*-gene product (v-Src) was associated with protein kinase activity[3] prompted the question of whether other tumour virus gene products might also possess phosphorylating activities and that this might underlie cell transformation. It became apparent that infection with the polyoma virus induces extensive phosphorylation of cellular protein, but that the transforming protein, middle T-antigen itself, also becomes phosphorylated. After proteolytic digestion of ^{32}P-labelled protein, the labelled

Polyoma virus: Originally the Ludwik Gross parotid virus, found by Stewart and Eddy to induce a wide spectrum of tumours after inoculation into immunologically immature newborn mice. This, and later the monkey tumour virus SV40, revolutionized the field of tumour virology.

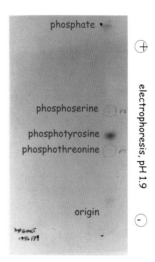

Figure 11.1 **Separation of phosphotyrosine from phosphoserine from phosphothreonine by paper electrophoresis.** Courtesy of Tony Hunter.

residues, presumed to be phosphoserines and phosphothreonines, were separated by electrophoresis. Unexpectedly, all of the label was confined to a new spot, now known to be due to phosphotyrosine (Figure 11.1).[4] In a sense this discovery was accidental. It is common practice when carrying out electrophoretic procedures to re-use the buffers on subsequent occasions. Eventually, the pH must alter, the anodic buffer becoming more acidic and the cathodic buffer becoming more alkaline. Had the pH 1.9 electrophoresis buffer been freshly prepared, the separation of phosphotyrosine would not have occurred. In the event, it had become more acidic, pH 1.7, so the phosphotyrosine migrated more slowly and was separated from the phosphothreonine.

Since the phosphate-tyrosine bond is comparatively resistant to alkali, the detection of tyrosine phosphorylation was simplified by treating ^{32}P-labelled cellular extracts with 1 mol/l NaOH. More recently antibodies have been developed that specifically recognize a phosphate-tyrosine epitope and this of course makes detection of tyrosine phosphorylation a very clean affair. The creation of antibodies having specificity for individual phosphoprotein-specific epitopes is now a business in its own right.

■ v-Src and other protein tyrosine kinases

With electrophoresis now carried out intentionally at pH 1.7, various other labelled protein digests were tested and it was found, unexpectedly, that v-Src can phosphorylate tyrosine residues on a range of quite unrelated proteins. This identified v-Src as a protein tyrosine kinase. Phosphorylation on tyrosine represents an authentic physiological process. This was confirmed by the finding that whereas labelling in non-transformed cells occurs almost exclusively on serines and threonines, in ^{32}P-labelled Rous sarcoma virus-transformed cells, phosphorylated threonine, serine and tyrosine residues are present in almost equal amounts.

It was also found that the transforming protein of the Abelson murine leukaemia virus (v-Abl) becomes labelled on a tyrosine residue when incubated *in vitro* with ^{32}P-ATP.[5] The target in this case is the tyrosine kinase itself, an example of autophosphorylation. Importantly, a second look at the phosphorylation status of the EGF receptor (see Chapter 10) showed that the labelling due to stimulation occurs on tyrosine, not threonine residues as previously reported.[6] A link between tyrosine phosphorylation and cell proliferation/transformation had been made.

■ Other processes mediated through tyrosine phosphorylation

Tyrosine phosphorylation is not limited to the actions of the transforming viruses or growth factors. It regulates a number of important signalling processes including:

- cell–cell and cell–matrix interactions through integrin receptors and focal adhesion sites[7] (see Chapter 14)
- stimulation of the respiratory burst in phagocytic cells, such as neutrophils and macrophages[8]
- activation of B lymphocytes by antigen binding to the B cell receptor[9]
- activation of T lymphocytes by antigen-presenting cells through the T cell receptor complex (TCR)[10] (see Chapter 12)
- the receptor for interleukin-2[11,12]
- the high-affinity receptor for immunoglobulin E (IgE) on mast cells and basophils (see Chapter 12).[13]

Here, we focus on the signal transduction pathway initiated by binding of the growth factors EGF and PDGF to their receptors. We will describe a number of principles that also apply for other tyrosine kinase containing receptors.

■ Tyrosine kinase-containing receptors

■ Crosslinking of receptors causes activation

Tyrosine kinase containing receptors come in several different forms. They are unified by the presence of a single membrane-spanning domain and an intracellular tyrosine protein kinase catalytic domain. The extracellular chains vary considerably, as illustrated in Figure 11.2. A general feature is that ligand binding results in dimerization of the receptors. In addition to activation by peptide ligands, some (but not all) of the functions of the EGF receptor can be elicited by crosslinking with antibodies.[14,15] Crosslinking of receptors by growth factors can be achieved in a number of ways. Platelet derived growth factor (PDGF) is itself a disulphide-linked dimeric ligand which crosslinks its receptor upon binding. When it binds to its receptor, crosslinking is automatic. EGF, a monomeric ligand, changes the receptor conformation in the extracellular domain allowing the occupied monomers to recognize each other. The activation signal is, of course, more complicated than this. For instance, the insulin receptor is a dimeric molecule, already crosslinked by default, yet it still requires the attachment of a ligand to become activated. For activation of all the receptor functions, not only must the receptor molecules be brought together as dimers, but they must also be oriented correctly in relation to each other.[16,17]

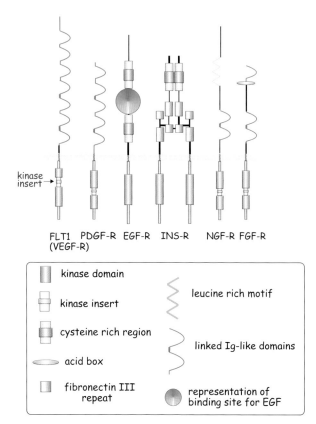

kinase
insert →

FLT1   PDGF-R EGF-R INS-R   NGF-R FGF-R
(VEGF-R)

kinase domain

kinase insert   leucine rich motif

cysteine rich region

                                        linked Ig-like domains
acid box

fibronectin III   representation of
repeat   binding site for EGF

Figure 11.2 Classification of receptors containing tyrosine protein kinase. All these receptors possess a single membrane-spanning segment and all of them incorporate a kinase catalytic domain, in some cases interrupted by an 'insert'. The extracellular domains vary as indicated. Some of these receptors exist in various isoforms: FLT1, fms-related tyrosine kinase receptor for vascular endothelial growth factor (VEGF); PDGF-R, platelet-derived growth factor receptor; EGF-R, epidermal growth factor receptor; INS-R, insulin receptor; NGF-R, nerve growth factor receptor; FGF-R, fibroblast growth factor receptor. Adapted and extended from Kavanaugh and Williams.[18]

Dimerization allows the kinase activity of both intracellular chains to encounter target sequences on the other, linked receptor molecule. This enables the intermolecular cross-phosphorylation of several tyrosine residues (Figure 11.3). The phosphorylated dimer then constitutes the active receptor. It possesses an array of phosphotyrosines which enable it to bind proteins (adapters and enzymes) bearing SH2 domains (see below and Chapter 18) to form *receptor signalling complexes*[19] (Figure 11.4). Additionally, the dimerized and phosphorylated receptor has the potential of phosphorylating its targets.

▪ Assembly of receptor signalling complexes

The formation of signalling complexes has been studied in a number of ways.

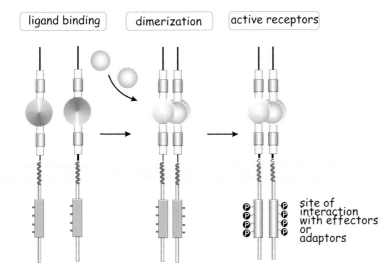

ligand binding | dimerization | active receptors

site of interaction with effectors or adaptors

Figure 11.3 Activation of the EGF receptor. On occupation by its ligand EGF, the EGF-R forms a dimer and this induces a change in the conformation of the cytoplasmic domain that reveals its latent tyrosine protein kinase activity. This phosphorylates the tyrosine residues on the linked receptor molecule (interphosphorylation). The dimerized, phosphorylated molecule constitutes the catalytically active receptor.

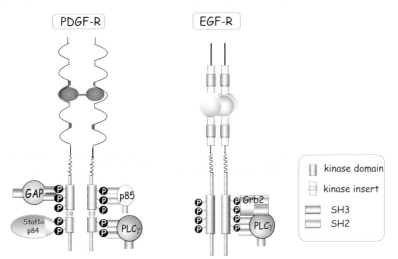

PDGF-R | EGF-R

GAP | p85 | Stat1α p84 | PLCγ

Grb2 | PLCγ

kinase domain
kinase insert
SH3
SH2

Figure 11.4 Formation of receptor signalling complexes. Activated EGF or PDGF receptors (EGF-R or PDGF-R) associate with effectors, including enzymes (PLCγ, GAP, etc.) or adaptor proteins (p85[PI-3kinase], Grb2, etc.) to form receptor signalling complexes.

By measurement of enzyme activity, the generation of second messengers and analysis of tyrosine phosphorylated substrates

The activated receptors for EGF and PDGF stimulate PLCγ. This results in the generation of DAG and IP_3, leading within seconds to the activation of protein kinase C and a rise in the concentration of intracellular free Ca^{2+}.[20–22] All of these can be measured. Furthermore, PLCγ itself becomes phosphorylated on tyrosine residues (see Chapter 5), indicating that it interacts directly with the catalytic domain of the receptor. Activation of the receptors for PDGF and insulin also causes activation of phosphatidylinositol 3-kinase (PI 3-kinase). This phosphorylates phophatidylinositol-4,5-bisphosphate (PIP$_2$) forming phosphatidylinositol-3,4,5-trisphosphate (PIP$_3$), a signalling lipid (see Chapter 13).[23] In addition, a number of serine-threonine kinases also become activated. These include ribosomal S6-kinase (implicated in protein synthesis), Raf-1 kinase (see below) and mitogen activated protein kinase (MAP kinase) (see below). Most importantly, the monomeric GTPase, Ras, becomes activated.[24–26] It changes from the GDP-bound state (inactive) to the GTP-bound state (active).

■ *By detecting the association of proteins with activated receptors*

To investigate the specific interactions of activated receptors, cells pre-labelled with ^{35}S-methionine are stimulated and then solubilized with detergent. The receptors, together with any associated proteins, are precipitated using an anti-receptor antibody. The associated proteins are detected by gel electrophoresis and autoradiography. Identification is achieved by microsequencing, immunoblotting and other techniques. This was how the associations of protein tyrosine kinase receptors with Ras-GAP, PLCγ and PI-3-kinase were originally demonstrated.[27,28] Using a similar approach, but with cell lysates and purified receptors, it was shown that a subunit (p85) of PI-3 kinase binds to the PDGF receptor.[29]

■ *By studying protein associations in a cell-free system: cloning of receptor targets*

Proteins expressed by a lambda-phage library in a bacterial host are screened for binding to the cytoplasmic domain of a receptor, labelled with ^{32}P-phosphate. The relevant bacterial clones are identified using autoradiography and the DNA sequence of the phage insert is determined. This was the way in which the associations of the EGF receptor with Grb2 (growth factor receptor binding protein-2) and the p85-subunit of PI 3-kinase were discovered.[30]

■ Src homology domains and the formation of receptor signalling complexes

Having established the formation of receptor signalling complexes, it was important to establish how these proteins interact with the tyrosine phosphorylated receptor. The answer came from p47[gag-crk], a transforming protein identified as a gene product of a chicken sarcoma virus.

Although its amino acid sequence fails to reveal any catalytic centre, p47[gag-crk] still appears to enhance the extent of tyrosine phosphorylation.[31,32] Its structure is dominated by the presence of domains closely resembling the SH2 and SH3 domains present in the Src protein (the structures of the Src homology domains are described in Chapter 18). When isolated from cell lysates, p47[gag-crk] was recovered in association with phosphorylated proteins and non-receptor tyrosine protein kinases.[33] In combination with the tyrosine protein kinase v-Abl and a substrate, p130[CAS], it increases the degree of phosphorylation from one to

Gag <u>G</u>lycosylated <u>a</u>nti<u>g</u>en, the gene encoding the internal capsid of the viral particle.

Crk <u>C</u>10 <u>r</u>egulator of <u>k</u>inase. This is a good example of how viruses mess about with genes, resulting in chimaeric proteins in which viral Gag sequences are fused to cellular protein CrkL. The mammalian cells

several residues (Figure 11.5). The finding that the enhancement of phosphorylation fails if the SH2 domain of p47$^{gag\text{-}crk}$ is removed gave the first indication that SH2 domains function to regulate protein interactions in a manner dependent on the presence of phosphotyrosine.[32] We understand that p47$^{gag\text{-}crk}$ acts as an adaptor molecule, linking the phosphorylated substrate p130^{CAS} to the protein tyrosine kinase and promoting further, multiple phosphorylations. The human homologue of the viral p47$^{gag\text{-}crk}$ protein is CrkL. Sequence analysis of proteins that bind to tyrosine phosphorylated receptors has shown that many, but not all of them, contain SH2 domains. Others contain protein tyrosine-phosphate binding domains, PTB, quite distinct from SH2 but equally able to recognize tyrosine phosphorylated residues situated in particular amino acid sequences.

The assembly of signalling complexes depends on the recruitment by tyrosine phosphorylated receptors of other proteins, adaptors and enzymes, having such SH2 or PTB domains.

Further evidence for a role of SH2 domains in transmitting the signals due to PTKs came from the finding that only the γ isoforms of PLC are directly activated by these receptors.[19] Significantly, PLCγ, but not the β and δ isoforms, possesses SH2 domains (see Chapter 5 for a discussion of the PLC family). There are many other proteins containing SH2 domains that associate with receptor PTKs in the formation of signalling complexes and a selection of these is illustrated in Figure 11.6. Some of these proteins themselves become phosphorylated as a result of this association though it is not clear whether this is always necessary for their activation. In the case of PLCγ, phosphorylation is certainly necessary.

Of the variety of adapters and enzymes that interact with EGF and PDGF receptors, some appear to bind more tightly than others, exhibiting sensitivity to the amino acid residues in the immediate vicinity of the phosphotyrosines (Figure 11.7). Thus, a particular receptor might transmit its

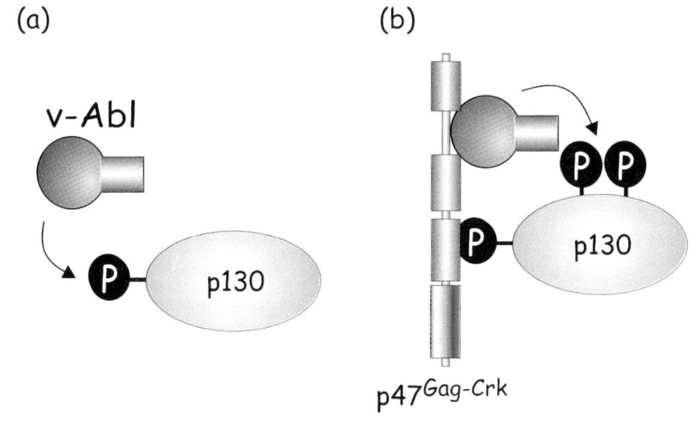

express the gene product once it is inserted into their genome. CrkL is an SH2 and SH3 domain-containing adaptor protein and is implicated in pathogenesis of chronic myelogenous leukaemia.

CAS Crk-associated substrate.

(a)

(b)

p47$^{Gag\text{-}Crk}$

Figure 11.5 Gag-Crk is an adaptor protein: (a) In the absence of Gag-Crk, v-Abl phosphorylates the p130 protein on a single tyrosine residue. (b) The association of phosphorylated p130 with p47$^{gag\text{-}crk}$ facilitates the further phosphorylation of this same protein at a number of additional sites. This adaptor function of p47$^{gag\text{-}crk}$ explains how a protein lacking a catalytic domain can nevertheless promote tyrosine phosphorylation.

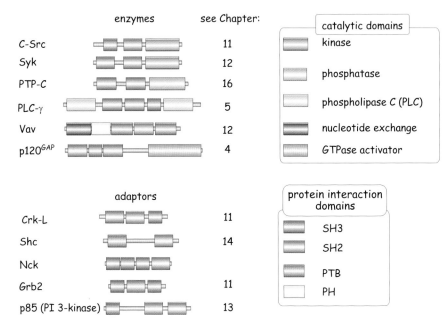

Figure 11.6 Domain organization of proteins that associate with phosphorylated tyrosine kinase-containing receptors. Many proteins that associate with tyrosine-phosphorylated receptors contain SH2 domains. These recognize specific amino acid stretches in the vicinity of phosphorylated tyrosine residues. Unlike the enzymes, the adaptors lack intrinsic catalytic activity but serve to link phosphorylated receptors with other effector proteins. Some of the proteins presented in this figure are discussed elsewhere in this book.

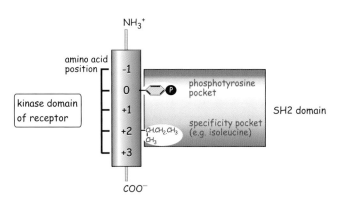

Figure 11.7 Recognition of phosphotyrosine and adjacent amino acids by the SH2 domain. Selectivity of recognition between different targets containing SH2 domains is conferred by the sequence of amino acids, particularly the third residue immediately adjacent on the C-terminal side of the phosphorylated tyrosine. As examples:

PI 3-kinase	-x-pY-x-x-M-
Grb2	-x-pY-x-N-x-
Src	-x-pY-x-x-I-.

signal through a panel of SH2-containing proteins. It remains unclear however, if two or more intracellular proteins can bind to a single receptor molecule simultaneously.

Branching of the signalling pathway

A number of signal transduction pathways branch out from the receptor signalling complex (Figure 11.8). Two such branches are described in detail in the next paragraphs and others are discussed in different chapters (PI 3-kinase, Chapter 13; STATS, Chapters 12 and 17).

The PLCγ-protein kinase C signal transduction pathway

Among the activities set in train by activation of the EGF and PDGF receptors is the generation of DAG and IP_3 by PLCγ. The DAG remains in the membrane and acts as a stimulus for PKC. The consequence is the transformation of a phosphotyrosine signal through activation of PLCγ into a phosphoserine/phosphothreonine signal. One of the first substrates of PKC is the EGF receptor itself. This becomes phosphorylated on a serine residue very close to the transmembrane domain and has the effect of inactivating the receptor.

Although numerous proteins have proved to be substrates of the PKC enzymes, we still lack full understanding of how these kinases determine the changes in gene transcription that occur after stimulation. As discussed earlier (Chapter 9), one consequence of the activation of PKC is _de_-phosphorylation of the transcription factor c-Jun, a component of the inducible transcription factor AP-1 (Figure 9.13, page 209).

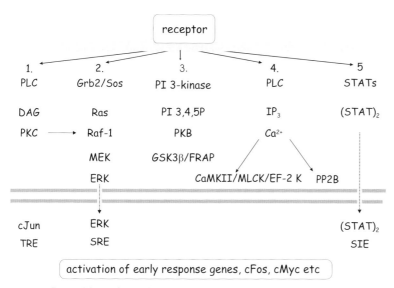

Figure 11.8 Branching of the signal transduction pathways. Following activation of receptor PTK, several signal transduction pathways can be activated. Five of these are indicated. Further details and abbreviations feature in the following paragraphs and figures.

■ The Ras signalling pathway

▨ *From the tyrosine kinase to Ras*

For almost 40 years it has been known that infection of rats with murine leukaemia viruses can provoke the formation of a sarcoma.[34,35] A major advance was the discovery that the Harvey murine sarcoma virus encodes a persistently activated form of the H-ras gene in which valine is substituted for glycine at position 12. Expression of this mutant in quiescent rodent fibroblasts resulted in altered cell morphology, stimulation of DNA synthesis and cell proliferation.[36] When over-expressed, normal H-c-Ras also induces oncogenic transformation[37] as does micro-injection of the mutant protein.[38] Conversely, injection of neutralizing antibodies to inhibit normal Ras function reverses cell transformation.[39] Stimulation of quiescent cells with serum or with growth factors promotes the binding of GTP to Ras.[40] It became apparent that Ras is an important component in the signalling pathways regulating cell proliferation, but how this would fit into the known pathways emanating from growth factor receptors remained unclear for a considerable time. The first clues came from genetic analysis of signal transduction pathways that operate in the invertebrates *Drosophila melanogaster* and *Caenorhabditis elegans.*

▨ *Photoreceptor development in the fruit fly*

The compound eyes of insects are formed of a hexagonal array of small units, ommatidia: in the case of the fruit fly, there are approximately 800 'small eyes'. Each is composed of 8 photoreceptor cells (R1–R8) and 12 accessory cells. On the basis of their morphology, order of development, axon projection pattern and spectral sensitivity, the photoreceptor cells can be classified into three functional classes: R8, the first to appear, followed by R1–R6 and then R7. The photosensitive pigment resides in a microvillus stack of membranes, the rhabdomere. The larger rhabdomeres of cells R1–R6 are arranged as a trapezoid surrounding the rhabdomeres of cells R7 and R8, the R8 rhabdomere being located below R7 (Figure 11.9). The development of R7 requires the products of two genes, *sevenless* (*sev*) and *bride-of-sevenless* (*boss*). The phenotypes generated by loss-of-function mutations in either of these genes are identical, R7 failing to initiate neuronal development. These mutations are readily detected in a behavioural test. Given a choice between a green and a UV light, normal (WT) flies will move rapidly towards the UV source.[41] Failure to develop cell R7, the last of the photoreceptor cells to be added to the ommatidial cluster, correlates with the lack of this fast phototactic response, and the flies move towards the green light.[41]

The *sev* product is required only in the R7 precursor, but the *boss* function must be expressed in the developing R8. Cloning revealed the *boss* product as a 100 kDa glycoprotein having seven transmembrane spans and an extended N-terminal extracellular domain.[42] Although ultimately expressed on all of the photoreceptor cells, at the time that R7 is being specified it is only present on the oldest, R8.[43] The product of the *sev* gene is a receptor protein tyrosine kinase.[44] Evidence for direct interaction

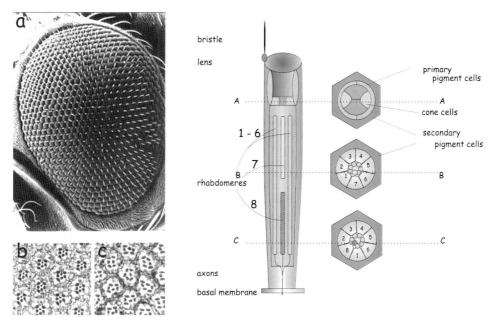

bristle

lens

A .... A cone cells

primary pigment cells

1 - 6

7

B
rhabdomeres

8

secondary pigment cells

B .... B

C C

axons

basal membrane

Figure 11.9 **The *sevenless* mutation in fly eyes.** The events leading to the development of cell R7 in eyes of *Drosophila* have provided a key to understanding the pathway downstream of receptor PTKs. Genes acting downstream of the *sevenless* receptor were revealed by screening for mutations that affect the development of cell R7. The eye of the fly is built up of ommatidia, groups of eight photoreceptor cells each covered by a single lens. The drawing illustrates the basic anatomy of a single ommatidial unit in longitudinal section. Sections cut at a, b and c are shown in transverse section on the right. Since two of the cells, R7 and R8, do not extend the full length of the ommatidial unit, the transverse sections b–b and c–c only reveal seven cells, not all eight. The scanning electron microscope image shows the geometrical arrangement of ommatidia. The thin sections B and C are both representative of cuts through section b–b in the drawing. Note that in B, taken from a wild-type fly, seven cells are evident, whereas in C, taken from a fly having the *sevenless* mutation, there are only six. From Dickson and Hafen.[48]

between the products of these two genes came from the demonstration that cultured cells expressing the *boss* product tend to form aggregates with cells expressing *sev*.[43]

It is now understood that the binding of Boss (the ligand) to Sev (the receptor kinase) leads to the activation of kinase activity and that this ultimately determines the fate of R7 as a neuronal cell. Since a reduction in the gene dosage of the fly *Ras1* impairs signalling by Sev, and persistent activation of *Ras1* obviates the need for the *boss* and *sev* gene products, it follows that the activation of Ras is an early consequence of Sev activity.[45] Further genetic screens of flies expressing constitutively activated Sev led to the identification of two intermediate components of this pathway as *Drk* (*downstream of receptor kinases*) and *Sos* (*son of sevenless*; see column 1 of Figure 11.10). The Sos protein shows substantial homology with the yeast CDC25 gene product, a guanine nucleotide exchange catalyst for RAS.[46] Although a reduction in the gene dosages of *Drk* and *Sos* impairs the signal from constitutively activated Sev, there is no effect on signalling from constitutively activated Ras. In the pathway of activation, this places the functions of the *Drk* and *Sos* products into a position intermediate between Sev and Ras. The *Drk* gene codes for a small protein consisting

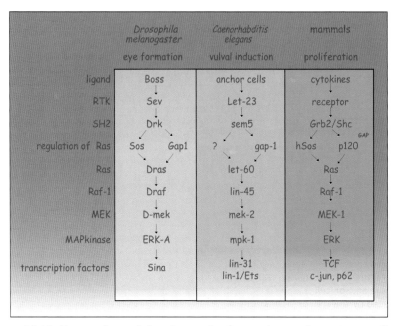

	Drosophila melanogaster eye formation	*Caenorhabditis elegans* vulval induction	mammals proliferation
ligand	Boss	anchor cells	cytokines
RTK	Sev	Let-23	receptor
SH2	Drk	sem5	Grb2/Shc
regulation of Ras	Sos Gap1	? gap-1	hSos p120 GAP
Ras	Dras	let-60	Ras
Raf-1	Draf	lin-45	Raf-1
MEK	D-mek	mek-2	MEK-1
MAPkinase	ERK-A	mpk-1	ERK
transcription factors	Sina	lin-31 lin-1/Ets	TCF c-jun, p62

Figure 11.10 Comparison of signal transduction pathways downstream of a tyrosine protein kinase receptor in species of three separate phyla. The striking homologies that exist between the genes coding for proteins operating downstream of receptor PTKs in very distant phyla enabled the sequence of events downstream of the EGF receptor to be elucidated.

exclusively of Src homology domains, two SH3 flanking a single SH2 domain (see Chapter 18). Having no catalytic activity of its own, Drk acts as an adaptor. It binds to the tyrosine phosphorylated receptor and links it to the proline-rich domains of Sos.[47]

Vulval cell development in nematode worms

In the nematode *C. elegans*, a similar pathway of activation involving autophosphorylation of a tyrosine kinase receptor leads to activation of the GTPase Let-60, a homologue of Ras (column 2 of Figure 11.10). This determines the development of vulval cells (Figure 11.11). Again, these mutants were first identified from genetic analysis of lethal mutations (let, <u>let</u>hal mutants), morphological changes in vulval development (sem, <u>sex</u> <u>m</u>uscle mutants) or alterations in cell lineage (lin, <u>lin</u>eage mutants).[49] They constitute the components of a signal transduction pathway based on Lin-3 (a product of the anchor cell), Let-23 (a tyrosine kinase receptor of the p5.p cell) and Sem-5 that associates with a (Sos-like) guanine nucleotide exchange protein. This brings about nucleotide exchange on Let-3. (Figure 11.10).

In both nematode and fly, the Ras protein acts as a switch that determines cell fate. In *C. elegans*, the activation of Ras determines the formation of vulval as opposed to hypodermal (skin) cells. In *Drosophila* photoreceptors, the activation of Ras determines the development of R7

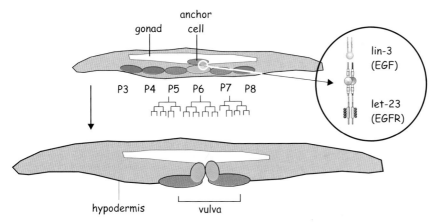

Figure 11.11 **Vulval development in *C. elegans*.** Because it is a relative simple structure, formed from just a few cells, the vulva is well suited for the genetic analysis of cell differentiation during embryological development. It is the product of just three cell lineages, the descendants of cells p5.p, p6.p and p7.p. Development is initiated by a signal from the anchor cell that lies adjacent to p6.p. The ligand Lin-3 (a homologue of EGF), produced by the anchor cell binds its receptor Let-23 (homologous to the EGF-R) on the surface of cell p6.p. Cell p6.p in turn releases signals to its neighbours, p5.p and p7.p. This initates a sequence of events involving the MAP kinase pathway that determines the fate of these cells as components of vulval tissue.

as a neuronal as opposed to a cone cell. In both cases, Ras proteins operate downstream of receptor tyrosine kinases that are activated by cell–cell interactions.

Regulation of Ras in vertebrates

The elucidation of the Ras pathway in vertebrates was based on the identification of proteins having sequence homologies with those present in *Drosophila* and *C. elegans* (column 3 of Figure 11.10).[50–52] Expression or microinjection of these proteins (and appropriate reagents such as peptides, antibodies, etc.) were used to restore or modulate the activity of this pathway in cells derived from mammals, flies or worms, bearing loss-of-function mutations. A vertebrate protein Grb2 (growth factor receptor binding protein 2), lacking catalytic activities but having SH2 and SH3 domains, was found to be capable of restoring function in Sem-5 deficient mutants. In addition, Grb2 was found to associate with a protein that is recognized by an antibody raised against the *Drosophila* protein, Sos. In this way the sequence of events became apparent. Grb2 is an adaptor protein, linking the phosphorylated tyrosine kinase receptor to the guanine nucleotide exchanger in vertebrates (Figure 11.12). The mammalian Sos homologue, hSos, is likewise a guanine nucleotide exchange factor which interacts with Ras.[53] Grb2 is composed exclusively of Src homology domains, one SH2 flanked by two SH3 domains. Because of the nature of the interaction of SH3 with proline-rich sequences (see Chapter 18) it is likely that Grb2 and Sos remain associated even under non-stimulating conditions. The main effect of receptor activation is to ensure the recruitment of the Grb2/Sos complex to the plasma membrane.[54]

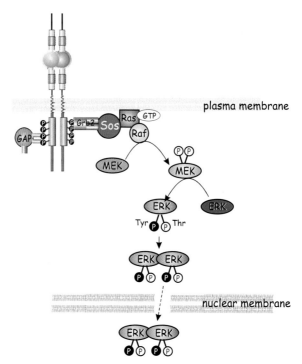

plasma membrane

nuclear membrane

Figure 11.12 Regulation of the Ras-MAP kinase pathway by receptor protein tyrosine kinases. The adaptor protein Grb2, in association with the guanine exchange factor Sos, attaches to the tyrosine phosphorylated receptor through its SH2 domains. This brings the Grb2/hSos complex into the vicinity of the membrane where it catalyses guanine nucleotide exchange on Ras. The activated Ras associates with the serine/threonine protein kinase Raf-1. Its localization at the membrane results in activation and subsequent phosphorylation of the dual specificity kinase MEK which phosphorylates ERK on both a tyrosine and a threonine residue. Dimerization exposes a signal peptide that allows this MAP kinase to interact with proteins that guide it into the nucleus (translocation).

As already pointed out (see Chapter 4), the Ras-GTPase activating protein p120[GAP] also contains two SH2 domains. It too binds to phosphotyrosines on activated receptors and is a component of the signalling complex that assembles on activated PDGF receptors (Figure 11.13). It is unclear what role the association of GAP plays in signal transduction. For instance, cells that express a mutant of the PDGF receptor that fails to bind GAP manifest normal activation of Ras.[55]

From Ras to MAP kinase and the activation of transcription

The events following the activation of Ras lead to the activation the extracellular signal regulated protein kinase, ERK. ERK was originally recovered as a serine/threonine phosphorylating activity present in the cytosol of EGF-treated cells and given the name mitogen activated protein kinase, MAP kinase.[56]

It enters the nucleus and is an activator of early response genes. This pathway operates quite independently of second messengers and, in some cell types, cAMP actually opposes it. There are two intermediate steps and both of these involve a phosphorylation (Figure 11.12). The immediate activator of ERK is MEK, MAP kinase-ERK kinase). This most unusual enzyme phosphorylates ERK on both a threonine (T) and a tyrosine (Y) residue. These are in the target sequence LTEYVATRWYR*APE* (Table 11.1), seven residues on the N-terminal side of the conserved motif APE, present in the catalytic centre of the kinase. To date, ERK appears to be the unique substrate for phosphorylation by MEK, indicating a particularly high level of specificity.

Moving further upstream, the first kinase in the cascade is Raf-1. This was initially identified as an oncogene product. The subsequent finding that activated Ras recruits Raf-1 to the membrane and in consequence brings about kinase activation, links ERK with the Ras pathway.[57–59] In the activation of Raf-1, it is its recruitment to the plasma membrane, not its actual association with activated Ras, that is necessary. Of course, the association with Ras is essential under normal conditions.[60,61] However, a mutant form of Raf-1 possessing a C-terminal -Caax box that acts as a site for prenylation (see Chapter 4), and which is therefore permanently associated with the plasma membrane, instigates the downstream events independently of Ras. Accordingly, the role of Ras in the physiological situation can be regarded as that of a membrane-located recruiting sergeant.

Beyond ERK: activation of early response genes

As a result of the double phosphorylation, ERK undergoes dimerization and the exposure of a signal peptide which enables it to interact with proteins that promote its translocation into the nucleus. Within, it catalyses the phosphorylation of its substrates on Ser-Pro and Thr-Pro motifs. In the case of stimulation by EGF and PDGF, the activation of ERK is an absolute requirement for cell proliferation. The early response genes become activated within an hour of receptor stimulation. Their activation is transient and it can occur under conditions in which protein synthesis is inhibited. Activation of the EGF receptor results in the rapid induction of the transcription factor c-Fos, one of the first cytokine-inducible transcription factors to be discovered.[62] It occupies a central position in the regulation of gene expression. Other early response genes include c-myc and c-jun. The promoter region of the c-fos gene contains a serum response element (SRE), a DNA domain that binds the transcription factors p67[SRF] (serum response factor) and p62[TCF] (ternary complex factor). Phosphorylation of p62[TCF] by ERK increases the formation of a complex of both transcription factors with the DNA to promote transcription of the c-fos gene[63] (Figure 11.13).

In the case of stimulation through the EGF receptor, c-Jun is also activated by phosphorylation. This occurs at sites quite distinct from those that are dephosphorylated as an indirect consequence of PKC activation (see above). Expression of persistently activated (oncogenic) forms of Ras (v-Ras) activates c-Jun-mediated gene transcription. However, since the presence of a persistently activated form of Raf-1 is without effect on c-Jun, this does not appear to occur through the Raf-1–MEK–ERK pathway. Moreover,

MAP kinase Mitogen activated protein kinase; since cloning, referred to as **ERK**, extracellular signal regulated kinase.

Raf Rat fibrosarcoma, also called MAP kinase-kinase-kinase, MAP-KKK.

p62[TCF] was first identified as part of a complex of three components, together with **p67[SRF]** and DNA. It was therefore referred to as 'ternary complex factor'.

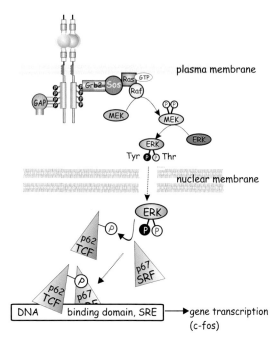

Figure 11.13 Activation of transcription by ERK. Inside the nucleus ERK phosphorylates p62^{TCF} which then associates with p67^{SRF} to form an active transcription factor complex. This binds to DNA at the SRE.

c-Jun is poorly phosphorylated by ERK and its activation is more likely to be a consequence of phosphorylation by its own, Jun N-terminal kinase (JNK, another member of the mitogen activated kinases) (Figure 11.15). c-Jun and c-Fos together form the activator protein complex-1 (AP-1) which promotes expression of yet another series of genes.

Fos From feline osteosarcoma virus.
Myc The cellular counterpart of the transforming gene of the avian leukosis retrovirus MC29.
Jun From avian sarcoma virus-17: we are told that *junana* is 17 in Japanese. Note: abbreviations for genes are presented in lower case: fos, myc, etc. The same abbreviations with the first letter capitalized indicates their respective protein gene products.

Regulation of protein synthesis

It is must be evident that the mitogenic signal initiated by growth factors requires an increased rate of protein synthesis and probably also the selective translation of specific mRNAs (since not all mRNAs are translated at the same rate). Here again, ERK plays a role, initiating ribosomal protein synthesis by regulating the binding of the initiation factor-4E (eIF-4E) to the cap of the mRNA and to the initiation factor complex (Figure 11.14). Activation of ERK results in phosphorylation of eIF-4E at Ser-209, though this site is not itself accessible to ERK. Instead, it is Mnk1 (MAP kinase-integrating kinase-1), which acts downstream of ERK, that links it with the initiation of protein synthesis. Mnk1 does not interact directly with eIF-4E, but uses a docking site in its partner, eIF-4G. Association of eIF-4E with the other components of the initiation complex has the effect of ironing out a hairpin loop close to the 5′ cap of the RNA and this facilitates the association of eIF-2.GTP with the small ribosomal subunit (40S) and the messenger[64,65] (Figure 11.14).

Interestingly, Mnk1 is the target of some viruses when hijacking cellular protein synthesis. A gene product of adenovirus (protein p100) binds

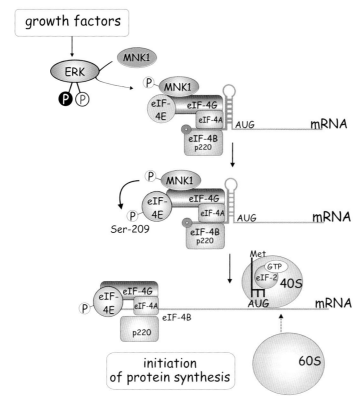

Figure 11.14 Activation of protein synthesis by ERK. ERK induces the phosphorylation and activation of Mnk1 which in turn phosphorylates and activates eIF-4E. This associates with the initiation complex at the poly-A tail of the mRNA and effectively irons out the folds in the mRNA, allowing the attachment of the ribosome at the start codon, AUG.

eIF-4G and displaces Mnk1 so that it is no longer able to phosphorylate and activate eIF-4E. As a result, the cellular mRNA remains untranslated. However, the viral mRNA is unaffected and the cell becomes a machine for the manufacture of viral proteins.[66]

Termination of the ERK response
The termination of the signals initiated by growth factors is critical. The pathways that terminate the activity of ERK are discussed in Chapter 17 (page 378).

A family of MAP kinase-related proteins
Once it was cloned, it was apparent that ERK is a member of a substantial family of proteins, referred to as MAP kinases. These may be classified into three main functional groups. The first of these mediate mitogenic and differentiation signals. The other two are associated with cellular responses to stress and inflammatory cytokines. They operate in three pathways:

- The ERK pathway. ERK1 and ERK2 are the prototypic MAP kinases described in the previous paragraphs. There are seven members of the ERK family (ERK1–7). However, most of the higher numbered isoforms do not appear to function in the mitogenic pathway.

- The JNK/SAPK pathway. SAPK stands for <u>s</u>tress <u>a</u>ctivated <u>p</u>rotein <u>k</u>inase. Within this class, the Jun N-terminal kinases (JNK) form a subfamily (SAPK/JNK1–3).

- The p38/HOG pathway. HOG indicates <u>h</u>igh <u>o</u>smolarity <u>g</u>lycerol, which causes cellular stress in yeast (*S. cerevisiae*) resulting in the activation of this protein kinase. The p38MAP-kinases form another subfamily (four members).

Each of these pathways, shown in Figure 11.15, involves a kinase cascade resulting in the phosphorylation and activation of the MAP-kinase family member. Each contains a dual phosphorylation site, TEY, TPY or TGY, the central residue in this motif being characteristic of the class, as shown in Table 11.1 It is evident that cells are endowed with parallel pathways of activation, and that these may operate individually or in combination to initiate specific patterns of gene expression. Additionally, cross-talk between the pathways undoubtedly occurs.

■ *MAP kinases in other organisms*

Pathways regulated by MAP kinases are widely distributed and can be found in all eukaryotic organisms.[67] In *S. cerevisiae*, physiological

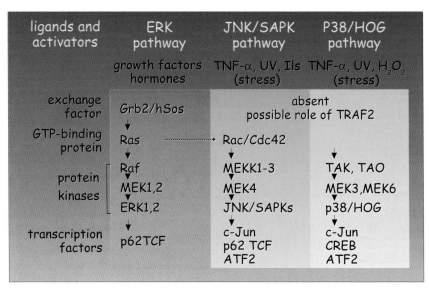

Figure 11.15 Parallel pathways to transcription and the MAP kinase family. The MAP kinases can be classified into three groups, based on the identity of the intermediate residue in their dual phosphorylation motifs (TEY, TGY or TPY). This classification also defines three distinct signal transduction pathways indicated as the ERK, the JNK/SAPK and the P38/HOG pathways, each having unique protein kinases acting upstream.

Table 11.1 Dual phosphorylation sites in MAP-kinase family members

Species	Kinase	Domain VII	Linker L12	Domain VIII (catalytic loop)
			TEY motif	
Human	ERK1	DFGLAR	IADPEHDHTGF	LTEYVATRWYRAPEIMLNSK
Rat	ERK1	DFGLAR	IADPEHDHTGF	LTEYVATRWYRAPEIMLNSK
Human	ERK2	DFGLAR	VADP–HDHTGF	LTEYVATRWYRAPEIMLNSK
Rat	ERK2	DFGLAR	VADP–HDHTGF	LTEYVATRWYRAPEIMLNSK
			TGY motif	
Mouse	P38/HOG	DFGLAR	HTDDE------	MTGYVATRWYRAPEIMLNWN
			TPY motif	
Rat	SAPKα	DFGLAR	TACTN------FM	MTPYVVTRYYRAPEVILGMG
Rat	SAPKβ	DFGLAR	TAGTS------FM	MTPYVVTRYYRAPEVILGMG
Rat	SAPKγ	DFGLAR	–AGTS------FM	MTPYVVTRYYRAPEVILGMG
Human	JNK1	DFGLAR	TAGTS------FM	MTPYVVTRYYRAPEVILGMG

processes regulated by MAP kinases include mating, sporulation, maintenance of cell wall integrity, invasive growth, pseudohyphal growth and osmoregulation. MAP kinase is a regulator of the immune response and embryonic development in *Drosophila*. It has also been implicated as a regulator in slime moulds, plants and fungi.

Other Ras activators and effectors

Guanine nucleotide exchange factors other than hSos have also been found to activate Ras and other effectors have also been found (see Figure 4.15, page 92). These may interact with unique sequences in the effector loop. The question remains, how many different effectors can attach to activated Ras and what determines the level of their priority?[68]

The Ca²⁺/calmodulin pathway

The elevation of cytosol Ca^{2+} following activation of PLCγ results in widespread protein phosphorylation by serine/threonine protein kinases. These include the broad spectrum Ca^{2+}-calmodulin-dependent protein kinase II (CaM-kinase II) (see Chapter 8), myosin light chain kinase (MLCK), phosphorylase kinase and elongation factor-2 kinase (EF-2 kinase). All of these are activated by the Ca^{2+}/calmodulin complex.

The level of phosphorylation of a particular substrate at any time must be determined by the rates of both phosphorylation and dephosphorylation. Ca^{2+}/calmodulin can affect this balance through activation of calcineurin, a protein phosphatase (see Chapter 8). This leads to the activation of transcription factors that play essential roles in the activation of T lymphocytes (Chapter 17). Clearly, Ca^{2+} is an extremely versatile second messenger modulating numerous intracellular signals (Figure 8.1, page 174).

There is still an awful lot to learn about Ras.

As our knowledge of Ras and the pathways that it regulates have advanced, so its involvements have become more complex and entangled. Ras is not merely an activator of cell growth. Indeed, in some cells it causes growth inhibition and differentiation. In others it blocks differentiation. It has become apparent that depending on the cell, the state of the cell, the activation state of other GTPases, possibly the identity of the Ras isotype (N-, K- or H-), it is a regulator of multiple cell functions. Nor is its relationship to cell transformation a matter of simple cause and effect. As examples:

• Oncogenic Ras only induces transformation if the recipient cell already has endured a number of mutations in its tumour suppressor genes such as p53 or Rb. Otherwise its introduction results in apoptosis (see Chapter 10, 'Cancer and transformation', page 246).

• Most of the experimental work on Ras has been carried out on rodent fibroblast cell lines that are far more susceptible to transformation than the epithelial cells in which most (human) Ras-related tumours occur. More than this, the phenotypes of the transformed cells are very different. Ras transformed (human breast) epithelial cells are characterized by disruption of the adherens junctions and the appearance of stress fibres and focal adhesions (see Chapter 14), but transformed fibroblasts are associated with a loss of stress fibres. Although constitutively activated mutants of Ras or its effector Raf can both induce the transformed phenotype in mouse fibroblasts, only Ras can induce transformation of rat intestinal epithelial cells.

• Due to the availability of reagents (antibodies, mutants, etc.), most investigations have concentrated on H-Ras. However, although the human Ras genes N, H and K are indeed very similar, and in many experimental situations appear to function in the same way, there is no reason to believe that their actions are identical. The conservation of three ras genes in vertebrate evolution begs the question whether or not the gene products

■ Activation of PI 3-kinase

Association of the adaptor p85$^{PI 3-kinase}$ with the tyrosine phosphorylated receptor, positions the attached catalytic subunit, p110$^{PI 3-kinase}$ at the membrane where it phosphorylates inositol phospholipids at the 3 position of the inositol ring. This results in the activation of PKB, a protein kinase of prime importance in cell survival, proliferation, motility and glucose metabolism. This pathway is considered in Chapter 13.

■ Direct phosphorylation of transcription factors

The simplest way in which a plasma membrane receptor could alter gene expression would be by direct phosphorylation of transcription factors. The activation of transcription by the interferons is an example. Transcription factors known as STATs were recognized as targets for interferon receptors, but it is now apparent that they also mediate the signals of EGF and PDGF receptors[69,70] (see Chapter 12). Following phosphorylation of p84^{Stat1a} and p91^{Stat1b} by the activated growth factor receptor, they combine to form a dimeric complex and, as a consequence, they translocate to the nucleus. Here they promote transcription of early response genes such as c-fos (Figure 11.16).

The STAT dimer, formed after tyrosine phosphorylation by the PDGF receptor, was originally described as Sis-inducible factor (SIF), a transcription factor complex activated by the viral oncogene, v-Sis. This viral oncogene codes for the precursor of PDGF and activates a similar signal transduction pathway.[71,72]

A switch in receptor signalling: activation of ERK by 7TM receptors

■ Pathway switching mediated by receptor phosphorylation

As described in Chapter 4, G-protein-linked receptors are themselves substrates not only for PKA and PKC but also for receptor specific kinases, which preferentially target occupied (and therefore activated) receptors (Figure 4.10, page 85). On the other hand, phosphorylation by PKA and PKC, triggered by an increase in second messenger production, affects occupied and unoccupied receptors alike. The result of phosphorylation is to switch the attention of the receptors to alternative G-proteins. As described earlier (see Chapter 9) β-adrenergic receptors phosphorylated by PKA now speak to the G_i proteins instead of G_s and this opens the door to the ERK pathway (Figure 9.5, page 196). Because receptor-specific kinases, such as βARK, target only receptors that are occupied, they are only called into action under conditions of more robust stimulation. The phosphorylation sites, though also present in the C-terminal region, are distinct from those targeted by PKA.

The β-receptor phosphorylated by βARK recruits β-arrestin (see page 84) to a region on its third intracellular loop that would normally interact

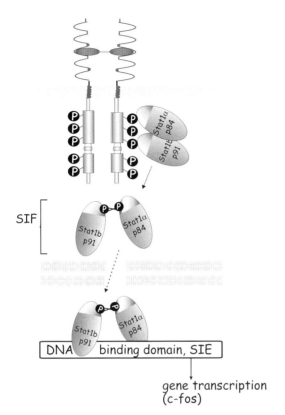

SIF

DNA binding domain, SIE

gene transcription
(c-fos)

Figure 11.16 Direct phosphorylation of the STAT class of transcription factors. Through their SH2 domains, the p84^{Stat1a} and p91^{Stat1b} associate with the receptor and become phosphorylated on tyrosine residues. They form a dimer (SIF) which translocates to the nucleus where it binds to an Sis-inducible element (SIE) and activates transcription of, for example, the c-*fos* gene.

have specific functions? Although the data are so far very sketchy, the embryonic lethality of K-ras (but not of N- or H-ras) supports the idea of non-redundancy of function.

In this chapter we have indicated the role of Ras as an activator of the ERK pathway; but this is not the only pathway implicated in its regulation of cellular proliferation. In Chapter 13 we consider Ras as one of a number of activators of PI 3-kinase and the consequent activation of protein kinase B. There are several additional (and also potential) effector pathways through which the effects of activated Ras can operate.

For a comprehensive discussion of these questions, see Shields *et al.*[68]

STAT Signal transducer and activator of transcription.

with G-proteins, so blocking the transmission of signals through G_s (Figure 11.17). Importantly, the bound β-arrestin now acts as a docking site for the tyrosine protein kinase Src (pp60^{src}), through its SH3 domain, and this also initiates a signalling pathway resulting in the activation of ERK.

The steps following the recruitment of Src are not yet clear. It is possible that it initiates a series of events similar to those that occur in focal adhesion complexes, in which it phosphorylates the adaptor protein Shc-1 which, in turn, binds to the Grb2/Sos complex (for details, see Chapter 14, Figure 14.15, page 330). This ultimately results in the activation of Ras, Raf, MEK and lastly ERK.[73]

Other 7TM receptors may employ different pathways to reach Ras. For instance, the lysophosphatidic acid receptor, LPA, also activates ERK but in this case, dominant negative Src is without effect. Here the signal appears to involve PI 3-kinase, an unidentified tyrosine kinase and docking protein and finally Grb2/Sos.[74]

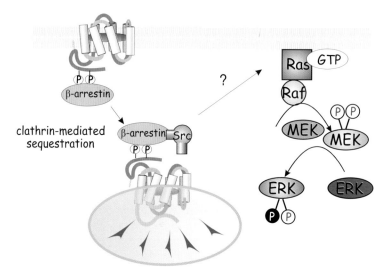

Figure 11.17 Phosphorylation of the β-adrenergic receptor by βARK (see Figure 4.10, page 85) allows binding of β-arrestin, an adaptor protein. This terminates communication with G-proteins but signals the Ras-ERK pathway by recruitment of Src (via its SH3-domain). The arrestin targets the receptor to clathrin-coated pits, there to be removed from the cell surface and directed towards lysosomes.

■ Pathway switching by transactivation

An alternative manner of switching, also allowing signals emanating from G-protein linked receptors to activate the Ras/ERK pathway, arises from a process that has been called transactivation.[75,76] As an example, activation of receptors responding to carbachol (a stable muscarinic cholinergic analogue) or thrombin, appears to cause the release of an EGF-like growth factor (HB-EGF) from its membrane-bound precursor.

This then acts in an autocrine/paracrine manner to stimulate its own receptor pathway, resulting in the activation of ERK. The switch arises as a result of activation of a metalloproteinase that cleaves the inactive precursor form of the growth factor (proHB-EGF)[76,77] (Figure 11.18). In effect,

HB-EGF Heparin-binding EGF-like growth factor which exists as a membrane-bound precursor, proHB-EGF.

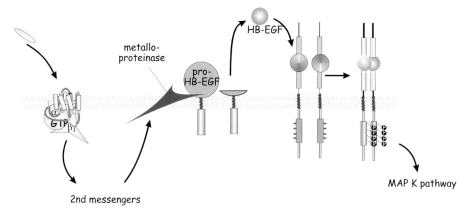

Figure 11.18 Transactivation of receptors. Activation of the 7TM receptor results in activation of a signal transduction pathway that activates an extracellular endoproteinase releasing the membrane bound precursor of HB-EGF. This binds to the EGF receptor and induces yet another series of signal transduction pathways.

the first ligand induces the expression of a second, quite unrelated ligand, which in turn sets in train its own distinct signal transduction pathways. Such transactivation expands enormously the repertoire of signalling systems that a cell can apply to integrate its responses to the great diversity of stimuli. In the case described, the mechanism of transactivation involves the initial generation of intracellular signals that have the effect of activating a transmembrane enzyme (metalloproteinase) that has its catalytic site situated in the extracellular domain. The mature growth factor then stimulates its characteristic intracellular signal pathways.

■ References

1 Arkinstall, S., Payton, M., Maundrell, K. Activation of phospholipase Cλ in *Schizosaccharomyces pombe* by coexpression of receptor or nonreceptor tyrosine kinases. *Mol. Cell Biol.* 1995; 15(3): 1431–8.

2 Skorokhod, A., Gamulin, V., Gundacker, D., Kavsan, V., Muller, I.M. Origin of insulin receptor-like tyrosine kinases in marine sponges. *Biol. Bull.* 1999; 197: 198–206.

3 Collett, M.S., Erikson, R.L. Protein kinase activity associated with the avian sarcoma virus src gene product. *Proc. Natl. Acad. Sci. USA* 1978; 75: 2021–4.

4 Eckhart, W., Hutchinson, M.A., Hunter, T. An activity phosphorylating tyrosine in polyoma T antigen immunoprecipitates. *Cell* 1979; 18: 925–33.

5 Witte, O.N., Dasgupta, A., Baltimore, D. Abelson murine leukaemia virus protein is phosphorylated in vitro to form phosphotyrosine. *Nature* 1980; 283: 826–31.

6 Ushiro, H., Cohen, S. Identification of phosphotyrosine as a product of epidermal growth factor-activated protein kinase in A-431 cell membranes. *J. Biol. Chem.* 1980; 255: 8363–5.

7 Giancotti, F.G. Integrin signaling: specificity and control of cell survival and cell cycle progression. *Curr. Opin. Cell Biol.* 1997; 9: 691–700.

8 Naccache, P.H., Gilbert, C., Caon, A.C. *et al.* Selective inhibition of human neutrophil functional responsiveness by erbstatin, an inhibitor of tyrosine protein kinase. *Blood* 1990; 76: 2098–104.

9 Burg, D.L., Furlong, M.T., Harrison, M.L., Geahlen, R.L. Interactions of Lyn with the antigen receptor during B cell activation. *J. Biol. Chem.* 1994; 269: 28136–42.

10 Cantrell, D.A. T cell antigen receptor signal transduction pathways. *Cancer Surv.* 1996; 27: 165–75.

11 Williamson, P., Merida, I., Greene, W.C., Gaulton, G. The membrane proximal segment of the IL-2 receptor β-chain acidic region is essential for IL2-dependent protein tyrosine kinase activation. *Leukemia* 1994; 8 (Suppl 1): S186–9.

12 Kirken, R.A., Rui, H., Evans, G.A., Farrar, W.L. Characterization of an interleukin-2 (IL-2)-induced tyrosine phosphorylated 116-kDa protein associated with the IL-2 receptor β-subunit. *J. Biol. Chem.* 1993; 268: 22765–70.

13 Li, W., Deanin, G.G., Margolis, B., Schlessinger, J., Oliver, J.M. Fcε R1-mediated tyrosine phosphorylation of multiple proteins, including phospholipase Cγ1 and the receptor βγ2 complex, in RBL-2H3 rat basophilic leukemia cells. *Mol. Cell Biol.* 1992; 12: 3176–82.

14 Defize, L.H., Moolenaar, W.H., van der Saag, P.T., de Laat, S.W. Dissociation of cellular responses to epidermal growth factor using anti-receptor monoclonal antibodies. *EMBO J.* 1986; 5: 1187–92.

15 Spaargaren, M., Defize, L.H., Boonstra, J., de Laat, S.W. Antibody-induced dimerization activates the epidermal growth factor receptor tyrosine kinase. *J. Biol. Chem.* 1991; 266: 1733–9.

16 Syed, R.S., Reid, S.W., Li, C.W. *et al.* Efficiency of signalling through cytokine receptors depends critically on receptor orientation. *Nature* 1998; 395: 511–16.

17 Ortega, E., Schweitzer, S.R., Pecht, I. Possible orientational constraints determine secretory signals induced by aggregation of IgE receptors on mast cells. *EMBO J.* 1988; 7(13): 4101–9.

18 Kavanaugh, W.M., Williams, L.T. Signaling through receptor tyrsoinse kinases. In: Heldin, C.H., Purton, M. (Eds) *Signal transduction*. Chapman & Hall, London, 1996; 3–18.

19 Anderson, D., Koch, C.A., Grey, L. *et al.* Binding of SH2 domains of phospholipase Cγ1, GAP, and Src to activated growth factor receptors. *Science* 1990; 250: 979–82.

20 Pandiella, A., Beguinot, L., Velu, T.J., Meldolesi, J. Transmembrane signalling at epidermal growth factor receptors overexpressed in NIH 3T3 cells. Phosphoinositide hydrolysis, cytosolic Ca2+ increase and alkalinization correlate with epidermal-growth-factor-induced cell proliferation. *Biochem. J.* 1988; 254: 223–8.

21 Gilligan, A., Prentki, M., Knowles, B.B. EGF receptor down-regulation attenuates ligand-induced second messenger formation. *Exp. Cell Res.* 1990; 187: 134–42.

22 Gonzalez, F.A., Gross, D.J., Heppel, L.A., Webb, W.W. Studies on the increase in cytosolic free calcium induced by epidermal growth factor, serum, and nucleotides in individual A431 cells. *J. Cell Physiol.* 1988; 135: 269–76.

23 Higaki, M., Sakaue, H., Ogawa, W., Kasuga, M., Shimokado, K. Phosphatidylinositol 3-kinase-independent signal transduction pathway for platelet-derived growth factor-induced chemotaxis. *J. Biol. Chem.* 1996; 271: 29342–6.

24 Liu, X.Q., Pawson, T. The epidermal growth factor receptor phosphorylates GTPase-activating protein (GAP) at Tyr-460, adjacent to the GAP SH2 domains. *Mol. Cell Biol.* 1991; 11: 2511–16.

25 Medema, R.H., Vries Smits, A.M. de, Zon, G.C. van der, Maassen, J.A., Bos, J.L. Ras activation by insulin and epidermal growth factor through enhanced exchange of guanine nucleotides on p21ras. *Mol. Cell Biol.* 1993; 13: 155–62.

26 Muroya, K., Hattori, S., Nakamura, S. Nerve growth factor induces rapid accumulation of the GTP-bound form of p21ras in rat pheochromocytoma PC12 cells. *Oncogene* 1992; 7: 277–81.

27 Meisenhelder, J., Suh, P.G., Rhee, S.G., Hunter, T. Phospholipase Cγ is a substrate for the PDGF and EGF receptor protein-tyrosine kinases in vivo and in vitro. *Cell* 1989; 57: 1109–22.

28 Koch, C.A., Anderson, D., Moran, M.F., Ellis, C., Pawson, T. SH2 and SH3 domains: elements that control interactions of cytoplasmic signaling proteins. *Science* 1991; 252: 668–74.

29 Escobedo, J.A., Navankasattusas, S., Kavanaugh, W.M. *et al.* cDNA cloning of a novel 85 kd protein that has SH2 domains and regulates binding of PI3-kinase to the PDGF β-receptor. *Cell* 1991; 65: 75–82.

30 Skolnik, E.Y., Margolis, B., Mohammadi, M. *et al.* Cloning of PI3 kinase-associated p85 utilizing a novel method for expression/cloning of target proteins for receptor tyrosine kinases. *Cell* 1991; 65: 83–90.

31 Mayer, B.J., Hamaguchi, M., Hanafusa, H. A novel viral oncogene with structural similarity to phospholipase C. *Nature* 1988; 332: 272–5.

32 Matsuda, M., Mayer, B.J., Fukui, Y., Hanafusa, H. Binding of transforming protein, P47gag-crk, to a broad range of phosphotyrosine-containing proteins. *Science* 1990; 248: 1537–9.

33 Mayer, B.J., Hanafusa, H. Association of the v-crk oncogene product with phosphotyrosine-containing proteins and protein kinase activity. *Proc. Natl. Acad. Sci. USA* 1990; 87: 2638–42.

34 Harvey, J.J. An unidentified virus which causes the rapid production of tumours in mice. *Nature* 1964; 204: 1104–5.

35 Simons, P.J., Dourmashkin, R.R., Turano, A., Phillips, D.E., Chesterman, F.C. Morphological transformation of mouse embryo cells in vitro by murine sarcoma virus (Harvey). *Nature* 1967; 214: 897–8.

36 Sweet, R.W., Yokoyama, S., Kamata, T. *et al.* The product of ras is a GTPase and the T24 oncogenic mutant is deficient in this activity. *Nature* 1984; 311: 273–5.

37 Chang, E.H., Furth, M.E., Scolnick, E.M., Lowy, D.R. Tumorigenic transformation of mammalian cells induced by a normal human gene homologous to the oncogene of Harvey murine sarcoma virus. *Nature* 1982; 297: 479–83.

38 Feramisco, J.R., Gross, M., Kamata, T., Rosenberg, M., Sweet, R.W. Microinjection of the oncogene form of the human H-ras (T-24) protein results in rapid proliferation of quiescent cells. *Cell* 1984; 38: 109–17.

39 Feramisco, J.R., Clark, R., Wong, G. *et al.* Transient reversion of ras oncogene-induced cell transformation by antibodies specific for amino acid 12 of ras protein. *Nature* 1985; 314: 639–42.

40 Satoh, T., Endo, M., Nakafuku, M., Nakamura, S., Kaziro, Y. Platelet-derived growth factor stimulates formation of active p21ras.GTP complex in Swiss mouse 3T3 cells. *Proc. Natl. Acad. Sci. USA* 1990; 87: 5993–7.

41 Harris, W.A., Stark, W.S., Walker, J.A. Genetic dissection of the compound eye of *Drosophila melanogaster. J. Physiol.* 1976; 256: 415–39.

42 Hart, A.C., Kramer, H., Vactor, D.L. Van, Paidhungat, M., Zipursky, S.L. Induction of cell fate in the Drosophila retina: the bride of sevenless protein is predicted to contain a large extracellular domain and seven transmembrane segments. *Genes Dev.* 1990; 4: 1835–47.

43 Kramer, H., Cagan, R.L., Zipursky, S.L. Interaction of bride of sevenless membrane-bound ligand and the sevenless tyrosine-kinase receptor. *Nature* 1991; 352: 207–12.

44 Hafen, E., Basler, K., Edstroem, J.E., Rubin, G.M. Sevenless, a cell-specific homeotic gene of Drosophila, encodes a putative transmembrane receptor with a tyrosine kinase domain. *Science* 1987; 236: 55–63.

45 Fortini, M.E., Simon, M.A., Rubin, G.M. Signalling by the sevenless protein tyrosine kinase is mimicked by Ras1 activation. *Nature* 1992; 355: 559–61.

46 Jones, S., Vignais, M.-L., Broach, J.R. The CDC25 protein of *S. cerevisiae* promotes exchange of guanine nucleotides bound to ras. *Mol. Cell Biol.* 1991; 11: 2641–6.

47 Lowenstein, E.J., Daly, R.J., Batzer, A.G. *et al.* The SH2 and SH3 domain-containing protein GRB2 links receptor tyrosine kinases to ras signaling. *Cell* 1992; 70: 431–42.

48 Dickson, B., Hafen, E. Genetic dissection of eye development in *Drosophila.* In: Bate, M., Martinez-Arias, A. (Eds) *Development of Drosophila melanogaster.* Cold Spring Harbor Press, New York, 2000; 1327–62.

49 Kornfeld, K. Vulval development in *Caenorhabditis elegans. Trends Genet.* 1997; 13: 55–61.

50 Gale, N.W., Kaplan, S., Lowenstein, E.J., Schlessinger, J., Bar-Sagi, D. Grb2 mediates the EGF-dependent activation of guanine nucleotide exchange on Ras. *Nature* 1993; 363: 88–92.

51 Stern, M.J., Marengere, L.E., Daly, R.J. *et al.* The human GRB2 and *Drosophila* Drk genes can functionally replace the *Caenorhabditis elegans* cell signaling gene sem-5. *Mol. Biol. Cell* 1993; 4: 1175–88.

52 Li, N., Batzer, A., Daly, R. *et al.* Guanine-nucleotide-releasing factor hSos1 binds to Grb2 and links receptor tyrosine kinases to Ras signalling. *Nature* 1993; 363(6424): 85–8.

53 Aronheim, A., Engelberg, D., Li, N, al-Alawi, N. Schlessinger, J., Karin, M. Membrane targeting of the nucleotide exchange factor Sos is sufficient for activating the Ras signaling pathway. *Cell* 1994; 78: 949–61.

54 Rozakis-Adcock, M., Fernley, R., Wade, J., Pawson, T., Bowtell, D. The SH2 and SH3 domains of mammalian Grb2 couple the EGF receptor to the Ras activator mSos1. *Nature* 1993; 363: 83–5.

55 Burgering, B.M., Freed, E., van der Voorn, L., McCormick, F., Bos, J.L. Platelet-derived growth factor-induced p21ras-mediated signaling is independent of platelet-derived growth factor receptor interaction with GTPase-activating protein or phosphatidylinositol-3-kinase. *Cell Growth Differ.* 1994; 5(3): 341–7.

56 Ray, L.B., Sturgill, T.W. Rapid stimulation by insulin of a serine/threonine kinase in 3T3-L1 adipocytes that phosphorylates microtubule-associated protein 2 in vitro. *Proc. Natl. Acad. Sci. USA* 1987; 84: 1502–6.

57 Moodie, S.A., Willumsen, B.M., Weber, M.J., Wolfman, A. Complexes of Ras.GTP with Raf-1 and mitogen-activated protein kinase kinase. *Science* 1993; 260: 1658–61.

58 Warne, P.H., Viciana, P.R., Downward, J. Direct interaction of Ras and the amino-terminal region of Raf-1 in vitro. *Nature* 1993; 364: 352–5.

59 Koide, H., Satoh, T., Nakafuku, M., Kaziro, Y. GTP-dependent association of Raf-1 with Ha-Ras: identification of Raf as a target downstream of Ras in mammalian cells. *Proc. Natl. Acad. Sci. USA* 1993; 90: 8683–6.

60 Leevers, S.J., Paterson, H.F., Marshall, C.J. Requirement for Ras in Raf activation is overcome by targeting Raf to the plasma membrane. *Nature* 1994; 369: 411–14.

61 Stokoe, D., Macdonald, S.G., Cadwallader, K., Symons, M., Hancock, J.F. Activation of Raf as a result of recruitment to the plasma membrane. *Science* 1994; 264: 1463–7.

62 Kruijer, W., Cooper, J.A., Hunter, T., Verma, I.M. Platelet-derived growth factor induces rapid but transient expression of the c-fos gene and protein. *Nature* 1984; 312: 711–16.

63 Treisman, R. Regulation of transcription by MAP kinase cascades. *Curr. Opin. Cell Biol.* 1996; 8: 205–15.

64 Lin, T.A., Kong, X., Haystead, T.A. *et al.* PHAS-I as a link between mitogen-activated protein kinase and translation initiation (see comments). *Science* 1994; 266: 653–6.

65 Sonenberg, N., Gingras, A.C. The mRNA 5′ cap-binding protein eIF4E and control of cell growth. *Curr. Opin. Cell Biol.* 1998; 10: 268–75.

66 Cuesta, R., Xi, Q., Schneider, R.J. Adenovirus-specific translation by displacement of kinase Mnk1 from cap-initiation complex eIF4F. *EMBO J.* 2000; 19: 3465–74.

67 Whitmarsh, A.J., Davis, R.J. Structural organization of MAP-kinase signaling modules by scaffold proteins in yeast and mammals. *Trends Biochem. Sci.* 1998; 23: 481–6.

68 Shields, J.M., Pruitt, K., McFall, A., Shaub, A., Der, C.J. Understanding ras: 'It ain't over 'til it's over'. *Trends Cell Biol.* 2000; 10: 147–54.

69 Vignais, M.L., Sadowski, H.B., Watling, D., Rogers, N.C., Gilman, M. Platelet-derived growth factor induces phosphorylation of multiple JAK family kinases and STAT proteins. *Mol. Cell Biol.* 1996; 16: 1759–69.

70 Zhong, Z., Wen, Z., Darnell, J.E.J. Stat3: a STAT family member activated by tyrosine phosphorylation in response to epidermal growth factor and interleukin-6. *Science* 1994; 264: 95–8.

71 Wagner, B.J., Hayes, T.E., Hoban, C.J., Cochran, B.H., Doolittle, R.F. The SIF binding element confers sis/PDGF inducibility onto the c-fos promoter. *EMBO J.* 1990; 9: 4477–84.

72 Doolittle, R.F., Hunkapiller, M.W., Hood, L.E. *et al.* Simian sarcoma virus onc gene, v-sis, is derived from the gene (or genes) encoding a platelet-derived growth factor. *Science* 1983; 221: 275–7.

73 Hall, R.A., Premont, R.T., Lefkowitz, R.A. Heptahelical receptor signaling: beyond the G protein paradigm. *J. Cell Biol.* 1999; 145: 927–32.

74 Kranenburg, O., Verlaan, I., Hordijk, P.L., Moolenaar, W.H. Gi-mediated activation of the Ras/MAP kinase pathway involves a 100 kDa tyrosine-phosphorylated Grb2 SH3 binding protein, but not Src nor Shc. *EMBO J.* 1997; 16: 3097–105.

75 Daub, H., Wallasch, C., Lankenau, A., Herrlich, A., Ullrich, A. Signal characteristics of G protein-transactivated EGF receptor. *EMBO J.* 1997; 16: 7032–44.

76 Prenzel, N., Zwick, E., Daub, H. *et al.* EGF receptor transactivation by G-protein-coupled receptors requires metalloproteinase cleavage of proHB-EGF. *Nature* 1999; 402: 884–8.

77 Zwick, E., Daub, H., Aoki, N. *et al.* Critical role of calcium-dependent epidermal growth factor receptor transactivation in PC12 cell membrane depolarization and bradykinin signaling. *J. Biol. Chem.* 1997; 272: 24767–70.

Signalling pathways operated by non-receptor protein tyrosine kinases

■ The non-receptor protein tyrosine kinase family

There is an important family of receptors that induce responses similar to those of the receptor tyrosine kinases yet possess no intrinsic catalytic activity. Instead, they recruit catalytic subunits from within the cell in the form of one or more non-receptor protein tyrosine kinases (nrPTKs). These proteins exist within the cytosol as soluble components, or they may be membrane-associated. They can be divided into nine main families with four additional enzymes that do not appear to belong to any of the defined families (Figure 12.1).[1] Recruitment of non-receptor PTKs and the consequent tyrosine phosphorylations are usually the first steps in the assembly of a substantial signalling complex consisting of a dozen or more proteins that bind and interact with each other. Further details about non-receptor tyrosine kinases are given in 'Non-receptor protein tyrosine kinases and their activation' (page 294).

Examples of the class of receptors that recruit non-receptor PTKs include those that mediate immune and inflammatory responses:

- The T-lymphocyte receptor (TCR) is involved in detection of foreign antigens, presented together with the major histocompatibility complex (MHC). Subsequently it regulates the clonal expansion of T cells.[2]

- The B lymphocyte receptor for antigen is important in the first line of defence against infection by micro-organisms.[3]

- The interleukin-2 receptor (IL-2R). The cytokine IL-2, secreted by a subset of T-helper cells, enhances the proliferation of activated T- and B-cells, increases the cytolytic activity of natural killer (NK) cells and the secretion of IgG.

- Immunoglobulin receptors, such as the high affinity receptor for IgE (IgE-R), present on mast cells and blood-borne basophils.[4] This plays an important role in hypersensitivity and the initiation of acute inflammatory responses.

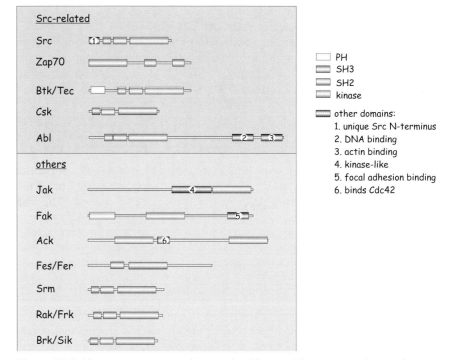

Figure 12.1 Non-receptor protein tyrosine kinases. These protein kinases form a large family. Most of them contain SH2 and SH3 domains. Several were originally discovered as transforming genes of a viral genome, hence names like src or abl, derived from Rous sarcoma virus or Abelson murine leukaemia virus. Adapted from Hunter.[1]

- Erythropoietin receptors. The cytokine erythropoietin plays an important role in the final stage of maturation of erythroid cells into mature red blood cells. For this reason it has been used by athletes to boost their performance in endurance sports such as cycling. The erythropoietin receptor is present on a number of cell types in addition to erythroid progenitor cells, suggesting that erythropoietin may have specific effects on other tissues, still to be discerned.

- Prolactin receptors. The pituitary hormone prolactin plays a pivotal role in the regulation of lactation. In addition, it has been implicated in modulation of immune responses. For instance, it regulates the level of NK-mediated cytotoxicity. It has also gained attention as a potential male contraceptive.

Other non-receptor PTKs are discussed in Chapter 14, where non-receptor protein tyrosine kinases are shown to play a pivotal role in cell survival and proliferation.

▪ T cell receptor signalling

T lymphocytes have a central role in cell-mediated immunity. When activated, they proliferate and differentiate to become either cytotoxic or helper T cells. Cytotoxic T cells kill specific targets, most commonly virus-infected cells, while helper T cells assist other cells of the immune system. Effective T cell activation requires the coordinated interactions of several different ligands with their receptors. This is not the only unusual aspect. The primary ligands that activate T cells are not classical first messengers (i.e. blood-borne soluble molecules), but proteins presented on the membranes of neighbouring cells. For example, the stimulation of T lymphocytes that are naive to any previous form of activation involves the T cell receptor (see below). This detects foreign antigenic protein in the form of a short peptide that is proffered on the surface of the target cell. An infected cell, expressing viral proteins, presents fragments bound in a groove on the MHC, a plasma membrane protein. Class I MHC molecules are expressed on virtually all nucleated cells and are characteristic of the host so that they are recognized by the immune system as 'self'. The task of the T cell is to kill the infected cell.

Circulating antigens are processed by specialized cells such as B-lymphocytes and macrophages which then present fragments on their surfaces, again attached to an MHC molecule (class II). Here, the function of the T cell is to help the target cell to make antibodies or to ingest invading micro-organisms.

More than one lymphocyte receptor must be engaged to ensure activation

The cell–cell interaction necessary for T cell activation requires intimate contact, calling specialized adhesion molecules into play (Chapter 14), as well as three different ligand–receptor interactions (Figure 12.2). It is the combination of antigen and MHC that initiates the T cell response, but there are two possible outcomes. Only in the presence of a second, 'co-stimulatory' signal do the cells become fully activated. The full response comprises transcription of early response genes, followed by synthesis and release of the cytokine IL-2, entry into the cell cycle and differentiation into an effector or memory cell. In the absence of a co-stimulatory signal, the T cell becomes unresponsive or anergic.

The signalling events are outlined in Figure 12.2. Initial cell–cell contact involves low specificity interaction between B7 (on the antigen presenting cell) and CD28 (on the T-cell). The α and β chains of the T cell receptor (TCR) may then bind to the peptide presented by the antigen-presenting cell. The TCR is associated with the CD3 molecule to comprise a complex of eight polypeptides, all of which span the membrane (Figure 12.2). In spite of having no intrinsic catalytic domains, activation of this complex results in tyrosine phosphorylations due to the recruitment of non-receptor PTKs. In addition, there is activation of PLCγ with production of IP$_3$ and DAG and elevation of cytosol Ca^{2+}. Thus, the consequences of receptor ligation are not dissimilar from those induced by the receptors for EGF or PDGF. An early candidate explaining the induction of tyrosine kinase activity emerged with the discovery of the PTK, Lck (p56^{lck}), a T cell-specific member of the Src family[5,6] (see Figure 12.1). Lck is associated with the cytosolic tail of CD4 (in helper T cells) or CD8 (in cytotoxic T cells).[7] The extracellular domains of these molecules bind to the MHC protein and this not only strengthens the rather weak interaction established

(a)

(b)

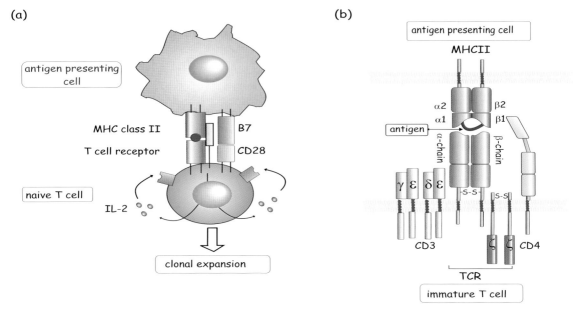

Figure 12.2 Clonal expansion of naïve T-lymphocytes: (a) Clonal expansion is signalled through binding of an antigen presented in the groove of the MHC II on an antigen-presenting cell to the T-cell receptor. This, and the additional interaction of a number of co-stimulatory adhesion molecules, such as B7 and CD28, result in the production of IL-2 together with its receptor on the T cell. This induces a proliferation signal that is responsible for the clonal expansion of those T lymphocytes that exclusively recognize this particular antigen. As a result, the host can eliminate the antigen with great efficiency. (b) The TCR possesses a disulphide-linked heterodimer of α and β chains. These have hypervariable regions that detect the antigen, presented as a short peptide in the groove of an MHC molecule. This heterodimer, along with two ζ-chains, forms a complex of four polypeptides (γ, ε and δ, ε) of the CD3 molecule. A CD4/CD8 molecule is also associated with the TCR in helper/cytotoxic T cells respectively. This binds to the MHC and brings Lck, a non-receptor PTK, into the vicinity of the ζ-chains.

between the TCR and antigen, but it also brings CD4 (or CD8) into the vicinity of the TCR complex, leading Lck to its targets on the ζ-chains. However, as with other Src family kinases, Lck is inactive until specific residues have been dephosphorylated. This is accomplished by yet another transmembrane protein, CD45, which possesses protein tyrosine phosphatase activity. Its function is described more fully in Chapter 17.

Activation of Lck results in the phosphorylation of the ζ-chains.[8] The target tyrosines are confined to ITAM motifs (immunoreceptor tyrosine-based activation motifs). ITAMs are also present in the γ, δ and ε chains of CD3 and are targets of another Src family kinase Fyn (p59[fyn]) associated with the ε chain.[9] Fyn is also activated by dephosphorylation. Both Fyn and Lck are needed for efficient TCR signalling. Phosphorylation of ITAMs provides docking sites for SH2 domain-bearing molecules with the resulting recruitment of the PTK, ZAP-70 (ζ-chain-associated protein-tyrosine kinase of 70 kDa).[10] Once bound, this in turn becomes phosphorylated and thereby activated, causing phosphorylation of multiple substrates. The sequence of events then follows a pattern in which phosphotyrosines bind SH2- or PTB-domain-containing proteins that may themselves be PTKs that can phosphorylate other proteins in succession. At each stage

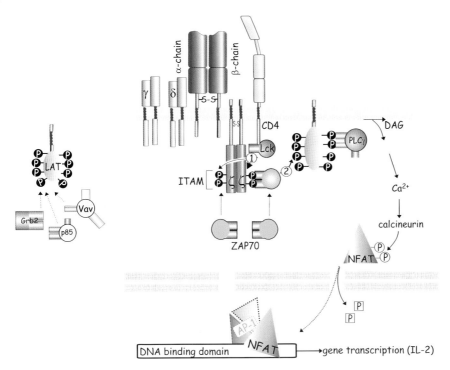

Figure 12.3 Signals from the activated TCR complex are transduced by cytosolic tyrosine protein kinases. The TCR activates Lck that phosphorylates the two ζ-chains in the ITAM motif. The phosphotyrosine residues form a docking site for the SH2 domain of ZAP70, another cytosolic PTK, which in turn, phosphorylates several (maximally nine) tyrosine residues on the transmembrane adaptor protein LAT. Various proteins attach to LAT. These include the guanine exchange factor Vav, the adaptor Grb2, the adaptor subunit of PI 3-kinase and PLCγ. All of these play important roles in the activation of the IL-2 gene. The elevation of intracellular Ca^{2+} activates calcineurin which dephosphorylates NF-AT, the nuclear factor of activated T cells. Together with the AP-1 complex, NF-AT drives the transcription of the IL-2 gene.

there is the opportunity for branching, through a range of effectors. By successive recruitment, an extensive signalling complex is assembled that includes multiple effector enzymes (Figure 12.3). An important branch-point is offered by the integral membrane protein LAT (linker for activation of T-cells) that presents no less than nine substrate tyrosine residues.[11] When phosphorylated, these recruit a broad range of signalling molecules, all through interaction with SH2-domains. These include: Grb2, SLP76 and PLCγ, PI 3-kinase (through its p85 regulatory subunit) and the guanine nucleotide exchange factors Dbl and Vav.

The formation of a signalling complex around the TCR and the branching pathways that emanate from it, resemble the mechanisms used by the growth factors. However, the destinations of these pathways are not all clear. The PLCγ pathway (DAG, IP_3 and elevation of Ca^{2+}) leads to the activation of the phosphatase calcineurin which activates the transcription factor NF-AT (nuclear factor of activated T cells).[8,12] This is essential for clonal expansion of T cells because of its pivotal role in the induction of

IL-2 expression (see Chapter 17). NF-AT requires the assistance of the activator protein-1 complex (AP-1), in order to drive expression of IL-2.

As already mentioned, full T cell activation does not take place in the absence of a co-stimulatory signal, obtained through the engagement of a separate receptor. For some helper T cells, this must be provided by CD28 on the lymphocyte which binds to the B7 molecule on an antigen-presenting cell. The signalling pathway activated by CD28 involves PI-3 kinase (Chapter 13), but the details of the mechanism and how it interacts with the events set in train by the TCR are unclear.

■ The IgE receptor and a signal for exocytosis

Tissue mast cells and circulating basophils are of haematopoietic lineage. Best known for their roles in allergy, they mediate both immediate and delayed hypersensitivity reactions.[4] They also help to defend the body against bacterial[13] and parasitic infections and take part in inflammatory responses. Their immunological stimulus is provided by polyvalent antigens that bind and crosslink IgE that is itself bound to a high-affinity immunoglobulin receptor, IgE-R (specifically, FcεR1). Initially, the signalling mechanism has similarities with that of lymphocytes, in that it involves the successive recruitment of tyrosine kinases and SH2-domain containing proteins (adaptors and effectors).

IgE-receptor aggregation sets in train a series of events. The acute consequence is the secretion of preformed products stored in secretory granules, and this takes place within a few minutes. The released substances include vasoactive agents and mediators of inflammation (histamine, proteoglycans, neutral proteases, acid hydrolases). Then, over minutes to hours, the cells synthesize and secrete cytokines, among others IL-2 and IL-6, and arachidonate-derived inflammatory mediators, such as the leukotriene LTB_4.[14] The formation of new granules and recovery of cell morphology then continues over a period extending from hours to weeks.[15]

The initial events that follow receptor aggregation involve the recruitment of Src-family tyrosine kinases, including Lyn and Syk (Figure 12.4). Like the T cell receptor, the IgE-R is located together with the scaffold protein LAT in a lipid raft (microdomain: see below).[16] Phosphorylation of LAT by Syk provides a docking site for a number of SH2 domain-containing proteins. Of these, Vav is of importance because it regulates activation of members of the Rho family of GTPases. Vav is endowed with numerous domains that enable it to integrate diverse incoming and outgoing signals. These include one SH2 domain, two SH3 domains, a Dbl homology (DH) domain, a pleckstrin homology (PH) domain, a leucine-rich region and a cysteine-rich region (see Chapter 18). The DH domain, in particular, is characteristic of the guanine nucleotide exchange factors that catalyse GTP/GDP exchange on Rho family GTPases. These control diverse cellular responses, including the reorganization of the cytoskeleton[17] and the regulation of the Jun N-terminal kinases (JNK) that mediate differentiation[18] (see also Figure 11.15, page 274).

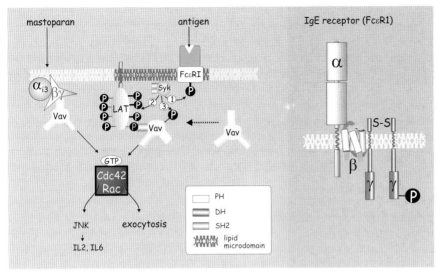

Figure 12.4 Signal pathways emanating from the high-affinity receptor for IgE. Antigen binding causes the recruitment and activation of the cytosolic tyrosine protein kinase Syk which phosphorylates the FcεRI and LAT. Subsequent recruitment and phosphorylation of Vav causes GTP-exchange on Cdc42 and Rac, activating pathways for degranulation and cytokine production.

Intracellular signalling pathways generally overlay one another and, although they may culminate in different effects, they share common elements. This can make it hard to map out the sequence of events that lie between the binding of a ligand and the final cellular response. In mast cells and other related cells of haematopoietic origin (neutrophils, eosinophils), the signalling events that lead to exocytosis of secretory granules are not easy to separate from those that direct the other cell responses. Activation of phospholipase C produces an elevation of cytosol Ca^{2+}, but the fusion of secretory granules with the plasma membrane is not, in itself, dependent on Ca^{2+}.[19–21] Instead, unlike most other secretory cells, in which exocytosis follows directly from an elevation of cytosol Ca^{2+}, exocytosis is dependent on the presence of GTP. Activation of Cdc42 and Rac is the key determining step committing cells of this class to undergo exocytosis, though the steps linking these GTPases to the proteins that regulate membrane fusion remain unknown.[22–24]

A second pathway of activation in mast cells is triggered by agents such as the wasp venom peptide mastoparan. Rather than interacting with cell surface receptors, such 'receptor-mimetic' agents are able to insert into the membrane to cause direct activation of heterotrimeric G-proteins of the G_i class.[25,26] Here, it is the βγ-subunits of the G-protein that provide the signal for exocytosis.[27] As in the pathway from the IgE-R, it is possible that Vav participates in the integration of these signals since it possesses a PH domain (binds βγ-subunits) and has guanine nucleotide exchange activity.

The lipid raft hypothesis

A feature of the signal transduction mechanisms in cells that are stimulated through immune recognition receptors is the involvement of heterogeneous regions in the plasma membrane. These microdomains (or rafts) are characterized by their lipid composition. The restriction of certain membrane-associated or transmembrane signalling proteins to particular lipid domains is especially apparent in T and B lymphocytes and mast cells. The existence of lipid domains is widespread, but knowledge of their wider involvement in signalling mechanisms is lacking and their study has not been without disagreement. Initially, the discovery of detergent-resistant fractions in membranes obtained from epithelial cells[28] led to the postulation of membrane regions enriched in glycosphingolipids and cholesterol. These have also been referred to as detergent-resistant (or detergent-insoluble) membrane domains, or more simply as lipid rafts. They are thought to be phase-separated regions, which, because of the saturated nature of the acyl chains of the component sphingolipids and the presence of cholesterol, possess a higher order of rigidity than the remainder of the membrane. Direct visualization of lipid microdomains in a smooth muscle cell by 'single molecule' fluorescence microscopy indicates dimensions ranging between 0.2 and 2 μm at ambient temperatures, but they may be smaller at 37°C.[29] Both their size and density are likely to vary with cell type.

Certain signalling molecules appear to be confined to lipid rafts. These include CD4/CD8 in T cells and the dual acylated (*N*-myristoylated and palmitoylated) non-receptor tyrosine kinases of the Src family (see 'Non-receptor protein tyrosine kinases and their activation', page 294). The non-acylated Src kinase Lck is also raft-associated, presumably through association with CD4/CD8. Another important signalling molecule located in rafts is LAT (see above). Other membrane proteins such as the transferrin receptor and the tyrosine phosphatase CD45 are generally absent from these domains. The primary receptors, the TCR, the BCR and the IgE-R, were thought to exist outside rafts until crosslinked during activation. However, this is now less clear cut and it seems that there is a population of raft-associated receptors that increases upon stimulation. Exactly how lipid rafts function as platforms for the assembly of signalling complexes is unclear.

Interferons and their effects

Classification of the interferons, a group of cytokines known originally for their antiviral properties, has been difficult. They may be divided into two main classes: type 1 (IFNα, β and ω) and type 2 (IFNγ). Type 1 interferons, produced by monocytes, macrophages, activated lymphocytes and certain epithelial cells, have mostly antiviral activity.

Type 2 interferons are released by T cells and natural killer cells and they are potent immunomodulators and inhibitors of proliferation. For example, IFNγ not only induces the production of MHC class II molecules, but also elevates the expression of HLA-G (a form of MHC I) and

CD4 in T lymphocytes. All this increases the ability of T cells to detect and respond to viral protein fragments offered by antigen-presenting cells, and helps B cells to mount an antibody response. IFNγ also promotes the expression of receptors for IgG on macrophages and the secretion of TNFα. Although the mechanisms set in train by IFNγ are ultimately antiviral, they are not specific for any particular virus.

The processes activated by the type 1 interferons are less easy to summarize. IFNα, perhaps the best understood, has important actions in the control of differentiation and growth. A number of cancers are susceptible to IFNα and it is used in adjuvant therapy as an anticancer drug. In particular, it has been applied in the treatment of hairy cell leukaemia (a chronic B cell leukaemia[30]) and to treat certain metastasizing cancers, such as renal carcinoma and Kaposi's sarcoma, which occurs in a quarter of AIDS sufferers.

Interferon-α receptor and STAT proteins

The interferon-α receptor (type 1) is composed of three subunits, IFNαR1, IFNαR2b and IFNαR2c. IFNαR2c binds the latent forms of the cytosol proteins STAT1 and STAT2. To date, seven members of the mammalian family of STAT proteins have been identified (see Chapter 11).

The IFNαR1 subunit is complexed with the non-receptor PTKs, Tyk2 and Jak1 (Table 12.3) which are also in an inactive state. Stimulation of cells through the IFNαR-1 leads to subunit trimerization followed by activation of its associated PTKs (Figure 12.5), which phosphorylate the receptor subunits. The phosphorylation of IFNαR1 (Y466) allows binding of STAT2 (p113) through its SH2 domain, bringing the transcription factor close to the activated tyrosine kinases. Once phosphorylated, STAT2 promotes the phosphorylation of STAT1 (p91). The two STAT proteins interact through mutual binding of their SH2 domains to phosphotyrosine residues, then dissociate from the receptor as a heterodimer which combines with IRF (interferon regulatory factor). This trimeric complex, ISGF3 (interferon-α stimulated-gene factor 3). translocates to the nucleus and binds to DNA at the interferon-stimulated response element (ISRE).[31]

The STATs also convey signals issuing from several of the interleukin receptors. The specificity of the intracellular signal is determined by the particular cytosolic tyrosine protein kinase associated with the receptor and by the combinations of STATs that are phosphorylated and activated (see Table 12.1).

Table 12.1 Diversity in cytokine-induced signalling. The spectrum of recruited non-receptor tyrosine kinases and downstream STATs and associated proteins (IRS) determine the outcome of the cellular response. Adapted from Ihle.[32]

Cytokines	IL-2		IL-4	IL-9	IL-10	IL-12
Non-receptor tyrosine kinase	Jak1	and	Jak3		Jak1/Tyk2	Jak2/Tyk2
STATS	STAT5		STAT6	STAT3	STATs1 and 3	STAT 4
Associated protein			IRS	IRS		

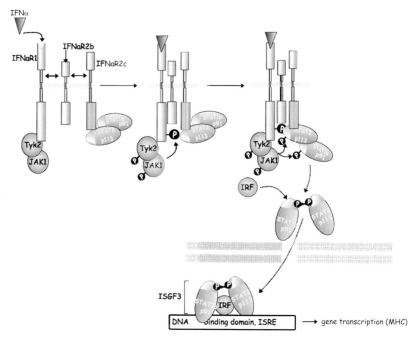

Figure 12.5 Activation of STATS by the interferon receptor. Binding of IFNα causes trimerization of the subunits of the IFNα receptor (IFNαR1, IFNαR2b and IFNR2c). This activates the associated cytosolic protein kinases Tyk2 and Jak1, which phosphorylate each other and phosphorylate the IFNαR1 subunit of the receptor complex. STAT2, complexed with the IFNαR2c receptor, binds to the phosphorylated IFNαR1 and in doing so becomes phosphorylated by the cytosolic protein kinases. The phosphorylated STAT2 catalyses the phosphorylation of STAT1 and these detach from the receptor complex, dimerize and are accompanied by an interferon regulatory factor of 48 kDa (IRF). The ternary complex ISGF3 enters the nucleus, binds to DNA at the ISRE, inducing expression of genes such as the MHC.

Oncogenes, malignancy and signal transduction

Viral oncogenes

Infection by viruses carrying oncogenes can cause malignant cell growth. Although viruses were first recognized as causative agents in avian cancers 90 years ago, for much of the twentieth century there was doubt that any human cancers were initiated in this way. Even now, almost all the information in this area refers to non-human animals. There are a number of problems here. First of all, as was already apparent in the first decade of the twentieth century,[33–35] demonstration of a viral mode of transmission depends on the induction of disease by transfer of tissue filtrates from animal to animal. Some viruses only become oncogenic as a consequence of multiple passages and through different animal species. Secondly, while there are many human cancers that are certainly associated with viral infection, it is far from certain in most cases whether the virus initiates the condition or whether it is merely permissive of induction by another agent, such as a chemical carcinogen. In general, the

Phosphoinositide 3-kinases, protein kinase B and signalling through the insulin receptor

At a viva voce examination at one of the older universities, a student was questioned about insulin and how it works. After some embarrassed hesitation, he assured the examiners that that he had known all about it, but unfortunately he had forgotten. 'What a pity', was the response 'This means that now, nobody knows'.

Insulin receptor signalling; it took some time to discover

A lack of knowledge about how a particular drug or therapy works has never been an impediment to its application in the clinic (see Chapter 1). Indeed, for most practitioners, this is generally a very secondary consideration. A well-known case in point is aspirin, first introduced by the Bayer Company in 1898. With the wholehearted encouragement of the medical profession, it has been consumed by the tonne and with good cause too. Yet it was more than 80 years before its most sensitive target was identified as cyclo-oxygenase, preventing the synthesis of prostaglandins.[1,2] Another example is the treatment of childhood (type I) diabetes with insulin.[3,4]

Several glycogenolytic hormones (adrenaline, glucagon, vasopressin, growth hormone) can mobilize the metabolic stores and elevate the concentration of blood glucose, but only insulin has the reverse effect. It increases the net uptake of glucose from the blood and it increases its conversion to glycogen and triglyceride, at the same time inhibiting their breakdown. Since the acute effects of the glycogenolytic hormones require elevations of cAMP or cytosol Ca^{2+}, it follows that the actions of insulin must be mediated through a third route, one that escapes the

The association of diabetes mellitus with the pancreas was established by the work of Oscar Minkowski.[47] A dog from which he had removed the pancreas began to suffer an uncontrollable polyuria. The following day, there being no cage large enough to accommodate the animal (which had been well house-trained), it was kept tied up in the laboratory. According to the story (subsequently denied by Minkowski), the assistant noticed that flies settled wherever it had passed urine. Regardless of this, the animal passed 12% of sugar and was suffering from diabetes mellitus.

It is customary to credit the Canadians, Banting and Best with the discovery of insulin, their colleagues McLeod and Collip somehow standing close by (or not quite so close by in the case of McLeod, who may have spent some of the critical months on a fishing holiday on the Isle of Skye). For sure, were it not for Banting's certainty and indeed obsession, it is unlikely that patients would have been successfully treated within a few months of the commencement of the experimental trials. However, it is legitimate to ask whether the Canadians were actually the first discoverers of insulin. Here, we are not considering the widely canvassed claim of Nicholas Paulescu,[3,4] who was certainly in the race at about the same time, but that of Eugene Gley who was working 25 years earlier. Sadly, his right to scientific immortality fails on two counts. First, he was never in a position to apply his preparation in the treatment of patients. This is the importance of Banting and Best's contribution. Whether he had the means to make the repeated and rapid analyses of blood glucose concentration, necessary to monitor its anti-glycosuric effect, is uncertain. More likely, he just lacked the audacity to inject his preparation into people, a restraint to which Banting was quite insensitive. Anyway, Gley was working in 1895 and all this had to wait for another quarter century,

attentions of either PKA or PKC. Only in 1980 did it become apparent that the mechanism involves autophosphorylation of the receptor.[5] The discovery of protein kinase B (PKB) and the 3-phosphorylated inositol lipids eventually provided the key.[6]

PI 3-kinase

Until recently, it was normal to consider the reactions leading up to and including the synthesis of the polyphosphoinositides as constitutive (unregulated) processes. Attention was focused primarily on the generation of the breakdown products DAG and IP_3. It is now clear that matters are far more complex. For example, the small GTPase Rho regulates the inositol lipid 5-kinase PI(4,5)K, responsible for the conversion of phosphatidylinositol-4-phosphate, PI(4)P, to phosphatidylinositol-4,5-bisphosphate.[7] Although it is far from clear that $PI(4,5)P_2$ qualifies as a signalling molecule, inositol lipids that are phosphorylated in the 3-position certainly do. They are the products of the PI 3-kinases that phosphorylate PI and its derivatives PI(4)P and $PI(4,5)P_2$ position to give PI(3)P, $PI(3,4)P_2$ and $PI(3,4,5)P_3$ (Figure 13.1).

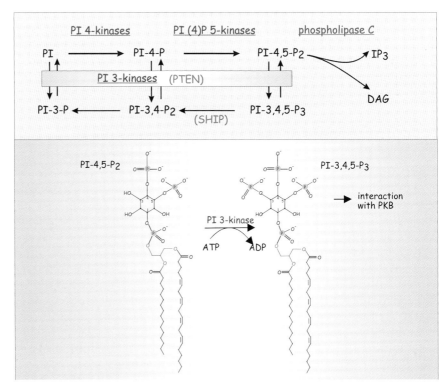

Figure 13.1 Phosphoinositide 3-kinases and the generation of 3-phosphorylated lipids. The PI 3-kinases phosphorylate the 3 OH-position in the inositol ring of the phosphatidylinositol lipids. The 3-OH phosphorylated inositol lipids are not substrates for PLC. The phosphatases PTEN and SHIP reverse the reaction. The PH domain of PKB interacts preferentially with the $PI(3,4,5)P_3$ product.

PI(3,4,5)P$_3$ was first detected in stimulated human neutrophils.[8] Although initially contemplated as a new substrate for PLC, and thus the source of a water-soluble second messenger (inositol-1,3,4,5-phosphate or IP$_4$), it soon became clear that the lipid itself is a signalling entity. Phosphatidylinositol occupies a special place among phospholipids since its headgroup offers such a multiplicity of phosphorylation patterns. Although inositol lipids have been detected in analyses of bacteria, there is no indication that any of them plays a regulatory role. However, it is clear that some bacteria rely on the inoside metabolism of host cells to regulate their invasiveness.[9] PI is present in all eukaryotic cells, though its metabolism in unicellular organisms such as yeast is restricted, since they lack the means to generate either PI(4)P or PI(4,5)P$_2$. The substrate specificity of the yeast PI 3-kinase (coded by the gene *VPS34*) is appropriately limited to PI and therefore it only generates the monophosphorylated derivative PI(3)P (which can then be converted to PI(3,5)P$_2$, essential for Golgi membrane recycling[10,11]). The enzymes catalysing the formation of PI(3,4)P$_2$ and PI(3,4,5)P$_3$ probably evolved with the need for more complex forms of metabolic regulation following the emergence of the first metazoans.

Unlike the more familiar polyphosphoinositides, the inositol phospholipids that have a phosphate group at the 3 position do not serve directly as substrates for phospholipase-C. They are metabolized by the hydrolysis of the phosphate groups at the 3- and 5- positions (Figure 13.1) (see Chapter 17). Both the 3-kinases and the 3-phosphatases are under the control of receptor-mediated processes and it is apparent that the 3-phosphorylated inositol phospholipids are fully fledged second messengers. The difficulty has been in assigning the processes that they regulate.

but we do know from his report that there can be little doubt that he had successfully isolated an antiglycosuric agent from extracts of pancreas. He provides sufficient detail, rare in his day, for the modern reader to have some confidence that his preparation was as good as he claimed. But where was his report all this time? Instead of going public, Gley sealed it in an envelope which he placed into the hands of the secretary of the Société de Biologie de Paris, with the firm instruction that it was not to be opened until directed by him.[5] With this bizarre act he waived all claims to be credited as the discoverer of insulin and so it lay hidden until word came through from Canada in 1921. The application of insulin was certainly one of the key milestones of modern medical practice, yet its mechanism of action as a glucose-lowering hormone has been widely regarded as a 'mystery', even into the last decade of the twentieth century.

A family of PI 3-kinases

Cloning and screening strategies have revealed a number of enzymes having 3-phosphorylating activity (Figure 13.2). The PI 3-kinases comprise a family of enzymes subdivided into three classes (Figure 13.2). They have distinct substrates and various forms of regulation. They all have four homologous regions, the kinase domain being the most conserved.

Class I PI 3-kinases

Class I PI 3-kinases phosphorylate PI, PI(4)P and PI(4,5)P$_2$ (the preferred substrate). These enzymes comprise two subunits, a regulatory (p55 or p85) and a catalytic subunit (p110), each existing in various forms (Figure 13.2). The multidomain structure of the regulatory subunit, in particular p85, strongly suggests that they should be able to interact with other signalling proteins. The SH2 domains enable them to bind to phosphotyrosine residues. Similarly, the SH3 domains allow interaction with proline-rich sequences, present for instance in the adaptor molecule Shc, the GTPase-activating protein Cdc42GAP or the regulator of TCR signalling, Cbl.[12] In addition, the p85-subunit contains a BCR homology domain[13] that interacts with members of the Rho family of GTPases, Rac and Cdc42, providing yet further opportunities for regulation.[14]

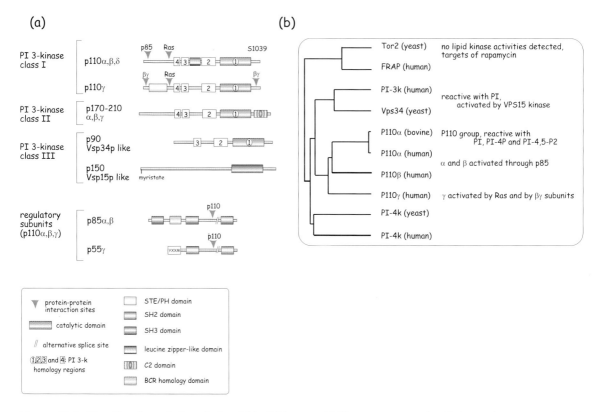

Figure 13.2 Classification of phosphoinositide 3-kinases: (a) The enzymes are classified into three groups based on the molecular structure of the subunit that contains the kinase domain. The class I is subdivided into groups A and B. Group A comprises α, β and δ which interact with the regulatory subunits, p85 or p55. Group B has one member, p110γ, regulated by G-protein βγ-subunits. It is also found associated with a p101 protein of unknown identity. Class II comprises three isoforms, none of which requires a regulatory subunit for its activation. Class III has only one member. The mode of regulation of activity of classes II and III remains to be discovered. (b) Dendrogram showing the evolutionary relationships of the PI 3-kinases.

There are four isoforms of the p110-subunit. All contain a kinase domain and a Ras interaction site.[15] In addition the α, β and δ isoforms possess an interaction site for the p85-subunit. The class I enzymes can be further subdivided. Class IA enzymes interact through their SH2 domains with phosphotyrosines present on either protein tyrosine kinases or to docking proteins such as insulin receptor substrates (IRS, see below) or LAT (see Chapter 12). Uniquely, the class IA enzymes activate protein kinase B (PKB, see below).

The p110γ catalytic subunit, found only in mammals, is the single member of class IB. It has its own regulatory subunit, p101, a protein without apparent sequence homologies. Importantly, p110γ can be regulated by G-protein βγ-subunits. It is not clear whether or not the association with p101 is required for this regulation (see Chapter 14).[16]

The class I PI 3-kinase enzymes may respond to different upstream signals and have different functions. This is apparent in macrophages in which distinct isotypes modulate separate cellular responses. Here, mitogenic signalling (DNA synthesis) induced by colony stimulating factor-1 is mediated by p110α whereas actin organization and cell migration require the β or δ isoforms.[17]

Class II PI 3-kinases

The three members of this group, PI 3-kinase II α, β, and γ have substrate specificity for PI and PI(4)P. They are all monomeric (170–210 kDa) with a C-terminal C2 domain. Their mode of activation is unclear.

Class III PI 3-kinases

These are represented by the human homologues of the yeast gene product VPS34 (vacuolar protein sorting mutant). They only phosphorylate PI to form PI(3)P. Like the yeast enzyme, the human enzyme is tightly coupled to a regulatory subunit (homologue of VPS15), a protein Ser/Thr kinase which both phosphorylates and recruits the catalytic unit to membranes. Since the monophosphorylated product PI(3)P is ubiquitous, and its level does not appear to alter when cells are stimulated, it is likely that the class III enzymes serve a housekeeping rather than a signalling role.

Studying the role of PI 3-kinase

Wortmannin, an inhibitor of PI 3-kinase

It might be thought that the availability of an inhibitor of the PI 3-kinases, in this case wortmannin, would have provided an unambiguous key to the understanding of the pathways and cellular functions that they control. This antifungal antibiotic isolated from *Penicillium wortmannii* was originally identified as a toxic agent causing acute necrosis of lymphoid tissues, severe myocardial haemorrhage and haemoglobinuria.[18] At an appropriate dose, it is also a powerful anti-inflammatory agent.[19,20]

Used judiciously, and in combination with other inhibitors (e.g. Lilly inhibitor, LY294002) and independent approaches, wortmannin certainly has its place in the battery of techniques for investigating signal transduction. At concentrations above 10^{-9} mol/l, it associates covalently with the p85 regulatory subunit of PI 3-kinase.[21] In the quest for inhibitory effects, however, it has often been applied at high concentrations, leading to false positive reports of inhibition. Furthermore, due to the existence of multiple isoforms of PI 3-kinase, not all of which are targets of this inhibitor, a negative result does not necessarily rule out a role for the products of 3-phosphorylation.

Unfortunately, it has all too readily been assumed that the actions of **wortmannin** are specific to the 3-kinases, even though it was in use as an inhibitor of myosin light chain kinase for at least a couple of years before its first use in the field of PI 3-kinases.[20] At concentrations above 100 nmol/l wortmannin also inhibits a form of PI 4-kinase and in the micromolar range, in addition to its effects on myosin light chain kinase, it possibly inhibits other protein kinases as well. As a result of its uncritical use, the 3-phosphorylated lipids were considered to be implicated in several processes that are quite innocent of any such relationship.

Measuring the lipid product

More directly, a role for 3-phosphorylation can be tested by measuring the levels of the 3-phosphorylated lipids in stimulated cells. Although PI(3)P is a normal lipid component, the amounts of PI(3,4)P$_2$ and PI(3,4,5)P$_3$ in resting cells are vanishingly low. Unless there is an increase in their levels following stimulation, their roles as second messengers can be ruled out.[22]

This could be taken as an object lesson in the use of inhibitors in general. Experience advises that the terms 'potent' and 'highly specific', frequently used to promote pharmacological agents on their first outing, may wear a bit thin after a year or two. As the evidence of side-effects accumulates, new 'potent' and 'highly specific' agents come to take their place. The application of 'potent' and 'highly specific' inhibitors of calmodulin and later of protein kinase C impeded progress in an earlier generation.

■   Pathways of activation for PI 3-kinase

The process by which receptors for growth factors activate the class IA PI 3-kinases through interaction with the SH2 domain of the p85 regulatory subunits is well established (Figure 13.3). PI 3-kinase can also be recruited and activated by non-receptor tyrosine kinases such as those of the Src family through the same mechanism.[23] In platelets, following stimulation by thrombin, it appears that activation of PI 3-kinase is linked to focal adhesion kinase (FAK: see Chapter 14). This protein, associated with integrin signalling and cytoskeletal organization in focal adhesion sites, contains proline-rich regions that interact with the SH3 domain on p85.[24] In addition, phosphorylation of FAK allows interaction with the SH2 domain of p85 and this appears to be the route of activation following attachment of cells to solid substrates.

Apparently, there are two routes to the activation of PI 3-kinase. One of these is direct and the other indirect through the involvement of focal adhesion sites (for further discussion concerning the link between FAK and the cytoskeleton, see Chapter 14). The small GTPase Ras also activates PI 3-kinase through direct interaction with p110α (Figure 13.3).[15] Other GTPases, particularly those of the Rho family, are also involved as regulators (and downfield effectors) in pathways regulated by PI 3-kinase.[16]

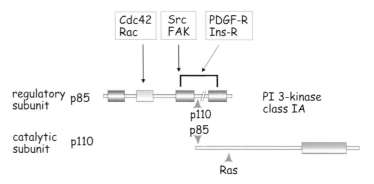

Figure 13.3 Multiple pathways to activate PI 3-kinase class IA. Several proteins interact with the p85 regulatory subunit. The Rho family GTPases, Rac and Cdc42, interact at the BCR domain. PDGF-R, rPTKs, the nRPTKs Src and FAK, and the insulin receptor substrate (IRS) all interact at the SH2 domains. The catalytic subunit can interact directly with the monomeric GTPase Ras.

■   Protein kinase B and activation through PI(3,4,5)P₃

akt Acutely transforming retrovirus AKT8 in rodent T-cell lymphoma.

The viral oncogene v-*akt* encodes a fusion product of a cellular serine–threonine protein kinase and the viral structural protein Gag. This kinase, PKB, is similar to both PKCε (73% identity to the catalytic domain) and protein kinase A (68%). It differs from other protein kinases since it contains a PH domain. These can bind to polyphosphoinositide head groups and also G-protein βγ-subunits (see Chapter 18). To date there are three subtypes, α, β and γ all of which show a broad tissue distribution

(Figure 13.4). Shortly after its discovery, it was found that PI(3,4,5)P$_3$ is the activator of PKB. The mechanism of this activation has turned out to be far from simple, with the phospholipid playing two distinct roles. One of these is direct, the lipid head-group binding to a PH domain in the N-terminal segment of PKB.[25] The other interaction is indirect, involving a soluble protein kinase, PDK1 (3-phosphoinositide dependent protein kinase), also endowed with a PH-domain.[26] Binding of PI(3,4,5)P$_3$ is crucial since it enables PDK1 and PKB to come together (Figure 13.5). Unlike PKB (and many other protein kinases involved in signal transduction), PDK1 is constitutively active. Specificity in this signalling pathway is refined by PI(3,4,5)P$_3$ acting both as a recruiting sergeant and then as an activation signal. PDK1 phosphorylates PKB in its catalytic loop but the full activation signal requires a second phosphorylation in the C-terminal domain. This reaction is dependent on yet another protein kinase, PDK2, which has yet to be properly identified. Unless both buttons are pressed, nothing happens. Double phosphorylation of PKB causes its detachment from the membrane and this enables it to interact with its substrates elsewhere in the cell. The viral oncogene product, v-Akt, has a lipid anchor (myristoyl group), which means that the protein kinase is already located at the membrane and this may facilitate its activation.

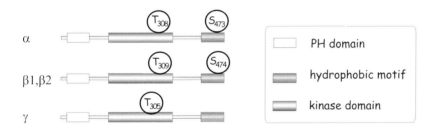

Figure 13.4 Classification of protein kinase B. There are multiple isoforms of PKB, designated α, β1, β2, and γ. They all contain a PH domain and a hydrophobic motif at the C-terminus. They differ slightly in the localization of the regulatory phosphorylation sites.

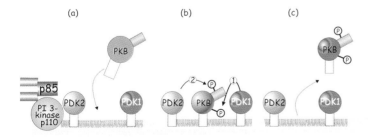

Figure 13.5 Mechanism of activation of protein kinase B. Generation of PI(3,4,5)P$_3$ (brown) serves as a membrane recruitment signal for PKB. Associated with the membrane it is (a) firstly phosphorylated in its catalytic domain by PDK1; (b) then by PDK2 in the hydrophobic motif and (c) the activated PKB then detaches from the membrane.

Insulin: the role of IRS, PI 3-kinase and PKB in the regulation of glycogen synthesis

From the insulin receptor to PKB

The stimulation by insulin of glucose uptake, glycogen synthesis and protein synthesis are all sensitive to wortmannin.[27,28] In contrast, insulin-mediated activation of ERK is without significant effect on glucose transport or the activation of glycogen synthesis.[29,30] Binding of insulin to its receptor also results in prolonged activation of PI 3-kinase, although there is no direct interaction between these two components. Instead, the activation of PI 3-kinase is mediated through its interaction with a molecule that is a substrate of the insulin receptor. Since the receptor is a dimer, the possibility exists that ligand binding induces a domain movement that brings about autophosphorylation.[31] This is followed by phosphorylation of a number of insulin receptor substrates (IRS). To date four such proteins have been identified (IRS-1 to IRS-4), all possessing PH and PTB domains. The PTB domain binds directly to the tyrosine phosphorylated region of the insulin receptor that is immediately proximal to the membrane and this results in phosphorylation of the substrate at multiple sites. The p85 regulatory subunit of PI 3-kinase (class IA) can then bind to IRS-1 through its SH2 domains[32] (Figure 13.6).

Phosphorylation on the **IRS** occurs at tyrosine residues that are part of the Tyr-X-X-Met motif where X stands for any amino acid. IRS-1 contains nine such motifs and is thus phosphorylated at a number of sites.

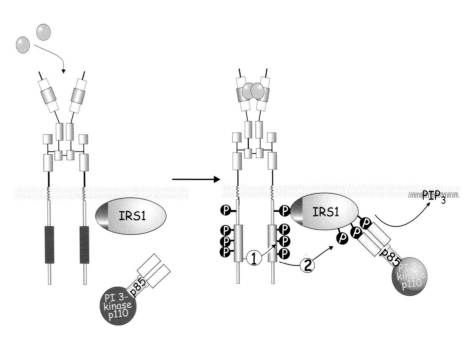

Figure 13.6 Activation of PI 3-kinase by the insulin receptor. The IRS-1 binds and is phosphorylated by the activated insulin receptor. It then serves as a docking site for the SH2-domains of the p85 regulatory subunit of PI 3-kinase leading to the generation of PI(3,4,5)P$_3$ (marked in brown).

Following prolonged activation of PI 3-kinase and the production of 3-phosphorylated polyphosphoinositides, a number of Ser/Thr kinases associate with the plasma membrane. Important among these is PKB. The intervention of PI 3-kinase in the activation of PKB is well supported by a variety of observations.[33] Its activity is quiescent in serum-starved fibroblasts, but it becomes active shortly after the addition of insulin or platelet derived growth factor (PDGF). This does not occur in cells that contain a mutant form of PKB lacking the PH domain and it depends on the presence of phosphotyrosines on IRS-1 or the PDGF receptor. Notice that it is $PI(3,4,5)P_3$, not $PI(3,4)P_2$ that activates PKB and that this correlates with the ability of the isolated PH domain (and the intact kinase) to bind to this particular phospholipid.[34] Similar to insulin, PDGF is able to activate PKB but this has no immediate effect on glucose metabolism. Other factors must be in place to direct the signal transduction pathway downstream of the insulin and the PDGF receptors. For instance, the proximity of phosphorylated IRS-1 to the insulin receptor relays the activation of PKB in the direction of glucose metabolism. This connection is absent in the case of the PDGF receptor.

■   From PKB to glycogen synthase

The minimum sequence motif required for efficient phosphorylation of small peptides by PKB is RxRxx(S/T)(F/L) and a number of substrates have been found that possess this. With respect to insulin signalling, one such substrate is glycogen synthase kinase, GSK-3β, phosphorylation of which suppresses its activity[23] (Figure 13.7).

Inactivation of GSK-3β necessarily reduces the phosphorylation at regulatory serine residues on glycogen synthase. Alone, however, this is not sufficient to cause the abrupt onset of glycogen synthesis. In order for this to occur, it is also necessary to remove the phosphate groups from the glycogen synthase by activation of the serine–threonine protein phosphatase-1G (PP1G, see Figure 13.7). This is mediated through the action of the insulin-stimulated protein kinase (ISPK), a protein kinase homologous to the ribosomal S6 kinase-II. The insulin-stimulated cell is now fully engaged in the synthesis of glycogen (see figure 17.10).

Note that the activity of glycogen synthase is enhanced when it is in the dephospho form.

■   The role of PI 3-kinase in activation of glucose transport and protein synthesis

Insulin activation of glucose transport in muscle and fat occurs through the transfer of glucose transporter molecules (GLUT4) from intracellular vesicles to the plasma membrane. PI 3-kinase and phosphorylated IRS-1 are co-localized on the surface of these vesicles. The mechanism by which PI 3-kinase mediates transfer to and fusion with the plasma membrane is not clear, but it is likely that PH domains play a role.

PI 3-kinase is also implicated in regulation of protein synthesis through the eukaryotic translation initiator factor-4E (eIF-4E) and p70 S6-kinase. eIF-4E is the limiting initiation factor in most cells and its activity plays a principal role in determining global translation rates. It is regulated by

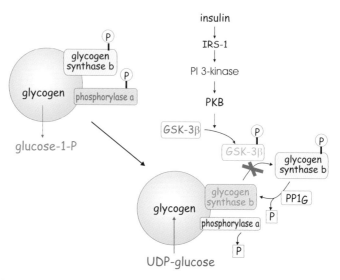

Figure 13.7 Insulin-mediated activation of glycogen synthase. Binding of insulin to its receptors leads to the activation of PKB. One of the substrates of PKB is glycogen synthase kinase-3β, which becomes inactivated. In addition, the phosphatase PP1G ensures rapid dephosphorylation and activation of glycogen synthase allowing glycogen synthesis to recommence.

phosphorylation, for instance through the ERK pathway, but also by binding to translational repressor proteins, 4E-BPs. These are inactivated by phosphorylation (see Chapter 11). The S6 protein is a component of the 40S ribosome subunit and its phosphorylation increases the rate of translation resulting in enhanced protein synthesis.[35] The S6 component is phosphorylated by S6 kinase, of which several isoforms have been identified.[36] Their activities are regulated by insulin, growth factors or glucagon.[37,38] Both 4E-BP1 and p70 S6-kinase are under the control of PKB, but this is indirect, involving yet another protein kinase, FRAP/mTOR (Figure 13.8).

This was initially recognized as the target of rapamycin, an immunosuppressant and an inhibitor of protein synthesis (structurally related to FK506, see Chapter 17).[39] FRAP/mTOR phosphorylates 4E-BP1 and this causes the release of eIF-4E, which can now participate in the initiation of protein synthesis.[40] It also phosphorylates p70 S6-kinase.[41]

There exists another and more direct way to activate p70 S6-kinase. Generation of PI(3,4,5)P$_3$ translocates PDK1 to the plasma membrane where it phosphorylates p70 S6-kinase at a site in the catalytic loop. As with the activation of PKB, a second phosphorylation has to occur in the C-terminal domain before it becomes fully activated. This is most likely brought about by one of the atypical PKCs, since growth factor-mediated activation of p70 S6-kinase is prevented in cells that express large amounts of inactive forms of PKCγ and ζ.[42] However, this mode of activation is insensitive to rapamycin and we conclude that under physiological conditions it is not operational.

FRAP FKBP-rapamycin associated protein, a protein kinase that is the human homologue of the yeast *TOR* gene.

TOR, target of rapamycin, a phosphoinositide kinase homologue that phosphorylates proteins, see figure 13.2

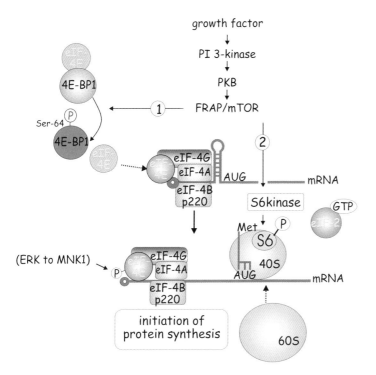

Figure 13.8 Regulation of protein synthesis by PI 3-kinase. Through activation of PKB, PI 3-kinase controls initiation and translation in protein synthesis. Activated PKB phosphorylates and activates the protein kinase FRAP/mTOR. This phosphorylates 4E-BP1, an inhibitor of the initiation factor eIF-4E. The liberated eIF-4E attaches to the cap-structure of mRNA and by ironing out a hairpin it facilitates the association of eIF2.GTP and the 40S ribosomal subunit. FRAP/mTOR also phosphorylates and activates S6 kinase, which in turn phosphorylates the S6 protein of the 40S ribosomal subunit. Phosphorylated S6 increases the efficiency of protein translation. Phosphorylation of 4E-BP1 and S6 plus the addiitional phosphorylation eIF-4E (by MAP kinase) are essential for the increase of protein synthesis mediated by growth factors or insulin.

Other processes mediated by the 3-phosphorylated inositol phospholipids

A number of seemingly disparate functions are modified in cells and organisms in which the synthesis of 3-phosphorylated lipids is chronically altered (Figure 13.3). These include

- The induction of haemangiomas in chickens and transformation of fibroblasts in cells containing a constitutively activated retrovirus-encoded PI 3-kinase.

- Inhibition of PI 3-kinase prevents T-cell activation by preventing the translocation of NF-AT.

- In vascular endothelial cells subjected to shear stress, nitric oxide synthase (NOS) is activated by PKB. The released NO relaxes the vascular smooth muscle.[43]

- The GTPases, Rac and Cdc42 in cooperation with PI 4-P 5-kinase and PI 3-kinase, play a role in assembly of the submembranous actin filament system leading to particle internalization.[44]

- PKB phosphorylates and activates a transcription factor of the forkhead/winged-helix family.[45] This has a function in resisting apoptosis and may regulate the lifespan of *C. elegans*. Mutants of this nematode having reduced activity of the insulin/IGF-1-receptor homologue DAF-2 and therefore unable to activate PI 3-kinase, live twice as long as normal.

- In epithelial cells, activation of the apoptosis pathway is suppressed as a consequence of PI 3-kinase activation. We return to the consideration of this in Chapter 14.

Multiple kinases and multiple phosphorylation sites, PDK1 as an integrator of multiple inputs

The search for pathways leading to the activation of PKB and the discovery of the role of PDK1 has brought to light a general theme for activation of the serine–threonine protein kinases of the AGC family.

AGC kinases, so named because they have sequence similarity with the first well-characterized serine/threonine protein kinases A (cAMP dependent), G (cGMP dependent) and C (Ca^{2+} and diacylglycerol dependent).

A threonine residue in a consensus sequence present in the catalytic loop of all these enzymes (Table 13.1) must be phosphorylated to achieve catalytic activity (see page 404). Not only this, but a number of these enzymes also have a conserved sequence at the C-terminus that contains a second phosphorylation site (serine or threonine). When phosphorylated, this allows for maximal catalytic activity. PDK1 phosphorylates several of the AGC protein kinase family (in addition to PKB, also p70 S6 kinase, PKCs α, β1, β2, δ, γ and ζ) in their catalytic loops. Thus, it acts as a central regulator of protein kinases. The atypical protein kinases C γ and ζ can phosphorylate PKCδ and p70 S6 kinase at the second phosphorylation site in their C-terminal domains (see Table 13.1). They may therefore also play a role in regulation of protein kinase activity.

We begin to see that the idea of a simple linear sequence of phosphorylations (an enzyme cascade) is no longer tenable.[46] Instead, it now appears that particular protein kinases cooperate, perhaps as a complex, to phosphorylate each other and themselves, with the encouragement of appropriate second messenger(s) (Figure 13.9). These phosphorylations determine the catalytic activity of these protein kinases and so integrate the converging pathways on to PKB.

So, who did discover insulin?

Probably 'nobody discovered insulin'. It existed first as an idea (Minkowski), then as a proven hypothesis (Gley and others) and finally as a practical way of alleviating diabetes (Paulescu and Banting). (Adapted from Henderson.[4])

Table 13.1 Alignment of conserved sequences in the family of AGC protein kinases

Protein kinase	Activation loop (PDK1 site)	C-terminal domain (PDK2 site)
PKB	KTFCGTPEY	FPQFSY
p70S6K	HTFCGTIEY	FLGFTY
PKCδ	HTFCGTIEY	FLGFTY
PKCα	STFCGTPDY	FAGFSF
PKCβI	RTFCGTPDY	FEGFSY
PKCβII	KTFCGTPDY	FAGFSY
PKCγ	RTFCGTPDY	GQGFTY
cAMP dependent protein kinase (no PDK2 site)		
PKAα	WTLCGTPEY	FSEF (no PDK2 site)
aPKC or PKC-related protein kinases (no PDK2 site)		
PKCζ	STFCGTPNY	FEGFEY
PKCλ	STFCGTPNY	FEGFEY
PRK1	STFCGTPEF	FLDFDF
PRK2	STFCGTPEF	FRDFDF

The catalytic loops and C-terminal domains of these proteins are phosphorylated by different kinases. Because the catalytic loops are always phosphorylated by PDK1, these are referred to as PDK1-sites. The phosphorylated residues in the C-terminal domains are known as 'PDK2' sites. PDK2 has yet to be identified. From Peterson and Schreiber.[46]

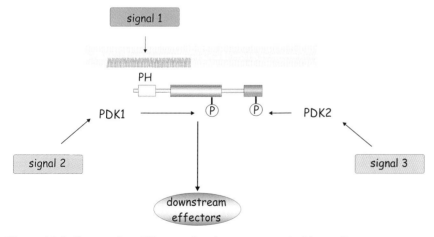

Figure 13.9 Converging different signals on to protein kinase B.

■ References

1 Vane, J.R. Inhibition of prostaglandin synthesis as a mechanism of action for aspirin-like drugs. *Nat. New Biol.* 1971; 231: 232–5.

2 Roth, G.J., Stanford, N., Majerus, P.W. Acetylation of prostaglandin synthase by aspirin. *Proc. Natl. Acad. Sci. USA* 1975; 72: 3073–6.

3 Murray, I. Paulescu and the isolation of insulin. *J. Hist. Med.* 1971; 26: 150–7.

4 Henderson, J.R. Who really discovered insulin? *Guy's Hosp. Gaz.* 1971; 85: 314–18.

5 Gley, E. Action des extraits de pancréas sclérosé sur des chiens diabétiqués (par extirpation du pancréas). *C.R. Soc. Biol. (Paris)* 1922; 87: 1322–5.

6 Kasuga, M., Karlsson, F.A., Kahn, C.R. Insulin stimulates the phosphorylation of the 95,000-dalton subunit of its own receptor. *Science* 1982; 215: 185–7.

7 Chong, L.D., Traynor-Kaplan, A., Bokoch, G.M., Schwartz, M.A. The small GTP-binding protein Rho regulates a phosphatidylinositol 4-phosphate 5-kinase in mammalian cells. *Cell* 1994; 79: 507–13.

8 Traynor-Kaplan, A.E., Harris, A.L., Thompson, B.L., Taylor, P., Sklar, L.A. An inositol tetrakisphosphate-containing phospholipid in activated neutrophils. *Nature* 1988; 334: 353–6.

9 Ireton, K., Payrastre, B., Chap, H. *et al.* A role for phosphoinositide 3-kinase in bacterial invasion. *Science* 1996; 274: 780–2.

10 Gary, J.D., Wurmser, L.S., Emr, S.D. Fap1p is essential for PtdIns(3)P 5-kinase activity and the maintenance of vacuolar size and membrane homeostasis. *J. Cell Biol.* 1998; 143: 65–79.

11 Herman, P.K., Stack, J.H., DeModena, J.A., Emr, S.D. A novel protein kinase homolog essential for protein sorting to the yeast lysosome-like vacuole. *Cell* 1991; 64: 425–37.

12 Rudd, E., Schneider, H. Lymphocyte signaling: Cbl sets the threshold for autoimmunity. *Curr. Biol.* 2000; 10: R344.

13 Musacchio, A., Cantley, L.C., Harrison, S.C. Crystal structure of the breakpoint cluster region-homology domain from phosphoinositide 3-kinase p85 α subunit. *Proc. Natl. Acad. Sci. USA* 1996; 93: 14373–8.

14 Diekmann, D., Brill, S., Garrett, M.D. *et al.* Bcr encodes a GTPase-activating protein for p21rac. *Nature* 1991; 351: 400–2.

15 Rodriguez-Viciana, P., Warne, P.H., Dhand, R. *et al.* Phosphatidylinositol-3-OH kinase as a direct target of Ras. *Nature* 1994; 370: 527–32.

16 Wymann, M.P., Pirola, L. Structure and function of phosphoinositide 3-kinases. *Biochim. Biophys. Acta* 1998; 1436: 127–50.

17 Haesebroeck, B. Van, Jones, G.E., Allen, W.E. *et al.* Distinct PI(3)Ks mediate mitogenic signalling and cell migration in macrophages. *Nat. Cell Biol.* 1999; 1: 69–71.

18 Gunther, R., Abbas, H.K., Mirocha, C.J. Acute pathological effects on rats of orally administered wortmannin-containing preparations and purified wortmannin from *Fusarium oxysporum. Food Chem. Toxicol.* 1989; 27: 173–9.

19 Closse, A., Haefliger, W., Hauser, D. *et al.* 2,3-Dihydrobenzofuran-2-ones: a new class of highly potent antiinflammatory agents. *J. Med. Chem.* 1981; 24: 1465–71.

20 Nakanishi, S., Kakita, S., Takahashi, I. *et al.* Wortmannin, a microbial product inhibitor of myosin light chain kinase. *J. Biol. Chem.* 1992; 267: 2157–64.

21 Wymann, M.P., Bulgarelli-Leva, G., Zvelebil, M.J. *et al.* Wortmannin inactivates phosphoinositide 3-kinase by covalent modification of Lys-802, a residue involved in the phosphate transfer reaction. *Mol. Cell Biol.* 1996; 16: 1722–33.

22 Kaay, J. Van der, Batty, I.H., Cross, D.A., Watt, P.W., Downes, C.P. A novel, rapid, and highly sensitive mass assay for phosphatidylinositol 3,4,5-trisphosphate (PtdIns(3,4,5)P3) and its application to measure insulin-stimulated PtdIns(3,4,5)P3 production in rat skeletal muscle in vivo. *J. Biol. Chem.* 1997; 272: 5477–81.

23 Cross, D.A., Alessi, D.R., Cohen, P., Andjelkovich, M., Hemmings, B.A. Inhibition of glycogen synthase kinase-3 by insulin mediated by protein kinase B. *Nature* 1995; 378: 785–9.

24 Guinebault, C., Payrastre, B., Racaud, S.C. *et al.* Integrin-dependent translocation of phosphoinositide 3-kinase to the cytoskeleton of thrombin-activated platelets involves specific interactions of p85 α with actin filaments and focal adhesion kinase. *J. Cell Biol.* 1995; 129: 831–42.

25 Bellacosa, A., Testa, J.R., Staal, S.P., Tsichlis, P.N. A retroviral oncogene, akt, encoding a serine-threonine kinase containing an SH2-like region. *Science* 1991; 254: 274–7.

26 Corvera, S., Czech, M.P. Direct targets of phosphoinositide 3-kinase products in membrane traffic and signal transduction. *Trends Neurosci.* 1998; 8: 442–7.

27 Sakaue, H., Noguchi, T., Matozaki, T. *et al.* Ras-independent and wortmannin-sensitive activation of glycogen synthase by insulin in Chinese hamster ovary cells. *J. Biol. Chem.* 1995; 270: 11304–9.

28 Hara, K., Yonezawa, K., Sakaue, H. *et al.* 1-Phosphatidylinositol 3-kinase activity is required for insulin-stimulated glucose transport but not for RAS activation in CHO cells. *Proc. Natl. Acad. Sci. USA* 1994; 91: 7415–19.

29 Yamamoto-Honda, R., Tobe, K., Kaburagi, Y. *et al.* Upstream mechanisms of glycogen synthase activation by insulin and insulin-like growth factor-I. Glycogen synthase activation is antagonized by wortmannin or LY294002 but not by rapamycin or by inhibiting p21ras. *J. Biol. Chem.* 1995; 270: 2724–9.

30 Dorrestijn, J., Ouwens, D.M., Berghe, N. Van den, Bos, J.L., Maassen, J.A. Expression of a dominant-negative Ras mutant does not affect stimulation of glucose uptake and glycogen synthesis by insulin. *Diabetologia* 1996; 39: 558–63.

31 Garrett, T.P., McKern, N.M., Lou, M. *et al.* Crystal structure of the first three domains of the type-1 insulin-like growth factor receptor. *Nature* 1998; 394: 395–9.

32 Valverde, A.M., Lorenzo, M., Pons, S., White, M.F., Benito, M. Insulin receptor substrate (IRS) proteins IRS-1 and IRS-2 differential signaling in the insulin/insulin-like growth factor-I pathways in fetal brown adipocytes. *Mol. Endocrinol.* 1998; 12: 688–97.

33 Franke, T.F., Yang, S.I., Chan, T.O. *et al.* The protein kinase encoded by the Akt proto-oncogene is a target of the PDGF-activated phosphatidylinositol 3-kinase. *Cell* 1995; 81: 727–36.

34 Franke, T.F., Kaplan, D.R., Cantley, L.C., Toker, A. Direct regulation of the Akt proto-oncogene product by phosphatidylinositol-3,4-bisphosphate. *Science* 1997; 275: 665–8.

35 Pearson, R.B., Thomas, G. Regulation of p70s6k/p85s6k and its role in the cell cycle. *Prog. Cell Cycle Res.* 1995; 1: 21–32.

36 Grove, J.R., Banerjee, P., Balasubramanyam, A. *et al.* Cloning and expression of two human p70 S6 kinase polypeptides differing only at their amino termini. *Mol. Cell Biol.* 1991; 11: 5541–50.

37 Novak-Hofer, I., Thomas, G. Epidermal growth factor-mediated activation of an S6 kinase in Swiss mouse 3T3 cells. *J. Biol. Chem.* 1985; 260: 10314–19.

38 Smith, C.J., Rubin, C.S., Rosen, O.M. Insulin-treated 3T3-L1 adipocytes and cell-free extracts derived from them incorporate 32P into ribosomal protein S6. *Proc. Natl. Acad. Sci. USA* 1980; 77: 2641–5.

39 Brown, E.J., Albers, M.W., Shin, T.B. *et al.* A mammalian protein targeted by G1-arresting rapamycin-receptor complex. *Nature* 1994; 369: 756–8.

40 Manteuffel, S.R. von, Gingras, A.C., Ming, X.F., Sonenberg, N., Thomas, G. 4E-BP1 phosphorylation is mediated by the FRAP-p70s6k pathway and is independent of mitogen-activated protein kinase. *Proc. Natl. Acad. Sci. USA* 1996; 93: 4076–80.

41 Burnett, P.E., Barrow, R.K., Cohen, N.A., Snyder, S.H., Sabatini, D.M. RAFT1 phosphorylation of the translational regulators p70 S6 kinase and 4E-BP1. *Proc. Natl. Acad. Sci. USA* 1998; 95: 1432–7.

42 Romanelli, A., Martin, K.A., Toker, A., Blenis, J. p70 S6 kinase is regulated by protein kinase Czeta and participates in a phosphoinositide 3-kinase-regulated signalling complex. *Mol. Cell Biol.* 1999; 19: 2921–8.

43 Dimmeler, S., Fleming, I., Fisslthaler, B. *et al.* Activation of nitric oxide synthase in endothelial cells by Akt-dependent phosphorylation. *Nature* 1999; 399: 601–5.

44 Kwiatkowska, K., Sobotka, A. Signaling pathways in phagocytosis. *Bioessays* 1999; 21: 422–31.

45 Kops, G.J., Ruiter, N.D. de, Vries-Smits, A.M. de *et al.* Direct control of the Forkhead transcription factor AFX by protein kinase B. *Nature* 1999; 398: 630–4.

46 Peterson, R.T., Schreiber, S.L. Kinase phosphorylation: Keeping it all in the family. *Curr. Biol.* 1999; 9: R521.

47 von Mering, J. and Minkowski, O. *Arch. Exp. Pathol. Pharmakol.* 1890; 26: 371–87

Signal transduction to and from adhesion molecules

In this chapter we consider the adherence of cells to surfaces or to other cells and ask how this affects their responses to soluble agonists such as growth factors. The molecules that effect adhesion behave both as targets for signals that are generated within cells (hence, inside-out signalling) and as receptors for extracellular signals (outside-in signalling). These two aspects are well exemplified in the regulation of survival, proliferation, differentiation and in leukocyte trafficking, discussed below. This list is far from complete. Adhesion molecules are also of prime importance in the functioning of synapses and in neuronal cell outgrowth, differentiation of keratinocytes, gene expression in mammary epithelial cells, thymic selection and the activation of T lymphocytes.

■ Adhesion molecules

The binding of cells to the extracellular matrix and their attachment to other cells occurs through specific adhesion molecules. Initially these were described innocently as a sort of glue, sticky molecules having no functional implications. We now recognize that they also act as signalling molecules and are properly described as receptors (Figure 14.1). However, the ligands that interact with adhesion molecules are generally insoluble, frequently adhesion molecules themselves, presented by adjacent cells or components of the extracellular matrix.

The adhesion molecules first came to light as a result of investigations of brain development around 1970. It was realized that the very precise organization of neural cells in the central nervous system must require a dynamic process of cell guidance and cell adhesion. This would drive the direction-seeking processes of neurite outgrowth and synapse formation. Two main ideas were presented.[1] The first suggested that during development, in order to establish precise cell–cell contacts, the interacting cells must each present unique adhesion molecules that fit each other as a lock and key (chemoaffinity hypothesis). The second idea was that the set of adhesion molecules is limited, but their binding capacity could be modulated over time. For instance, developing neuronal cells would all offer the same molecule, and during neurite outgrowth this would be in a

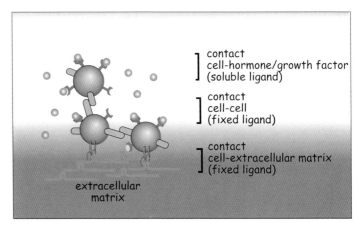

extracellular
matrix

Figure 14.1 Three levels of communication. Cells are in contact with soluble ligands (hormones and cytokines), with each other and with the extracellular matrix. The latter two contacts are mediated by adhesion molecules which, in many instances, should be regarded as receptors which convey signals into the cells.

low affinity state. The cell might then convert the adhesion molecule into its high affinity state, to promote binding to its counterpart on a nearby cell.

It now appears that there is truth in both these propositions. The number of adhesion molecules is certainly limited and their capacity to interact with counter-receptors is regulated by their levels of expression (presence or absence) and also by their binding state (affinity). In the realm of immunology, the set of adhesion molecules expressed on a cell surface and their state of activation has been called the 'area code'.[2]

Clear evidence for specific adhesion interactions came from studies of the re-aggregation of disaggregated tissues. Tissue cells dispersed by treatment with trypsin, which strips proteins from the surface, only recover their adherent properties after a period in culture. The need for a recovery period suggested that new adhesion molecules must be expressed on the cell surface. The re-aggregation can be prevented by monovalent Fab$_1$ fragments prepared from antibodies raised against the cell membranes which block cell surface epitopes (Figure 14.2).

Because such binding sites could never be saturated by a few micrograms of 'blocking' antibodies the possibility of weak non-specific interactions was precluded and it followed that there is a limited number of specialized molecules that determine cell–cell interactions. These are the adhesion molecules, and the first to be discovered was the neuronal cell adhesion molecule (NCAM).[3]

Early attempts to identify individual adhesion molecules were hampered by the lack of specific antibodies. A major advance was the advent of monoclonal antibodies and their application to questions regarding cell adhesion. This led to the identification of a molecule expressed on the surface of macrophages, Mac-1, which plays a pivotal role in the binding of leukocytes to the vascular endothelium.[4] It also determines the binding of the serum complement factor C3bi, which, together with the Fc recep-

Divalent antiserum would have the effect of aggregating the cells. The use of **monovalent Fab** fragments also ensures that no signals ensue from the crosslinking of surface molecules that might follow the use of intact divalent antiserum.

tor, mediates activation of the respiratory burst (generation of toxic oxygen metabolites).

Later, a monoclonal antibody that recognizes lymphocyte function-associated antigen-1 (LFA-1) was used to show that the binding of cytotoxic T-lymphocytes to their target cells is also mediated by specific adhesion molecules.[5]

Adhesion molecules not only link cells to surfaces but they also make the connection between the extracellular matrix and the cytoskeleton. Because it was perceived that they integrate intracellular and extracellular events, they were called integrins.[6] These proteins can bind to a wide range of extracellular matrix molecules, including fibronectin, vitronectin, osteopontin, thrombospondin, fibrinogen and von Willebrand factor. A

Complement The term originally used to refer to the heat-labile factor in serum that causes immune cytolysis, the lysis of antibody-coated cells. It now refers to the entire functionally related system, comprising at least 20 distinct serum proteins, that is the effector not only of immune cytolysis but also of other functions.

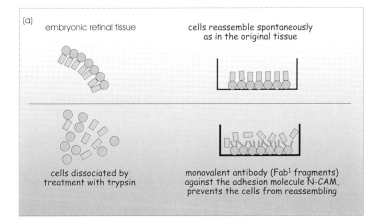

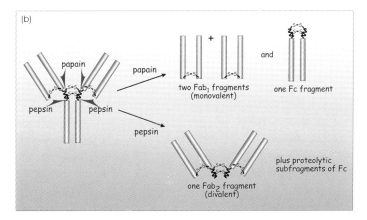

Figure 14.2 **Reassembly of retinal dispersed cells: discovery of NCAM:** (a) The re-assembly of dispersed cells is prevented by antibodies that recognize adhesion molecules. By this approach, the role of the adhesion molecule NCAM in the maintenance of tissue integrity was discovered. Since intact bivalent antibodies would have forced the cells to aggregate it was essential to use monovalent Fab$_1$ fragments that are unable to form crosslinks. (b) The generation of the fragments Fc, Fab$_1$ (monovalent) and Fab$_2$ (divalent) by enzyme digestion of immunoglobulin.

number of integrins bind these proteins through the recognition of short motifs such as Arg-Gly-Asp (RGD) or Glu-Ile-Leu-Asp-Val (EILDV).[7] The adhesion molecules Mac-1 and LFA-1 share substantial sequence homology with the integrins, and it was confirmed that they too are members of this family of adhesion molecules.

Naming names

Adhesion molecules have been given names that express function (intercellular adhesion molecule-1, ICAM-1); location and function (endothelium leukocyte adhesion molecule-1, ELAM-1); the time of induction of expression during T-cell activation (very late antigen-4, VLA-4); or in recognition of their integration of the extracellular matrix with the intracellular cytoskeleton (integrins). Of course, there are other names, often bestowed after cloning (αM integrin) or through recognition by specific monoclonal antibodies (CD11b, cluster of differentiation 11b). If this is not all perfectly clear, then one should appreciate that CD18/CD11b is synonymous with Mac-1, which is synonymous with integrin αMβ2 and that CD62E is synonymous with ELAM-1 which is synonymous with E-selectin. Such are the problems of nomenclature when molecular cloning rubs shoulders with immunology, pharmacology and all the rest.

Immunoglobulin superfamily

NCAM is a member of a large family of cell surface proteins that express repeated immunoglobulin-like domains at their extracellular N-termini. These Ig domains are globular loop-like structures resistant to proteases, stabilized by sulphydryl bridging (Figure 14.3). The proteins of the Ig-family

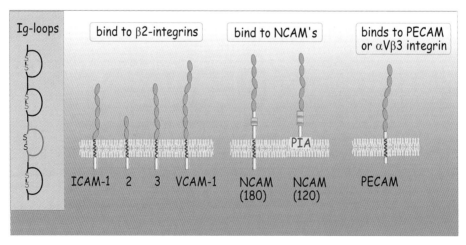

Figure 14.3 Cellular adhesion molecules belonging to the immunoglobulin superfamily. These adhesion molecules are characterized by the presence of repeated loop-like structures that are homologous to those present in immunoglobulins (Ig-loops). There are several members of this family, all having a single membrane-spanning domain. They interact with different ligands (or counter-receptors). PIA, phosphatidyl inositol anchor.

of adhesion molecules include NCAM, platelet endothelial cell adhesion molecule (PECAM, CD31), ICAM-1[8] and vascular cell adhesion molecule-1 (VCAM-1).[9] They all have a single membrane-spanning domain. The subfamily of ICAMs can be subdivided into ICAM-1 and ICAM-3 (previously CD50), and ICAM-2, the shortest, which has only two Ig-repeats and is constitutively expressed on all endothelial cells. There are two splice variants of VCAM-1 which express either five or seven Ig-domains. Thus far, no signal transduction pathway has been discovered that emanates from the adhesion molecules of the Ig class. Nor has any intracellular signalling pathway been discovered that can alter their affinity. They are best regarded as ligands, not receptors. They bind to integrin molecules which act as true receptors on other cells. The regulation of cell–cell binding by the Ig-class is determined by their level of expression.

■ Integrins

The integrins are the most dynamic and versatile of the adhesion molecules. They are composed of two subunits (α and β), linked non-covalently (Figure 14.4). There exist at least 12 α- and 7 β-subunits (Figure 14.5). The extracellular chains possess binding sites for divalent cations. Depending on the particular $\alpha\beta$ combination, they may bind to either ICAMs, VCAM-1 or MadCAM (on mucosal cells). They may also bind to components of the extracellular matrix and to the blood proteins, fibrinogen or von Willebrand Factor. The intracellular domain of the β-subunit interacts with the cytoskeleton. It can be phosphorylated, possibly affecting its

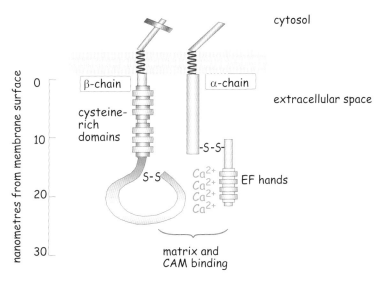

Figure 14.4 **Structure of integrins.** These adhesion molecules are composed of two non-covalently bound subunits (α and β) and they link the extracellular matrix to the intracellular cytoskeleton. The regulation of integrin activation, an inside-out cellular signal transduction pathway, remains poorly defined.

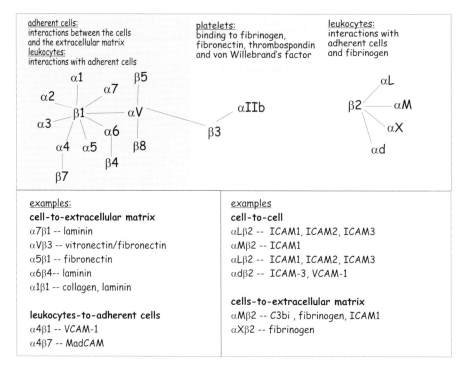

adherent cells:
interactions between the cells
and the extracellular matrix
leukocytes:
interactions with adherent cells

platelets:
binding to fibrinogen,
fibronectin, thrombospondin
and von Willebrand's factor

leukocytes:
interactions with
adherent cells
and fibrinogen

examples:	examples
cell-to-extracellular matrix	**cell-to-cell**
α7β1 -- laminin	αLβ2 -- ICAM1, ICAM2, ICAM3
αVβ3 -- vitronectin/fibronectin	αMβ2 -- ICAM1
α5β1 -- fibronectin	αLβ2 -- ICAM1, ICAM2, ICAM3
α6β4-- laminin	αdβ2 -- ICAM-3, VCAM-1
α1β1 -- collagen, laminin	
	cells-to-extracellular matrix
leukocytes-to-adherent cells	αMβ2 -- C3bi , fibrinogen, ICAM1
α4β1 -- VCAM-1	αXβ2 -- fibrinogen
α4β7 -- MadCAM	

Figure 14.5 **Mixing and matching integrin subunits.** Various combinations of α- and β-subunits create integrins having specific ligand binding characteristics. The β2-integrins are restricted to blood cells, whereas β1 are also present in tissue cells. Platelets have the unique combination of αIIbβ3, a vital component in the blood clotting cascade. Integrins bind to specific sequences in proteins of the extracellular matrix such as Arg-Gly-Asp (RGD) or Glu-Ile-Leu-Asp-Val (EILDV).

binding activity and this may be of importance for circulating cells such as leukocytes. Here, chemokines (chemotactic cytokines) and their receptors (G-protein linked) play an important role. In contrast, for tissue cells (growing on a matrix) the integrins appear to be in a permanently active state.

■ Cadherins

Specific adhesion molecules regulating intercellular recognition are fundamental to the process of animal morphogenesis. In the formation of the early embryo, the compaction of the morula is mediated through Ca^{2+}-dependent adhesion molecules[10] (Figure 14.6). Uvomorulin, one of the first to be identified, is instrumental in the transition from a grape-like to a mulberry-like embryo.

Further studies, using embryonal carcinoma cells (teratocarcinoma F9 cells), revealed a whole range of Ca^{2+} dependent adhesion molecules, now collectively called cadherins.[11]

The cadherins comprise a large family of proteins (at least 11 members in humans). They have been classified on the basis of the sequence similarities of their extracellular domains, in particular the Ca^{2+}-binding

Latin, *uva*: grape; *morum*: mulberry; hence **uvomorulin.**

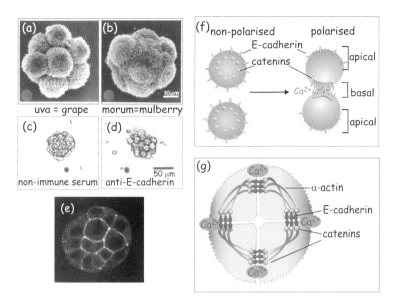

Figure 14.6 The role of cadherin in the compaction of the eight-cell stage mouse embryo: (a and b) Scanning electron microscopy reveals that after three cell divisions (eight-cell stage) mouse embryos change from a uva (grape)-like to a morula (mulberry)-like aggregate. This process, called compaction, enables the cells to attach firmly to each other, manifesting the first signs of polarization. Their morphology now presents distinct basal (contact) and apical (peripheral) membrane surfaces. (c and d) Light microscope images showing the inhibition of the compaction process by anti-E-cadherin. The antiserum prevents the attainment of the compact form though cell division continues unimpeded. The non-immune serum allows compaction to proceed normally. This type of experiment demonstrated that contact between cells is mediated by a limited set of membrane adhesion molecules. (e) Immunofluorescence microscopy using anti-cadherin antibodies indicates that E-cadherin is located at the cell–cell boundaries of the mouse embryo at the morula stage. (f) Schematic presentation showing the changes in distribution of E-cadherin and its interaction with β-catenin following the formation of cell contacts. The polarization creates distinct basal and apical membrane surfaces. (g) The organization of actin as a purse-string that draws β-catenin and E-cadherin together, causing compaction of the embryo. Panels c, d and e courtesy of Rolf Kemler.

domains[12] (Figure 14.7). The physiological significance of their striking Ca^{2+} dependence observed *in vitro* remains unclear. They were initially named after the tissue in which they were discovered (epithelial, placental, neuronal, etc.), but these labels have little meaning since they are widely expressed. Uvomorulin is present in epithelial cells and proved to be the same as E-cadherin.[13] In developmental processes, the expression of each cadherin subclass is regulated both spatially and temporally and this is associated with individual morphogenic events. Thus, the tendency of cells to segregate or aggregate correlates with the expression of particular subclasses of cadherins. This is particularly apparent during gastrulation and the formation of the neural tube. In addition, regulation can also occur through activation, as occurs at the onset of compaction (Figure 14.6).

Cadherins generally mediate homotypic cell–cell adhesion, acting as both receptor and ligand. They are present in cell–cell junctions, adherens junctions and desmosomes (Figure 14.8) where they are associated with bundles of cortical actin or intermediate filaments, via the intermediates of catenins or desmoplakin, respectively. By forming cell–cell contacts, the cadherins appear to act as tumour suppressors. When these fail, as in metastasizing epithelial tumours, there is a loss of baso-lateral localization (see below).

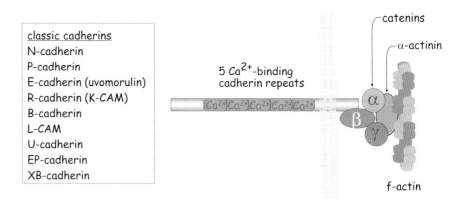

Figure 14.7 Molecular structure of cadherins. Cadherins form a large family of adhesion molecules characterized by five Ca^{2+}-binding domains. In the cytosol they interact with the catenins (α, β and γ) which, with α-actinin, make the bridge to the actin cytoskeleton.

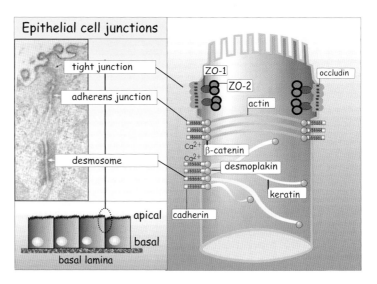

Figure 14.8 Epithelial cells are firmly attached to each other by a set of junctions which resist physical stress. Tight junctions form a barrier for most solutes; adherent junctions are connected to the cellular adhesion belt and maintain tissue integrity; desmosomes are linked to intermediate filaments and also act to maintain cellular integrity. Cadherins make the connection in both adherent junctions and desmosomes through homophilic intercellular interactions.

Selectins

Selectins are present on the surface of white blood cells, platelets and also the cells lining the blood vessels (endothelial cells). They mediate the initial low-affinity adhesion sites for lymphocytes and for leukocytes such as neutrophils. This prepares the way for cell migration into the lymph nodes and tissues. Selectins contain a number of recognized domains. At the N-terminus there is a C-type (Ca^{2+}-binding) lectin domain. This enables binding to particular carbohydrate residues (galactose, N-acetyl glucosamine and fucose) present on cell surface glycoproteins and glycolipids. Selectins also contain a single EGF-like domain and a series of repeats (complement regulatory protein, CRP) (Figure 14.9). The retention of lymphocytes in lymph nodes can be prevented by monosaccharides (l-fucose and d-mannose), and by the polysaccharide fucoidin (rich in l-fucose), giving an indication that carbohydrate residues may be involved in the homing of lymphocytes.[14] This was confirmed by the finding of a lymph-node-specific homing receptor which possesses a C-type lectin domain.[15] Various selectins, mediating intercellular interactions, have since been identified in vascular endothelial cells (E), platelets (P) and leukocytes (L). L-selectin, responsible for lymphocyte homing, is expressed on all circulating leukocytes except for a subpopulation of memory cells. It recognizes CD34 (a heavily glycosylated mucin) on endothelial cells. E- and P-selectins are expressed on endothelial cells and recognize, respectively, sialyl LewisX and P-selectin glycosylated ligand-1 (PSGL-1) on leukocytes.

Little is known about the role of the selectins in signal transduction. However, we cannot let them pass without remarking on their remarkable property by which their affinity is modulated by the imposition of shear forces.[16]

Selectin Homologous to the Ca^{2+}-dependent (C-type) lectin. Lectin is derived from the word *select*, and originally applied to plant proteins that bind to specific carbohydrate residues present on nitrogen-fixing bacteria. They have been used in cell biological work because they bind to specific glycosidic residues present in the Golgi system and on cell surfaces. Only later was it found that animal cells also possess similar proteins, generally on the surfaces of endothelial cells.

sialyl Lewis Sialylated sugars expressed on blood cells that serve as antigens in blood group typing as developed by Lewis.

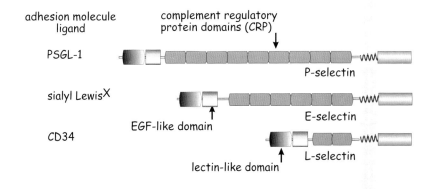

Figure 14.9 Molecular structure of selectins. The selectins are characterized by the presence of an N-terminal lectin-homology domain and a variable number of CRPs. Through their lectin domain, selectins interact with sugar residues present in cell surface glycoproteins and glycolipids.

■ Cartilage link proteins

The glycosaminoglycan hyaluronan, a high molecular weight polysac-cacharide (with the repeating unit D-glucuronic acid (1-β-3) N-acetyl-D-glucosamine), is present in the tissue matrix and body fluids of all vertebrates. It plays a fundamental role in regulating cell migration and differentiation.[17] The majority of hyaluronan-binding proteins belong to the cartilage link-protein superfamily. All of these contain a conserved disulphide-linked link module (approximately 100 amino acid residues) that has a three-dimensional structure resembling the sugar-binding domain of E-selectin. They can be further subdivided in two groups, those that are true cellular adhesion molecules such as CD44[18,19] and those that are components of the extracellular matrix, such as cartilage-link protein itself, aggrecan, versican, brevican and TNF-stimulated gene-6 (TSG-6).[20] This second group plays an important role in the architecture of the extracellular matrix, bringing together the pressure resistant qualities of glycosaminoglycans and the tension resistant qualities of the extracellular matrix proteins, an essential quality of cartilage in articulated joints.

CD44, originally discovered as the homing receptor of lymphocytes, is the best studied of the cartilage link proteins. It is required when lymphocytes bind to high endothelial venules and leave the circulation to enter lymph nodes.[21] It is a single membrane-spanning protein encoded by 19 exons of which 9 in the extracellular domain (v2–v10) are variably spliced (Figure 14.10). Because of the extensive glycosylation and the variable regions, the molecular mass of CD44 can vary between 85 and 200 kDa.

Glycosaminoglycans and the **extracellular matrix proteins** share with wood and reinforced concrete the valuable qualities of resistance to the forces of tension and compression. These are two important requirements for the construction not only of organisms but also of buildings.

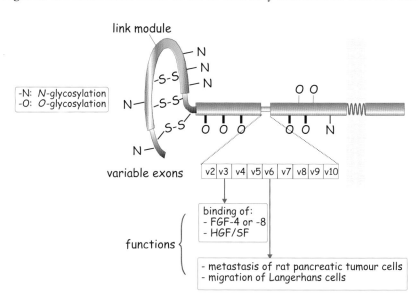

Figure 14.10 **Molecular structure of CD44.** This adhesion molecule is a member of the family of cartilage link proteins, characterized by a link-domain which serves as a binding site for the glycosaminoglycan, hyaluronan. CD44 has a number of exons that are variably spliced, v2–v10, and a large number of glycosylation sites. This heterogeneity has functional consequences, determining its capacity to bind FGFs or HGF/SF and its role in cell migration and metastasis of tumours.

The link-module is implicated in the binding of hyaluronan, but the splicing extends the range of binding capacities. For instance, the inclusion of exon v3 allows the attachment of heparan or chondroitin sulphate, enabling it to interact with fibroblast growth factor (FGF)-4 or -8 and with hepatocyte growth factor/scatter factor (HGF/SF). Other potential ligands are osteopontin, fibronectin and ankyrin. Homophilic intercellular interactions with CD44 are also reported.

The splice variants stimulated great interest when it appeared that CD44v4–7 might determine metastasis in rat pancreatic tumour cells. In short, it seemed at first, to have the quality of a metastasis gene product.[22] More than this, a monoclonal antibody raised against v6 prevents metastasis. Unfortunately, however, although human tumours often express CD44-splice variants, and although in certain instances this is a marker for poor prognosis, there seem to be no functional implications in the process of metastasis. The signal transduction pathway emanating from CD44 remains uncertain. A number of membrane proteins, including ezrin, radixin, and moesin, that associate with CD44, belong to the group of erythrocyte band 4.1-related proteins. These are proposed to function as links between the membrane and the cytoskeleton,[23] but it remains to be shown that they actually relay signals from CD44 into the cell. Another point of interest is, of course, the capacity of CD44v3 to bind growth factors of the FGF and HGF/SF families. This may mean that it is involved in the recruitment of the respective receptors, thereby mediating intracellular signalling. In the following sections, signalling to and from CD44 will not be further discussed.

Adhesion molecules and cell survival

The survival of endothelial and epithelial cells depends critically upon the contacts they make with each other and with the extracellular matrix. Without contact, they die through the controlled process of apoptosis.[24,25] This mechanism protects the organism against dysplastic growth (meaning wrongly formed, abnormal: see 'Cancer and cell transformation', page 246), preventing stray cells from colonizing inappropriate locations. Cells have an intrinsic drive to self-destruct but are prevented from doing this by signals emanating from specific 'rescue' pathways (Figure 14.11). One such signal (outside-in) follows from the attachment of the integrin α5β1 to the extracellular matrix.

When fibroblasts spread on an extracellular matrix having fibronectin as an abundant component, the integrins (mainly α5β1 and αVβ3) form multimeric clusters that attach to the cytoskeleton at focal adhesion sites. These are composed of a number of proteins, some having structural, others signalling roles. Together they form a focal adhesion complex as depicted in Figure 14.12a. The structural components vinculin and talin form a binding site for the actin cytoskeleton and thus direct the formation of stress fibres and actin structures within the cortical region of the cell. Talin also forms the site of attachment for the tyrosine kinase FAK (focal adhesion kinase). Attachment, resulting in activation and

unattached cell
susceptible to apoptosis

attached cell
protected against apoptosis

Figure 14.11 **The basics of apoptosis.** Attachment of tissue cells to the extracellular matrix protects them against apoptosis (programmed cell death). When tissue cells, such as epithelial or endothelial cells, are brought into suspension they become susceptible to self-destruction, even in the presence of sufficient growth and nutrient factors.

autophosphorylation (Tyr397) enables FAK to act as a docking site for the SH2 domain of the p85-regulatory subunit of PI 3-kinase (Figure 14.12a) leading to the generation of phosphatidyl inositide 3-phosphate lipids.

Downstream in the pathway of rescue, PKB effects a number of phosphorylations[26,27] which, if inhibited, prevents rescue from apoptosis.[28] PKB promotes rescue through at least two pathways. One is through direct phosphorylation and inactivation of components of the apoptotic machinery, including Bad and Caspase-9.[29] The other is by phosphorylation of FKHRL1, a transcription factor (member of the *Drosophila* forkhead/winged-helix family). When phosphorylated, FKHRL1 is retained in the cytosol and is so prevented from activating genes critical for induction of factors that promote cell death (such as the Fas ligand).[30]

The importance of FAK is underlined by the finding that cells expressing a constitutively active form survive in suspension, even though they are effectively 'homeless'.[31] The situation normally provoking the death of cells following detachment has been called *anoikis* (meaning homeless). Here, the protein kinase is active regardless of the failure to make contact with an extracellular matrix. Rescue from apoptosis also occurs when cells express constitutively activated oncogenic forms of Ras or Src, and thus activate PI-3 kinase and the ERK pathway. Unlike FAK, these not only prevent apoptosis but also promote proliferative signals that result in tumour formation[32] (see Chapter 11).

One might expect that circulating cells, such as leukocytes, must have other pathways to protect them from apoptosis. For sure, some white blood cells such as neutrophils are very short-lived (just a few hours in the blood), but others, such as the lymphocytes responsible for immunological memory, survive for years. Whether memory cells need to be continually reminded to survive by occasional encounters with other cells or the extracellular matrix remains to be established.

The **basal lamina** is a thin layer made up of laminin and entactin (glycoproteins), type IV collagen, and perlecan (heparan sulphate proteoglycan).

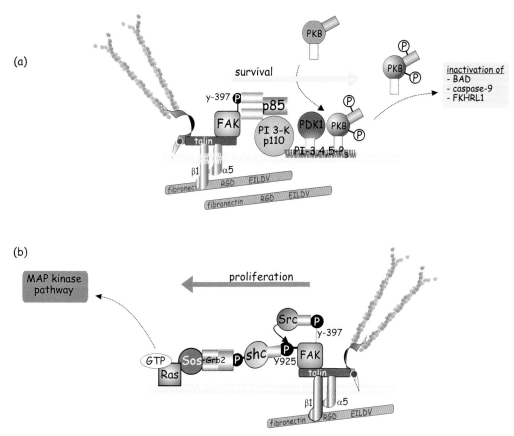

Figure 14.12 Survival and proliferation: (a) The focal adhesion site promotes cell survival signals through activation of protein kinase B (PKB). As tissue cells spread out on an extracellular matrix, focal adhesion sites are formed. These are composed of clustered β1-integrins associated with talin, vinculin and the actin cytoskeleton. The focal adhesion kinase FAK attaches to talin, autophosphorylates on a tyrosine residue (Y397) and provides the activation signal for PI 3-kinase. Production of PI(3,4,5)P_3 provides a binding site for the PH domains of PDK1 and PKB. PKB is phosphorylated on two serine/threonine residues and detaches from the membrane to phosphorylate and inactivate substrates that would otherwise sensitize the cells to apoptosis. These include Bad, caspase-9 and the transcription factor FKHRL-1. (b) The focal adhesion site promotes cell proliferation signals through activation of Ras. Autophosphorylation of FAK (Y397) also generates a docking site for Src which phosphorylates FAK at a second tyrosine residue, Y925. This acts as the docking site for the adaptor Shc which itself becomes phosphorylated and binds Grb2. This initiates the activation of the Ras-ERK pathway, necessary for initiation of the cell cycle.

■ Jun-N-terminal kinase (JNK) induces apoptosis in detached epithelial cells

While FAK and PKB together promote rescue, the pathway that drive detached cells into apoptosis involves the JNK pathway (described in Figure 11.15, page 274). Cells containing dominant negative components of the JNK pathway are resistant to apoptosis.[33,34] However, there is an important difference in the mechanism. The activation of MEKK-1 occurs, not through phosphorylation, but through limited proteolysis by caspase-3 (the caspases are sulphydryl enzymes that cleave peptides at specific aspartate residues, see 'Apoptosis', page 335). The consequent activation of JNK prepares cells for apoptosis. How the FAK/PKB and JNK

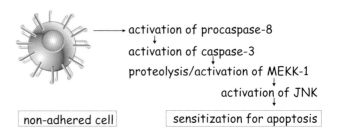

activation of procaspase-8
↓
activation of caspase-3
↓
proteolysis/activation of MEKK-1
↓
activation of JNK
↓

| non-adhered cell | | sensitization for apoptosis |

Figure 14.13 **Stress-induced signals for apoptosis.** Non-adherent cells lack the FAK-mediated survival signal and, moreover, facilitate apoptosis by activation of the JNK pathway. Caspase-3, activated by procaspase-8, cleaves MEKK-1, causing its activation. This is followed by phosphorylation and activation of JNK. JNK phosphorylates and activates the Jun family of transcription factors which induce transcription of genes implicated in the induction of apoptosis.

pathways are involved in the downstream processes that determine survival or death remains uncertain.

■ Scatter factor rescues endothelial cells from apoptosis

In the process of wound healing, damaged tissues become revascularized. The newly formed endothelial cells migrate under the guidance of a matrix of collagen and hyaluronan already deposited by the tissue fibroblasts and macrophages. In this way, they move into the wound. However, to do this, they must detach, at least partially, from the extracellular matrix and this could make them vulnerable. Their migration is induced by an angiogenic factor, scatter factor (also known as hepatocyte growth factor) which also protects them against apoptosis.

■ Adhesion molecules and regulation of the cell cycle

An essential requirement for the proliferation of tissue cells, driven by growth factors, is their attachment to a suitable surface. If EGF or PDGF is added to suspended fibroblasts, the activation of the ERK pathway is merely transient. The cells fail to proliferate and die through apoptosis.[35] Proliferation only proceeds under the influence of two independent stimuli, one due to a growth factor, the other from adhesion molecules.

   The assembly of focal adhesion complexes that occurs when fibroblasts spread on a fibronectin matrix involves the clustering of integrins which attach to the cytoskeleton. This assembly also requires monomeric GTPases of the Rho family, in particular Rac and RhoA.[36] However, the signal directing exchange of guanine nucleotides, which precedes clustering, remains uncertain. One possibility is offered by the opposing effects of the α-subunits of G_{12} (negative) and G_{13} (positive) on the guanine nucleotide exchange activity of RhoGEF.[37] It is evident that RhoA mediates its effect in the assembly process through the activation of PI-4P 5-kinase, generating PIP_2 (Figure 14.14). As a result, vinculin attaches to the plasma membrane and unmasks cryptic binding sites that allow it to form a cross bridge between talin (which is linked to the integrin) and actin (the cytoskeleton).[38]

The formation of integrin clusters allows the binding of FAK which undergoes autophosphorylation (Tyr397) and then recruits Src (or Fyn) kinases, to cause further phosphorylation (Tyr925) and the formation of an activated tyrosine protein kinase complex. The phosphorylated FAK, residue Tyr925, now binds the adaptor protein Shc, which binds Grb2 and activates the Ras pathway (Figure 14.12). This may serve to augment the signal from the growth factor receptor and results in prolonged activation

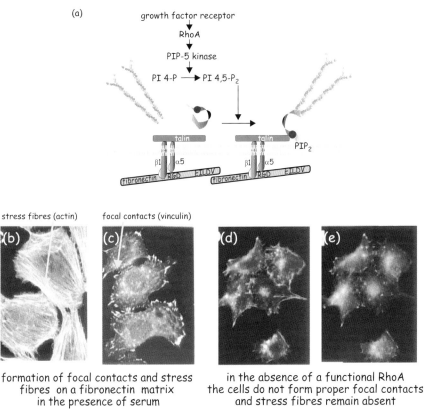

(a) growth factor receptor

RhoA

PIP-5 kinase

PI 4-P $\longrightarrow$ PI 4,5-P$_2$

PIP$_2$

talin

$\beta 1$ $\alpha 5$

fibronectin RGD EILDV

stress fibres (actin)   focal contacts (vinculin)

(b)   (c)   (d)   (e)

formation of focal contacts and stress fibres on a fibronectin matrix in the presence of serum

in the absence of a functional RhoA the cells do not form proper focal contacts and stress fibres remain absent

Figure 14.14 **Formation of focal adhesion sites:** (a) Growth factors play an essential role in the formation of focal adhesion sites through activation of the GTPase RhoA. This promotes formation of PIP$_2$ through activation of PI-4P5 kinase. The PIP$_2$ attaches to vinculin, causing it to open up so that it can attach to both talin and actin, making the link between the integrins and the cytoskeleton. The role of RhoA in this process is nicely illustrated in the micrographs of cells stained with (b) fluorescent phalloidin that binds to actin or with (c) antibodies against vinculin. In the presence of matrix proteins and serum, the fibroblasts spread out, forming stress fibres (b) and focal adhesion sites (c). However, when a non-functional N19-RhoA is introduced (a so-called dominant negative mutant), this is prevented. Even in the presence of abundant growth factors these cells are unable to enter the cell cycle. Adapted from Clark et al.[39]

of ERK (Figure 14.15).[40,41] The sustained signal ensures progression from G_0 to G_1 and entry into the cell cycle.

Downstream from the focal adhesion complex, GTPases of the Rho family are again at the centre of events, relaying both inside-out and outside-in signals, this time initiating entry into the cell cycle (Figure 14.16). The molecular components linking these GTPases with the integrins and then with the subsequent processes needed for entry into the cycle have not yet been fully elucidated. However, prolonged activation of Ras through the concerted actions of growth factors and integrins not only triggers the ERK pathway, but also stimulates a series of events culminating in the activation of RhoA. This then plays a crucial role in cell cycle progression through its effect on the expression of the inhibitor p21[waf/cip] (see Policing the drivers of the cell cycle, page 240). The presence of a mutated, dominant negative RhoA, prevents initiation of DNA synthesis by constitutively activated forms of Ras because the cells continue to express p21[Waf/Cip] at a high level, thus preventing the protein kinase activity of the cyclinD/CDK4 or cyclinE/CDK2 complexes. When Rho is active, expression of p21[waf/cip] is suppressed and DNA synthesis proceeds. The picture that emerges is that Ras emits at least two signals. One of these is a growth promoting signal mediated through the activation of ERK, the other activates RhoA thereby allowing cells to pass through the G_1 phase.[42,43]

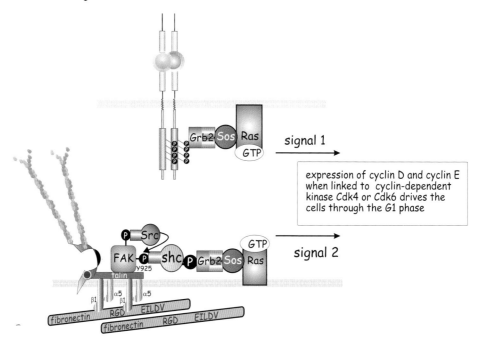

signal 1

expression of cyclin D and cyclin E when linked to cyclin-dependent kinase Cdk4 or Cdk6 drives the cells through the G1 phase

signal 2

Figure 14.15 Two signals required for induction of cell proliferation. The signals due to the receptors for growth factors and from focal adhesion sites converge at Ras and ensure sustained activation of ERK and its translocation to the nucleus. Only under these conditions does expression of cyclins D and E ensue. These, together with the cyclin-dependent protein kinases (CDKs) then drive the cells through the G_1 phase of the cell cycle to induce a round of proliferation.

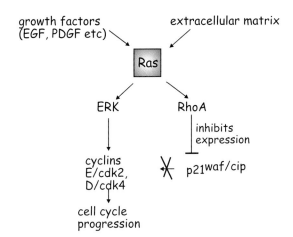

Figure 14.16 **The activation of RhoA downstream of Ras is essential for proper induction of the cell cycle.** Sustained activation of Ras not only results in the sustained activation of ERK, but also the activation of RhoA. This inhibits the expression of the cell cycle inhibitor p21$^{waf/cip}$, an important regulator of the protein kinase activity of the cyclinD/CDK4 and cyclinE/CDK2 complexes.

Adhesion molecules as tumour suppressors

Adherens junctions and desmosomes in endothelial and epithelial cells

The surfaces, separating the interior from the exterior, including the skin, the linings of the body cavities and the gut are formed from tight sheets of epithelial cells. Similarly, the lining of the vasculature, including the heart, blood vessels and lymphatics, is formed from endothelial cells which are specialized epithelial cells. The extracellular matrix is minimal, comprising a thin sheet, the basal lamina. It is the cells themselves that bear the brunt of physical stresses such as the peristalsis of the gut and the pulsatile movement of the vasculature. Tissue structure is maintained by an elaborate series of cellular attachments, tight junctions (also known as zonula occludens), adherent junctions and desmosomes (Latin *occludo* to block-up; *desmo* meaning comb, hence bundle) (Figure 14.8).

Tight junctions form the impermeable barrier between the exterior and the interior. The selective passage of solutes through the cellular sheet is mediated by transporter proteins in the plasma membrane and by trafficking vesicles within the cells. Both of these are particularly abundant in epithelial and endothelial cells. At tight junctions the plasma membranes of two cells are brought into tight apposition by homotypic interactions of several molecules of the membrane protein occludin, which is complexed with zonula occludens (ZO) -1 and -2. The adherent junctions and desmosomes ensure resistance to physical stress. The adhesion molecule cadherin, attached to the cytoskeleton, plays an important role. In the desmosome it is linked to desmoplakin, the site of attachment of the intermediate filaments. In epithelial cells these are formed from polymers of keratin. Malformed keratin, as occurs in the rare inherited condition epidermolysis bullosa, is associated with a tendency for skin chafing and blistering due to rupture of epidermal cells in the basal layer. The adherent junctions are the sites at which cadherin is linked to catenins, to which actin filaments attach. This structure forms the cellular adhesion belt, essential for tissue integrity (a role comparable to the steel rods in reinforced concrete).

■ Loss of adherent junctions induces de-differentiation

If the adherent junctions stabilizing a cell layer are artificially destabilized, for instance by antibodies against cadherin, there is a loss of cell polarity, the particular spatial distribution of the membrane proteins becoming randomized. The insulin binding proteins (normally disposed on the basal surface), and the glucose transporters (normally disposed on the apical surface), are now distributed all over the membrane. The tight junctions disassemble through detachment of the ZO-proteins. The link between the cadherins and the catenins is broken, and, like the ZO-proteins, these are now scattered into the cytosol (Figure 14.17).[44] The cells produce more fibronectin (extracellular matrix), and switch from the production of keratin to vimentin (both forming intermediate filaments). In short, they begin to resemble fibroblasts. This phenomenon, also referred to as the epithelial–mesenchymal transition, plays an important role in the reorganization of tissue in the developing embryo. It allows tight sheets of cells to loosen up, detach and then migrate, to establish

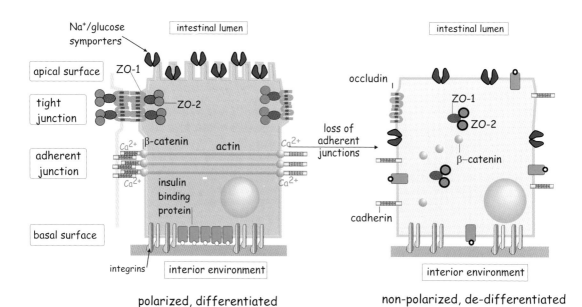

polarized, differentiated   non-polarized, de-differentiated

Figure 14.17 Loss of cell–cell contact and the dissipation of cell polarity. The polarized state is well illustrated by the selective distribution of insulin binding proteins (basolateral surface) and the Na/glucose symporters (apical surface). With the loss of adherent junctions this polarization of functions dissipates, tight junctions disappear. The Na/glucose symporters and insulin binding proteins are distributed randomly. The cells become de-differentiated.

new tissues elsewhere (organ development). A similar, but uncontrolled transition occurs in the formation of cancer cells as a consequence of transformation of epithelial cells.[45]

■ β-Catenin plays a crucial role in the de-differentiation of epithelial cells

β-Catenin is important in the stabilization of adherent junctions but when detached from cadherin its role changes radically. On liberation into the cytosol it translocates to the nucleus where it associates with the ternary transcription factor, Tcf.[46]

The β-catenin/Tcf complex induces expression of the genes for cyclin-D1, c-myc and fibronectin, all favouring entry into the cell cycle.[47,48] Moreover, it also represses apoptosis[49] (Figure 14.18). In an experimental model, ectopic (over-) expression of β-catenin was found to promote cell de-differentiation.[50] Clearly, the uncontrolled liberation of β-catenin can have sinister consequences (Table 14.1). Under normal circumstances the concentration of cytosolic β-catenin is strictly limited by its association with a complex from adenomatous polyposis coli (APC), conductin and glycogen synthase kinase-3β (GSK 3β). Once associated in the complex, β-catenin becomes phosphorylated on several residues and,

Tcf T-cell factor, initially cloned as a lymphoid cell transcription factor. Members of the Tcf family are now recognized as activators and repressors of genes implicated in regulation of the cell cycle in a number of cell types. The T-cell factor is quite distinct from the ternary complex factor p62[TCF], linked with the MAP kinases (page 271).

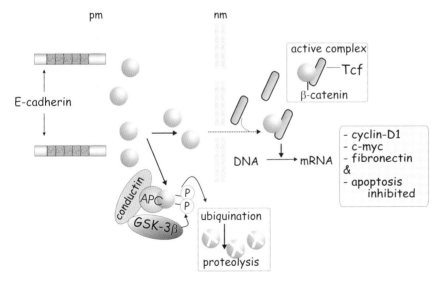

Figure 14.18 Loss of cadherin-mediated cell–cell contact induces proliferation. This occurs through the creation of β-catenin/Tcf transcription factor complexes. Depolarized and de-differentiated epithelial cells (as illustrated in Figure 14.17) acquire the capacity to proliferate and a badly organized multilayer of epithelial cells gradually develops. β-Catenin liberated from the adherent junctions associates with the transcription factor Tcf to induce the expression of genes (cyclinD1, c-myc and fibronectin) that promote the cell cycle. In addition, they establish a survival pathway that renders these cells less prone to apoptosis. By forming a complex with APC (bound to conductin), β-catenin is brought into the vicinity of GSK-3. Phosphorylation at serine/threonine residues enables ubiquination of β-catenin, marking it for destruction. Under non-pathological conditions, the liberation of β-catenin is always transient.

Table 14.1 Mutations of β-catenin and APC in human cancers

Mutated genes	Consequence	Type of cancer
β-catenin	Insensitive to APC-mediated degradation	Invasive colon cancer
Cadherin	Non-functional truncated protein	Colon cancer
APC	Incapable of promoting degradation of β-catenin	Adenomatous polyposis coli

as a consequence, it is targeted for destruction through the ubiquitin-proteasome pathway.[51]

■ Mutations of β-catenin and APC in human cancers

Loss-of-function mutations in genes coding for cadherin and the APC protein, and gain-of-function mutations in the gene coding for β-catenin, are implicated in carcinogenesis, particularly in the colon.[52] These mutations and their consequences give evidence for the important regulatory roles of these proteins. In some patients with colon cancer the gene for E-cadherin expresses a truncated product. This shortened protein is unable to participate in junctional complexes so that the adherent junctions are improperly formed. Over time the cells detach and become invasive. In a culture of highly invasive epithelial cells (of kidney or mammary gland origin), the ectopic expression of functional E-cadherin was found to reduce the invasive phenotype. The cells reformed as a single epithelial sheet and re-established their polarity.[53] A mutated form of β-catenin, insensitive to degradation, cannot be properly cleared from the cytosol and as a result it translocates to the nucleus and associates with Tcf. Here it activates gene expression, resulting in invasive tumour development.[54] Finally, in the condition adenomatous polyposis coli, the APC protein is mutated and incapable of promoting the destruction of cytosolic free β-catenin.[55]

■ A role of cadherin in contact inhibition?

Contact inhibition is the term used to describe the abrupt arrest of the cell cycle that occurs in cultures of rapidly proliferating epithelial cells at the point when a confluent monolayer forms. The phenomenon has been recognized for many years but, even now, a full description of the molecular mechanism remains elusive. It was clear from the start that the arrest is not a consequence of accumulation of inhibitory factors but is mediated by an adhesion molecule.[56] Cadherin plays an important role. As the density of the cells increases, so enabling them to make multiple contacts, the forming adherent junctions sequester the free β-catenin and in this way reduce the proliferation promoting action of Tcf/β-catenin. A programme of cell differentiation and complete cell cycle arrest ensues. There are certainly other molecules involved in this process, among them very likely, receptor protein tyrosine phosphatases[57] (see Chapter 17).

Essay: Apoptosis

Cell death is a normal, indeed essential, aspect of organ development and it is manifest throughout adult life.[25] As an extreme example, neutrophils are short lived, surviving in the circulation for about 4 hours. Millions die and are replaced by new cells every minute. Nerve cells also die, but are not replaced. In the human brain, the number of neurones is maximal at age 3 months. It appears that neuronal cells have access to a limited amount of a survival factor, produced by the neighbouring cells with which they are in contact at synapses. Neuronal cells that have inadequate or incorrect connections, not fully integrated into the neuronal network, become starved of survival signals. Cells dying for these reasons (as opposed to cells damaged by injury, which undergo necrosis) commit a form of controlled suicide, called programmed cell death or apoptosis.[25,58–62] It is apparent that all cells have continuously to face up to the possibility of suicide. Indeed, in the absence of survival factors this must be counted their normal destiny.

Characteristics of an apoptotic cell

A cell undergoing apoptosis can be recognized by the gradual disintegration of its nuclear envelope and the presence of highly condensed chromatin in the nucleus. The plasma membrane forms small blebs and vesicles. It loses contact with the extracellular matrix, shrinks and gradually disintegrates, without, however, releasing its content into the environment (Figure 14.19). The detritus is scavenged within hours by neighbouring cells or by professional macrophages without the slightest sign of inflammation. This contrasts with cells dying as a result of tissue injury (necrosis) in which they undergo lysis, releasing their contents into the environment to induce an inflammatory response that serves to initiate the process of tissue remodelling and repair (wound healing).

apoptosis From the Greek απω (from, away, detached); πτῶσις (falling). Hence, falling away.

Caspases, cellular proteases cause apoptosis

A good understanding of the mechanism of apoptosis came from genetic studies using the nematode *C. elegans*. It was found that deletion of *CED* genes (cell death abnormal) prevent the programme of cell death that is integral to the process of development. The effect was to alter the sculpture of the embryo. Homologous genes exist in mammals. These code for proteolytic enzymes of the family of caspases that contain a cysteine residue in the catalytic site and cleave their substrates at a consensus motif, Asp-Glu-Val-Asp (hence, cysteine-aspartate-proteases).[63] Procaspases are activated by cleavage at the same consensus site, either by themselves or by other caspases. There exist two classes of caspases, those that exclusively cleave and activate other caspases, (initiator caspases), and those that cleave other proteins (effector caspases) (Figure 14.20). Of course, the initiator caspases themselves have to be activated. It appears that the activity of these enzymes is suppressed through binding to inhibitory proteins that prevent their dimerization, thus preventing their

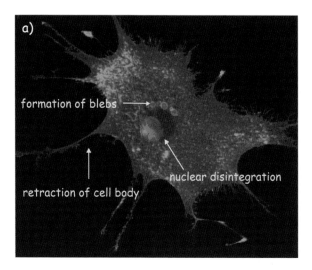

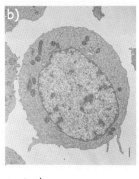
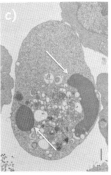

control   apoptotic cell

Figure 14.19 **Induction of apoptosis – morphology:** (a) Apoptotic cells are characterized by retraction of the cell body, formation of membrane blebs and disintegration of the nucleus. The fibroblast cell surface was stained with anti-CD44 antibodies (FITC green) and the nucleus with DAPI (red). (b, c) Transmission electron micrographs illustrating the apoptotic process in a mouse T-cell lymphoma. The characteristics of apoptosis, nuclear fragmentation (arrows) and formation of cytosolic vacuoles are evident. The cells expressed a human Fas-antigen (see page 339) and apoptosis was induced by treatment with anti-Fas antibodies. (b) and (c), Courtesy of Shin Yoneraha and Masazumi Sameshima.

mutual activation.[64] Only when a sufficient amount of initiator caspases is activated does the process of cell destruction become inevitable.

■ Cellular targets of caspases

Different caspases have different roles (Figure 14.21):

- inactivation or destruction of inhibitors of apoptosis
- inactivation of the inhibitory protein (ICAD) of the caspase activated DNAase (CAD). The inactivation of ICAD, through proteolysis, allows CAD to degrade strands of DNA

- inactivation of Bcl-2, a protein that maintains caspases in their inactive state (see below)
- destruction of cellular compartments and signal transduction pathways
- degradation of proteins that direct the organization of the cytoskeleton, such as gelsolin, p21-activated protein kinase (PAK), FAK or lamin
- destruction of the machinery that repairs and replicates DNA
- degradation of repair enzymes, DNA-dependent protein kinase (DNA-PK), the splicing enzyme U1-70K and the replication factor C.

PAK-2 is cleaved during apoptosis through the action of a caspase. This activates PAK2 which effects changes in the morphology of the apoptotic cell.
DNA-PK, DNA-dependent protein kinase holoenzyme, acts as a serine/threonine protein kinase that phosphorylates many DNA-binding proteins and transcription factors. Defective DNA-PK activity is linked to defective DNA-repair.
Replication factors A and C (RF-A and RF-C) Cellular proteins essential for complete elongation of DNA during replication and transcription.

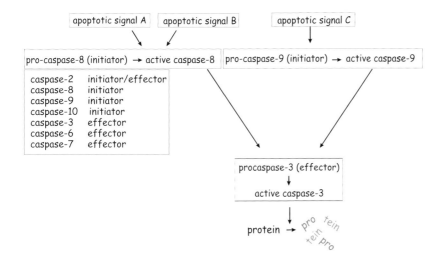

Figure 14.20 **Initiator and effector caspases.** Cascades of caspase activation. The cascades are initiated by apoptotic signals that activate the initiator caspases. These in turn activate effector caspases that cleave their specific substrates causing apoptosis.

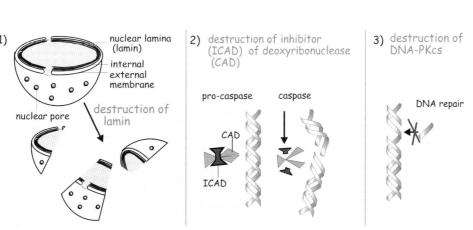

Figure 14.21 **Different apoptosis targets of effector caspases.**

Bcl-2 B-cell lymphoma due to a chromosomal translocation that recombines the candidate transforming gene *bcl-2*, with the Ig heavy chain.
The human **Bcl-2** protein is remarkably similar to Ced-9 and it can suppress apoptosis in *C. elegans* when artificially introduced. Apparently the apoptosis machinery is well conserved throughout evolution.

■ Regulation of caspases

These examples illustrate the role of various regulatory proteins implicated in the induction of apoptosis under different conditions.

■ *Regulation by Bcl-2*

Apoptosis can be induced by an intracellular signal as a response to an insufficiency of external survival factors. Alternatively, there is the possibility that it may be controlled by an internal clock. Bcl-2, a protein originally implicated in the generation of B-cell lymphomas, plays an important role.

These tumours arise as a consequence of a chromosomal translocation that places a number of genes in the vicinity of a hyperactive IgH locus. One of these, *bcl-2*, becomes heavily transcribed (Figure 14.22) and by inhibiting apoptosis, it acts to increase the lifespan of the B-lymphocytes and contributes to tumour formation.

Bcl-2 binds to Apaf, an adaptor for initiator procaspases, and in this way prevents their dimerization and mutual activation (Figure 14.23).

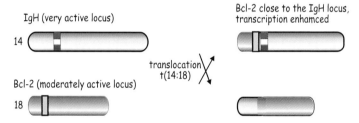

Figure 14.22 Translocation of Bcl-2. Translocation between chromosome 14 and 18 brings the *bcl-2* locus under regulation of the very active IgH locus. This augments the expression of Bcl-2 with the consequence of enhanced protection against apoptosis.

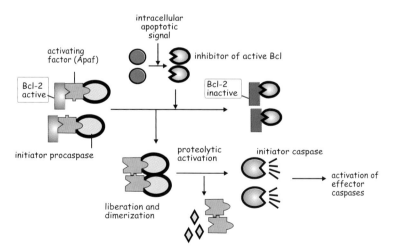

Figure 14.23 Bcl-2 and the activation of effector caspases. The activated inhibitor of Bcl-2 competes for binding with Apaf. This liberates the Apaf/initiator procaspase complex which now dimerize and activate each other.

Bcl-2 is present on the surface of organelle membranes (mitochondria, endoplasmic reticulum, nucleus). When an apoptotic signal is presented, the cells activate an inhibitor of Bcl-2, Bad for instance, that competes with the binding of Apaf. The Apaf/initiator procaspase complex is thus dislodged and Bcl-2 remains attached to the inhibitor, so enabling initiator pro-caspases to approach and cleave each other. In this way, active initiator caspases are generated, and the cascade of proteolytic reactions commences, resulting in cell death.

Induction of apoptosis by Fas ligand

Apoptosis can be induced by external signals through activation of the Fas receptor by the Fas ligand.[65] Fas, a member of the TNF/NGF family of receptors (see Chapter 15), is expressed on most cells but its ligand is presented only on the surface of cytotoxic T cells. Of the many different types of lymphocytes that are generated when the immune system is provoked by a virus or tumour-specific antigen, these are the most important in the mediation of apoptosis. Through the T-cell receptor (Chapter 12) they recognize virus infected and tumour cells presenting their specific antigens in the context of MHC-class I. As a result, the cytotoxic T cells present the Fas-ligand which binds and aggregates its receptor, Fas, as a trimeric complex on the target cell. (Figure 14.24). The activated receptors each bind a complex of adaptor proteins, FADD and FLASH, associated with the initiator pro-caspase-8. This brings the pro-caspases-8 together, resulting in their mutual cleavage and activation.

Activation of caspases through cytochrome c

The mitochondrion is a source and amplifier of apoptotic signals and represents an important conduit to cell death. Signals from receptors (e.g. TNFα), and also some cytotoxic drugs elicit the release of cytochrome c into the cytoplasm where it binds and dimerizes two adaptor subunits, Apaf, that are linked to the initiator pro-caspase-9. Dimerization causes mutual cleavage, liberating the active caspase and so a proteolytic

Apaf <u>A</u>poptotic protease <u>a</u>ctivating <u>f</u>actor; binds initiator procaspases and is involved in their dimerization. Binding of Bcl-2 to Apaf prevents this.

The **Bcl-2** proteins are important regulators of cell death and survival. They have pro- and anti-apoptotic roles:

• pro-apoptotic are Bid, Bax, Bad, Bak, Bok and Bim

• anti-apoptotic are Bcl-2, Bcl-X$_L$, Bcl-w, Mcl-1 and A1.

They all possess one or more Bcl-2-homology domains (BH1, 2, 3 or 4). Some members of this family interact with each other, forming homo- or heterodimers, and these interactions appear to modulate their function. For instance, Bid inhibits the anti-apoptotic action of Bcl-2 and stimulates the pro-apoptotic action of Bax.

Bad Bcl-associated dimer.

Bax Bcl-associated partner containing six exons.

Bid Bcl-2 interacting death agonist.

Bok Bcl-2-related ovarian killer

Bak bcl-2 homologous antagonist/killer.

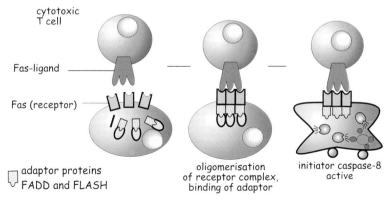

cytotoxic T cell

Fas-ligand

Fas (receptor)

adaptor proteins
FADD and FLASH

oligomerisation of receptor complex, binding of adaptor

initiator caspase-8 active

Figure 14.24 **Induction of apoptosis by Fas ligand.** Expression of Fas ligand on the cytotoxic T cell causes oligomerization of the Fas receptor, binding of the adaptor/procaspase-8 complex and activation.

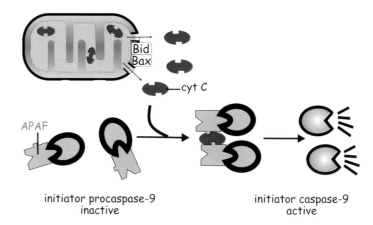

initiator procaspase-9
inactive

initiator caspase-9
active

Figure 14.25 Cytochrome c in the activation of initiator caspase-9. Cytotoxic drugs or certain receptors activate the apoptosis agonists Bid and Bax. These open a pore in the mitochondrial outer membrane to allow leakage of cytochrome c. In the cytoplasm it dimerizes and activates procaspase-9.

cascade is inititated (Figure 14.25). The leakage of cytochrome *c* occurs through the permeability transition pore complex (PTPC), a structure that bridges both the inner and outer mitochondrial membranes.[66] Bcl-2 family proteins associate with the PTPC to reduce (Bcl-2 and Bcl-SL) or promote (Bax in combination with Bad or Bid) permeability. Phosphorylation of Bad by PKB causes its sequestration in the cytoplasm preventing it from binding to the mitochondrial membrane and assisting Bax in opening the PTPC. In addition, mitochondria contain caspases and an apoptosis inducing factor, AIF, that promote apoptosis directly after their release into the cytoplasm.

References

1 Edelman, G.M. Cell adhesion molecules. *Science* 1983; 219: 450–7.
2 Springer, T.A. Traffic signals for lymphocyte recirculation and leukocyte emigration: the multistep paradigm. *Cell* 1994; 76: 301–14.
3 Brackenbury, R., Thiery, J.P., Rutishauser, U., Edelman, G.M. Adhesion among neural cells of the chick embryo. I. An immunological assay for molecules involved in cell–cell binding. *J. Biol. Chem.* 1977; 252: 6835–40.
4 Springer, T., Galfre, G., Secher, D.S., Milstein, C. Mac-1: a macrophage differentiation antigen identified by monoclonal antibody. *Eur. J. Immunol.* 1979; 9: 301–6.
5 Davignon, D., Martz, E., Reynolds, T., Kurzinger, K., Springer, T.A. Lymphocyte function-associated antigen 1 (LFA-1): a surface antigen distinct from Lyt-2,3 that participates in T lymphocyte-mediated killing. *Proc. Natl. Acad. Sci. USA* 1981; 78: 4535–9.
6 Tamkun, J.W., DeSimone, D.W., Fonda, D. *et al.* Structure of integrin, a glycoprotein involved in the transmembrane linkage between fibronectin and actin. *Cell* 1986; 46: 271–82.
7 Ruoslahti, E., Pierschbacher, M.D. New perspectives in cell adhesion: RGD and integrins. *Science* 1987; 238: 491–7.
8 Dustin, M.L., Rothlein, R., Bhan, A.K., Dinarello, C.A., Springer, T.A. Induction by IL-1 and interferon-γ: tissue distribution, biochemistry, and function of a natural adherence molecule (ICAM-1). *J. Immunol.* 1986; 137: 245–54.
9 Osborn, L., Hession, C., Tizard, R. *et al.* Direct expression cloning of vascular cell

adhesion molecule 1, a cytokine-induced endothelial protein that binds to lymphocytes. *Cell* 1989; 59: 1203–11.

10 Ducibella, T., Anderson, E. Cell shape and membrane changes in the eight-cell mouse embryo: prerequisites for morphogenesis of the blastocyst. *Dev. Biol.* 1975; 47: 45–58.

11 Yoshida, C., Takeichi, M. Teratocarcinoma cell adhesion: identification of a cell-surface protein involved in calcium-dependent cell aggregation. *Cell* 1982; 28: 217–24.

12 Takeichi, M. Cadherins in cancer: implications for invasion and metastasis. *Curr. Opin. Cell Biol.* 1993; 5: 806–5.

13 Boller, K., Vestweber, D., Kemler, R. Cell-adhesion molecule uvomorulin is localized in the intermediate junctions of adult intestinal epithelial cells. *J. Cell Biol.* 1985; 100: 327–32.

14 Stoolman, L.M., Rosen, S.D. Possible role for cell-surface carbohydrate-binding molecules in lymphocyte recirculation. *J. Cell Biol.* 1983; 96: 722–9.

15 Lasky, L.A., Singer, M.S., Yednock, T.A. *et al.* Cloning of a lymphocyte homing receptor reveals a lectin domain. *Cell* 1989; 56: 1045–55.

16 Alon, R., Chen, S., Fuhlbrigge, R., Puri, K.D., Springer, T.A. The kinetics and shear threshold of transient and rolling interactions of L-selectin with its ligand on leukocytes. *Proc. Natl. Acad. Sci. USA* 1998; 95: 11631–6.

17 Laurent, T.C., Fraser, J.R. Hyaluronan. *FASEB J.* 1992; 6: 2397–404.

18 Aruffo, A., Stamenkovic, I., Milnick, M., Underhill, C.B., Seed, B. CD44 is the principal cell surface receptor for hyaluronate. *Cell* 1990; 6: 1303–13.

19 Banerji, S., Ni, J., Wang, S.X. *et al.* LYVE-1, a new homologue of the CD44 glycoprotein, is a lymph-specific receptor for hyaluronan. *J. Cell Biol.* 1999; 144: 789–901.

20 Day, A.J. The structure and regulation of hyaluronan-binding proteins. *Biochem. Soc. Trans.* 1999; 27: 115–21.

21 Goldstein, L.A., Zhou, D.F., Picker, L.J. *et al.* A human lymphocyte homing receptor, the hermes antigen, is related to cartilage proteoglycan core and link proteins. *Cell* 1989; 56: 1063–72.

22 Gunthert, U., Hofmann, M., Rudy, Y. *et al.* A new variant of glycoprotein CD44 confers metastatic potential to rat carcinoma cells. *Cell* 1991; 65: 13–24.

23 Tsukita, S., Oishi, K., Sato, N., Sagara, J., Kawai, A. ERM family members as molecular linkers between the cell surface glycoprotein CD44 and actin-based cytoskeletons. *J. Cell Biol.* 1994; 126: 391–401.

24 Frisch, S.M., Francis, H. Disruption of epithelial cell-matrix interactions induces apoptosis. *J. Cell Biol.* 1994; 124: 619–26.

25 Raff, M.C. Cell suicide for beginners. *Nature* 1998; 396: 119–22.

26 King, W.G., Mattaliano, M.D., Chan, T.O., Tsichlis, P.N., Brugge, J.S. Phosphatidylinositol 3-kinase is required for integrin-stimulated AKT and Raf-1/mitogen-activated protein kinase pathway activation. *Mol. Cell Biol.* 1997; 17: 4406–18.

27 Khwaja, A., Rodriguez, V.P., Wennstrom, S., Warne, P.H., Downward, J. Matrix adhesion and Ras transformation both activate a phosphoinositide 3-OH kinase and protein kinase B/Akt cellular survival pathway. *EMBO J.* 1997; 16: 2783–93.

28 Kennedy, S.G., Wagner, A.J., Conzen, S.D. *et al.* The PI 3-kinase/Akt signaling pathway delivers an anti-apoptotic signal. *Genes Dev.* 1997; 11: 701–13.

29 Downward, J. Mechanisms and consequences of activation of protein kinase B/Akt. *Curr. Opin. Cell Biol.* 1998; 10: 262–7.

30 Brunet, A., Bonni, A., Zigmond, M.J. *et al.* Akt promotes cell survival by phosphorylating and inhibiting a Forkhead transcription factor. *Cell* 1999; 96: 857–68.

31 Frisch, S.M., Vuori, K., Ruoslahti, E., Chan-Hui, P.Y. Control of adhesion-dependent cell survival by focal adhesion kinase. *J. Cell Biol.* 1996; 134: 793–9.

32 Lin, H.J., Evinerm, V., Prendergast, G.C., Whitem, E. Activated H-ras rescues E1A-induced apoptosis and cooperates with E1A to overcome p53-dependent growth arrest. *Mol. Cell Biol.* 1995; 15: 4536–44.

33 Frisch, S.M., Ruoslahti, E. Integrins and anoikis. *Curr. Opin. Cell Biol.* 1997; 9: 701–6.

34 Cardone, M.H., Salvesen, G.S., Widmann, C., Johnson, G., Frisch, S.M. The regulation of anoikis: MEKK-1 activation requires cleavage by caspases. *Cell* 1997; 90: 315–23.

35 Assoian, R.K. Anchorage-dependent cell cycle progression. *J. Cell Biol.* 1997; 136: 1–4.

36 Hotchin, N.A., Hall, A. The assembly of integrin adhesion complexes requires both extracellular matrix and intracellular rho/rac GTPases. *J. Cell Biol.* 1995; 131: 1857–65.

37 Hart, M.J., Jiang, X., Kozasa, T. *et al.* Direct stimulation of the guanine nucleotide exchange activity of p115 RhoGEF by Gα13. *Science* 1998; 280: 2112–14.

38 Gilmore, A.P., Burridge, K. Regulation of vinculin binding to talin and actin by phosphatidyl-inositol-4-5-bisphosphate. *Nature* 1996; 381: 531–5.

39 Clark, E.A., King, W.G., Brugge, J.S., Symons, M., Hynes, R.O. Integrin-mediated signals regulated by members of the rho family of GTPases. *J. Cell Biol.* 1998; 142: 573–86.

40 Bottazzi, M.E., Zhu, X., Bohmer, R.M., Assoian, R.K. Regulation of p21(cip1) expression by growth factors and the extracellular matrix reveals a role for transient ERK activity in G1 phase. *J. Cell Biol.* 1999; 146: 1255–64.

41 Lenormand, P., Sardet, C., Pages, G. *et al.* Growth factors induce nuclear translocation of MAP kinases (p42mapk and p44mapk) but not of their activator MAP kinase kinase (p45mapkk) in fibroblasts. *J. Cell Biol.* 1993; 122: 1079–88.

42 Olson, M.F., Paterson, H.F., Marshall, C.J. Signals from Ras and Rho GTPases interact to regulate expression of p21Waf1/Cip1. *Nature* 1998; 394: 295–9.

43 Olson, M.F., Ashworth, A., Hall, A. An essential role for Rho, Rac, and Cdc42 GTPases in cell cycle progression through G1. *Science* 1995; 269: 1270–2.

44 Fialka, I., Schwarz, H., Reichmann, E. *et al.* The estrogen-dependent c-JunER protein causes a reversible loss of mammary epithelial cell polarity involving a destabilization of adherens junctions. *J. Cell Biol.* 1996; 132: 1115–32.

45 Thiery, J.P., Chopin, D. Epithelial cell plasticity in development and tumor progression. *Cancer Metastasis Rev.* 1999; 18: 31–42.

46 Roose, J., Clevers, H. TCF transcription factors: molecular switches in carcinogenesis. *Biochim. Biophys. Acta* 1999; 1424: 23–7.

47 Tetsu, O., McCormick, F. β-Catenin regulates expression of cyclin D1 in colon carcinoma cells. Nature 1999; 398: 422–6.

48 He, T.C., Sparks, A.B., Rago, C. *et al.* Identification of c-MYC as a target of the APC pathway. *Science* 1998; 281: 1509–12.

49 Ahmed, Y., Hayashi, S., Levine, A., Wieschaus, E. Regulation of armadillo by a *Drosophila* APC inhibits neuronal apoptosis during retinal development. *Cell* 1998; 93: 1171–82.

50 Orford, K., Orford, C.C., Byers, S.W. Exogenous expression of β-catenin regulates contact inhibition, anchorage-independent growth, anoikis, and radiation-induced cell cycle arrest. *J. Cell Biol.* 2000; 146: 855–68.

51 Orford, K., Crockett, C., Jensen, J.P., Weissman, A.M., Byers, S.W. Serine phosphorylation-regulated ubiquitination and degradation of β-catenin. *J. Biol. Chem.* 1997; 272: 24735–8.

52 Hirohashi, S. Inactivation of the E-cadherin-mediated cell adhesion system in human cancers. *Am. J. Pathol.* 1998; 153: 333–9.

53 Vleminckx, K., Vakaet, L., Mareel, M., Fiers, W., van Roy, F. Genetic manipulation of E-cadherin expression by epithelial tumor cells reveals an invasion suppressor role. *Cell* 1991; 66: 107–19.

54 Morin, P.J., Sparks, A.B., Korinek, V. *et al.* Activation of β-catenin-Tcf signaling in colon cancer by mutations in β-catenin or APC. *Science* 1997; 275: 1787–90.

55 Gumbiner, B.M. Carcinogenesis: a balance between β-catenin and APC. *Curr. Biol.* 1997; 7: 443–6.

56 Schutz, L., Mora, P.T. The need for direct cell contact in 'contact' inhibition of cell division in culture. *J. Cell. Physiol.* 1968; 71: 1–6.

57 Zondag, G.C.M., Moolenaar, W.H. Receptor protein tyrosine phosphatases: involvement in cell-cell interaction and signaling. *Biochimie* 1997; 79: 477–83.

58 Ashkenazi, A., Dixit, V.M. Death receptors: signaling and modulation. *Science* 1998; 281: 1305–8.

59 Thornberry, N.A., Lazebnik, Y. Caspases: enemies within. *Science* 1998; 281: 1312–16.

60 Green, D.R., Reed, J.C. Mitochondria and apoptosis. *Science* 1998; 281: 1309–12.

61 Adams, J.M., Cory, S. The Bcl-2 protein family: arbiters of cell survival. *Science* 1998; 281: 1322–6.

62 Evan, G., Littlewood, T. A matter of life and cell death. *Science* 1998; 281: 1317–22.

63 Miura, M., Zhu, H., Rotello, R., Hartwieg, E.A., Yuan, J. Induction of apoptosis in fibroblasts by IL-1 β-converting enzyme, a mammalian homolog of the *C. elegans* cell death gene ced-3. *Cell* 1993; 75: 653–60.

64 Salvesen, G.S., Dixit, V.M. Caspase activation: The induced-proximity model. *Proc. Natl. Acad. Sci. USA* 1999; 96: 10964–7.

65 Suda, T., Takahashi, T., Golstein, P., Nagata, S. Molecular cloning and expression of the Fas ligand, a novel member of the tumor necrosis factor family. *Cell* 1993; 75: 1169–78.

66 Kroemer, G. Mitochondrial control of apoptosis: an overview. *Biochem. Soc. Symp.* 1999; 66: 1–15.

Adhesion molecules and trafficking of leukocytes

▪ Inflammation and its mediators

The first indications that white cells adhere to the walls of the finer vessels and then emigrate into the tissues under conditions of inflammation were recorded more than 150 years ago (Figure 15.1).[1–3] Cohnheim's detailed histological description of 1882[4] remained the basis of most text book accounts of inflammation for a further 80 years. The first electron microscopic investigation of leukocyte adherence and migration provided much needed detail but failed to reveal how cells adhere to the inflamed endothelium.[5] Nor were any ideas forthcoming about how the cells penetrate an apparently coherent endothelial cell layer.

It is essential that migratory cells can tolerate changes in their environments. They must be able to move from the bone marrow (where they are attached) into the blood and then into the tissues, particularly the lymph nodes (where they become attached again). To do all this they must be able to switch their integrins 'on' and 'off'. If the integrins on a leukocyte were permanently 'on', it could never leave the bone marrow. Once in the blood, the integrin function should be switched 'off'. Leukocytes leave the circulation when they encounter an appropriate signal of inflammation or infection (which is common experience), or more generally in the continuous process of immune surveillance. For monocytes and neutrophils, departure from the circulation is a one-way journey, their destiny with death in the tissues. By contrast, lymphocytes move from the blood into the tissues, then through the lymphatic system and then back into blood. We focus on the regulation of leukocyte adhesion and extravasation under inflammatory conditions (see Figure 15.2).

At sites of inflammation (caused by injury or infection), mediators are released that affect the expression and affinity of adhesion molecules and that affect the release of chemokines from endothelial cells. Though technically cytokines, chemokines were originally identified and characterized as chemotactic agents. By attracting the host phagocytic cells to sites of tissue breakdown, these peptides could mediate an essential first step in tissue repair and healing. They arrest leukocyte circulation and

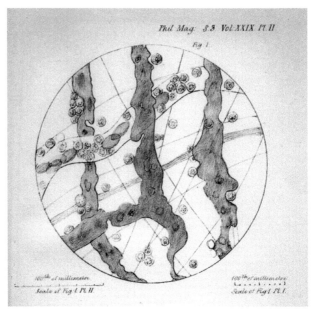

Figure 15.1 **Augustus Waller's microscopic examination of cell adherence and extravasation from the vessels of an inflamed frog tongue.**

"The blood, as we are aware, consists of a transparent fluid holding in suspension numerous particles, most of which are red and of a flattened shape, while a few others are colourless, and spherical in form. . . . The peculiar manner in which the lymph-globules, or corpuscles, conduct themselves when in the capillaries, when in an organ in a state of irritation, has of late engaged much attention. The experiments of Mr W Addison of Malvern, have greatly contributed to show these important functions in inflammation. In the tongue of the frog and toad they may be frequently seen circulating with the red particles in the vessels, down to the minutest capillaries. As it has already been pointed out, these spherules are generally found, when they come into contact with the parieties of the vessels, to retain their adherence with greater force than is manifested by the red particles in the like circumstances; as in the figure, where the current was observed to continue for many minutes without displacing the globules near the sides of the vessel. Thus we frequently see a lymph-globule remain in the same place, notwithstanding the current of red particles sweeping and pushing by it. The appearance in the larger vessels of these spherules, adherent to their inner surface, has been very aptly compared to so many pebbles or marbles over which a stream runs without disturbing them. . . . The corpuscles, which are transparent, are occasionally seen to be granulated. . . . Let us now examine the admirable manner in which nature has solved the apparent paradox, of eliminating, from a fluid circulating in closed tubes, certain particles floating in it, without causing any rupture or perforation in the tubes, or allowing the escape of the red particles, which are frequently the smaller of the two, or that of the fluid part of the blood itself. . . . After the observation had continued for half an hour, numerous corpuscles were seen outside the vessels, together with a very few blood discs in the proportion of about one to ten of the former. No appearance of rupture could be seen in any of the vessels. The corpuscles were generally distant about $0^{mm}03$ from their parieties. After the experiment had lasted about two hours, thousands of these corpuscles were seen scattered over the membrane, with scarcely any blood discs. . . . No trace of the corpuscular extravasation could be seen, except the presence of the corpuscles themselves. . . . I consider therefore as established,-1st, the passage of these corpuscles 'de toute pièce' through the capillaries; 2ndly, the restorative power in the blood, which immediately closes the aperture thus formed. . . . In endeavouring to account for the fact of the passage of the corpuscles through the vessels we find considerable difficulties. It cannot be referred to the influence of vitality, as it is observed likewise to take place after death. It may be surmized, either that the corpuscle, after remaining a certain time in contact with the vessel, gives off by exudation from within itself some substance possessing a solvent power over the vessel, or that the solution of the vessel takes place in virtue of some of those molecular actions which arise from the contact of two bodies; actions which are now known as exerting such extensive influence in digestion, and are referred to what is termed the catalytic power."

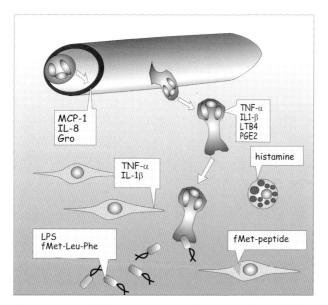

Figure 15.2 Generation of inflammatory mediators at sites of infection. Tissue damage and bacterial infection induce the release of inflammatory mediators from various cell types. Bacterial lipopolysaccharide (LPS), which acts on local fibroblasts, vascular endothelial cells, mast cells and resident leukocytes induces release of histamine, IL-1β and TNFα, but also MCP-1, IL-8 or Gro. These mediators are responsible for the up-regulation of adhesion molecules on endothelial cells and the activation of integrins on leukocytes. This results in extravasation followed by migration to the site of damage/infection. Formylmethionyl peptides released from bacteria facilitate the migratory response (chemotaxis). These processes allow the rapid accumulation of leukocytes, in particular neutrophils, at sites of infection as a first line of defence. It also initiates the process of tissue remodelling and repair (wound healing).

cause these cells to accumulate at sites of inflammation (as in pus). Inflammatory mediators are derived from a number of sources, as listed in Table 15.1.

Leukocytes are capable of sensing concentration differences of chemokines as small as 1% over a distance of 40 μm, approximating the linear dimensions of a cell. They respond by moving in the direction of the source, in the process called chemotaxis.[6]

In the next sections we describe the signalling pathway that emanates from the TNFα receptor in vascular endothelial cells. This results in elevated cell surface expression of adhesion molecules and the release of other inflammatory mediators and chemokines. It also plays a crucial role in regulating the extravasation of circulating leukocytes (also known as transendothelial cell migration).

Table 15.1 Sources of inflammatory mediators

- Release from invading micro-organisms. These include lipopolysaccharide (LPS or endotoxin), shed from the surface. Also N-formylmethionyl peptides (fMet-peptides) derived from the initiator sequence of bacterial protein synthesis
- Release from resident cells. These include TNFα and IL-1β from activated macrophages or fibroblasts, histamine from mast cells and chemokines (MCP-1, IL-8, Gro) from endothelial cells or resident leukocytes
- Release from the mitochondria of damaged cells. Peptides derived from the amino-terminus of proteins (such as NAD$^+$ dehydrogenase) coded by the mitochondrial genome. These share with eubacteria the characteristic N-formylmethionyl initiator sequence
- Components of complement such as C5a and of fibrinolysis such as thrombin

■ Tumour necrosis factor-α; potential antitumour agent or inflammatory cytokine?

The observation that the size of a tumour occasionally diminishes after bacterial infection has a long history. As early as 1848, LeGrand noted two cases of long-standing scrofulous lymphoma which appeared to regress after infection with erysipelas.[7] In 1888 P. Bruns reported on the spontaneous regression of human tumours following infection with *S. erysipelas*:[8]

Erysipelas An acute superficial form of cellulitis involving the dermal lymphatics, usually caused by infection with group A streptococci. Formerly called St Anthony's fire.

> The frequently observable fact that new formations, particularly those of a malign nature, may be caused to regress or to disappear by a concurrent erysipelas is of very great interest both in theory and practice. In respect of the former this fact merits particular attention, especially at the present time, given the current preoccupation with the study of the aetiology of new formations, the results of which may well be capable of shedding new light upon the beneficial effects of erysipelas. On the practical front these observations have given rise to experiments in which curative erysipelas is artificially induced in order to bring about the healing of inoperable new formations. However, the admissibility of such experiments remains an open question . . .

W. B. Coley[9] reported several cases of tumour regression and even disappearance following repeated inoculation of erysipelas. Speculating on possible mechanisms, he considered

Die öfters beobachtete Thatsache, dass Neubildungen, namentlich malingner Natur, durch ein interkurrentes Erysipel zur Verkleinerung oder zum Verschwinden gebracht werden, ist von hervorragendem theoretischem und praktischem Interesse. In ersterer Hinsicht verdient diese Thatsache gerade gegenwärtig besondere Beachtung, wo die Untersuchung der Aetiologie

> 1) that the erysipelas coccus has a direct destructive action upon the cell elements of the new growth; 2) that the high temperature (induced by the infection) alone is sufficient; 3) that sarcoma and carcinoma are both of bacterial origin and the erysipelas germ has a direct antagonistic effect upon the cancer bacillus . . . it seems in the light of present knowledge not improbable that all of the three theories may contain an element of truth, and that a larger theory combining all these elements is necessary to explain the curative action of erysipelas.

In the firm belief that the cure of cancer was at hand, Coley and others applied cell-free filtrates of streptococci as 'Coley's toxins' over a period of about 40 years, apparently with some success.[9–11] The effect of these bacterial products is associated with severe haemorrhagic reaction mainly confined to the core of the tumour, which rapidly sloughs off. However, it generally leaves a ring of viable tissue which, unfortunately, continues to grow. Not surprisingly, such treatments also tend to cause widely disseminated systemic effects and, all too frequently, these can lead to circulatory collapse and death. Early attempts to separate the haemorrhagic from the cytotoxic components in culture filtrates of *Serratia marcescens* led to the isolation of endotoxin (also called LPS, a surface component of Gram-negative bacteria).[12,13] Among its many biological activities, this could elicit haemorrhagic necrosis of both experimental and primary subcutaneous tumours.

All this provoked the hunt for immune mediators, cytotoxic for transformed cells but not causing septic shock like the endotoxin. The first of these, TNFα and TNFβ and lymphotoxin, products released by cytotoxic T lymphocytes, attracted immense interest.[14,15] Instead, however, TNF was also found to provoke severe systemic toxicity and to possess general pro-inflammatory properties. The focus of interest shifted from tumour necrosis to mechanisms of inflammation.

◼ TNFα and regulation of adhesion molecule expression in endothelial cells

The binding of TNFα to its receptor results in the formation of homotrimeric clusters.[16] In endothelial cells this leads to the enhanced expression of VCAM-1 and the ICAMs,[17] and release of chemokines. As a result, the adherence of neutrophils is greatly amplified.[18] This effect of TNFα and those induced by other inflammatory stimuli are mediated through the nuclear transcription factor known as <u>n</u>uclear <u>f</u>actor κB (NF-κB), a protein complex that attaches to the immunoglobulin κ light chain gene[19] (Figure 15.3). The factor is made up of two proteins, p50 and p65, both members of the NF-κB/Rel family of transcriptional regulators. At first, NF-κB was believed to be restricted to lymphocytes but it is now clear that it is widely distributed although generally retained in the cytosol by a high affinity inhibitory factor, inhibitor κB (I-κB).[20,21] Activation of NF-κB requires that it is first released from its association with the inhibitor. This occurs through serine-phosphorylation of I-κB, which marks it for ubiquitination and destruction by the proteasome. NF-κB then translocates to the nucleus, where it promotes the transcription of many genes, including the adhesion molecules (already mentioned), chemokines (IL-8, MCP-3, MIP-1a), cytokines (IL-1β, IL-6, TNFα, GM-CSF) and enzymes (iNOS, cyclo-oxygenase-2, cytosolic PLA$_2$).

The trimerization of the TNF receptor initiates the formation of an unusually large signalling complex (approximately 700 kDa) (Figure 15.3). There is still uncertainty about the functions of its individual components, which have been given macabre names such as <u>T</u>NF <u>r</u>eceptor-<u>a</u>ssociated

der malignen Neubildungen auf der Tagesordnung steht, deren Ergebnisse vielleicht geeignet sind, auch über die Art und Weise der salutären Wirkung des Erysipels Licht zu verbreiten. In praktischer Beziehung haben jene Beobachtungen zu Versuchen mit der kunstlichen Erzeugung eines kurativen Erysipels Veranlassung gegeben, um inoperable bösartige Neubildungen zur Heilung zu bringen. Allein die Zulässichkeit.solcher Versuche ist noch eine offene Frage ...

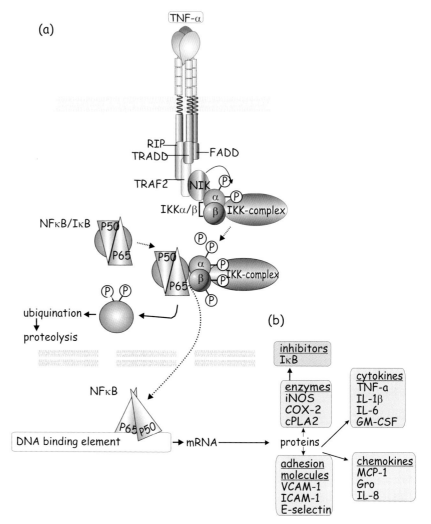

(a)

(b)

Figure 15.3 NF-κB and the expression of adhesion molecules. (a) The binding of TNFα to its receptor results in trimerization followed by formation of a truly massive receptor signalling complex. This in turn associates with the IKK-complex, of which the two kinases, IKKα and β, are essential components. NIK phosphorylates and activates these kinases, allowing them to dissociate and cause phosphorylation of IκB, the inhibitor of NF-κB. The phosphorylated inhibitor is ubiquinated and destroyed by proteolysis. The liberated NF-κB translocates to the nucleus and activates transcription of a large number of genes as illustrated. (b) NF-κB plays an important role in the TNFα-mediated expression of adhesion molecules, enzymes, chemokines and cytokines in endothelial cells. Among these is also the gene for the inhibitory factor I-κB itself, so that when a sufficient number of copies of this protein have been attained, it will bind to NF-κB, sequestering it in the cytoplasm. This negative feedback loop plays an important role in the termination of the inflammatory response.

death domain (TRADD), Fas-associated death domain (FADD), TNF receptor-associated factor (TRAF), receptor inhibitor protein (RIP) and NF-κB-inducing kinase (NIK).[22] All this results in the activation of the serine/threonine protein kinase, NIK, a member of the MEKK family of protein kinases. Its position in the pathway is equivalent to that of Raf in the activation of ERK1/2 (Figure 11.15, page 274) and it phosphorylates two further protein kinases, IKKα and IKKβ (Inhibitor κB Kinases-α and -β), which are the equivalent to MEK1. These detach from the signalling complex and bind to the inhibitor IκB (itself tightly bound to NF-κB). There they phosphorylate IκB on two residues, rendering I-kBα prone to ubiquitination and subsequent degradation, and releasing NF-κB p50/p65. The dissociation of IκB reveals a targeting signal on NF-κB, directing it to the nucleus where it promotes transcription of the genes for VCAM-1, ICAM, etc.[23] (Figure 15.3).

NF-κB also enhances the expression of IκB, thereby initiating a negative feedback loop that attenuates the TNFα signal. This secures the return to basal levels of cell surface expression of VCAM-1 and ICAM-1. Their expression is therefore transient, allowing termination of the inflammatory response. Without this, the influx of leukocytes would be prolonged causing chronic tissue damage, as occurs in the formation of ulcers.

TNFα (or histamine, IL-1β, LPS, thrombin, etc.: see Table 15.1), causes transient up-regulation of VCAM-1 and ICAM-1 and also of E-and P-selectins on vascular endothelial cells. P-selectin, stored in specialized secretory granules, is rapidly transferred to the plasma membrane. Expression of E-selectin, ICAM-1 and VCAM-1 occurs more slowly and then persists for a few days. The increase in the density of these adhesion molecules on the surface of endothelial cells increases their 'avidity' (the product of density × affinity) for leukocytes.

■ Chemokines and activation of integrins on leukocytes

Integrins play crucial roles in the arrest, flattening and the transmigration of leukocytes through the endothelial cell layer (extravasation, see Figure 15.2). However, in circulating cells, they remain quiescent unless activated by chemokines provided by endothelial cells (Table 15.2).

The chemokines are small proteins having four conserved cysteines that form two essential intramolecular disulphide bonds (Cys1 with Cys3 and Cys2 with Cys4: see Figure 15.4). They are classified as CC, CXC and CXXXC chemokines, according to the spacing of the first two cysteines in the N-terminal segment.

Table 15.2 Chemokines

Receptor family	Receptor subtype	Ligand
CXC	CXCR1	IL-8
	CXCR2	IL-8, GROα,β,γ, NAP-2, ENA78,
	CXCR3	GCP-2
	CXCR4	ID10, Mig
		SDF-1
CC	CCR1	RANTES, MIP-1α, MCP-2, MCP-3
	CCR2a/b	MCP-1, MCP-2, MCP-3, MCP-4
	CCR3	Eotaxin, RANTES, MCP-3, MCP-4
	CCR4	RANTES, MIP-1α, MCP-1
	CCR5	RANTES, MIP-1α, MIP1β

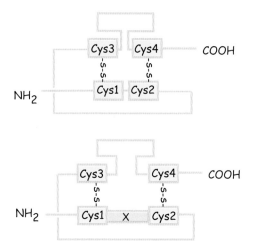

Figure 15.4 **Representation of intramolecular disulphide bonding in CC and CXC cytokines.**

Chemokines became a focus of attention when it was shown that the envelope glycoprotein of human immunodeficiency virus (HIV) competes for binding with chemokines. Together with CD4, the chemokine receptors are the port of entry of HIV-1 into T-lymphocytes. Occupation of chemokine receptors, for instance of CXCR-4, prevents the penetration of cells by HIV-1.[24] In some rare individuals, because of a homozygous mutation, the cytokine receptor CCR-5 is not exposed at the cell surface. The virus cannot gain entry and these fortunate people appear to be immune to infection by HIV.[25] The mutation does not appear to induce any deleterious phenotype, which indicates that there is some redundancy among these receptors. Essential functions are well covered in the absence of CCR-5.

The chemokine receptors interact with pertussis-toxin-sensitive G-proteins (G_i/G_o). Of the 13 receptors that have been characterized, most recognize more than one chemokine (Table 15.2). They elicit a number of signal transduction events which are illustrated in Figure 15.5.[26]

The integrins are activated by a separate pathway in which PKC is implicated (they can be activated by phorbol ester[27]). However, there is the problem that chemokine-induced integrin activation is not prevented when PKC is inhibited.[28] On the other hand, activation is sensitive to inhibitors of protein tyrosine kinases and of PI 3-kinase.[29] It is possible that $\beta\gamma$-subunits derived from the activation of the G-protein G_i provide the link between the receptor and the Ras-ERK kinase pathway (see Figure 9.5, page 196).

In leukocytes, the $\beta\gamma$-subunits interact with the p110γ-catalytic subunit of PI 3-kinase causing local generation of PI(3,4,5)P_3 (see Chapter 13) that recruits proteins having PH domains to the plasma membrane. These include PKB and also guanine nucleotide exchange factors that cause the activation of Ras, Rac and RhoA. The activation of Rac is essential for the induction of cell migration, through the formation of filopodia[28] and activation of the respiratory burst. RhoA is essential for the activation of

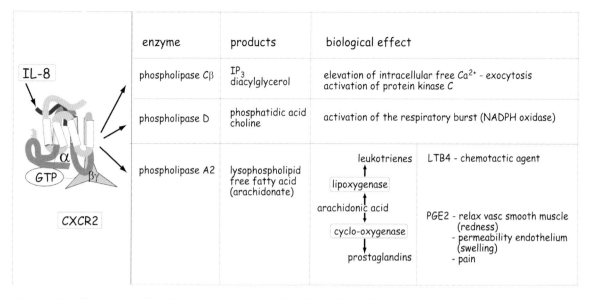

	enzyme	products	biological effect		
	phospholipase Cβ	IP$_3$ diacylglycerol	elevation of intracellular free Ca^{2+} - exocytosis activation of protein kinase C		
	phospholipase D	phosphatidic acid choline	activation of the respiratory burst (NADPH oxidase)		
	phospholipase A2	lysophospholipid free fatty acid (arachidonate)	leukotrienes ↑ lipoxygenase ↑ arachidonic acid ↑ cyclo-oxygenase ↓ prostaglandins	LTB4 - chemotactic agent	PGE2 - relax vasc smooth muscle (redness) - permeability endothelium (swelling) - pain

Figure 15.5 Signal transduction pathways emanating from chemokine receptors in leukocytes. Binding of the chemokine IL-8 to its receptor, CXCR2, results in the activation of a number of signal transduction pathways through the activation of phospholipases Cβ, A$_2$ and D.

the integrins[28] (Figure 15.6). These bind to ICAM-1 or VCAM-1 on the endothelial cells with high affinity, causing spreading and subsequent transmigration.[30] In this way, the suspended leukocytes take on the characteristics of adherent cells. Once in the tissue, the activated integrins have further important roles in augmenting the neutrophil responses such as phagocytosis, exocytosis and the generation of reactive oxygen metabolites, necessary for the killing and clearing of micro-organisms.

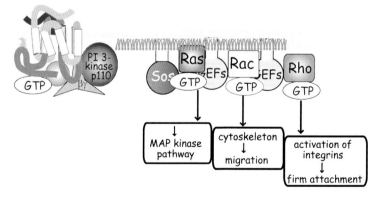

Figure 15.6 Activation of PI 3-Kγ by chemokine receptors is essential for integrin activation. Chemokine receptors are linked to heterotrimeric GTP-binding proteins. Here it is the βγ-subunits which are of importance as activators. These bind directly to the PH domain of the p110γ catalytic subunit of PI 3-kinase resulting in the formation of PI(3,4,5)P$_3$ (marked in brown) and the activation of exchange factors (GEFs) for Ras (Sos), Rac2 (Tiam) and RhoA (Dbl). In leukocytes, RhoA is essential for the activation of integrins, resulting in firm endothelial attachment and extravasation.

In addition to the chemokines, the synthetic tripeptide agonist formyl-Met-Leu-Phe (fMLP) and the serum complement component C5a have been used widely in experimental investigations. They also interact with characteristic 7TM G-protein linked receptors on leukocytes to initiate a wide range of biological responses including calcium mobilization, activation of integrins, migration, release of reactive oxygen metabolites, lysosomal enzyme secretion, etc.

■ The three-step process of leukocyte adhesion to endothelial cells

The processes described form part of a more general sequence of events that is now referred to as the three-step process of leukocyte adhesion to endothelial cells.

Depending on the type of leukocyte, different integrins participate in the firm arrest to endothelium. For instance, in monocytes $\alpha M\beta 2$ is predominantly involved in arrest, whereas in neutrophils this role is taken by $\alpha L\beta 2$. Lymphocytes rely mainly on $\alpha 4\beta 1$ (see Figure 14.5, page 320).

Leukocytes make contact with the vascular endothelium from time to time, by the presentation of surface ligands (L-selectin, sialyl Lewis-X or PSGL-1: see above) to receptors on the surfaces of the endothelial cells, (CD34, E or P selectin). These low-affinity interactions tether the cells in a manner by which they can attach, then roll on the surfaces of the endothelial cells and finally detach (in Figure 15.7a). However, the presence of chemokines causes the leukocyte integrins to become activated so that they bind to ICAM-1 or VCAM-1 present at elevated levels due to the presence of inflammatory mediators. The leukocytes are arrested at this point. They flatten and are, in principle, competent to transmigrate through the tight layer of endothelial cells.[30] Not all bound cells transmigrate. Some merely perch on the endothelial cell surface and then detach. Those cells that do transmigrate, pass through the basal membrane and make it right through into the tissues.

A mutation in the gene coding for the $\beta 2$ integrins is associated with an inherited tendency to frequent and persistent infections (leukocyte adhesion deficiency). This is due to the failure of leukocytes to reach sites of inflammation in sufficient numbers.[31–34]

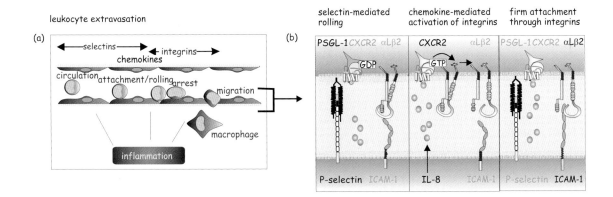

(a) leukocyte extravasation

selectins — integrins —
chemokines

circulation — attachment/rolling — arrest — migration

macrophage

inflammation

(b)

selectin-mediated rolling | chemokine-mediated activation of integrins | firm attachment through integrins

PSGL-1 CXCR2 $\alpha L\beta 2$ CXCR2 $\alpha L\beta 2$ PSGL-1 CXCR2 $\alpha L\beta 2$

GDP   GTP

P-selectin ICAM-1 | IL-8   ICAM-1 | P-selectin ICAM-1

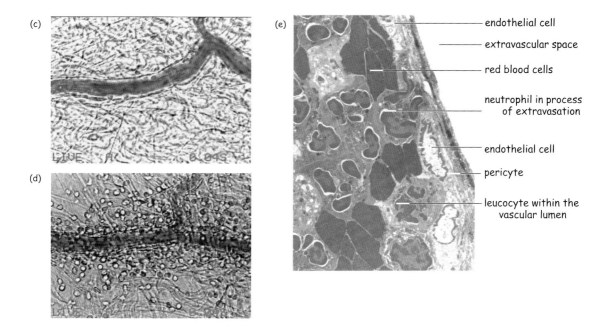

Figure 15.7 **The three-step model of leukocyte transendothelial migration:** (a) Representation of the three-step model of leukocyte transendothelial migration through the wall of a venule. This comprises (1) a selectin-mediated rolling of leukocytes on the vascular endothelial cell surface, followed by (2) a chemokine-mediated activation of integrins on the leukocytes resulting in (3) arrest and migration of the leukocytes through the vascular endothelial cell layer. It should be noted that a large proportion of leukocytes are normally attached and roll on the surface of the endothelium. It is the firm arrest and subsequent migration that is triggered by local inflammation as a consequence of the release of chemokines and an enhanced expression of members of the selectins, VCAM and ICAM. (b) A molecular representation of the same three steps, with purple (on top) representing the leukocyte and red (at the bottom) representing the endothelial cell. *Left panel:* Initial contact between the two cells is made through P-selectin interacting with the P-selectin-glycosylated ligand-1 (PSGL-1). This interaction must have a rapid turn-off rate allowing the cells to roll. *Centre panel:* The close proximity to the endothelial cells allows detection of IL-8 by its receptor CXCR2. This causes the activation of GTP-binding proteins that in turn activate the integrin αLβ2. *Right panel:* The activated integrin now binds ICAM-1 causing firm attachment of the leukocyte to the endothelial cell. This interaction is a prerequisite for further flattening and migration of leukocytes through the endothelial cell layer. (c) Video-micrograph illustrating a rat mesenteric venule (~ 40 μm diameter). The blurring within the venule is due to the flow of blood. (d) Video-micrograph illustrating leukocyte accumulation into the mesenteric tissue after upstream administration of the chemokine LTB4 (from Sussan Nourshargh, Imperial College London). (e) Thin section electron micrograph, showing a leukocyte (three-lobed nucleus) migrating through a layer of endothelial cells into rat mesenteric tissue (from Sussan Nourshargh, Imperial College London).

References

1   Addison, W. Experimental and practical researches on the structure and function of blood corpuscles; on inflammation; and on the origin and nature of tubercles in the lungs. *Trans. Prov. Med. Surg. Assoc.* 1843; 11: 233–305.

2   Waller, A. Microscopic examination of some of the principal tissues of the animal frame, as observed in the tongue of the living frog, toad etc. *Phil. Mag.* 1846; 29: 271–87.

3   Waller, A. Microscopic observations on the perforation of the capillaries by the corpuscles of the blood, and on the origin of mucus and pus-globules. *Phil. Mag.* 1846; 29: 397–405.

4 Cohnheim, J. *Lectures on General Pathology*. The New Sydenham Society, London, 1882.

5 Marchesi, V.T., Florey, H.W. Electron micrographic observations on the emigration of leucocytes. *Quart. J. Exp. Physiol.* 1960; 45: 343–8.

6 Zigmond, S.H. Ability of polymorphonuclear leukocytes to orient in gradients of chemotactic factors. *J. Cell Biol.* 1977; 75: 606–16.

7 Legrand, A. De l'analogie et des différences entre les tubercules et les scrofules: Influence des maladies eruptives sur le developpement et la marche des scrofules et des tubercules. *Rev Médicale française et étrangère* 1848; 2: 392–448.

8 Bruns, P. Die Heilwirkung des Erysipels auf Geschwülste (The healing effect of erysipelas on tumours). *Beitr. Klin. Chir.* 1868; 3: 443–66.

9 Coley, W.B. Contribution to the knowledge of sarcoma. *Ann. Surg.* 1891; 14: 199–220.

10 Nauts, H.C., Fowler, G.A., Bogatko, F.H. A review of the influence of bacterial infection and of bacterial products (Coley's toxins) on malignant tumors in man. *Acta Med. Scand. Suppl.* 1953; 275: 5–103.

11 Pearl, R. Cancer and tuberculosis. *Am. J. Hyg.* 1929; 9: 97–159.

12 Shear, M.J., Andervont, H.B. Chemical treatment of tumors. III. Separation of hemorrhage-producing fraction of *B. coli* filtrate. *Proc. Soc. Exp. Biol. Med.* 1936; 34: 323–5.

13 Shear, M.J., Turner, F.S. Chemical treatment of tumors: Isolation of the hemorrhage-producing fraction from Serratia marcescens (Bacillus prodigiosus) culture filtrate. *J. Natl Cancer Inst.* 1943; 4: 81–97.

14 Kolb, W.P., Granger, G.A. Lymphocyte in vitro cytotoxicity: characterization of human lymphotoxin. *Proc. Natl. Acad. Sci. USA* 1968; 61: 1250–5.

15 Ruddle, N.H., Waksman, B.H. Cytotoxicity mediated by soluble antigen and lymphocytes in delayed hypersensitivity. 3. Analysis of mechanism. *J. Exp. Med.* 1968; 128: 1267–79.

16 Vandevoorde, V., Haegeman, G., Fiers, W. Induced expression of trimerized intracellular domains of the human tumor necrosis factor (TNF) p55 receptor elicits TNF effects. *J. Cell Biol.* 1997; 137: 1627–38.

17 Osborn, L., Hession, C., Tizard, R. *et al.* Direct expression cloning of vascular cell adhesion molecule 1, a cytokine-induced endothelial protein that binds to lymphocytes. *Cell* 1989; 59: 1203–11.

18 Pohlman, T.H., Stanness, K.A., Beatty, P.G., Ochs, H.D., Harlan, J.M. An endothelial cell surface factor(s) induced in vitro by lipopolysaccharide, interleukin 1, and tumor necrosis factor-α increases neutrophil adherence by a CDw18-dependent mechanism. *J. Immunol.* 1986; 136: 4548–53.

19 Sen, R., Baltimore, D. Multiple nuclear factors interact with the immunoglobulin enhancer sequences. *Cell* 1986; 46: 705–16.

20 Baeuerle, P.A., Baltimore, D. IκB: A specific inhibitor of the NF-κB transcription factor. *Science* 1988; 242: 540–6.

21 Verma, I.M., Stevenson, J.K., Schwarz, E.M. Rel/NF-κB/IκB family: Intimate tales of association and dissociation. *Genes Dev.* 1995; 9: 2723–35.

22 Stancovski, I., Baltimore, D. NF-κB activation: The IκB kinase revealed? *Cell* 1997; 91: 299–302.

23 Cross, D.A., Alessi, D.R., Cohen, P., Andjelkovich, M., Hemmings, B.A. Inhibition of glycogen synthase kinase-3 by insulin mediated by protein kinase B. *Nature* 1995; 378: 785–9.

24 Oberlin, E., Amara, A., Bachelerie, F. *et al.* The CXC chemokine SDF-1 is the ligand for LESTR/fusin and prevents infection by T-cell-line-adapted HIV-1. *Nature* 1996; 1(382): 833–5.

25 Liu, R., Paxton, W.A., Choe, S. *et al.* Homozygous defect in HIV-1 coreceptor accounts for resistance of some multiply-exposed individuals to HIV-1 infection. *Cell* 1996; 86: 367–77.

26 Baggiolini, M. Chemokines and leukocyte traffic. *Nature* 1998; 392: 565–8.

27 Wright, S.D., Silverstein, S.C. Tumor-promoting phorbol esters stimulate C3b and C3b' receptor-mediated phagocytosis in cultured human monocytes. *J. Exp. Med.* 1982; 156: 1149–64.

28 Laudanna, C., Campbell, J.J., Butcher, E.C. Role of Rho in chemoattractant-activated leukocyte adhesion through integrins. *Science* 1996; 271: 981–3.

29 Bokoch, G.M. Chemoattractant signaling and leukocyte activation. *Blood* 1995; 86: 1649–60.

30 Springer, T.A. Traffic signals for lymphocyte recirculation and leukocyte emigration: the multistep paradigm. *Cell* 1994; 76: 301–14.

31 Arnaout, M.A., Dana, N., Gupta, S.K., Tenen, D.G., Fathallah, D.M. Point mutations impairing cell surface expression of the common β subunit (CD18) in a patient with leukocyte adhesion molecule (Leu-CAM) deficiency. *J. Clin. Invest.* 1990; 85: 977–81.

32 Kishimoto, T.K., Hollander, N., Roberts, T.M., Anderson, D.C., Springer, T.A. Heterogeneous mutations in the β subunit common to the LFA-1, Mac-1, and p150,95 glycoproteins cause leukocyte adhesion deficiency. *Cell* 1987; 50: 193–202.

33 Arnaout, M.A., Pitt, J., Cohen, H.J. *et al.* Deficiency of a granulocyte-membrane glycoprotein (gp150) in a boy with recurrent bacterial infections. *N. Engl. J. Med.* 1982; 306: 693–9.

34 Anderson, D.C., Springer, T.A. Leukocyte adhesion deficiency: an inherited defect in the Mac-1, LFA-1, and p150,95 glycoproteins. *Annu. Rev. Med.* 1987; 38: 175–94.

Signalling through receptor bound protein serine/threonine kinases

■ The TGFβ family of growth factors

TGFβ1, first identified as a transforming factor for mesenchymal cells, then as an inhibitor of proliferation in epithelial cells (see Chapter 10, page 231), is a member of a family of structurally related proteins, most of which have little or nothing to do with cell transformation. Loss-of-function mutations in the related genes of *Drosophila* (*Dpp*) and *Xenopus* (*Vg1*) give rise to developmental defects. On the basis of sequence comparisons, the TGFβ family as identified in mammals can be divided into a number of subfamilies that include bone morphogenetic protein (BMP), activin and TGFβ (Table 16.1). The list of the TGFβ family of growth factors and knowledge of their role in development is still expanding.[1]

■ Two signalling receptors for TGFβ; type I and type II

TGFβ1 came to notice in a search for transforming growth factors, and initially it was anticipated that its receptor would be linked, either directly or indirectly, to tyrosine phosphorylations.

When it was later found that it inhibits proliferation of epithelial cells, interest shifted to the possibility that it prevents growth factor signalling. Neither of these ideas proved very fruitful. TGFβ1 has no effect on tyrosine phosphorylation, nor (at least when measured on a time scale of minutes) does it affect the early events in EGF or PDGF receptor signalling. So it remained until 1990, when a receptor for activin was cloned and found to contain a putative transmembrane Ser/Thr-protein kinase domain.[2] Similar domains were then found in two of the TGFβ1 receptors.[3–5]

On the basis of their structural and functional properties, the receptors for this family of ligands are divided into two subfamilies, type I and type II (TβR-I and TβR-II: Figure 16.1; Table 16.2).[6] They are very similar. Both receptors are glycoproteins having molecular weights of 55 kDa and 70 kDa respectively. Both have a single membrane-spanning segment and an

Note that **TGFα**, a totally unrelated growth factor despite its similar name, acts conventionally through a tyrosine kinase pathway by binding to the EGF receptor.

Table 16.1 Some activities of the TGFβ family. The factors listed have been identified in human and/or mouse. Equivalents in other species are indicated (D, *Drosophila melanogaster*; c, *Caenorhabditis elegans*; x, *Xenopus laevis*).

BMP2 subfamily			
BMP2 (Dpp Daf-7 (c))	D, C	Gastrulation, neurogenesis, chondrogenesis, interdigital apoptosis	x: mesoderm patterning D: dorsalization, eye/wing development c: dauer larva formation
BMP4			
BMP5 subfamily			
BMP5 (60 A)	D	Participates with BMP2 and 4 in the development of nearly all organs	
BMP6/Vgr1			
BMP7OP1			
BMP8/OP2			
GDF5 subfamily			
GDF5/CDMP1		Chondrogenesis in developing limbs	
GDF6/CDMP2			
GDF7			
vg1 subfamily			
GDF1 (Vg1)	X		x: Vg1: axial mesoderm formation (also in fish)
GDF3/VG32			
bmp3 subfamily			
BMP3/osteogenin		Osteogenic differentiation, endochondral bone formation, monocyte chemotaxis	
GDF10			
Activin subfamily			
Activin βA		FSH production, erythroid cell differentiation	x: mesoderm induction
Activin βB			
Activin βC			
Activin βE			
TGFβ subfamily			
TGFβ1		Cell cycle control in haematopoietic and epithelial cells, control of mesenchymal cell proliferation and differentiation, production of extracellular matrix, immunosuppression	
TGFβ2			
TGFβ3			
Distant members			
Inhibin α		Inhibition of FSH production, mesoderm induction	
MIS/AMH		Müllerian duct regression	

intrinsic serine-threonine protein kinase domain in the C-terminal segment. What distinguishes the type I receptor is a highly conserved region of 30 amino acids, immediately preceding the protein kinase domain, containing the sequence TTSGSGSGLP. This is the GS domain, which becomes phosphorylated after binding of TGFβ. It plays a crucial role in the activation of TβR-I. The type I receptor also has a binding site for the

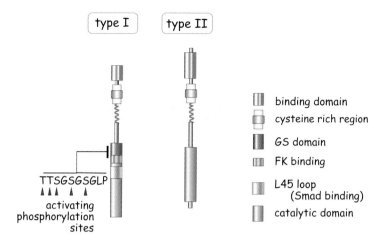

type I   type II

binding domain
cysteine rich region
GS domain
FK binding
L45 loop
(Smad binding)
catalytic domain

TTSGSGSGLP
activating
phosphorylation
sites

Figure 16.1 Domain structures of the TGFβ receptors. The type I and II receptors contain a serine/threonine protein kinase domain, a cysteine-rich region and a TGFβ-binding domain. The type I receptor contains an additional GS domain and a number of phosphorylation sites that are involved in its activation by the type II receptor which is constitutively activated.

Table 16.2 The type I and II families of receptors of the TGFβ-family of growth factors. Different members of the TGFβ1-family of growth factors bind to different type I and II receptors and form different hetero-tetrameric complexes. This list is not complete; numerous receptors still have to be identified. Listed members are of vertebrate origin unless otherwise indicated (D, *D. melanogaster*; C, *C. elegans*).

ALK, activin receptor-like kinases. These are human genes recognized by their sequence similarity with the activin receptor type II and Daf-1.

Ligand	Type I receptor	Type II receptor
BMP2	BMPR-IA (ALK-3)	BMPR-II
BMP7/8	BMPR-IB (ALK-6)	ActR-II
Dpp (D)	Tkv (D)	Punt (D)
	Sax (D)	
Daf-7 (C)	Daf-1 (C)	Daf-4 (C)
GDF5	ActR-I	ActR-IIB
Activin	ActR-I (ALK-2)	ActR-IIB
	ActR-IB (ALK-4)	ActR-II
TGFβI	TβR-I (ALK-5)	TβR-II

12 kDa immunophilin FK506-binding protein (FKBP12), a molecule th... may act as a negative regulator of the receptor signalling function. Importantly, the protein kinase activity of TβR-II is constitutively active.

The mammalian receptors are homologous with the punt (*put*), thick-veins (*tkv*) and saxophone (*sax*) genes of *Drosophila* and with the dauer phenotypes-1 and 4 (*daf-1* and *daf-4*) genes of *C. elegans*. The downstream signalling pathway from the TGFβ1-receptors was revealed by searching for mammalian homologues of their counterparts in these organisms. It became apparent that the mammalian TGFβ1 receptors transmit signals into cells through a unique set of transcription factors, the Smad proteins (see below).[7]

Here we outline some of the mechanisms by which members of the TGFβ family of receptors elicit their effects on target cells. We concentrate on the pathway activated by TGFβ1 through TβR-I and TβR-II receptors and in particular on the pivotal role of the Smad proteins in relaying signals from cell-surface receptors to the nucleus. It will become apparent that these pathways are rather similar to those already described for the activation of the STATS (see Chapter 11), through the tyrosine kinase-containing receptors such as those for EGF and PDGF and for interferon. The main theme is that a receptor complex phosphorylates a transcription factor. This forms an oligomeric complex that translocates to the nucleus to interact with DNA-response elements in promoter regions of genes.

■ Accessory receptors: betaglycan and endoglin

The search for cell surface TGFβ-binding proteins revealed a third set of receptors, type III. These differ from types I and II, their intracellular domains lacking any sequence motif that could be involved in signal transduction. One of the type III binding proteins, betaglycan, is a trans-membrane proteoglycan with heparan and chondroitin sulphate chains.[8] It functions as a co-receptor with ActRII to bind inhibin, a member of the TGFβ superfamily. Inhibin and activin inhibit and activate the secretion of FSH from porcine pituitary glands.[9] This functional antagonism may be explained by the finding that they both bind ActRII. In the case of activin, binding is followed by recruitment of the type I receptor, ActR-IB, to initiate the signal. Inhibin recruits betaglycan so that there is a loss of the activin signal due to a loss of available type II receptors. This puts a block on further proceedings (Figure 16.2).[10] The other accessory receptor is endoglin, a transmembrane protein[11] that may play a role in the binding and presentation of TGFβ to the type II receptor.[12]

■ TGFβ-mediated receptor activation

TGFβ1 is a disulphide-linked homodimer (Figure 16.3). It brings together pairs of type I and II receptors to form heterotetrameric receptor complexes. Ligand-independent homo-oligomers may exist, but have no signalling capacity. TGFβ1 can bind to TβR-II in the absence of TβR-I, but not vice versa. TGFβ1, however, cannot signal into the cell in the absence of TβR-I. These findings indicate that the most likely sequence of events is that TGFβ1 first binds to TβR-II, then subsequently alters its conformation so that it can be recognized by TβR-I (sequential binding). When the ligand brings the two receptors in close proximity, the type II receptor, which is constitutively active, phosphorylates the three serine and two threonine residues in the GS domain of TβR-I (Figure 16.4). With this, the dormant serine-threonine protein kinase of TβR-I becomes activated and we speak of an active receptor complex. As with the protein

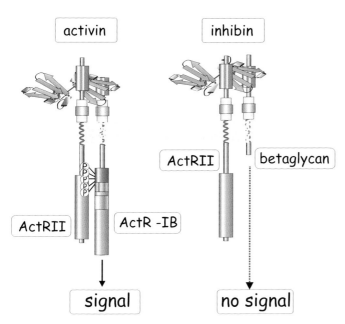

Figure 16.2 Accessory receptor type III binds inhibin. Both activin and inhibin bind first to ActRII but then recruit different co-receptors. Activin recruits the type I receptor and transmits signals. Inhibin recruits betaglycan, which is incapable of transmitting signals and counteracts the activin response.

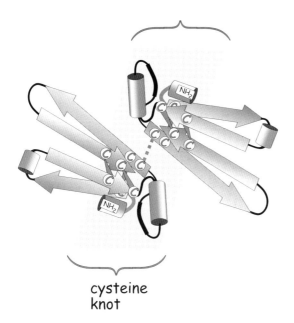

Figure 16.3 Disulphide-linked dimer structure of TGFβ1. The cysteine knot is also present in PDGF and NGF. Adapted from McDonald and Hendrickson.[13]

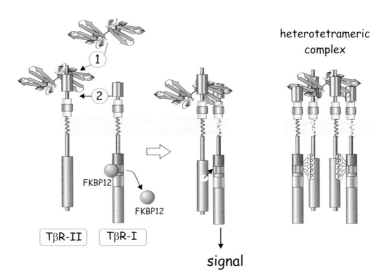

heterotetrameric
complex

FKBP12

FKBP12

TβR-II   TβR-I

signal

Figure 16.4 Activation of the TGFβ-receptor. TGFβ1 binds to the type II receptor, causing a conformational change that allows it to be recognized by the type I receptor. The association of the two receptors results in detachment of the inhibitory FKBP12 protein and this is followed by phosphorylation of the type I receptor by the type II receptor. The activated type I receptor signals into the cell. Because TGFβ1 is itself a dimer, two receptor complexes can form around the growth factor.

tyrosine kinase-containing receptors, both oligomerization and phosphorylation constitute the receptor activation event.

■ Downstream signalling; *Drosophila, Caenorhabditis* and Smad

Genetic screening of accessible organisms such as *Drosophila* and *C. elegans* provided crucial pointers to the mechanism of TGFβ signalling in mammalian cells. A *Drosophila* gene, decapentaplegic (*dpp*), when mutated, causes pattern deficiencies and duplications in structures derived from one or more of the 15 major imaginal disks.[14]

Later it was found that Dpp is equivalent to BMP-2 and -4 of mammalian cells. In early development, Dpp is responsible for dorsal/ventral polarity. Later, as segments appear, Dpp functions in the definition of boundaries between the segmental compartments. As part of this process, Dpp defines the position of future limbs, including wings, legs and antennae. It also has a role in the structuring of the mesoderm (see web page http://sdb.bio.purdue.edu/fly/aimain/1aahome.htm). The Dpp gene product acts through three receptors that are homologous to the family of mammalian TGFβ receptors: thickveins, saxophone (TβRI) and punt (TβRII). To dissect the mechanism through which Dpp operates, genetic screening was carried out to identify a mutation that, when combined with Dpp mutations that produce a 'weak' phenotype, generates one that is more severely affected. This strategy is designed to identify two genes coding for proteins that operate in the same pathway. The mutant

Decapentaplegic
Paralysed at 15 sites.
Imaginal From the Latin *imago*, the final and perfect stage or form of an insect after its metamorphoses.

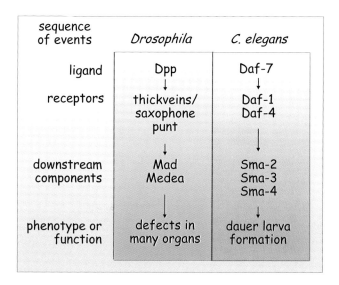

sequence of events	Drosophila	C. elegans
ligand	Dpp ↓	Daf-7 ↓
receptors	thickveins/ saxophone punt ↓	Daf-1 Daf-4 ↓
downstream components	Mad Medea ↓	Sma-2 Sma-3 Sma-4 ↓
phenotype or function	defects in many organs	dauer larva formation

Figure 16.5 Signal transduction pathways downstream of serine/threonine kinase receptors in two phyla. Homologies between the genes implicated in these pathways allowed the elucidation of the sequence of events downstream of the TGFβ receptor in mammals.

gene so obtained was named *mad* (mothers against Dpp) and the flies exhibit defects that resemble those of the Dpp mutants.[15] However, wild-type *dpp* cannot restore the defects induced by mutations in *mad* and this places *mad* downstream of *dpp* (Figure 16.5).[16]

The nematode *C. elegans* responds to conditions of overcrowding and starvation by arresting its development as a dauer (resilient, durable) larva. Genetic screening of mutants having the dauer larva phenotype revealed a number of genes, daf-1, -2, -3, -4, etc. (for dauer larva formation,) of which *daf-1* and *daf-4* code for serine/threonine receptor protein kinases.[17,18] *Daf-4* mutants are dauer-constitutive and the larvae are smaller than the wild-types. Screening for mutants with similar phenotypes revealed three more genes, *sma-2*, *sma-3* and *sma-4* (small).[19] The Sma gene products act downstream of Daf-4. DNA sequencing indicated that Mad (fly) and Sma (worm) proteins are homologous, and with these sequences to hand, nine human homologues coding for Smad proteins (amalgamation of Sma and Mad) were revealed.[7,20]

■ Smads have multiple roles in signal transduction

The Smad proteins have two regions of homology, MH1 and MH2 (Mad homology), at the N- and C-terminals. The nine members of this family have different functions which are controlled through their selective interactions with the two TGFβ receptors and with each other[21] (Figure 16.6). On the basis of their structures and functions they are divided into three groups.

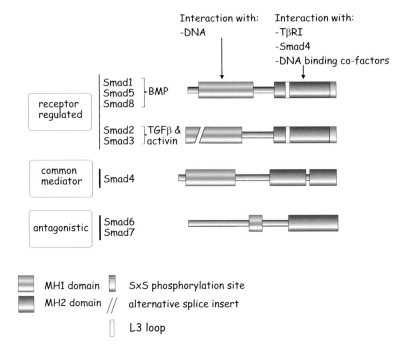

Figure 16.6 **Structure domains of the Smads.** These transcription factors are subdivided in three groups on the basis of functional and structural differences. The MH1 domain interacts with DNA, the MH2 domain interacts with the type I receptor, Smad4 and transcriptional partners. Adapted from Massagué and Wotton.[22]

Receptor-regulated Smads (Smad1, 2, 3, 5 and 8)

These are phosphorylated through the activated type-I receptors. For example, TGFβ1 induces the phosphorylation of Smad2 and Smad3, whereas the equivalent activated receptors for BMP-2 or BMP-4 (BMP-R1A) phosphorylate Smads 1, 5 and 8. Serine phosphorylation in the C-terminal MH2 domain, in the SxS sequence, causes a structural alteration that allows them to bind to Smad4 (Figure 16.7). Short structural elements in the type I receptor and in the Smads determine the specificity of interaction, the exposed L45 loop in the type I receptor kinase domain interacting with the L3 loop in the MH2 domain of the Smads. Exchanging residues in these domains is sufficient to switch the signalling specificity of the TGFβ and BMP pathways.[23,24]

The receptor-regulated Smads induce receptor-specific responses and they are for this reason also referred to as 'pathway-restricted' Smads. This is illustrated by over-expression of Smad1 in *Xenopus*. Here it induces ventral mesoderm formation (a typical BMP response), whereas over-expression of Smad2 induces formation of dorsal mesoderm (a typical activin response) (Figure 16.8).[23] Mutation of the serine in the SxS motif inhibits receptor-mediated phosphorylation of Smads1 and 2 and prevents their association with Smad4, their accumulation in the nucleus and interaction with DNA binding proteins. The *Drosophila* homologues are MAD and dSmad2 and the *C. elegans* homologues are Sma-2 and 3.

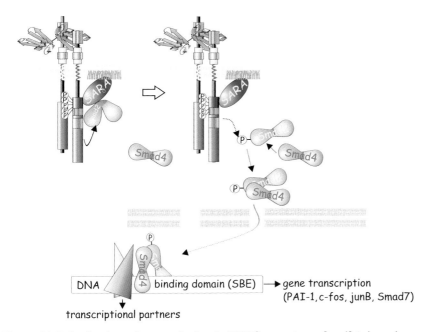

DNA /\ binding domain (SBE) → gene transcription
(PAI-1, c-fos, junB, Smad7)

transcriptional partners

Figure 16.7 Activation of transcription by TGFβ receptors. Smad2 is bound to the type I receptor in association with the Smad anchor for receptor activation (SARA). Upon activation of the receptor complex, Smad2 becomes phosphorylated, detaches from the receptor and associates with the common mediator Smad4. Together they translocate to the nucleus and bind, with their MH1 domains, to DNA at the Smad binding element (SBE). Binding also occurs through interaction of Smads with transcriptional partners (co-factors).

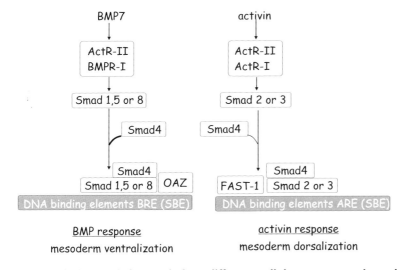

Figure 16.8 TGFβ growth factors induce different cellular responses through the recruitment of different combinations of Smads and transcriptional partners. The BMP response in *X. laevis* is due to the combination of Smads 1, 5 or 8 interacting with the transcriptional partner OAZ resulting in mesoderm ventralization. In contrast, activin recruits Smads 2 or 3 and together with Smad4 and FAST-1, as transcriptional partner, induces mesoderm dorsalization.

Common mediator Smad 4

This is required to form the hetero-oligomer complexes with the receptor-regulated Smads but, since it lacks the consensus substrate sequence (SXS), it is not phosphorylated by any of the receptors. In mammalian cells it binds to phosphorylated Smad1, 2, 3, 5 and 8 and forms complexes that translocate to the nucleus, there to act as transcription factors. The *Drosphila* and *C. elegans* homologues are Medea and Sma-4 respectively.

Antagonistic Smads (Smad6 and Smad7)

Lacking the C-terminal SxS phosphorylation site and possessing only a distantly related MH1 domain, these Smads diverge structurally from other members of the family. When bound to TβR-I they therefore prevent phosphorylation of Smads 1, 2, 3, 5, and 8. Expression of these antagonistic Smads offers a means of inhibiting the cellular responses of members of the TGFβ-family of cytokines. Since expression of Smad7 is induced by TGFβ1 it may form a part of a negative feedback loop (Figure 16.9).[25]

Receptor-binding proteins involved in modulation of receptor function

As with many other receptors (for example, the β$_2$-adrenergic receptor: see Chapter 3), over-expression of the receptors for TGFβ causes constitu-

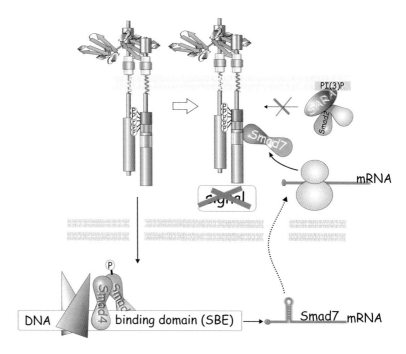

Figure 16.9 Induction of an inhibitor Smad. One of the transcripts induced by TGFβ1 is the mRNA coding for Smad7. The Smad7 protein associates with the type I receptor and prevents access of Smad2 and SARA. Downstream signalling is thus attenuated by a negative feedback loop.

tive activation. In this case, however, it is essential to over-express both TβR-I and TβR-II together in order to ensure the formation of the appropriate hetero-oligomers. It is believed the tendency to form such oligomers in normal cells is prevented by the presence of FKBP12, a small protein that binds to the TβR-I and impedes its phosphorylation by TβR-II. When TGFβ1 is bound to TβR-I, FKBP12 cannot associate and signalling can proceed[26] (Figure 16.4). SARA, another protein that binds to the TGFβ receptors, plays a role in the anchoring of Smads (in particular Smad2 and 3) in their non-phosphorylated state (Figure 16.7). It contains a FYVE finger domain that may serve to locate the protein at the membrane through interaction with PI(3)P.

Activation of the receptor and subsequent phosphorylation of Smad2 or 3 results in their detachment from SARA and the formation of complexes with Smad4. Mutation in SARA, the anchoring protein, causes misdirection of Smad2, preventing its association with the receptor, and inhibition of the TGFβ-dependent transcriptional responses. This indicates that Smad localization in the vicinity of the receptor is important for TGFβ signalling. Thus although Smads themselves bind to the type I receptor, through the L3 loop in the MH2 domain, SARA increases the efficiency, as well as the selectivity, of receptor signalling by favouring the phosphorylation of Smad2 or 3 and preventing unwanted cross-talk with other pathways.[26]

The **FYVE finger** (named after the first letter of four proteins found to contain it; Fab1p, YOTB, Vac1p and EEA1) is a double-zinc binding domain that is conserved in more than 30 proteins from yeast to mammals. It is found in several proteins involved in intracellular traffic.

■ Transcriptional regulation by Smads

Smad complexes can bind directly to DNA at the SBE sequence CAGAC, but optimal induction of transcription requires their association with other factors or 'transcriptional partners' (Figure 16.7). These bind to the Smads and to DNA at a separate binding element and strengthen the attachment to DNA.[22] Different transcriptional partners are involved when different types of receptor are activated. For instance, the transcription factor OAZ[27] binds selectively to the BMP-activated Smad1/Smad4 complex and plays a role in the ventralization of mesoderm, whereas FAST binds to complexes of Smad2 or 3 with Smad4, activated by activin (Figure 16.8) and promotes dorsalization.[28]

The association of additional partners may also depend on the presence of other extracellular signals and the specificity of signalling is determined through integration of different signals at the level of activation of transcription factors. For instance, one action of TGFβ is in the modification of the cell matrix through the induction of proteases and protease inhibitors. Induction of the plasminogen activator inhibitor-1 (PAI-1) requires a TRE site that overlaps with the SBE in the promoter for efficient signalling (see Fig 9.13, page 209). The Smads do not bind at this site but cooperate with members of the Jun and Fos family of transcription factors (AP-1), which do.[29] Transcriptionally active Jun and Fos must first be activated through the ERK or JNK pathways and thus require the presence of suitable stimuli like serum or one of its growth factors (see Chapter 9, page 208 and Chapter 11, page 271).

OAZ Olf-1 associated zinc finger protein, a transcriptional partner initially identified in olfactory epithelium.
FAST Forkhead activating signal transducer-1, a member of the winged-helix family of putative transcription factors, an early immediate gene in the signalling of activin.

Another action of TGFβ is the slowing down of the cell cycle in epithelial cells,[30] mediated through the induction of the cell cycle inhibitors p15^{INK4B} and p21$^{CIP/WAF}$. Both of these are transcriptionally regulated through the action of Smads.

■ Role of Smads in tumour suppression

Given the important role of TGFβ in the suppression of cell proliferation, the terminal differentiation of haematopoietic cells and the activation of cell death mechanisms, it is perhaps not surprising that mutations in the components of the signal transduction pathway can increase susceptibility to aberrant cell proliferation. Together with other mutations that favour proliferation, this may ultimately result in the formation of tumours. Interestingly, the common mediator Smad4 was initially identified as one of the mutated or deleted genes linked to pancreatic and other carcinomas.[31] Most of these mutations are present in the MH2 domain, which contains the phosphorylation site. They either prevent the trimerization of Smad4, decrease the stability of the protein or prevent its interaction with non-Smad transcriptional partners. A small group of mutations are in the MH1 domain and these prevent interaction with Smad2. There are also hereditary and somatic forms of colorectal cancer in which the TβR-II receptor is mutated and the cells therefore lack their normal growth-inhibitory signal mechanism.

■ References

1 Kingsley, D.M. The TGF-β superfamily: new members, new receptors, and new genetic tests of function in different organisms. *Genes Dev.* 1994; 8: 133–46.
2 Mathews, L.S., Vale, W.W. Expression cloning of an activin receptor, a predicted transmembrane serine kinase. *Cell* 1991; 65: 973–82.
3 Bassing, C.H., Yingling, J.M., Howe, D.J. *et al.* A transforming growth factor β type I receptor that signals to activate gene expression. *Science* 1994; 263: 87–9.
4 Lin, H.Y., Wang, X.-F., Ng-Eaton, E., Weinberg, R.A., Lodish, H.F. Expression cloning of the TGF-β type II receptor, a functional transmembrane serine-threonine kinase. *Cell* 1992; 68: 775–85.
5 ten-Dijke, P., Yamashita, H., Ichijo, H., Franzen, P., Laiho, M., Miyazono K., Heldin, C.H. Characterization of type I receptors for transforming growth factor-β and activin. *Science* 1994; 264: 101–4.
6 Luo, K., Lodish, H.F. Signaling by chimeric erythropoietin-TGF-β receptors: homodimerization of the cytoplasmic domain of the type I TGF-β receptor and heterodimerization with the type II receptor are both required for intracellular signal transduction. *EMBO J.* 1996; 15: 4485–96.
7 Liu, F., Hata, A., Baker, J.C. *et al.* A human Mad protein acting as a BMP-regulated transcriptional activator. *Nature* 1996; 381: 620–3.
8 Wang, X.-F., Lin, H.Y., Ng, E.E. *et al.* Expression cloning and characterization of the TGF-β type III receptor. *Cell* 1991; 67: 797–805.
9 Ling, N., Ying, S.Y., Ueno, N. *et al.* Pituitary FSH is released by a heterodimer of the β-subunits from the two forms of inhibin. *Nature* 1986; 321: 779–82.
10 Lewis, K.A., Gray, P.C., Blount, A.L. *et al.* Betaglycan binds inhibin and can mediate functional antagonism of activin signalling. *Nature* 2000; 404: 411–14.
11 Cheifetz, S., Bellon, T., Cales, C. *et al.* Endoglin is a component of the transforming growth factor-β receptor system in human endothelial cells. *J. Biol. Chem.* 1992; 267: 19027–30.

12 Cheifetz, S., Massagué, J. Isoform-specific transforming growth factor-β binding proteins with membrane attachments sensitive to phosphatidylinositol-specific phospholipase C. *J. Biol. Chem.* 1991; 266: 20767–72.

13 McDonald, N., Hendrickson, W.A. A structural superfamily of growth factor containing a cystine knot motif. *Cell* 1993; 73: 421–4.

14 Spencer, F.A., Hoffmann, F.M., Gelbart, W.M. Decapentaplegic: a gene complex affecting morphogenesis in *Drosophila melanogaster*. *Cell* 1982; 28: 451–61.

15 Raftery, L.A., Twombly, V., Wharton, K., Gelbart, W.M. Genetic screens to identify elements of the decapentaplegic signaling pathway in *Drosophila*. *Genetics* 1995; 139: 241–54.

16 Newfeld, S.J., Chartoff, E.H., Graff, J.M., Melton, D.A., Gelbart, W.M. Mothers against dpp encodes a conserved cytoplasmic protein required in DPP/TGF-β responsive cells. *Development* 1996; 122: 2099–108.

17 Estevez, M., Attisano, L., Wrana, J.L. *et al.* The daf-4 gene encodes a bone morphogenetic protein receptor controlling *C. elegans* dauer larva development. *Nature* 1993; 365: 644–9.

18 Georgi, L.L., Albert, P.S., Riddle, D.L. daf-1, a *C. elegans* gene controlling dauer larva development, encodes a novel receptor protein kinase. *Cell* 1990; 61: 635–45.

19 Savage, C., Das, P., Finelli, A.L. *et al. Caenorhabditis elegans* genes sma-2, sma-3, and sma-4 define a conserved family of transforming growth factor β pathway components. *Proc. Natl. Acad. Sci. USA* 1996; 93: 790–4.

20 Heldin, C.H., Miyazono, K., ten-Dijke, P. TGF-β signalling from cell membrane to nucleus through SMAD proteins. *Nature* 1997; 390: 465–71.

21 Kretzschmar, M., Massagué, J. SMADs: mediators and regulators of TGF-β signaling. *Curr. Opin. Genet. Dev.* 1998; 8: 103–11.

22 Massagué, J., Wotton, D. Transcriptional control by the TGF-/Smad signaling system. *EMBO J.* 2000; 19: 1745–54.

23 Chen, Y.G., Hata, A., Lo, R.S. *et al.* Determinants of specificity in TGF-β signal transduction. *Genes Dev.* 1998; 12: 2144–52.

24 Feng, X.H., Derynck, R. A kinase subdomain of transforming growth factor-β (TGF-β) type I receptor determines the TGF-β intracellular signaling specificity. *EMBO J.* 1997; 16: 3912–23.

25 Nakao, A., Afrakhte, M., Moren, A. *et al.* Identification of Smad7, a TGFβ -inducible antagonist of TGF-β signalling. *Nature* 1997; 389: 631–5.

26 Tsukazaki, T., Chiang, T.A., Davison, A.F., Attisano, L., Wrana, J.L. SARA, a FYVE domain protein that recruits Smad2 to the TGFβ receptor. *Cell* 1998; 95: 779–91.

27 Hata, A., Seoane, J., Lagna, G. *et al.* OAZ uses distinct DNA- and protein-binding zinc fingers in separate BMP-Smad and Olf signaling pathways. *Cell* 2000; 100: 229–40.

28 Chen, X., Rubock, M.J., Whitman, M. A transcriptional partner for MAD proteins in TGF-β signalling. *Nature* 1996; 383: 691–6.

29 Zhang, Y., Feng, X.H., Derynck, R. Smad3 and Smad4 cooperate with c-Jun/c-Fos to mediate TGF-β-induced transcription. *Nature* 1998; 394: 909–13.

30 Kramer, I.M., Patel, R., Spargo, D., Riley, P. Initiation of growth inhibition by TGF β1 is unlikely to occur in G1. *J. Cell. Sci.* 1994; 107: 3469–75.

31 Hahn, S.A., Schutte, M., Hoque, A.T. *et al.* DPC4, a candidate tumor suppressor gene at human chromosome 18q21. *Science* 1996; 271: 350–3.

Protein dephosphorylation and protein phosphorylation

■ The importance of dephosphorylation

The phosphorylation of proteins, at serine, threonine or tyrosine residues, serves multiple roles in the regulation of cell function. However, this is only half the story. If the transfer of phosphate groups to proteins is to serve as a precise and sensitive signalling mechanism, then necessarily it must operate against a low background. Dephosphorylation is thus as important as phosphorylation, and it follows that the phosphoprotein phosphatases are integral components of the signalling systems operated by protein kinases.[1] In a number of cases dephosphorylation serves as a true reset button, bringing proteins back to their resting state. A good example is the role of the serine/threonine phosphatase PP1G which dephosphorylates phosphorylase *a*, thereby terminating the breakdown of glycogen. There are other proteins (for example, glycogen synthase, Src, c-Jun, p56[Lck], NF-AT) that are phosphorylated under 'resting' conditions and then become active as a consequence of dephosphorylation (Figure 17.1). In particular, the transcription factor c-Jun requires both dephosphorylation of serine/threonine residues near the DNA binding site, and phosphorylation of serines at the N-terminal region in order to be fully active.

■ Protein tyrosine phosphatases

A soluble protein phosphatase specific for phosphotyrosines (PTP1B) was first isolated from human placenta.[2] Its amino acid sequence has stretches homologous with the tandem repeat domains present in the cytoplasmic portion of the leukocyte common antigen CD45.[3] This is a receptor-like protein expressed on the surface of cells of haematopoietic lineage. Using the DNA sequence that codes for the catalytic domain as a probe (the conserved 'signature motif' [I/V]HCXAGXXR[S/T]), many more protein tyrosine phosphatases were revealed.

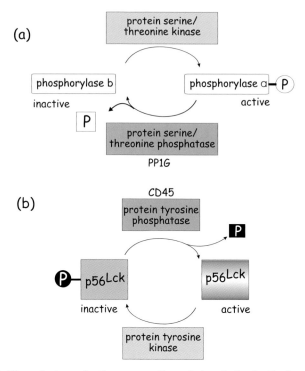

Figure 17.1 Phosphatases in the generation of signals for both deactivation and activation. (a) Protein phosphatases in the role of a 'reset button', deactivating their substrate. An example is the dephosphorylation of phosphorylase *a* by the serine/threonine phosphatase PP1G. (b) Protein phosphatase as activator. Illustrated is the example of activation of p56^Lck by CD45, a receptor-like tyrosine protein phosphatase.

Cloning data show the protein tyrosine phosphatases to be a family of multi-domain proteins having exceptional diversity (Figure 17.2).[4] They can be broadly divided into two groups, the transmembrane or receptor-like PTPs and the cytosolic PTPs. None of these are related to the serine-threonine-specific phosphatases. This is in contrast to the protein kinases (serine-threonine and tyrosine), which share common ancestry. Unlike the serine-threonine phosphatases, in which substrate specificity is determined by associated targeting subunits, the tyrosine phosphatases are all monomeric enzymes. How the diversity in structure reflects differences in substrate recognition remains unclear.

■ Transmembrane receptor-like PTPs

Nearly all the transmembrane PTPs are characterized by the presence of tandem repeats D1 and D2, expressing the catalytic signature motif. However, only the membrane proximal D1 domains are catalytically active. The transmembrane PTPs are classified on the basis of their extracellular segments (ectodomains). They range from very short chains having no clear function, to extended structures with putative ligand-binding domains, similar to those present in adhesion molecules (fibronectin

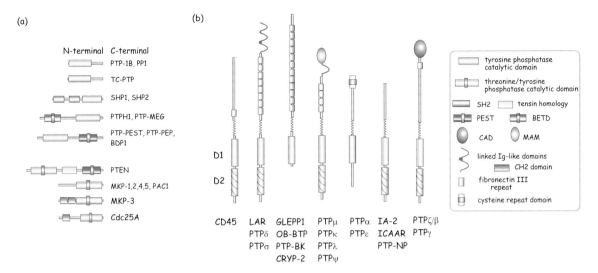

Figure 17.2 **Domain organization of tyrosine phosphatases:** (a) Cytosolic protein tyrosine phosphatases. These divide into two groups, those identified as tyrosine phosphatases and the dual-specific phosphatases, dephosphorylating serine, threonine and tyrosine residues. They are further subdivided by the presence of various homology domains, such as SH2 or PEST. (b) Receptor-like protein tyrosine phosphatases. All contain a tyrosine protein phosphatase motif, often in tandem, and are distinguished by differences in their extracellular domains. Some have elaborate extracellular structures resembling adhesion molecules or growth factor receptors, others appear rather sparse and ligand association is hard to imagine. No ligands have been identified for most of these 'receptor-like' phosphatases.

BETD	Band 4.1/ezrin/talin-homologous domains.	MAM	Meprin/A5/mu domain.
CAD	Carbonic anhydrase domain.	PEP	Proline, glutamate, serine and threonine-enriched protein tyrosine phosphatase.
CH2	Cdc25 homology domain-2.		
CRD	Cysteine repeat domain.	PEST	Proline, glutamate serine and threonine rich region.
CRYP	Chicken receptor tyrosine phosphatase.		
GLEPP	Glomerula epithelial protein 1.	PTEN	Phosphatase and tensin homologue deleted from chromosome 10.
IA	Islet cell auto-antigen.		
ICAAR	Islet cell auto-antigen receptor form.	PTP	Protein tyrosine phosphatase.
LAR	Leukocyte common antigen.	PTP-NPP	PTP-neuronal and pancreatic.
		STEP	Striatal enriched phosphatase.

repeats or immunoglobulin repeats) (Figure 17.2). Such diversity suggests a wide range of biological functions, which have, however, proved hard to pin down. This is largely because identification of their physiological substrates has been hampered by the non-specificity of these enzymes when assayed for activity *in vitro*. Moreover, although some are predicted to be receptors, the ligands have so far been elusive. PTPµ and PTPκ (see Figure 17.2) may take part in homotypic adhesion through their MAM domains.[5,6] When expressed in insect cells they induce Ca²⁺-independent cell aggregation, but even here the downstream pathways remain unclear.

■ Cytosolic PTPs

The cytosolic PTPs are also classified according to their domain structures which are understood to act as localization signals, directing the enzymes to the nucleus or cytoskeleton. An important subclass, SHP-1 and SHP-2, possess SH2 domains. Others are characterized by the presence of PEST sequences (Pro-Glu/Asp-Ser/Thr) in the vicinity of the C-terminus. Another subclass comprises dual-specific phosphatases, which can dephosphorylate at both tyrosine and threonine/serine residues. Since the first of these to be identified was VH1, encoded by vaccinia virus, they are also referred to as VH1-like phosphatases.[7] The dual-specific phosphatases also have homology with Cdc25, a regulator of mitosis in fission yeast (*Schizosaccharomyces pombe*). This activates cyclin-dependent kinase-2 (Cdc2/CDK1, see Chapter 10) by dephosphorylation of adjacent threonine and tyrosine residues.[8]

■ The role of PTPs in signal transduction

At first, interest in protein tyrosine phosphatases was driven by the perception that they might be antitumorigenic. They appeared to offer the possibility of counteracting the transforming effects of mutated, and therefore constitutively activated, protein tyrosine kinases. Thus it came as quite a surprise when it was found that some PTPs, rather than opposing the actions of the kinases, actually cooperate with them to reinforce their signals.[9] Several PTPs have now been described, many of them first cloned from haematopoietic cells.

■ Positive regulation through phosphotyrosine dephosphorylation

■ *CD45*

As with many other receptor-like PTPs, no specific activating ligand has been identified for CD45 with any certainty. Its importance became evident with the discovery that T lymphocytes lacking this antigen fail to become activated through the TCR, although they respond quite normally to stimulation through the IL-2 receptor (Figure 12.2, page 286). Normally, engagement of the TCR results in activation of the soluble protein tyrosine kinase Lck, resulting in phosphorylation of the ζ-chains. In cells lacking active CD45, Lck remains inactive and the subsequent tyrosine phosphorylations do not occur (see Chapter 12).

Where does the phosphatase activity of CD45 come into all this? The inactive (resting) state of Lck is maintained as a consequence of autoinhibition, through the intramolecular interaction of its own SH2 domain with a phosphotyrosine residue situated in its C-terminus (a 'closed conformation': Figure 17.3). Lck is activated first through dephosphorylation (Y505) by CD45, then, because this exposes the kinase active site, through autophosphorylation (Y394) in the catalytic domain.[10] The fully active Lck phosphorylates the ζ-chains of the engaged TCR. From this, it appears that CD45 must somehow be activated when the TCR binds to an MHC molecule on an antigen-presenting cell. Similar mechanisms of activation through dephosphorylation apply in the action of Fyn (associated with the ε-chain of CD3: see Chapter 12) and Src (dephosphorylation at Y527).[11]

How the phosphatase activity of CD45 is regulated remains unclear. Importantly, in cells expressing a truncated CD45, lacking the extracellular domain, Fyn and Lck are constitutively activated. With respect to regulation of phosphatase activity, there are some indications that, in contrast with the PTKs, dimerization may be the 'off' signal (note that in Figure 17.3 a ligand is indicated holding the two components of CD45 apart). In order to test this idea, cells have been generated that express a chimaeric molecule, comprising the intracellular phosphatase-containing portion of CD45 and the extracellular domain of the EGF receptor. Application of EGF, expected to dimerize the receptor, switches off the phosphatase activity.[12]

SHP-2

SHP-2 (Src homology phosphatase-2, initially named SH-PTP2 or Syp) has a broad tissue distribution. Targeted disruption of the gene in mice causes early embryonic death, but from this, other than concluding that it is important, little can be learned. Early indications about a possible role of SHP-2 in cell signalling came from two lines of evidence. First, the gene is homologous to *Drosophila* Corkscrew (Csw, a tyrosine phosphatase). This is involved in activation of the serine-threonine kinase Draf (equivalent to mammalian Raf) (Figure 17.4). Draf is activated through the Dras pathway and this implies that the phosphatase provides a positive signal downstream of receptor tyrosine kinase activation. The Corkscrew pathway directs the differentiation of the terminal, non-segmented regions of the fly embryo.[13]

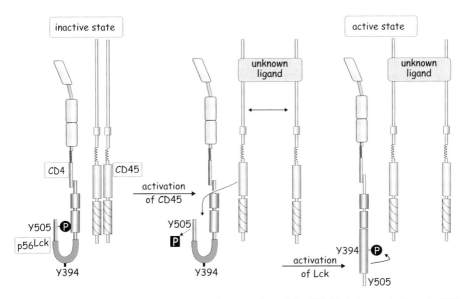

Figure 17.3 Activation of Lck by dephosphorylation mediated by CD45. Lck, attached to the TCR subunit CD4, is inactive as a result of the linking of its phosphotyrosine residue (505) with its own SH2 domain. Activation of CD45 by an unknown ligand (probably separating the dimerized protein), results in dephosphorylation of Y505. The Lck kinase catalytic domain unfolds, autophosphorylates (Y394) and is now able to phosphorylate substrates such as the ITAM domains in the ζ-subunits of the TCR.

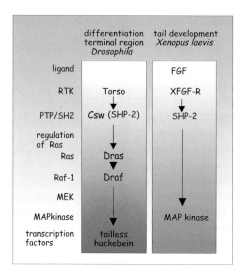

Figure 17.4 Tyrosine phosphatases in parallel pathways regulating transcription in *Xenopus* and *Drosophila*. Genetic analysis of phenotypes defective in differentiation of the terminal region of *Drosophila* or tail development in *Xenopus* provided a key to understand the role of SHP-2 in signalling downstream of protein tyrosine phosphatases.

Secondly, expression in *Xenopus* embryos of a dominant negative mutant form of SHP-2, lacking the catalytic domain, causes tail truncations and prevents animal cap elongation (*short toad*, composed mainly of head structures).[14] These processes are determined in part by the receptor for FGF. This also signals through the MAP kinase pathway and so the phosphatase SHP-2 must act between the FGF-R and the activation of MAP kinase. Similarly, fibroblasts expressing the catalytically inactive form of SHP-2 fail to activate MAP kinase in response to FGF, PDGF or insulin-like growth factor (IGF).[15] Although the substrate of SHP-2 has yet to be identified, it has become apparent that SHP-2 also functions as an adaptor.[16] In this way it can promote the activation of ERK with no involvement of its phosphatase activity (Figure 17.5). Following phosphorylation of SHP-2 by the PDGF receptor, it acts as a site for attachment for Grb2 and so sets in train the reactions of the Ras-MAP kinase pathway. The inability of the dominant negative mutants to activate the MAPK pathway can be understood if SHP-2 is required for both pathways and that transmission of signals down both pathways is required for an effective response. The loss of either the adaptor function or protein phosphatase activity results in insufficient activation, leading to the altered phenotypes.

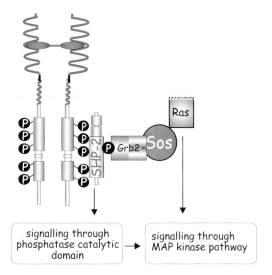

Figure 17.5 **SHP-2, protein tyrosine phosphatase and adaptor molecule.**
SHP-2 transmits two signals. It binds and is phosphorylated by the activated PDGF
receptor. It acts as a phosphatase, although specific ligands remain to be found. In addition
it acts as an adaptor, linking the receptor with Ras. Both pathways must be intact in
order to obtain a satisfactory signal for the activation of MAP kinase.

Negative regulation through dephosphorylation

MKP-1, dual-specific protein phosphatases and regulation of MAP kinase by
tyrosine protein phosphatases

A failure to dephosphorylate may have serious consequences, equivalent
to persistent phosphorylation by oncogenic PTKs. The substrates of the
dual-specific PTPases are predominantly members of the MAP kinase
family, which are phosphorylated in the TEY sequence by dual-specific
kinases such as the MEKs (see Chapter 11). One such PTPase, MKP-1
(MAP kinase phosphatase-1) is the product of a growth-factor-induced
early response gene.[17,18]

Addition of serum to quiescent cells induces rapid phosphorylation
and activation of MAP kinase. This is only transient, but it is sufficient to
induce gene transcription and to promote entry into the G_1 phase of the
cell cycle.[19] The activity of the dual-specific phosphatase is expressed
within 20 minutes and coincides with the dephosphorylation of the TEY
motif and the deactivation of MAP kinase (see Chapter 11). If synthesis of
MKP-1 is prevented, or if its catalytic activity is mutated, the duration of
the activated state of MAP kinase is greatly extended.[20,21] Conversely,
expression of a constitutively activated form of MKP-1 blocks G_1-specific
transcription and entry into S-phase.[20,21] Thus MKP-1 is a negative regulator of MAP kinase, serving to attenuate the growth factor signal (Figure
17.6). In addition, MKP-1 is a substrate of MAP kinase. When phosphorylated, MKP-1 becomes less sensitive to ubiquitin-directed proteolysis
and, as a result, it accumulates.[22]

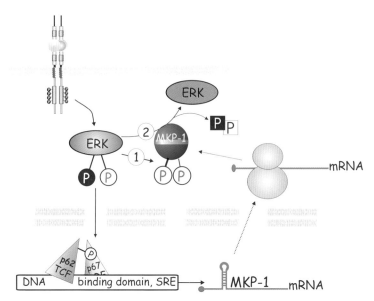

Figure 17.6 MKP-1, negative regulator of MAP kinases. One of the immediate early genes expressed following activation of ERK is the dual-specific protein phosphatase MKP-1. After translation, it becomes phosphorylated by ERK and in turn dephosphorylates and deactivates ERK, thereby terminating the growth factor signal.

There are indications of other mechanisms that restrain the signals from growth factor receptors. First, none of the mutations in MKP-1 predispose to tumour development in humans. If this pathway was absolutely essential, then one might expect that a tumorigenic mutation would have turned up by now. Secondly, MKP-1 deficient mice appear to develop normally. Their fibroblasts respond normally with respect to the extent and timing of the expression of c-*fos*, indicating that the control of MAP kinase is unperturbed.[23] Several MKPs have been cloned and, although they appear to have substrate-specificity (for instance, MKP-5 dephosphorylates JNK/SAPK and p38/HOG, but is an inefficient phosphatase for ERK), there may be redundancy of function.

■   SHP-1, Epo-R, STAT5 and JAK2

A role for the protein tyrosine phosphatase SHP-1 was first indicated in mice that had a somewhat moth-eaten appearance, with patches of inflamed skin and hair loss.[24] The phenotype is due to a systemic autoimmune condition caused by abnormalities in multiple cells of haemapoietic lineage, marked by a general over-expansion of cell numbers. These mice die within weeks of birth. SHP-1 has a down-regulatory effect on a number of signal transduction pathways, all related to the proliferation and differentiation of haemapoietic cells.

The receptor for erythropoietin, Epo-R, is a member of the family of haematopoietic growth factor receptors (Table 17.1), all characterized by a common binding domain and the absence of a cytoplasmic tyrosine

Table 17.1 Haematopoietic growth factor receptors

Stem cell factor receptor (c-kit)
(also known as the receptor for Steel-locus factor (SLF), mast cell growth factor and
Kit-ligand)

IL-2R, IL-3R

T-cell receptor (TCR)

Granulocyte macrophage-colony stimulating factor receptor (GM-SCF-R)

B-cell receptor (BCR)

Erythropoietin receptor (Epo-R),

Interferon α/β receptor (INF-α/β-R)

Colony stimulating factor-1 receptor (CSF-1R)

kinase domain. In order to transmit signals, these receptors have to recruit protein kinase activity. Following ligand-induced dimerization, Epo-R becomes phosphorylated by the action of the PTK JAK2, so generating a docking site for the transcription activator STAT5 (see Chapter 11). On docking, STAT5 is itself phosphorylated by the receptor-associated JAK2 and it then abandons the receptor to form a homodimer which translocates to the nucleus (Figure 17.7) (see also Chapter 12, page 291.

The phosphatase action of SHP-1 serves to attenuate the intracellular signal. The SH2 domain of SHP-1 binds to a phosphotyrosine residue (Y429) on Epo-R and this renders it susceptible to phosphorylation by JAK2. The phosphorylated, and hence activated, SHP-1 is thereby enabled to dephosphorylate and hence inactivate both the kinase JAK2 and the transcription activator STAT5. In this way it prevents the undocking, dimerization and translocation of STAT5 to the nucleus. In cells expressing an Epo-R mutant that lacks the Y429 site, SHP-1 is unable to bind to the Epo-receptor but STAT5 still can and, in consequence, Epo-induced STAT5 activation is greatly prolonged.[25]

Through its action as a negative regulator of transcription mediated by STAT5, SHP-1 serves an important role controlling the proliferation of haemopoietic progenitor cells. When its activity is suppressed there is a sustained level of tyrosine phosphorylation and this is coupled to enhanced cell proliferation. Conversely, its over-expression is coupled to suppression of cell growth.

▉ PTEN, a phosphatase for phosphoproteins and phospholipids

As indicated above, interest in protein tyrosine phosphatases was sparked off in the hope they might act as tumour suppressors. However, of the many tyrosine protein phosphatases investigated, none were found to exert such a role, until the discovery of PTEN (phosphatase and tensin homologue deleted from chromosome-10). This provides the best example of a gene coding for a PTP in which loss of function is linked with tumour development.[26,27] It plays an essential role in maintaining the

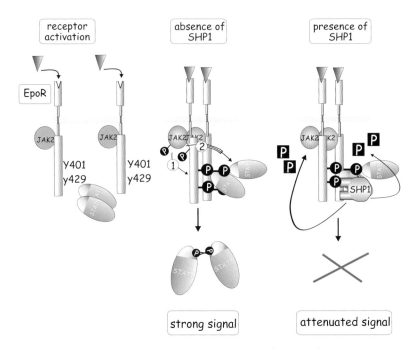

Figure 17.7 JAKs and STATs in the transduction of signals from haematopoietic growth factor receptors. Binding of Epo to its receptor causes dimerization and interphosphorylation of the associated cytosolic JAK2 tyrosine protein kinases. These phosphorylate the Epo receptors and generate docking sites for the SH2 domain of the STAT5 proteins. These become phosphorylated by JAK2, detach, dimerize and translocate to the nucleus to stimulate transcription. SHP-1 also binds to the tyrosine-phosphate docking sites and dephosphorylates both JAK2 and STAT5. As a consequence, the transcription factors cannot dimerize and remain in the vicinity of the membrane. The signal is attenuated.

balance between cell survival and cell proliferation (Figure 17.8). Thus it came as some surprise to find that it does so, not by controlling the level of protein tyrosine phosphorylation, but through its action as a phosphatase for PI-3,4,5-P3.[28]

PTEN is highly expressed in epithelial cells and it is classified as a tumour suppressor. Inactivating mutations have been detected in glioblastomas, melanomas, breast, prostate and endometrial carcinomas.[29] Ectopic expression in PTEN-deficient tumour cells results in arrest of the cell cycle in the G_1 phase, eventually followed by apoptosis. It also reduces cell migration, a finding that may explain why the loss of the gene product is frequently associated with late-stage metastatic tumours.[30]

The amino acid sequence of PTEN contains a tyrosine phosphatase signature motif (see above) and resembles most closely the dual-specificity phosphatases such as MKP-1. In spite of this, PTEN is generally a poor protein phosphatase, and it catalyses the dephosphorylation of PI-3,4,5-P3 with far greater alacrity.[28] The three-dimensional structure indicates the basis for this preference: its catalytic site, unlike that of other protein tyrosine phosphatases, is large enough to accommodate the

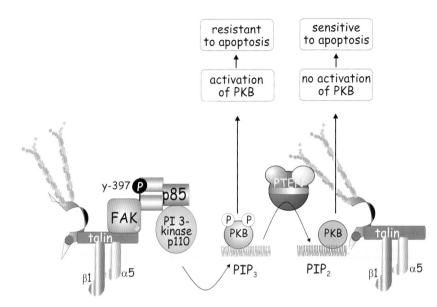

Figure 17.8 PTEN, a tyrosine phosphatase lookalike which dephosphorylates PIP$_3$. Formation of focal adhesion contacts leads to the activation of FAK which autophosphorylates at residue Y-397. This serves as a docking site for the SH2 domain of the p85 regulatory subunit of PI 3-kinase generating products that activate PKB and renders the cell resistant to apoptosis. The reaction is reversed by the lipid kinase PTEN, through dephosphorylation of the 3-phosphorylated inositide.

inositol ring. The presence of positively charged amino acids accounts for its preference for negatively charged substrates, such as the polyphosphoinositide head group.[31] Although protein substrates are not totally excluded (e.g. FAK and the adaptor protein Shc), the tumour-suppressive function of PTEN can be ascribed to its lipid phosphatase activity.[32]

Since PTEN catalyses the dephosphorylation of PI-3,4,5-P3, it is positioned at the head of two well-characterized signalling pathways, the one determining cell survival (signals from focal adhesion complexes), the other leading to cell proliferation (signals from receptor protein tyrosine kinases). In essence, PTEN serves to maintain PI-3,4,5-P3 at a low level. As a result, the recruitment and activation of PH-domain containing enzymes such as PKB and PDK1 are held in check.[33] PKB regulates cell survival through inhibition of activity of Bad and caspase-9 and through inhibition of expression of Fas (Chapter 14). It is also implicated in the regulation of expression of CyclinD1, which drives cells through the G$_1$ phase of the cell cycle. PDK1 regulates the activity, not only of PKB but also of ribosomal p70 S6 kinase, thereby controlling the rate of protein synthesis.

Serine-threonine phosphatases

The PTPs are all monomeric and their differences mainly determined by their domain structures, whereas the serine-threonine phosphatases are oligomeric and characterized by their association with targeting subunits. These direct them to particular locations, thus restricting their action to a limited range of substrates such as the phosphorylated form of myosin light chain or ribosomes. The first serine-threonine phosphatase to be discovered, PP1G, inactivates glycogen phosphorylase.[34,35]

Other serine-threonine phosphatases, initially classified in the four groups PP1, PP2A, PP2B and PP2C, have since been identified. Because they are relatively unspecific in their action, it was thought that this small number of enzymes would be sufficient to balance the effects of the numerous protein kinases.[36] Some functional diversity became apparent with the discovery of inhibitors that affect a restricted range of phosphatases. As examples, okadaic acid, a tumour promoter, is a potent inhibitor of PP1 and PP2A, whereas cyclosporin (used to inactivate T lymphocytes and prevent tissue rejection following organ transplantation) is a selective inhibitor of PP2B (calcineurin).

Real diversity came to light when it was realized that the purified activities only represent the catalytic subunits, but that in the cellular environment these enzymes are coupled with targeting or inhibitory subunits. The subcellular distributions, substrate selectivities and catalytic activities are largely determined by these subunits. For example, PP1 is potentially active against a wide range of peptide and protein substrates and so its activity must be carefully limited. Under normal conditions, it is associated with the glycogen-targeting G-subunit as the heterodimer PP1G, which has very high affinity for glycogen (K_{app} ~6 nmol/l). This ensures that its free form is kept at vanishingly low levels. In the situation of severe glycogen depletion, soluble inhibitor proteins (inhibitor 1, Inh1) ensure that its concentration in the cytosol is kept low (Figure 17.9). In these ways, PP1 is prevented from acting as a loose cannon, randomly dephosphorylating any phosphoprotein that might come in range. The association with particular targeting proteins narrows the range of available substrates. The other enzymes of glycogen metabolism, phosphorylase kinase, phosphorylase and glycogen synthase are also tightly bound to glycogen and all three are good substrates for dephosphorylation by PP1G.

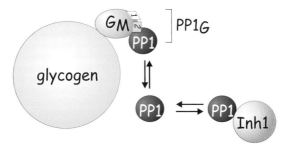

Figure 17.9 Serine phosphatases, targeting subunits and inhibitor proteins. PP1 never exists in a free soluble form. It is either bound to its regulator subunit G_m, associated with glycogen, forming the enzyme complex PP1G. Alternatively it is bound to the inhibitory subunit Inh1 which prevents uncontrolled dephosphorylation of other substrates.

◼ Classification of protein serine-threonine phosphatases

The serine-threonine phosphatases are classified in two superfamilies, PPP and PPM, listed in Table 17.2.[37,38] To date, the PPM comprise the Mg^{2+}-dependent PP2C and mitochondrial pyruvate dehydrogenase phosphatase (PDP). The PPP family is characterized by the presence of three invariant amino acid motifs (-GDxHG- . . . -GDxVxRG- . . . –GNH-) in the catalytic domain. The genome of yeast (*S. cerevisiae*) possesses genes coding for 12 catalytic subunits in the PPP family and it is likely that mammalian genomes will reveal 10 times more. Furthermore, the existence of a large number of targeting subunits will make for a huge number of functionally discrete serine-threonine phosphatases.

Table 17.2 Classification of protein serine-threonine phosphatases and their targeting subunits. The three-letter acronyms adhere to the conventions of the human genome nomenclature in the specification of a family. For the protein phosphatases, PPP indicates phosphoprotein phosphatase, PPM activation by magnesium.

Old	New	
Catalytic subunits (C)		
PPP family		
PP1 α, β, γ	PPP1CA, PPP1CB, PPP1CC	
PP2A α, β	PPP2CA, PPP2CB	Plus subunits either B or C
PP2B α, β, γ	PPP3CA, PPP3CB, PPP3CC	Plus subunits (calmodulin)
PP4	PPPC4	
PP5 (Ppt1)	PPP5C	
PP6	PPP6C	
PPM family		
PP2C α, β	PPM11CA, PPMC1B	Monomer Mg^{2+}-dependent
PDP	PPM2C	Mitochondrial pyruvate dehydrogenase phosphatase
Targeting or regulatory subunits (R)		
PPP1R family		
I-1	PPP1R1A	Inhibitor of PP1 when released from G-targeting subunit
DARPP32	PPP1R1B	Neuronal inhibitor of PP1 (dopamine and adenosine regulated)
PP11-2	PPP1R2	Inhibitor of PP1
G_M	PPP1R3	Targeting PP1 to glycogen and sarcoplasmic reticulum in skeletal muscle
G_L		Targeting PP1 to glycogen in liver
M110		Targeting PP1 to myofibrils in skeletal and smooth muscle
P53BP2		Targeting PP1 to the tumour suppressor nuclear protein p53
NIPP-1		Nuclear inhibitor of PP1

■ The role of PP1 in the regulation of glycogen metabolism

Site-specific phosphorylation of the G-subunit of PP1G enables it to generate the appropriate responses for adrenaline and Ca^{2+} on the one hand (glycogenolysis), and for insulin on the other (glycogen synthesis). In skeletal muscle this is catalysed by PKA at site 2, or by insulin-stimulated protein kinase (ISPK) at site 1 (Figure 17.10).[39] Both sites are located in the N-terminal regulatory domain. Phosphorylation at site 1 (insulin) brings about dephosphorylation of glycogen synthase and phosphorylase kinase. This increases the activity of the synthase and suppresses the activity of the kinase. Conversely, the effect of adrenaline is to cause phosphorylation at site 2, reducing the stability of the PP1G dimer by a factor of 10^4 and releasing the catalytic subunit to the cytosol, where it becomes associated with Inh1. The regulatory G-subunit remains associated with the glycogen. With the phosphatase activity suppressed, the phosphorylation of the glycogen metabolizing enzymes due to PKA persists, so that glycogen synthase is suppressed and phosphorylase kinase is activated. Dephosphorylation of PP1G at site 2, and hence its reactivation, is mediated by PP2A and the Ca^{2+}-dependent PP2B (the reset button)[39] (Figure 17.11).

The mechanisms that control phosphatase (PP1) activity in liver and muscle are different. In liver, the activity of the G_L subunit is not con-

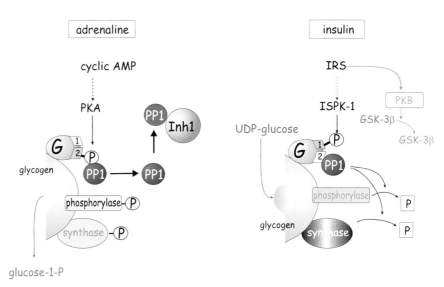

Figure 17.10 Decision-making in glycogen synthesis and breakdown: a balance of phosphorylation and dephosphorylation. (a) Adrenaline and insulin have opposing effects on the phosphatase activity of the PP1G complex. Adrenaline drives glycogenolysis through the activation of PKA, which leads to activation of phosphorylase. In addition, it ensures that phosphorylase remains activated by phosphorylating the regulatory subunit G_m at the 2 position. This causes detachment of PP1 which is sequestrated by Inh1 and renders the phosphatase inactive. PKA also phosphorylates and inactivates glycogen synthase. (b) Insulin, which drives gluconeogenesis, has the reverse effects. Through phosphorylation of IRS-1 it activates ISPK-1 and this phosphorylates the G_m subunit at the 1 position, so enforcing interaction with PP1. The phosphatase activates glycogen synthase and inactivates phosphorylase a.

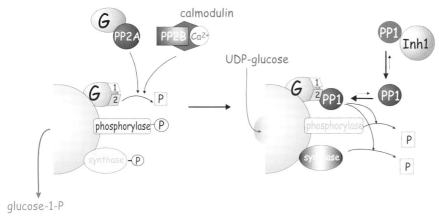

UDP-glucose

glucose-1-P

Figure 17.11 Rebalancing glycogen breakdown and synthesis: PP2A and PP2B.
When adrenaline is removed, glycogen metabolism has to be reset through
dephosphorylation of the site-2 serine of the G_m subunit of PP1G. Reactivation of PP1G
is mediated by the two phosphatases, PP2A and Ca^{2+}-dependent PP2B (calcineurin). In
this condition PP1G again becomes sensitive to adrenaline and insulin. Under these
'resting' conditions, it is likely that PP1 shuttles between G_m and Inh1, giving a low basal
production of glucose-1-P.

trolled by phosphorylation. Instead the inhibition of phosphatase activity,
necessary to suppress the action of glycogen synthase, is exerted by
the phosphorylated (active) form of the target enzyme phosphorylase[39]
(Figure 17.12). This acts as an allosteric inhibitor of PP1G at extremely low

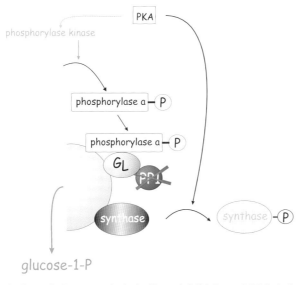

glucose-1-P

Figure 17.12 Regulation of glycogenolysis in liver: inhibition of PP1. Induction of
glycogenolysis differs in liver and muscle. In liver, the activated phosphorylase binds to the
regulatory subunit G_L (distinct from G_M in muscle) and this inhibits the activity of the
phosphatase PP1G. PKA also phosphorylates and inactivates glycogen synthase which
remains phosphorylated because of the inhibition of PP1G. Glycogen breakdown ensues.
Note that the displacement of phosphorylated glycogen synthase is intended to indicate
inhibition, not physical detachment from the glycogen.

concentrations. Again, this provides an effective means for coupling the activation of glycogenolysis to the suppression of glycogen synthesis, and vice versa.

The role of PP2B (calcineurin) in regulation of T-cell proliferation

Although we now recognize that it has a wide tissue distribution,[40] PP2B was originally identified as a calcium binding protein present in neural tissue, hence its alternative name, calcineurin.[41] Only later was it realized that it possesses phosphatase activity and that its regulatory subunit is the Ca^{2+}-binding protein calmodulin (see page 173). The whole assembly consists of three subunits: calcineurin A (the catalytic subunit), calcineurin B (a regulatory calmodulin-like subunit) and calmodulin itself. The B-subunit restricts the range of substrates. Calcineurin can be activated either by elevation of the cytosol concentration of Ca^{2+} or by phosphorylation of calcineurin B, which has the effect of increasing the Ca^{2+}-affinity so ensuring that activation can occur at resting Ca^{2+} concentrations.

Through the use of cyclosporin and FK506, used to suppress T lymphocyte-mediated immune responses, functional roles of calcineurin have come to light (see Chapter 8).[42] The potency and specificity of these drugs offered the possibility of new therapies to counter transplant rejection and to treat autoimmune diseases. Indeed, since its discovery, the use of cyclosporin A has enormously increased the survival of patients receiving kidney, heart and liver transplants. It also finds application in the treatment of asthma.[43] Both drugs prevent clonal expansion of T lymphocytes, predominantly by inhibiting IL-2 expression.[44] However, they bind to distinct intracellular receptors, cyclosporin to cyclophilin and FK506 to FK-binding protein (FKBP), known collectively as the immunophilins[45] (Figure 17.13). These act to prevent the access of calcineurin to NF-AT,[46] maintaining its phosphorylation state and so preventing its translation to the nucleus (Figure 17.14; see also Figure 12.3, page 287).[47] In addition to its well-established role in the transcription of IL-2, resulting in clonal expansion, NF-AT also induces the genes coding for IL-4, GM-CSF and TNFα. In this way, protein dephosphorylation plays a pivotal role in T-cell activation.[48,49]

Three steps of activation have been defined for NF-AT: dephosphorylation, nuclear translocation and increase in affinity for DNA (Figure 17.14).[50] In resting T-cells, NF-AT is phosphorylated on a number of serine residues and confined to the cytosol. The Ca^{2+} signal arising from activation of the T-cell receptor activates calcineurin and the resulting dephosphorylation exposes the nuclear localization sequence (NLS) and the DNA-binding region of NF-AT. Both molecules are transported, as a complex, to the nucleus. Members of the NF-AT family of transcription factors associate with the AP-1 complex (Jun and Fos, activated through the Ras/ERK pathway: see Chapter 11). The tight association of the three proteins on DNA drives the gene expression. The synergy between the

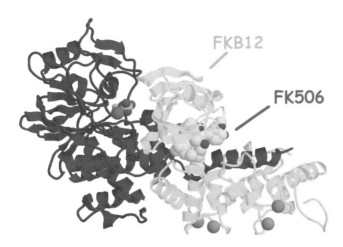

Figure 17.13 Immunophilins inhibit the phosphatase activity of calcineurin.
The drug FK506 binds to its intracellular receptor, the immunophilin FKBP12 (light green chains). The combination is immunosuppressant, binding to both the CnA- (catalytic, blue) and CnB- (lower green structure with four Ca^{2+} ions bound) subunits of the calcineurin heterodimer, inhibiting its action (see Figure 8.6, page 177).

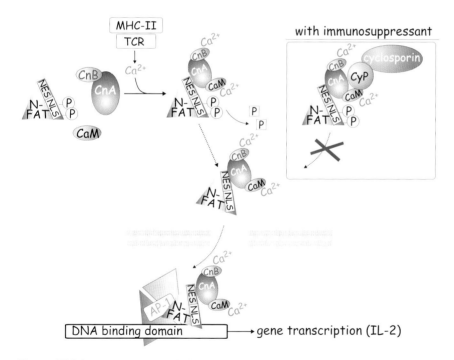

Figure 17.14 Calcineurin (PP2B), regulator of nuclear import of the transcription factor NF-AT. Activation of the TCR causes an increase in intracellular free Ca^{2+}. This induces the construction of the calcineurin complex comprising calmodulin (CaM), a regulatory subunit (CnB) and a phosphatase subunit (CnA). Calcineurin combines with the transcription factor NF-AT and the resulting dephosphorylation exposes the nuclear import signal. The calcineurin/NF-AT complex translocates to the nucleus and associates with a second transcriptionally active complex AP-1 to induce transcription of IL-2.

calcineurin and Ras/ERK pathways is necessary for successful T-cell proliferation.[51] To illustrate this synergy, the combination of constitutively activated Ras together with ionomycin (a Ca^{2+} ionophore) provides an effective stimulus for T cell activation; either stimulus applied alone is without effect.[52]

This pathway can be inhibited by the immunosuppressant drug cyclosporin. It binds cyclophilin (CyP) and the complex occupies the catalytic pocket of calcineurin, preventing access of NF-AT. The transcription factor remains phosphorylated. It cannot induce transcription of IL-2, and in this way it prevents clonal expansion of T lymphocytes.

Calcineurin also appears to play an important role in preventing the export of NF-AT from the nucleus.[53] In the absence of bound calcineurin, activated NF-AT becomes subject to futile cycling across the nuclear envelope. When intracellular free Ca^{2+} falls and calcineurin dissociates from the transcription factor, export prevails, thereby terminating the T-cell receptor signal.

References

1 Sun, H., Tonks, N.K. The coordinated action of protein tyrosine phosphatases and kinases in cell signaling. *Trends Biochem. Sci.* 1994; 19: 480–5.

2 Tonks, N.K., Diltz, C.D., Fischer, E.H. Purification of the major protein-tyrosine-phosphatases of human placenta. *J. Biol. Chem.* 1988; 263: 6722–30.

3 Charbonneau, H., Tonks, N.K., Kumar, S. *et al.* Human placenta protein-tyrosine-phosphatase: amino acid sequence and relationship to a family of receptor-like proteins. *Proc. Natl. Acad. Sci. USA* 1989; 86: 5252–6.

4 Fischer, E.H., Charbonneau, H., Tonks, N.K. Protein tyrosine phosphatases: a diverse family of intracellular and transmembrane enzymes. *Science* 1991; 253: 401–6.

5 Sap, J., Jiang, Y.P., Friedlander, D., Grumet, M., Schlessinger, J. Receptor tyrosine phosphatase R-PTP-κ mediates homophilic binding. *Mol. Cell Biol.* 1994; 14: 1–9.

6 Brady-Kalnay, S.M., Tonks, N.K. Identification of the homophilic binding site of the receptor protein tyrosine phosphatase PTPμ. *J. Biol. Chem.* 1994; 269: 28472–7.

7 Guan, K.L., Broyles, S.S., Dixon, J.E. A Tyr/Ser protein phosphatase encoded by vaccinia virus. *Nature* 1991; 350: 359–62.

8 Gautier, J., Solomon, M.J., Booher, R.N., Bazan, J.F., Kirschner, M.W. cdc25 is a specific tyrosine phosphatase that directly activates p34cdc2. *Cell* 1991; 67: 197–211.

9 Zheng, X.M., Wang, Y., Pallen, C.J. Cell transformation and activation of pp60c-src by overexpression of a protein tyrosine phosphatase. *Nature* 1992; 359: 336–9.

10 Yamaguchi, H., Hendrickson, W.A. Structural basis for activation of human lymphocyte kinase Lck upon tyrosine phosphorylation. *Nature* 1996; 384: 484–9.

11 Jove, R., Hanafusa, T., Hamaguchi, M., Hanafusa, H. In vivo phosphorylation states and kinase activities of transforming p60c-src mutants. *Oncogene Res.* 1989; 5: 49–60.

12 Desai, D.M., Sap, J., Schlessinger, J., Weiss, A. Ligand-mediated negative regulation of a chimeric transmembrane receptor tyrosine phosphatase. *Cell* 1993; 73: 541–54.

13 Perkins, L.A., Larsen, I., Perrimon, N. Corkscrew encodes a putative protein tyrosine phosphatase that functions to transduce the terminal signal from the receptor tyrosine kinase torso. *Cell* 1992; 70: 225–36.

14 Tang, T.L., Freeman, R.M., O'Reilly, A.M., Neel, B.G., Sokol, S.Y. The SH2-containing protein-tyrosine phosphatase SH-PTP2 is required upstream of MAP kinase for early *Xenopus* development. *Cell* 1995; 80: 473–83.

15 Shi, Z.Q., Lu, W., Feng, G.S. The Shp-2 tyrosine phosphatase has opposite effects in mediating the activation of extracellular signal-regulated and c-Jun NH2-terminal mitogen-activated protein kinases. *J. Biol. Chem.* 1998; 273: 4904–8.

16 Bennett, A.M., Tang, T.L., Sugimoto, S., Walsh, C.T., Neel, B.G. Protein-tyrosine-phosphatase SHPTP2 couples platelet-derived growth factor receptor β to Ras. *Proc. Natl. Acad. Sci. USA* 1994; 91: 7335–9.

17 Charles, C.H., Abler, A.S., Lau, L.F. cDNA sequence of a growth factor-inducible immediate early gene and characterization of its encoded protein. *Oncogene* 1992; 7: 187–90.

18 Sun, H., Charles, C.H., Lau, L.F., Tonks, N.K. MKP-1 (3CH134), an immediate early gene product, is a dual specificity phosphatase that dephosphorylates MAP kinase in vivo. *Cell* 1993; 75: 487–93.

19 Assoian, R.K. Control of the G1 phase cyclin-dependent kinases by mitogenic growth factors and the extracellular matrix. *Cytokine Growth Factor Rev.* 1997; 8: 165–70.

20 Brondello, J.M., McKenzie, F.R., Sun, H., Tonks, N.K., Pouyssegur, J. Constitutive MAP kinase phosphatase (MKP-1) expression blocks G1 specific gene transcription and S-phase entry in fibroblasts. *Oncogene* 1995; 10: 1895–904.

21 Sun, H., Tonks, N.K., Bar, S.D. Inhibition of Ras-induced DNA synthesis by expression of the phosphatase MKP-1. *Science* 1994; 266: 285–8.

22 Brondello, J.M., Pouyssegur, J., McKenzie, F.R. Reduced MAP kinase phosphatase-1 degradation after p42/p44MAPK-dependent phosphorylation. *Science* 1999; 286: 2514–17.

23 Dorfman, K., Carrasco, D., Gruda, M. *et al.* Disruption of the erp/mkp-1 gene does not affect mouse development: normal MAP kinase activity in ERP/MKP-1-deficient fibroblasts. *Oncogene* 1996; 13: 925–31.

24 Tsui, H.W., Siminovitch, K.A., de Souza, L., Tsui, F.W. Motheaten and viable motheaten mice have mutations in the haematopoietic cell phosphatase gene. *Nat. Genet.* 1993; 4: 124–9.

25 Klingmuller, U., Lorenz, U., Cantley, L.C., Neel, B.G., Lodish, H.F. Specific recruitment of SH-PTP1 to the erythropoietin receptor causes inactivation of JAK2 and termination of proliferative signals. *Cell* 1995; 80: 729–38.

26 Li, J., Yen, C., Liaw, D. *et al.* PTEN, a putative protein tyrosine phosphatase gene mutated in human brain, breast, and prostate cancer. *Science* 1997; 275: 1943–7.

27 Cantley, L.C., Neel, B.G. New insights into tumor suppression: PTEN suppresses tumor formation by restraining the phosphoinositide 3-kinase/AKT pathway. *Proc. Natl. Acad. Sci. USA* 1999; 96: 4240–5.

28 Maehama, T., Dixon, J.E. The tumor suppressor, PTEN/MMAC1, dephosphorylates the lipid second messenger, phosphatidylinositol 3,4,5-trisphosphate. *J. Biol. Chem.* 1998; 273: 13375–8.

29 Di Cristofano, A., Pandolfi, P.P. The multiple roles of PTEN in tumor suppression. *Cell* 2000; 100: 387–90.

30 Tamura, M., Gu, J., Matsumoto, K. *et al.* Inhibition of cell migration, spreading, and focal adhesions by tumor suppressor PTEN. *Science* 1998; 280: 1614–17.

31 Lee, J.O., Yang, H., Georgescu, M.M. *et al.* Crystal structure of the PTEN tumor suppressor: implications for its phosphoinositide phosphatase activity and membrane association. *Cell* 1999; 99: 323–34.

32 Myers, M.P., Pass, I., Batty, I.H. *et al.* The lipid phosphatase activity of PTEN is critical for its tumor supressor function. *Proc. Natl. Acad. Sci. USA* 1998; 95: 13513–18.

33 Alessi, D.R., Kozlowski, M.T., Weng, Q.P., Morrice, N., Avruch, J. 3-Phosphoinositide-dependent protein kinase 1 (PDK1) phosphorylates and activates the p70 S6 kinase in vivo and in vitro. *Curr. Biol.* 1998; 8: 69–81.

34 Cori, G.T., Cori, C.F. The enzymatic conversion of phosphorylase a to b. *J. Biol. Chem.* 1945; 158: 321–32.

35 Stralfors, P., Hiraga, A., Cohen, P. The protein phosphatases involved in cellular regulation. Purification and characterisation of the glycogen-bound form of protein phosphatase-1 from rabbit skeletal muscle. *Eur. J. Biochem.* 1985; 149: 295–303.

36 Cohen, P. The structure and regulation of protein phosphatases. *Annu. Rev. Biochem.* 1989; 58: 453–508.

37 Cohen, P.T. Novel protein serine/threonine phosphatases: variety is the spice of life. *Trends Biochem. Sci.* 1997; 22: 245–51.

38 Cohen, P.T.W. Nomenclature and chromosomal localization of human protein serine/threonine phosphatase genes. *Adv. Prot. Phosphatases* 1994; 8: 371–6.

39 Hubbard, M.J., Cohen, P. On target with a new mechanism for the regulation of protein phosphorylation. *Trends Biochem. Sci.* 1993; 18: 172–7.

40 Chantler, P.D. Calcium-dependent association of a protein complex with the lymphocyte plasma membrane: probable identity with calmodulin-calcineurin. *J. Cell Biol.* 1985; 101: 207–16.

41 Klee, C.B., Crouch, T.H., Krinks, M.H. Calcineurin: a calcium- and calmodulin-binding protein of the nervous system. *Proc. Natl. Acad. Sci. USA* 1979; 76: 6270–3.

42 Borel, J.F. Comparative study of in vitro and in vivo drug effects on cell-mediated cytotoxicity. *Immunology* 1976; 31: 631–41.

43 Sihra, B.S., Kon, O.M., Durham, S.R. *et al.* Effect of cyclosporin A on the allergen-induced late asthmatic reaction. *Thorax* 1997; 52: 447–52.

44 Bunjes, C., Hardt, D., Rollinghoff, M. and Wagner, H. Cyclosporin A mediates immunosuppression of primary cytotoxic T cell responses by impairing the release of interleukin 1 and interleukin 2. *Eur. J. Immunol.* 1981; 11: 657–61.

45 McKeon, F. When worlds collide: immunosuppressants meet protein phosphatases. *Cell* 1991; 66: 823–6.

46 Liu, J., Farmer, J.D., Lane, W.S., Friedman, J., Weissman. I. Schreiber, S.L. Calcineurin is a common target of cyclophilin-cyclosporin A and FKBP-FK506 complexes. *Cell* 1991; 66: 807–15.

47 Guerini, D. Calcineurin: not just a simple protein phosphatase. *Biochem. Biophys. Res. Commun.* 1997; 235: 271–5.

48 Flanagan, W.M., Corthesy, B., Bram, R.J., Crabtree, G.R. Nuclear association of a T-cell transcription factor blocked by FK-506 and cyclosporin A. *Nature* 1991; 352: 803–7.

49 Clipstone, N.A., Crabtree, G.R. Identification of calcineurin as a key signalling enzyme in T-lymphocyte activation. *Nature* 1992; 357: 695–7.

50 Rao, A., Luo, C., Hogan, P.G. Transcription factors of the NFAT family: regulation and function. *Annu. Rev. Immunol.* 1997; 15: 707–47.

51 Genot, E., Cleverley, S., Henning, S., Cantrell, D. Multiple p21ras effector pathways regulate nuclear factor of activated T cells. *EMBO J.* 1996; 15: 3923–33.

52 Chen, C.Y., Forman, L.W., Faller, D.V. Calcium-dependent immediate-early gene induction in lymphocytes is negatively regulated by p21Ha-ras. *Mol. Cell Biol.* 1996; 16: 6582–92.

53 Zhu, J., McKeon, F. NF-AT activation requires suppression of Crm1-dependent export by calcineurin. *Nature* 1999; 398: 256–9.

Protein domains and signal transduction

Structurally conserved protein modules

Many of the proteins involved in cellular signalling possess a multi-domain architecture (Table 18.1). Protein domains are regions within a protein molecule that exhibit structural homology.[1] More strictly, they are called 'structural domains', because it is their tertiary structures that are

Table 18.1 Examples of proteins with structurally conserved domains

Protein	Chapter	Structure
RasGAP	4	SH2 -SH3 -SH2 -PH —C2 -RasGAP
Sos	11	-DH -PH -RasGefN -RasGEF --polyPro-polyPro-polyPro-
PLCδι		PH —EF- EF- PI-PLC-X -PI-PLC-Y —C2 -
PLCγ1	5, 11	PH —EF- PI-PLC-X -PH, -SH2 -SH2 -SH3 -PH, -PI-PLC-Y —C2
Protein kinase B	13	PH -pkinase -
Grb2	11	SH3 -SH2 -SH3
Nck	11	SH3 -SH3 -SH3 -SH2
p85 subunit of PI3-kinase	13	SH3 ----SH2 ---SH2
β-Adrenergic receptor kinase	4	RGS -pkinase -PH
Btk	17	PH —BTK -SH3 -SH2 -pkinase
Vav	12	CH --DH -PH -C1 -SH3 -SH2 -SH3
Dbl	17	----DH -PH -
Shc	11	--PTB --SH2
Src	11, 12	--SH3 -SH2 -pkinase -

BTK, Btk motif; C1, homology with the DAG binding domain of PKC; C2, homology with the Ca^{2+} binding domain of PKC; DH, Dbl homology; EF, EF-hand pair; Ca^{2+} binding; PH, pleckstrin homology; PI-PLC-X or Y, phosphoinositide specific PLC; pkinase, protein kinase domain; PTB/PID, phosphotyrosine binding/interaction domain; Ras-GAP, Ras GTPase activating protein; Ras-GEF, Ras guanine nucleotide dissociation inhibitor; RGS, regulator of G protein signalling domain; SH, Src homology.

homologous and this may extend over a wide range of proteins. Their amino acid sequences tend to be poorly conserved. Each of the many types of domain is based on a compact, stable structure possessing a hydrophobic core. They consist typically of stretches of 40–100 amino acids and are encoded by discrete exons. A particular protein may possess several kinds of domains that have been brought together during evolution by the shuffling of exons. Because the tertiary structures of domains are compact, with N- and C-termini adjacent and exposed, they can insert as 'plug-in' modules into more extended structures.

Domains originate from a set of primordial globular structures that had the qualities of rigidity and stability. Where sequence homology exists, it preserves the basic properties; where the sequence is variable, it can allow variations of function, such as the specification of the target, localization or mode of activation. Thus, differences in sequence may reflect adaptations to meet special requirements. A particular protein may acquire a domain by gene fusion, allowing it to incorporate into its structure a subunit that was previously independent and that has been recruited for a particular purpose. Such insertion can occur most readily at loops in the parent structure, interrupting its linear sequence. By the same token, the linear sequence of a structural domain may itself be interrupted. For example, the chains that form one of the PH domains of PLCγ are separated by an insert of almost 350 residues containing three Src homology domains (Table 18.1).

■ Identification of domains

Since the level of sequence homology among domains of the same type is limited and, because of insertions, their identification from sequence data is difficult. Indeed, in the absence of structural information, their recognition has relied upon computer programs that make multiple comparisons of sequences and structures. As these methods have become more refined, domains that were not previously perceived have become evident in many proteins.

■ Domain function

While some globular proteins possess several different types of domains, or even multiple copies of a particular domain, there are others that possess only one. Many have no recognizable domains at all. It is common for signalling proteins to contain multiple domains. However, although our ability to detect the presence of domains has developed rapidly, our understanding of their functions is still in its infancy. Some domains have built-in enzyme functions (such as kinase and DH domains); others recognize specific peptide sequences (such as SH2 and SH3 domains); others (such as PH domains) recognize particular lipid head groups and yet others can bind phospholipids in a Ca^{2+}-dependent fashion (C2 domains). The non-catalytic interactions can enable the incorporation of signalling molecules into complex structures at specific membrane sites, all at the touch of a switch initiated by the binding of an activating ligand to its receptor.

In this chapter the basic principles of protein domains and their roles in signalling mechanisms are outlined. The discussion is limited to selected examples of domains and modules that have received mention elsewhere in this book. Although not strictly a domain, the EF-hand Ca^{2+} binding motif is also discussed.

Domains that bind oligopeptide motifs

SH2 domains

The SH2 domain is a region, separate and distinct from the catalytic domain in the non-receptor protein tyrosine kinase pp60[src] (reference 2; see Chapter 12), hence SH2, Src homology region 2. SH2 domains are present in all non-receptor PTKs, generally located immediately N-terminal to the kinase domain. They are also present in a large number of other proteins. They consist of about 100 residues and provide high-affinity binding sites for phosphorylated tyrosine residues ($K_d \sim 50–500$ nmol/l) (Figure 18.1). The tertiary structure consists typically of a central antiparallel β-sheet flanked on either side by two α-helices. The target phosphotyrosine is present in a four-residue motif and this binds by straddling the edge of the β-sheet. Residue 1 is the phosphotyrosine and residue 4 (i.e. pY+3) is usually hydrophobic. Each of these is held within a pocket, one on either side of the β-sheet. Five classes of SH2 domain have been recognized, distinguished by their binding specificities. These are determined by the residues that form the binding motif. For example, the SH2 domain of the Src family kinases binds optimally to the sequence

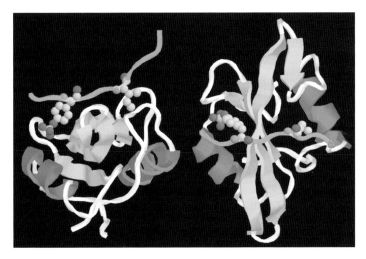

Figure 18.1 SH2 domains. Two orthogonal views of the structure of an SH2 domain (from the Src kinase Lck) with a bound phosphotyrosylpeptide ligand (EPQPYEEIPIYL). The backbone of the ligand (11 residues) is shown in green. The binding motif is a central pYEEI. The phosphotyrosine and isoleucine residues are shown at left and right respectively in each view. (Data source: 1lcj.pdb[13].)

pYEEI (see Figure 11.7, page 264). The affinity of SH2 domains for non-phosphorylated tyrosines or for phosphoserines or phosphothreonines is negligible.

SH2 domain-containing proteins may possess catalytic activity, such as Src (Chapter 12 and below) and PLCγ (Chapter 5). Alternatively they may possess no identifiable catalytic activities of their own, such as Grb2 (Chapter 11), which also possesses two SH3 domains (Table 18.1). These bind a motif that is quite different from that recognized by SH2 domains (see below). Thus, it can act as an adapter molecule, directing a tyrosine phosphorylation signal into a Ras pathway signal.

SH2 domains have been identified in animals, but not so far in the other main kingdoms of eukaryotic organisms, such as plants and fungi. In evolutionary terms, the most ancient form appears in the STAT protein (Chapter 12) of the slime mould *Dictyostelium discoideum*. This suggests a role for the SH2 domain in the evolution of multicellular animals. In response to an activating signal, STAT molecules dimerize through a mutual interaction involving an SH2 domain on each and a phosphotyrosine on the other. The dimeric form then enters the nucleus and binds to DNA to direct transcription, essentially as it does in higher animals.

◼ PTB/PID domains

Another type of domain that recognizes phosphotyrosine residues is the PTB (phosphotyrosine binding) domain (alternatively: phosphotyrosine interaction domain, PID). However, this has completely different molecular architecture from the SH2 domain. Also, in contrast with SH2 domains, the specificity of interaction is determined by the sequence of amino acids immediately on the N-terminal side of the phosphorylated tyrosine residue (NPXpY motif).[3,4] Curiously, PTB/PID domains are structurally very similar to PH domains, with a β-barrel structure and a long α-helix that packs against one end (compare Figure 18.2 with Figure 18.6). However, PH domains have quite different targets (see below), which bind either to the loops that join the β-chains or to the major helix. In PTB domains, by contrast, the target phosphotyrosine binds to one side of the β-barrel (Figure 18.3).

◼ SH3 domains

SH3 domains consist of 55–75 amino acids that form a twisted β-barrel structure (Figure 18.4). This stable core is conserved, but the loops that join the β-strands are variable. The motif on the target proteins, to which SH3 domains bind, consists of a proline-rich stretch of 8–10 residues. Such sequences form extended left-handed helices having three residues per turn (called a type II polyproline helix). The binding site is lined with hydrophobic aromatic amino acids.[5] The planar side-chains form ridges that fit into the grooves of the polyproline helix, rather like the threads of two adjacent screws that fit together (Figure 18.5). About three turns are involved in binding and the consensus motif is either RXLPPLPXX or

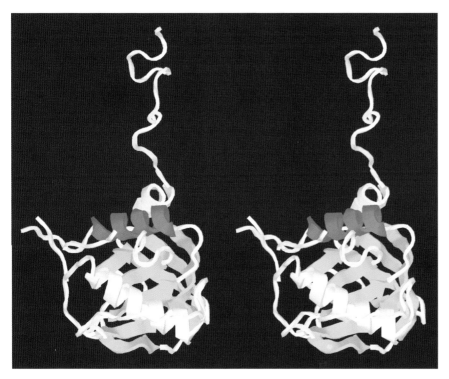

Figure 18.2 **PTB/PID domain structure.** Stereoscopic diagram of the structure of the PTB domain of Shc. The conserved secondary structure that forms the core is shown in colour (α-helix, magenta; β-structure, yellow). The helix closes off one end of a twisted β-barrel. (Data source: 1shc.pdb[14].)

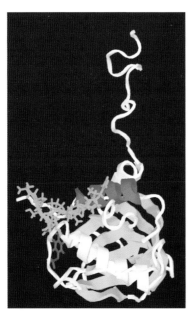

Figure 18.3 **PTB domain with bound phosphotyrosine-containing motif.** The target motif of the PTB domain is depicted in green and the phosphotyrosine is coloured red. (Data source: 1shc.pdb[14].)

How to view stereo images of molecular structures

There is much more information in a stereoscopic image than a conventional flat projection. Almost everyone can view stereo pictures with unaided eyes, but it does take a while to get used to it. Practice is the key, unless you are unfortunate enough to have one very weak eye.

There are two ways of seeing a three-dimensional image by observing a stereo pair. You may either cross your eyes, so that the left eye views the right-hand image and vice versa, or you may allow your eyes to diverge, so that each eye looks at the image in front of it. Rasmol-generated images are, by default, for convergent (cross-eyed) viewing and CHIME images are for divergent viewing. To view the images in this book (and most printed images), you should view the left image with the left eye and the right image with the right eye. If the two do not readily fuse into a single three-dimensional image, try the following:
• First touch your nose to the page between and below the stereo pair. The two images will now be superimposed, but the picture will be very blurred (although you may notice some three-dimensionality at this stage).
• Now move the page slowly away from you, but take care not to rotate it. Concentrate on the three-dimensional aspect and wait for your eyes to bring it into focus.
Note: if you attempt to view a divergent pair using the convergent strategy (crossed eyes) the image will be three-dimensional but inverted in a confusing way.

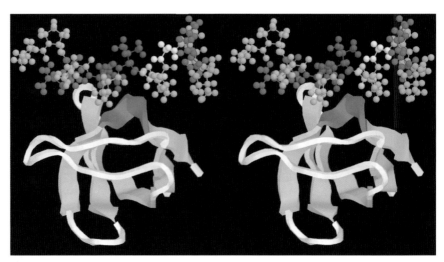

Figure 18.4 **Stereoscopic view of a left-handed proline-rich helix binding to the surface of an SH3 domain.** Structure, obtained by NMR, of the SH3 domain of the Src kinase Fyn, complexed with a synthetic peptide corresponding to residues 91–104 of the p85-subunit of PI3-kinase (PPRPLPVAPGSSLT). The peptide, in ball and stick format, is coloured to show Pro pink; Arg, Lys blue; Leu, Val green; Ala, Gly grey; Ser, Thr, orange. (Data source: 1azg.pdb[15].)

XXXPPLPXR, giving rise to two classes of SH3 domain-binding proteins. Those bearing the first sequence will align their motifs in the opposite direction to proteins with the second sequence. However, the binding in both cases is usually weak and these interactions are promiscuous. Such unreliability may account for the presence of two or more SH3 domains in adaptor proteins such as Grb2 (Chapter 11) and Nck (Table 18.1). In the case of Grb2, these bind to two neighbouring proline-rich regions on the target, thereby increasing the affinity and specificity of binding. The tandem arrangement of SH2 and SH3 domains found in a variety of signalling proteins may provide a conformational mechanism for regulating SH3-dependent interactions through tyrosine phosphorylation.[6]

Domains that bind proteins and lipids

PH domains

PH (pleckstrin homology) domains are a relatively recent addition to the inventory of known structural domains. Consisting of some 100 amino acid residues, sequence matches for them have been found in over 600 proteins, though detailed structures are known for only a few of these. Their function is less clear than that of the SH2 and SH3 domains and it varies among different PH domains. Many have been shown to mediate protein–protein or protein–lipid interactions. They were first identified as internal repeats in pleckstrin, the major substrate of protein kinase C in platelets.[7,8] Pleckstrin itself, rather unusually, possesses two PH domains, one at either end of the molecule.

The sequences which make up the various PH domains are much more

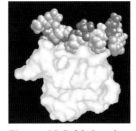

Figure 18.5 **Molecular surface of an SH3 domain bound to a proline-rich peptide.** The SH3 domain of Fyn depicted as a molecular surface (using <u>CHIME</u>). The target peptide is shown in spacefill format using the same colour scheme as Figure 18.4. (Data source: 1azg.pdb[15].)

variable than those of SH2 and SH3 domains. Like all domains, the few conserved residues tend to be immersed in the interior, where they maintain the core structure. Multiple sequence alignments of PH domains from different proteins reveal this lack of similarity, as illustrated in Table 18.2.

The conserved three-dimensional structures of PH domains are characterized by a β-barrel structure, consisting of seven antiparallel β-strands with a C-terminal α-helix packed against one end of the barrel. This forms a rigid frame. The process of evolution has adapted this structure, particularly in the interchain loop regions, to form PH domains with different binding specificities. It is noteworthy that the same fold forms the core of the PTB/PID domains (compare Figure 18.6 with Figure 18.2).

The assignment of function to PH domains is not straightforward. For signalling proteins, it is useful to identify specific ligands. Some are listed in Table 18.3. A number of proteins contain PH domains that can bind to the βγ-subunits of G-proteins. The best known of these is βARK, the kinase that catalyses phosphorylation of β-adrenergic receptors, but although it is known that the βγ-subunits bind to the C-terminal region of the PH domain, the structural details of the interaction are unclear. In contrast to βARK, the homologous protein rhodopsin kinase, which has no PH domain, possesses a farnesyl group at its C-terminus and this hydrophobic modification enables it to associate with membranes. This suggests that association of a PH domain with a membrane-tethered βγ-subunit also functions to attach the host molecule at a membrane site.

PH domains may also interact with membrane polyphosphoinositidcs. This occurs through binding of the 4- and 5-phosphate groups of the phospholipid headgroup to residues in a cleft between the loops connecting the β-strands. For instance, this is a property of the PH domain of PLCδ which binds to PI(4,5)P_2 (or IP$_3$) (Figure 18.6). This high-affinity interaction may direct the phospholipase to membrane regions containing PI(4,5)P_2, which is also the substrate for the catalytic domain. Another example is provided by the Btk/Tec family of non-receptor PTKs, except that in this case the PH domain is selective for PI(3,4,5)P_3, the product of PI3-kinase. It seems that the main function of these PH domains is to target proteins to specific membrane locations determined by the presence of individual polyphosphoinositides. The presence of IP$_3$, which might prevent this form of interaction, provides yet another possible link to signal transduction processes. The PH domain of βARK can bind to both PI(4,5)P_2 and βγ-subunits to ensure its translocation to a membrane site. A link with the regulation of heterotrimeric G-proteins and the activation status of 7TM receptors is then provided by the finding that the dissociation of the G$_{βγ}$/βARK complex is regulated by the level of PIP$_2$.

Finally, although we have good clues about the functions of a few important PH domains, we know little about the remainder. PH domains may use either or both of the two basic mechanisms described above for membrane translocation, but some use neither and much remains to be clarified.

Table 18.2 Sequence alignments of PH domains.

Protein		β1	Loop	β2	L	β3	Loop		β4
Dynamin	513	RKGWLTINNI	G.......IMKGG	SK.EYWFVLT	AE	NLSWY	KDDEE.....	.......KE	KKYMLS
PLCδ1	23	LKGSQLLKVK	S........SSW	RR.ERFYKLQ	ED	CKTIW	QESRKVMRTP	ESQLFSIEDIQ	EVRMGH
PLCβ2	22	SQGERFIKWD	D.......ETT	VA.SPV.ILR	VD	PKGYY	LYWTYQS..K	EMEFLDITSIR	DTRFGF
Pleckstrin1	7	REGYLVKKG.	.......SVFNT	WKPMWVVLL.	ED	GIEFY	KKKSDNS...	....	PKGMIP
β-Spectrin	2199	MEGFLNRKHE	WEAHNKK.ASSRS	WH.NVYCVIN	NQ	EMGFY	KDAKSAASGI	PY.......H	SEVPVS
Btk	6	LESIFLKRSQ	Q.....KKKTSPL	NFKKRLFLLT	VH	KLSYY	EYDFERGRR.	...GS..KKG	SIDVEK

Protein		L	β5	Loop	β6	Loop			β7	L	Alpha
Dynamin	558	VDN	LKLRDVE	KGFMS	SKHIF...	ALF	NTE..QR	NVYKDY	RQLELAC	ET	QEEVDSWKASFLRAG
PLCδ1	80	RT.	EGLEKFA	RDVPE	DRCFSI..	.VF	K......	...DQR	NTLDLIA	PS	PADAQHWVLGLHKII
PLCβ2	73	AKi	DVFNMii	DNSFL	LKTLTV..	.VS	G......	...PDM	iFHNFVS	YK	ENVGKAWAEDVLAlV
Pleckstrin1	51	LKG	STLTSPC	QDFGK	RMFVF...	KIT	T......	...TKQ	QDHFFQA	AF	LEERDAWVRDINKAI
β-Spectrin	2256	LKE	.AICEVA	LDYKK	KKHVF...	KLR	L......	....SD	GNEYFQA	KD	DEEMNTWIQAISSA.
Btk	61	ITC	VETVVPE	KNPiI	IERFPY..	.PF	QVV....	...YDE	GPLYVFS	PT	EELRKRWIHQLKNVI

Residues of PLC-δ1 and β-spectrin PH domains that specifically interact with ligand are in magenta. R → C mutation in the Src kinase Btk, found in immunodeficient mutant (Xid) mice, is in blue. *i* indicates an insertion. Modified from Ferguson et al.,[16] Runnels et al.[21]

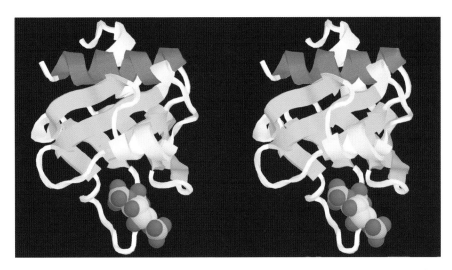

Figure 18.6 PH domain with bound IP₃. Stereoscopic diagram of the structure of the PH domain of PLCδ1 complexed with IP₃. The secondary structure forming the conserved core is indicated in colour (α-helix, magenta; β-structure, yellow). As for PTB domains, the major helix closes off one end of a twisted β-barrel. IP₃ is depicted as a ball and stick model in Rasmol cpk colours (red, oxygen; yellow, phosphorus). (Data source: 1mai.pdb[16].)

Table 18.3 Some physiological ligands of PH domains

Host protein	Ligands	Protein function
β-Adrenergic receptor kinases	βγ-subunits, PI(4,5)P₂	Down-regulation of receptors
Phospholipase Cδ1	PI(4,5)P₂, IP₃	Regulation of enzyme activity
Akt/PKB	PI(3,4)P₂, PI(3,4,5)P₃	A downstream effector of PI 3-kinase, regulation of enzyme activity
Btk	PI(3,4,5)P₃	Regulation of leukocyte activation
Ras-GAP	PI(3,4,5)P₃, I(1,3,4,5)P₄	Link between PI 3-kinase and Ras

Adapted from Lemmon et al.[9]

■ Polypeptide modules that bind Ca²⁺

■ The EF-hand motif

Common Ca^{2+}-binding regions on proteins include the EF-hand motif. E and F denote helical regions of the muscle protein parvalbumin, in which the motif was first identified. The Ca^{2+} binding site is formed by a loop of approximately 12 amino acids that links the two α-helices (a helix–loop–helix structure) (Figure 18.7). EF-hands generally occur as adjacent pairs. This arrangement allows them to fold compactly, and this is called an EF-hand unit. In general, the core of the EF-hand structure is reasonably conserved, but the outer regions vary and the Ca^{2+} dissociation constant can range from 10^{-7} to 10^{-5} mol/l. The important intracellular

Figure 18.7 EF-hand structure. Two views of the structure of one of the EF-hands of calmodulin showing a bound Ca^{2+} ion (green). The seven oxygen atoms that form the coordination shell are indicated in red. (Data source: 1cll.pdb[17].)

Ca^{2+}-sensor calmodulin possess two EF-hand units (i.e. four EF-hands). In each motif, the chemical ligands that coordinate each Ca^{2+} ion are oxygen atoms provided by aspartate and glutamate side-chains and by a peptide-bond carbonyl group, and also by a water molecule.[10,11] Conformational changes in calmodulin that result from Ca^{2+} binding are described in detail in Chapter 8.

C2 domains

Unlike the small EF-hand motifs, C2 domains are substantial structures that possess Ca^{2+} binding sites. They are also known as CaLB or Ca^{2+} and lipid-binding domains, and they exist in a wide range of intracellular proteins. Like the EF-hands their affinity for Ca^{2+} is variable. They are named for their homology with the so-called 'second conserved' regulatory domain of protein kinase Cβ (see Figure 9.8, page 200).

Structure of C2 domains

The binding of one ligand alters the affinity of other site(s). Positive cooperativity, as in this case, denotes an enhancement of the affinity at other sites.

C2 domains are made up of approximately 130 amino acid residues arranged in a rigid, eight-stranded, antiparallel β sandwich (Figure 18.8). Ca^{2+}-binding is confined to a region that is defined by three loops on one edge of the structure. In the C2 domain of PKCβ, depicted in Figure 18.8, five aspartate residues on two of the loops contribute coordinating oxygens that help form the Ca^{2+} binding pocket, in which up to three Ca^{2+} ions can be bound in a cooperative manner. Note, however, that the coordination sphere around each Ca^{2+} ion is not quite complete. C2 domains also bind negatively charged phospholipids (such as phosphatidylserine, phosphatidylinositol and polyphosphoinositides). Apart from the anionic residues, the side-chains in the loop regions are mostly positively charged. When calcium ions enter the pocket, the negative charges are neutralized and the protein can then bind electrostatically to the anionic lipid, which may complete the coordination spheres. Thus, phospholipid binding requires the binding of Ca^{2+}, which acts like an 'electrostatic switch', changing the surface potential of the protein.[12] This mechanism allows soluble proteins with C2 domains to become membrane-associated when intracellular Ca^{2+} becomes elevated to

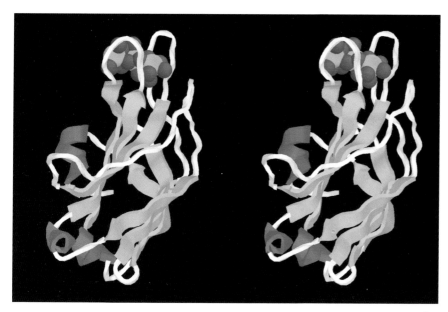

Figure 18.8 **C2 domain structure.** Stereo image showing secondary structure with Ca²⁺ in green. (Data source: 1a25.pdb[18].)

micromolar levels. It repeats a theme that we have encountered before: the recruitment of a signalling molecule through one of its domains to a specific membrane location.

In general, C2 domains do not necessarily exhibit all these characteristics. There are variations in the stoichiometry of binding (number of sites) and in the affinity for Ca^{2+}. Indeed, some do not bind Ca^{2+}, but can associate with other proteins, suggesting a possible alternative function.

Protein kinase domains

Protein kinases share a common domain

Protein kinases catalyse the transfer of a phosphate group from ATP to a hydroxyl residue on an amino acid side-chain. There are two principal classes: serine/threonine kinases and tyrosine kinases. In both, the catalytic activity is confined to a structurally conserved domain called a protein kinase domain. The basic architecture of kinase domains is typified by the catalytic subunit of the cAMP-dependent protein kinase A (PKA) and this is illustrated in Figure 18.9. The peptide chain is folded to form two lobes that are in close apposition, the cleft between them housing the catalytic site. There is also an N-terminal α-helical chain (the A helix) that binds to the surface of both lobes. It is sometimes referred to as the linker. The lobes are connected to each other by a single polypeptide chain which acts as a hinge.

The N-terminal lobe is the smaller of the two, possessing about 100 amino acids. At the N-terminus it is myristoylated, but there is no evidence that the myristoyl group is free to associate with membranes. Instead, it occupies a pocket and provides structural stability, helping to

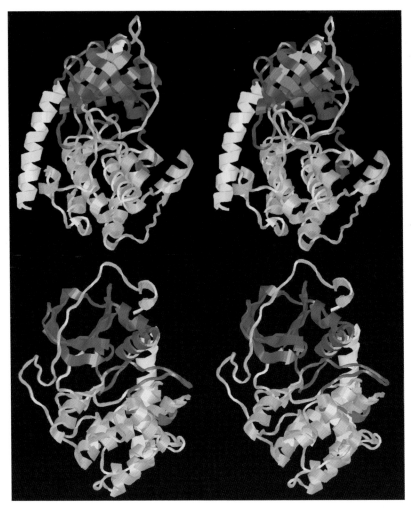

Figure 18.9 **Protein kinase domain architecture.** Stereo diagrams of the catalytic subunit of porcine PKA. The linker chain (A helix) is coloured cyan. The small lobe is blue and the large lobe is violet. The activation segment is shown in red. An N-terminal myristoyl group is not shown. (Data source: 1cdk.pdb[19].)

keep the linker in contact with the large lobe. The principal feature of the small lobe is an antiparallel β-sheet. There are also two α-helical chains: the B- and C-helices. The larger lobe consists of approximately 200 residues and it possesses mostly α-helical structure, arranged around a stable four-helix bundle.

Structural elements that regulate kinase activity

In broad terms, the principal function of the small lobe is to bind ATP, and that of the large lobe is to bind substrate and enable catalysis. However, the detailed interactions of the lobes with each other, as well as with ATP, Ca^{2+} and substrate, are complex. First, ATP must enter the cleft and bind in the correct orientation. To do this, it makes contact with residues on both lobes. Two Mg^{2+} ions are also bound, but for clarity these are not shown in the following figures. Next, the recognition and binding of a consensus motif on the target protein must take place. For PKA, the sequence of the motif is RXXT/Sh (where h indicates a hydrophobic side-chain). The target hydroxyl group on serine or threonine must be accurately aligned with the terminal phosphate of ATP and ultimately the ADP that is formed must be exchanged for new ATP.

All this requires the coordinated interaction of amino acid side-chains and peptide bonds in specific structures on both lobes. These are shown in **bold** below and indicated in Figure 18.10. The residues on the small lobe that form contacts with the nucleotide and align it, include the main chain nitrogens of a **glycine-rich loop**, between β-strands 1 and 2, and a lysine (K72) on β-strand 3 that interacts with a glutamate (D91) on the **C-helix**. In the large lobe, a 24 amino acid segment controls activation. It forms the loop following β-strand 8, termed the **activation loop**, and also part of the ensuing helix. On the activation loop is a crucial threonine (T197) that must itself be in a phosphorylated state for activation to take place. Once thought to be a stable post-translational modification, phosphorylation of T197 is now considered to be a regulatory event for PKA. Also following β-strand 8 are residues that contribute to the binding of the two Mg^{2+} ions (not shown in the figure) and, finally, following β-strand 6 is the **catalytic loop** that contains a glutamate (D166) which is thought to initiate the phosphate transfer reaction.

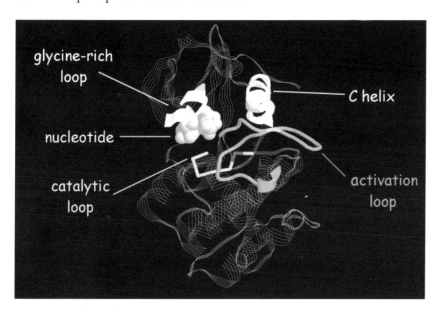

Figure 18.10 **Catalytic core of PKA.** Some of the important structures and residues that form the catalytic site. (Orientation as in Figure 18.9 (bottom), but the framework is faded and the linker and substrate are not shown). Key residues include K72 on β-strand 3, E91 in the C-helix, T197 in the activation loop and R165 in the catalytic loop (all coloured cyan). D166 in the catalytic loop is coloured red. The binding of ATP is indicated by the grey, spacefilled molecule. For experimental reasons this is an ATP analogue, adenylyl imidodiphosphate. (Data source: 1cdk.pdb[19].)

Effective catalytic activity depends on the completion of the active site, and this requires the binding of ATP and Mg^{2+}, the closure of the cleft between the two lobes and phosphorylation of T197. The position of the activation segment is critical for the correct alignment of the catalytic residues, and the contacts made by the phosphothreonine are particularly important. Also, ATP binding is very sensitive to the position of the C-helix. Thus, both the activation segment and the C-helix can act as regulators of kinase activity.

Binding of substrate in the correct orientation is also critical, and this is illustrated in Figure 18.11. In panel (a) the molecular surface of PKA in the

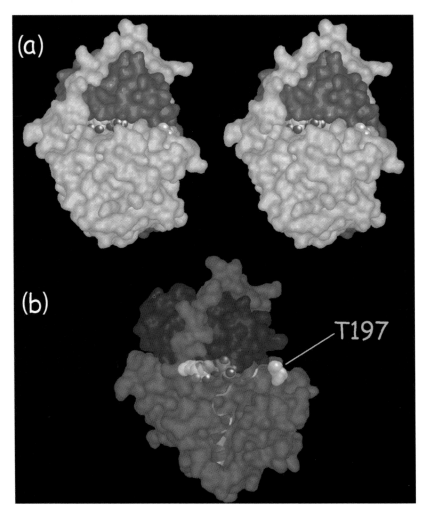

Figure 18.11 **Molecular surfaces of PKA showing ATP and substrate binding:** (a) Stereoscopic diagram of the calculated molecular surface of PKA (Van der Waals radii). The linker and small lobe are coloured blue; the large lobe is violet. The bound ATP (analogue) is just visible. T197 is coloured green. (b) The surface has been rendered transparent to reveal the nucleotide. The 'substrate' peptide is indicated in red (residues 5–24 of a protein called PKC inhibitor, PKI). (Data source: 1cdk.pdb[19].)

closed conformation is shown, with ATP bound deep in the cleft. In panel (b) the surface is rendered transparent to reveal the nucleotide and a pseudosubstrate peptide is shown in the substrate binding site.

Phosphorylation of T197 is important because it neutralizes the positive charges on a number of nearby cationic residues, including an arginine (R165) adjacent to D166. It also interacts with residues on the small lobe, helping to seal the cleft. T197 is present in the sequence RTWTL. This matches the PKA target consensus sequence and so indicates that kinase activity is regulated by autophosphorylation. A number of other kinases are also phosphorylated at positions equivalent to T197 on PKA. These include isoforms of PKC, ERK, MEK1, the cyclin-dependent kinases, CDK2 and CDK7, and Src family protein tyrosine kinases. However, although a phosphorylation in the activation loop may be conditional for kinase activation, autophosphorylation at this site does not occur in every case. For example, in PKC, ERK and the cyclin-dependent kinases, the sequence of the motif in the activation loop does not match the consensus sequence recognized by the kinase. This implies a requirement for an upstream kinase (for example, PDK1 for PKC, see Chapter 9).

Other protein kinases are not regulated by a phosphorylation (see Table 18.4). Instead, the positively charged arginine residue (equivalent to R165 in PKA) may be neutralised by an acidic (glutamate) residue (phosphorylase kinase), or substituted by a non-polar residue (myosin light chain kinase).

Table 18.4 Regulation of kinase activity by phosphorylation of the activation loop

Phosphorylated in the activation segment		Not phosphorylated in the activation segment	
Cyclic AMP-dependent kinase	PKA		
Cyclin-dependent kinase	p34^{cdc2} cdc2 CDK2 CDK7	Phosphorylase kinase	
MAP kinase	MAPK/ERK2	Casein kinase I	
MAP kinase kinase	MEK1	EGF receptor	EGFR
Raf1 kinase	Raf1	C-terminal Src kinase	Csk
Ca^{2+}/calmodulin kinase	CaMKI	Ca^{2+}/calmodulin kinase	CaMKII
Protein kinase C	PKC α, βII	Myosin light chain kinase	MLCK
Insulin-stimulated kinase	ISPK		
Glycogen synthase kinase	GSK3		
Insulin receptor kinase	IRK		
PDGF receptor	PDGFR		
c-Src family	Src, Yes, Fyn, Fgr, Lyn, Lck, Blk		

■ The regulatory domains of Src control protein kinase activity

For historical reasons, the numbering of the residues of human c-Src corresponds to the sequence of chicken c-Src. Thus 86–536 actually denotes residues 83–533.

The non-receptor protein tyrosine kinase c-Src provides an excellent example of the regulation of kinase activity by structural domains that can take part in protein–protein (or domain–domain) interactions. (The Src family of protein tyrosine kinases are described on page 294.

The three-dimensional structure of human c-Src (residues 86–836) is depicted in Figure 18.12. The linker chain and the two lobes of the kinase domain are evident. N-terminal to these structures there are the two Src homology domains (SH2 and SH3) and at the C-terminal there is a short, flexible segment that bears an inhibitory phosphotyrosine (pY527).

Regulation follows a general principle. The operational machinery that enables catalytic activity is contained within the kinase domain itself. The controls are located on adjacent domains and regulation depends on their interactions with the kinase domain. In the case of Src and related kinases, the elements that direct the regulation are the SH2 and SH3 domains. Src is held in an inactive state by the C-terminal chain phosphotyrosine. This provides an intramolecular binding site for the N-terminal SH2 domain. There is also an interaction between the SH3 domain and the linker chain, which adopts a left-handed helical conformation (resembling a type II polyproline helix, although there is only one proline residue). Together, these interactions force the molecule to adopt a compact configuration, distorting the small lobe, reducing access to the cleft and causing an outward rotation of the C-helix.

Activation of kinase activity follows events that destabilize the compact conformation. Lacking an isoleucine at pY+3, the amino acid sequence

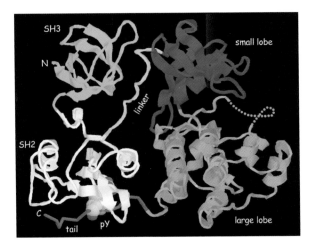

Figure 18.12 The structure of c-Src in its inactive state. This is the compact auto-inhibited conformation of c-Src. The kinase domain is to the right and the linker and lobes are coloured as in previous diagrams. The C-terminal tail region (red) contains pY527, which is shown in spacefill format, but is mostly hidden. The activation loop is sketched as a dotted line because it is disordered in this structure. (Data source: 1fmk.pdb[20].)

adjacent to the C-terminal pY527 does not match that of the motif that binds the SH2 domain of Src most strongly (see Figure 11.7, page 264 and Figure 18.1). Consequently, the binding is less tight, giving the opportunity for pY527 to become dephosphorylated. Displacement of pY527 by a higher-affinity phosphotyrosine might facilitate this. Removal of the phosphate group allows the molecule to relax into an active conformation. However, as in other kinases, phosphorylation of a key residue in the activation loop (Y416, equivalent to T197 in PKA) is also essential for activity.

A specific kinase, Csk (Figure 12.6, page 295), is responsible for phosphorylating the C-terminal tail of Src, inhibiting its activity. Mutation of tyrosine 527 to phenylalanine in avian c-Src produces a constitutively active oncogenic protein. Likewise the oncogenic viral form, v-Src, lacks a C-terminal tail.

References

1 Wetlaufer, D.B. Nucleation, rapid folding, and globular intrachain regions in proteins. *Proc. Natl. Acad. Sci. USA* 1973; 70: 697–701.

2 Koch, C.A., Moran, M., Sadowski, I., Pawson, T. The common src homology region 2 domain of cytoplasmic signaling proteins is a positive effector of v-fps tyrosine kinase function. *Mol. Cell Biol.* 1989; 9: 4131–40.

3 Kavanaugh, W.M., Williams, L.T. An alternative to SH2 domains for binding tyrosine-phosphorylated proteins. *Science* 1994; 266: 1862–5.

4 Kavanaugh, W.M., Turck, C.W., Williams, L.T. PTB domain binding to signaling proteins through a sequence motif containing phosphotyrosine. *Science* 1995; 268: 1177–9.

5 Yu, H., Rosen, M.K., Shin, T.B. *et al.* Solution structure of the SH3 domain of Src and identification of its ligand-binding site. *Science* 1992; 258: 1665–8.

6 Hu, K.Q., Settleman, J. Tandem SH2 binding sites mediate the RasGAP-RhoGAP interaction: a conformational mechanism for SH3 domain regulation. *EMBO J.* 1997; 16: 473–83.

7 Haslam, R.J., Koide, H.B., Hemmings, B.A. Pleckstrin domain homology. *Nature* 1993; 363: 309–10.

8 Imaoka, T., Lynham, J.A., Haslam, R.J. Purification and characterization of the 47,000-dalton protein phosphorylated during degranulation of human platelets. *J. Biol. Chem.* 1983; 258: 11404–14.

9 Lemmon, M.A., Falasco, M., Ferguson, K.M., Schlessinger, J. Regulatory recruitment of signalling molecules to the cell membrane by pleckstrin-homology domains. *Trends Cell Biol.* 2000; 7: 237–42.

10 Kretsinger, R.H. Calcium-binding proteins. *Annu. Rev. Biochem.* 1976; 45: 239–66.

11 Babu, Y.S., Sack, J.S., Greenhough, T.J. *et al.* Three-dimensional structure of calmodulin. *Nature* 1985; 315: 37–40.

12 Ubach, J., Zhang, X., Shao, X., Südhof, T.C., Rizo, J. Ca^{2+} binding to synaptotagmin: how many Ca^{2+} ions bind to the tip of a C2-domain? *EMBO J.* 1998; 17: 3921–30.

13 Eck, M.J., Shoelson, S.E., Harrison, S.C. Recognition of a high-affinity phosphotyrosyl peptide by the Src homology-2 domain of p56lck. *Nature* 1993; 362: 87–91.

14 Zhou, M.M., Ravichandran, K.S., Olejniczak, E.F. *et al.* Structure and ligand recognition of the phosphotyrosine binding domain of Shc. *Nature* 1995; 378: 584–92.

15 Renzoni, D.A., Pugh, D.J., Siligardi, G. *et al.* Structural and thermodynamic characterization of the interaction of the SH3 domain from Fyn with the proline-rich binding site on the p85 subunit of PI3-kinase. *Biochemistry* 1996; 35: 15646–53

16 Ferguson, K.M., Lemmon, M.A., Schlessinger, J., Sigler, P.B. Structure of the high affinity complex of inositol trisphosphate with a phospholipase C pleckstrin homology domain. *Cell* 1995; 83: 1037–46.

17 Chattopadhyaya, R., Meador, W.E., Means, A.R., Quiocho, F.A. Calmodulin structure refined at 1.7 A resolution. *J. Mol. Biol.* 1992; 228: 1177–92.

18 Sutton, R.B., Sprang, S.R. Structure of the protein kinase Cβ phospholipid-binding Ca^{2+} domain completed with Ca^{2+}. *Structure* 1998; 6: 1395–405.

19 Bossemeyer, D., Engh, R.A., Kinzel, V., Ponstingl, H., Huber, R. Phosphotransferase and substrate binding mechanism of the cAMP-dependent protein kinase catalytic subunit from porcine heart as deduced from the 2.0 A structure of the complex with Mn^{2+} adenylyl imidodiphosphate and inhibitor peptide PKI(5–24). *EMBO J.* 1993; 12(849): 859.

20 Xu, W., Harrison, S.C., Eck, M.J. Three-dimensional structure of the tyrosine kinase c-Src. *Nature* 1997; 595(602): 385.

21 Runnels, L.W., Jenco, J., Morris, A. Scarlata, S. Membrane binding of phospholipases C-β-1 and C-2 is independent of phosphatidylinositol 4,5-bisphosphate and the α and βγ subunits of G proteins. *Biochemistry* 1996; 35: 16824–32.

Index